METHODS IN MOLECULAR

Series Editor
John M. Walker
School of Life and Medical Sciences
University of Hertfordshire
Hatfield, Hertfordshire, AL10 9AB, UK

For further volumes:
http://www.springer.com/series/7651

3D Cell Culture

Methods and Protocols

Edited by

Zuzana Koledova

Department of Histology and Embryology, Masaryk University, Brno, Czech Republic

Humana Press

Editor
Zuzana Koledova
Department of Histology and Embryology
Masaryk University
Brno, Czech Republic

ISSN 1064-3745 ISSN 1940-6029 (electronic)
Methods in Molecular Biology
ISBN 978-1-4939-7019-3 ISBN 978-1-4939-7021-6 (eBook)
DOI 10.1007/978-1-4939-7021-6

Library of Congress Control Number: 2017939584

Printed on acid-free paper

This Humana Press imprint is published by Springer Nature
The registered company is Springer Science+Business Media LLC
The registered company address is: 233 Spring Street, New York, NY 10013, U.S.A.

Preface

In the last decade, there has been a boom in three-dimensional (3D) cell culture methods. This type of cell culture aims to completely recapitulate the 3D organization of cells and cell-cell and cell-matrix interactions in vitro to closely mimic a typical organ microarchitecture and microenvironmental signaling. Thereby, 3D cell cultures represent more accurate and physiologically more relevant models and thus bridge the gap between more traditional 2D cell cultures in Petri dish and live organism. They provide a malleable platform and a less costly and more ethical alternative to in vivo models. 3D cell cultures have revolutionized biological research: they have enabled studies of fundamental biological questions, improved drug efficacy and toxicity screening, and fuelled tissue engineering for clinical applications.

The intention of this book is to provide an overview of established 3D cell culture assays, presented with detailed methodology and useful tips. To this end, leaders in the field have been invited to share their protocols and know-how. Their contributions cover a wide spectrum of techniques and approaches for 3D cell culture, from organoid cultures through organotypic models to microfluidic approaches and emerging 3D bioprinting techniques, which are used in developmental, stem cell, cancer, and pharmacological studies. The first chapter provides an introduction to the 3D cell culture field and a guide to the methods presented in the book. Next, in the first part of the book, several chapters are devoted to the production of hydrogels and scaffolds for 3D cell culture, including biohybrid hydrogels, decelularized extracellular matrix, stiffness-tunable interpenetrating networks of reconstituted basement membrane matrix and alginate, and calcium phosphate foams. The second part of the book covers protocols for conventional, manually assembled 3D cell cultures. They include 3D mammary, prostate, lung, and intestinal organoid cultures, and organotypic assays, such as invasion assays, confrontational co-cultures, and a procedure for the production of full-thickness skin equivalents. The third part of the book presents techniques for 3D cell culture micropatterning, including photomask, micro-needle, soft lithography, and two-photon polymerization techniques. Microfluidic approaches for 3D cell culture are described in the fourth part of the book. They include several techniques for spheroid production and embedding into ECM as well as protocols for production of organs-on-a-chip. Another state-of-the-art approach to 3D cell culture, 3D bioprinting, is presented in the fifth part of the book. At last, but not least, techniques for imaging and image analysis of 3D cell cultures are provided in the sixth part of the book.

I hope that this volume on 3D cell culture will be useful to a wide readership of scientists, especially in the fields of developmental and cancer biology, pharmacology, medicine, and tissue engineering. I envision that the book will serve not only as a collection of protocols to be strictly followed but also as an instrumental guide for assay adaptation and establishment of new models. I believe that the book will inspire development of novel 3D cell culture techniques according to specific scientific needs and interests towards new generation of physiologically relevant and realistic 3D cell cultures.

Brno, Czech Republic *Zuzana Koledova*

Contents

Part V Bioprinting

Part VI Imaging and Image Analysis of 3D Cell Cultures

Contributors

MALIN ÅKERFELT • *High-Content Screening Laboratory (HCSLab), Institute of Biomedicine, University of Turku, Turku, Finland*
ANDREA ANFOSSO • *The Centenary Institute, Newtown, NSW, Australia; Sydney Medical School, University of Sydney, Sydney, NSW, Australia*
YOUMNA ATTIEH • *Institut Curie, PSL Research University, CNRS, UMR 144, Paris, France*
ANDREA BARBETTA • *Department of Chemistry, University of Rome "La Sapienza", Rome, Italy*
KIMBERLEY A. BEAUMONT • *The Centenary Institute, Newtown, NSW, Australia; Sydney Medical School, University of Sydney, Sydney, NSW, Australia*
DANIELA BELLONI • *Division of Experimental Oncology, San Raffaele Scientific Institute, Milan, Italy*
KAMBEZ H. BENAM • *Wyss Institute for Biologically Inspired Engineering, Boston, MA, USA*
ANGIOLA BERENZI • *Department of Clinical and Experimental Sciences, Institute of Pathological Anatomy, School of Medicine, University of Brescia, Brescia, Italy*
FABIEN BERTILLOT • *Institut Curie, PSL Research University, CNRS, UMR 144, Paris, France*
MARCUS BINNER • *Max Bergmann Center of Biomaterials Dresden, Leibniz Institute of Polymer Research Dresden, Dresden, Germany*
RAQUEL BLAZQUEZ • *Department of Internal Medicine III, University Hospital Regensburg, Regensburg, Germany*
JENNIFER BONIOTTI • *Laboratory of Tissue Engineering, Anatomy and Physiopathology Unit, Department of Clinical and Experimental Sciences, School of Medicine, University of Brescia, Brescia, Italy*
LAURA J. BRAY • *Max Bergmann Center of Biomaterials Dresden, Leibniz Institute of Polymer Research Dresden, Dresden, Germany; Science and Engineering Faculty and Institute of Health and Biomedical Innovation, Queensland University of Technology, Brisbane, QLD, Australia*
BRYAN N. BROWN • *Department of Bioengineering, University of Pittsburgh, Pittsburgh, PA, USA; Department of Obstetrics, Gynecology and Reproductive Sciences, Magee-Womens Research Institute, University of Pittsburgh, Pittsburgh, PA, USA; McGowan Institute for Regenerative Medicine, University of Pittsburgh, Pittsburgh, PA, USA*
MICHAEL J. BUCKENMEYER • *Department of Bioengineering, University of Pittsburgh, Pittsburgh, PA, USA*
MASSIMILIANO CAIAZZO • *Institute of Genetics and Biophysics, "A. Buzzati-Traverso", C.N.R., Naples, Italy; Laboratory of Stem Cell Bioengineering, Institute of Bioengineering, School of Life Sciences (SV) and School of Engineering (STI), Ecole Polytechnique Fédérale de Lausanne (EPFL), Lausanne, Switzerland; Department of Pharmaceutics, Utrecht Institute for Pharmaceutical Sciences (UIPS), Utrecht University, Utrecht, The Netherlands*

FERNANDO CALVO • *Tumour Microenvironment Team, Division of Cancer Biology, Institute of Cancer Research, London, UK*
LADISLAV CELKO • *Central European Institute of Technology (CEITEC), Brno University of Technology, Brno, Czech Republic*
GIULIO CERULLO • *Istituto di Fotonica e Nanotecnologie (IFN)-CNR and Department of Physics, Politecnico di Milano, Milano, Italy*
OVIJIT CHAUDHURI • *Department of Mechanical Engineering, Stanford University, Stanford, CA, USA*
YU-CHIH CHEN • *University of Michigan Comprehensive Cancer Center, Ann Arbor, MI, USA; Department of Electrical Engineering and Computer Science, University of Michigan, Ann Arbor, MI, USA*
DONG-WOO CHO • *Department of Mechanical Engineering, Pohang University of Science and Technology (POSTECH), Kyungbuk, Republic of Korea*
YOUNGJAE CHOE • *Wyss Institute for Biologically Inspired Engineering, Boston, MA, USA*
CRISTINA COLOSI • *Department of Chemistry, University of Rome "La Sapienza", Rome, Italy*
MARCO COSTANTINI • *Department of Chemistry, University of Rome "La Sapienza", Rome, Italy*
XIAOFENG CUI • *School of Chemistry, Chemical Engineering and Life Sciences, Wuhan University of Technology, Wuhan, China; Stemorgan Therapeutics, Albany, NY, USA; Clinic for Plastic Surgery and Hand Surgery, Klinikum Rechts der Isar, Technical University of Munich, Munich, Germany; Department of Biomedical Engineering, Rensselaer Polytechnic Institute, Troy, NY, USA*
J. LOWRY CURLEY • *AxoSim Technologies, LLC, New Orleans, LA, USA*
GUOHAO DAI • *Department of Biomedical Engineering, Rensselaer Polytechnic Institute, Troy, NY, USA*
MORGAN DELARUE • *Institut Curie, PSL Research University, CNRS, UMR 144, Paris, France; Physics Department, University of California, Berkeley, CA, USA*
MARIELLA DENTINI • *Department of Chemistry, University of Rome "La Sapienza", Rome, Italy*
STEPHANIE DESCROIX • *Institut Curie, PSL Research University, Institut Pierre Gilles De Gennes, CNRS UMR168, Paris, France; Sorbonne Universités, UPMC Univ Paris 06, Paris, France*
IRIS E. EDER • *Division of Experimental Urology, Department of Urology, Medical University of Innsbruck, Innsbruck, Austria*
THERESA EDER • *Division of Experimental Urology, Department of Urology, Medical University of Innsbruck, Innsbruck, Austria; Translational Radiooncology Laboratory, Department of Radiooncology and Radiotherapy, Charité University Hospital, Berlin, Germany; German Cancer Consortium (DKTK), German Cancer Research Center (DKFZ) Heidelberg, Partner Site Berlin, Berlin, Germany*
THOMAS C. FERRANTE • *Wyss Institute for Biologically Inspired Engineering, Boston, MA, USA*
MARINA FERRARINI • *Division of Experimental Oncology, San Raffaele Scientific Institute, Milan, Italy*

DAVIDE FERRARO • *Institut Curie, PSL Research University, Institut Pierre Gilles De Gennes, CNRS UMR168, Paris, France; Sorbonne Universités, UPMC Univ Paris 06, Paris, France*

ELISABETTA FERRERO • *Division of Experimental Oncology, San Raffaele Scientific Institute, Milan, Italy*

UWE FREUDENBERG • *Max Bergmann Center of Biomaterials Dresden, Leibniz Institute of Polymer Research Dresden, Dresden, Germany*

GUIFANG GAO • *School of Chemistry, Chemical Engineering and Life Sciences, Wuhan University of Technology, Wuhan, China; Stemorgan Therapeutics, Albany, NY, USA*

STEVEN C. GEORGE • *Department of Biomedical Engineering, Washington University in St. Louis, St. Louis, MO, USA*

MARIA-PAU GINEBRA • *Department of Materials Science and Metallurgical Engineering, Biomaterials, Biomechanics and Tissue Engineering Group, Technical University of Catalonia, Barcelona, Spain*

FLORIAN GROEBER • *Translational Center Wuerzburg, Fraunhofer Institute for Interfacial Engineering and Biotechnology (IGB), Wuerzburg, Germany*

PIYUSH B. GUPTA • *Whitehead Institute for Biomedical Research, Cambridge, MA, USA; Department of Biology and David H. Koch Institute for Integrative Cancer Research, Massachusetts Institute of Technology, Cambridge, MA, USA; Harvard Stem Cell Institute, Cambridge, MA, USA*

BASILE G. GURCHENKOV • *Institut Curie, PSL Research University, CNRS, UMR 144, Paris, France*

NIKOLAS K. HAASS • *Translational Research Institute, The University of Queensland Diamantina Institute, The University of Queensland, Brisbane, QLD, Australia; The Centenary Institute, Newtown, NSW, Australia; Discipline of Dermatology, University of Sydney, Sydney, NSW, Australia*

ALES HAMPL • *Department of Histology and Embryology, Faculty of Medicine, Masaryk University, Brno, Czech Republic; International Clinical Research Center – Center of Biomolecular and Cellular Engineering, St. Anne's University Hospital Brno, Brno, Czech Republic*

KAREN HUBBELL • *Stemorgan Therapeutics, Albany, NY, USA*

CHRISTOPHER C.W. HUGHES • *Department of Molecular Biology and Biochemistry, University of California, Irvine, CA, USA; Department of Biomedical Engineering, University of California, Irvine, CA, USA; Edwards Lifesciences Center for Advanced Cardiovascular Technology, University of California, Irvine, CA, USA*

DONALD E. INGBER • *Wyss Institute for Biologically Inspired Engineering, Boston, MA, USA; Harvard John A. Paulson School of Engineering and Applied Sciences, Cambridge, MA, USA; Vascular Biology Program, Boston Children's Hospital and Harvard Medical School, Boston, MA, USA*

ESMAIEL JABBARI • *Biomimetic Materials and Tissue Engineering Laboratory, Department of Chemical Engineering, University of South Carolina, Columbia, SC, USA; Department of Chemical Engineering, Swearingen Engineering Center, University of South Carolina, Columbia, SC, USA*

JOSEF JAROS • *Department of Histology and Embryology, Faculty of Medicine, Masaryk University, Brno, Czech Republic; International Clinical Research Center – Center*

of Biomolecular and Cellular Engineering, St. Anne's University Hospital Brno, Brno, Czech Republic
ZUZANA KOLEDOVA • *Department of Histology and Embryology, Faculty of Medicine, Masaryk University, Brno, Czech Republic*
ABRAHAM P. LEE • *Department of Biomedical Engineering, University of California, Irvine, CA, USA; Department of Mechanical and Aerospace Engineering, University of California, Irvine, CA, USA*
JELENA R. LINNEMANN • *Institute of Stem Cell Research, Helmholtz Center for Health and Environmental Research Munich, Neuherberg, Germany*
WENMING LIU • *College of Science, Northwest A&F University, Yangling, Shaanxi, China*
MATTHIAS LUTOLF • *Laboratory of Stem Cell Bioengineering, Institute of Bioengineering, School of Life Sciences (SV) and School of Engineering (STI), Ecole Polytechnique Fédérale de Lausanne (EPFL), Lausanne, Switzerland; Institute of Chemical Sciences and Engineering, School of Basic Science (SB), Ecole Polytechnique Fédérale de Lausanne (EPFL), Lausanne, Switzerland*
MARC MAZUR • *Wyss Institute for Biologically Inspired Engineering, Boston, MA, USA; University Medical Center Utrecht, Utrecht, CX, Netherlands*
GIOVANNA MAZZOLENI • *Laboratory of Tissue Engineering, Anatomy and Physiopathology Unit, Department of Clinical and Experimental Sciences, School of Medicine, University of Brescia, Brescia, Italy*
LISA K. MEIXNER • *Institute of Stem Cell Research, Helmholtz Center for Health and Environmental Research Munich, Neuherberg, Germany*
DANIEL H. MILLER • *Whitehead Institute for Biomedical Research, Cambridge, MA, USA; Department of Biology, Massachusetts Institute of Technology, Cambridge, MA, USA*
HARUKO MIURA • *Institute of Stem Cell Research, Helmholtz Center for Health and Environmental Research Munich, Neuherberg, Germany*
SEYEDSINA MOEINZADEH • *Biomimetic Materials and Tissue Engineering Laboratory, Department of Chemical Engineering, University of South Carolina, Columbia, SC, USA*
EDGAR B. MONTUFAR • *Central European Institute of Technology (CEITEC), Brno University of Technology, Brno, Czech Republic; Department of Materials Science and Metallurgical Engineering, Biomaterials, Biomechanics and Tissue Engineering Group, Technical University of Catalonia, Barcelona, Spain*
MICHAEL J. MOORE • *AxoSim Technologies, LLC, New Orleans, LA, USA; Department of Biomedical Engineering, Tulane University, New Orleans, LA, USA*
MICHELE M. NAVA • *Department of Chemistry, Materials and Chemical Engineering "Giulio Natta", Politecnico di Milano, Milano, Italy*
MATTHIAS NEES • *High-Content Screening Laboratory (HCSLab), Institute of Biomedicine, University of Turku, Turku, Finland*
RICHARD NOVAK • *Wyss Institute for Biologically Inspired Engineering, Boston, MA, USA*
PAOLA OCCHETTA • *Department of Electronics, Information and Bioengineering, Politecnico di Milano, Milano, Italy; Department of Biomedicine, University Hospital Basel, University Basel, Basel, Switzerland*
ROBERTO OSELLAME • *Istituto di Fotonica e Nanotecnologie (IFN)-CNR and Department of Physics, Politecnico di Milano, Milano, Italy*
FALGUNI PATI • *Department of Biomedical Engineering, Indian Institute Technology Hyderabad, Kandi, Telangana, India*

MICHAL PETROV • *TESCAN Brno, s.r.o., Brno, Czech Republic*
DUC T.T. PHAN • *Department of Molecular Biology and Biochemistry, University of California, Irvine, CA, USA*
TRAVIS A. PREST • *Department of Bioengineering, University of Pittsburgh, Pittsburgh, PA, USA*
TOBIAS PUKROP • *Department of Internal Medicine III, University Hospital Regensburg, Regensburg, Germany*
ANAS RABATA • *Department of Histology and Embryology, Faculty of Medicine, Masaryk University, Brno, Czech Republic*
MANUELA T. RAIMONDI • *Department of Chemistry, Materials and Chemical Engineering "Giulio Natta", Politecnico di Milano, Milano, Italy*
ROMANA E. RANFTL • *Tumour Microenvironment Team, Division of Cancer Biology, Institute of Cancer Research, London, UK*
MARCO RASPONI • *Department of Electronics, Information and Bioengineering, Politecnico di Milano, Milano, Italy*
CHRISTIAN REUTER • *Tissue Engineering and Regenerative Medicine, University Hospital Wuerzburg, Wuerzburg, Germany*
TOSHIRO SATO • *Division of Gastroenterology and Hepatology, Department of Internal Medicine, Keio University School of Medicine, Tokyo, Japan*
CHRISTINA H. SCHEEL • *Institute of Stem Cell Research, Helmholtz Center for Health and Environmental Research Munich, Neuherberg, Germany*
ARNDT F. SCHILLING • *Clinic for Plastic Surgery and Hand Surgery, Klinikum Rechts der Isar, Technical University of Munich, Munich, Germany*
ETHAN S. SOKOL • *Whitehead Institute for Biomedical Research, Cambridge, MA, USA; Department of Biology, Massachusetts Institute of Technology, Cambridge, MA, USA*
LOREDANA SPOERRI • *Translational Research Institute, The University of Queensland Diamantina Institute, The University of Queensland, Brisbane, QLD, Australia*
NATHALIE STEIMBERG • *Laboratory of Tissue Engineering, Anatomy and Physiopathology Unit, Department of Clinical and Experimental Sciences, School of Medicine, University of Brescia, Brescia, Italy*
SHINYA SUGIMOTO • *Division of Gastroenterology and Hepatology, Department of Internal Medicine, Keio University School of Medicine, Tokyo, Japan*
YOJI TABATA • *Laboratory of Stem Cell Bioengineering, Institute of Bioengineering, School of Life Sciences (SV) and School of Engineering (STI), Ecole Polytechnique Fédérale de Lausanne (EPFL), Lausanne, Switzerland*
MARKETA TESAROVA • *CEITEC BUT, Brno University of Technology, Brno, Czech Republic*
MERVI TORISEVA • *High-Content Screening Laboratory (HCSLab), Institute of Biomedicine, University of Turku, Turku, Finland*
DANIJELA MATIC VIGNJEVIC • *Institut Curie, PSL Research University, CNRS, UMR 144, Paris, France*
ROBERTA VISONE • *Department of Electronics, Information and Bioengineering, Politecnico di Milano, Milano, Italy*
LUCY VOJTOVA • *Central European Institute of Technology (CEITEC), Brno University of Technology, Brno, Czech Republic*

HEIKE WALLES • *Tissue Engineering and Regenerative Medicine, University Hospital Wuerzburg, Wuerzburg, Germany; Translational Center Wuerzburg, Fraunhofer Institute for Interfacial Engineering and Biotechnology (IGB), Wuerzburg, Germany*
JINYI WANG • *College of Science, Northwest A&F University, Yangling, Shaanxi, China*
XIAOLIN WANG • *Department of Micro/Nano Electronics, Shanghai Jiao Tong University, Shanghai, China*
CARSTEN WERNER • *Max Bergmann Center of Biomaterials Dresden, Leibniz Institute of Polymer Research Dresden, Dresden, Germany; Center for Regenerative Therapies Dresden, Technische Universität Dresden, Dresden, Germany*
KATRINA WISDOM • *Department of Mechanical Engineering, Stanford University, Stanford, CA, USA*
EUISIK YOON • *Department of Electrical Engineering and Computer Science, University of Michigan, Ann Arbor, MI, USA; Department of Biomedical Engineering, University of Michigan, Ann Arbor, MI, USA*
TOMMASO ZANDRINI • *Istituto di Fotonica e Nanotecnologie (IFN)-CNR and Department of Physics, Politecnico di Milano, Milano, Italy*

Chapter 1

3D Cell Culture: An Introduction

Zuzana Koledova

Abstract

3D cell culture is an invaluable tool in developmental, cell, and cancer biology. By mimicking crucial features of in vivo environment, including cell–cell and cell–extracellular matrix interactions, 3D cell culture enables proper structural architecture and differentiated function of normal tissues or tumors in vitro. Thereby 3D cell culture realistically models in vivo tissue conditions and processes, and provides in vivo like responses. Since its early days in the 1970s, 3D cell culture has revealed important insights into mechanisms of tissue homeostasis and cancer, and accelerated translational research in cancer biology and tissue engineering.

Key words 3D cell culture, Bioprinting, Extracellular matrix, Microenvironment, Microfluidics, Spheroid, Tissue architecture

1 3D Cell Culture: The Why

Tissue microenvironment, including extracellular matrix (ECM) and neighboring cells, is fundamental to tissue form and function and crucial in the control of tissue growth and development [1, 2]. However, traditional two-dimensional (2D) cell cultures, though easy and convenient to set up, lack proper environmental context and structural architecture and lead to changes in cell functions. When grown in 2D monolayers, normal epithelial cells often lose differentiation, become highly plastic and show characteristics displayed by tumor cells [3, 4], and malignant cells differ from their solid tumor counterparts [5].

Therefore, to provide a physiologically more relevant alternative to 2D cell culture, three-dimensional (3D) cell culture has been developed. By taking into account the crucial cell–cell and cell–ECM interactions, 3D culture allows the cells to replicate several critical features present in tissues, including morphology, differentiation, polarity [6], proliferation rate [7], gene expression [8–10] and genomic profiles [10, 11], and cellular heterogeneity and nutrient and oxygen gradients of tumors [12, 13].

Zuzana Koledova (ed.), *3D Cell Culture: Methods and Protocols*, Methods in Molecular Biology, vol. 1612,
DOI 10.1007/978-1-4939-7021-6_1, © Springer Science+Business Media LLC 2017

Importantly, 3D cell cultures span the gap between 2D cell cultures and animal models: 3D cell culture models incorporate physiologically relevant interactions while still being amenable to facile genetic manipulation, biochemical analysis, and imaging. By providing a reproducible, controlled microenvironment that mimics conditions in vivo, 3D cell cultures have become a valid alternative to the use of animal models. Moreover, due to the ability to create designer microenvironments in 3D [14], 3D cell cultures enable to study questions difficult to address in the organism. Among other findings, 3D cell culture models helped to demonstrate the fundamental rationale for 3D cultures themselves, i.e., that specific cellular and ECM microenvironment substantially influences cellular behavior. For example, when tumor cells were embedded in 3D collagen as single cells, they underwent individual cell migration and invasion; when embedded as small clusters, they invaded collectively; and when embedded as large aggregates, some of the cells invaded while some succumbed to necrosis [15, 16]. For another example, mouse mammary epithelial cells (MECs) were capable of in vivo-like structural organization and functional differentiation, including secretion of milk into lumen, only when cultured in reconstituted basement membrane matrix, but not when cultured in collagen I gels or in monolayers [17]. Subsequently, many studies corroborated on major effects of ECM composition and stiffness on cell signaling and behavior [18–23]. For example basal human MECs can form branching ducts with alveoli at their tips only when cultured in compliant matrix environment of floating collagen gels but not in rigid environment of adherent collagen gels [23], and increasing ECM stiffness alone induces malignant phenotypes in normal MECs [22].

2 3D Cell Culture: The How

Since their first employment in 1970s in the form of floating collagen gels, 3D cell culture systems have greatly evolved and are still undergoing development and refinements with the final goal to recreate entire organs in culture. Many fields of science have contributed tools and technologies, such as bioengineered materials, nanotechnology, microfluidics, and 3D bioprinting, to manipulate and control the cellular microenvironment (its chemistry, geometry, and mechanics) at multiple levels and scales with increased precision [24]. These advances have led to creation of myriad 3D cell culture systems designed according to specific scientific needs.

According to cellular input, cell culture format, and 3D microenvironment characteristics and means of production, several key types of 3D cell culture can be recognized [25]. This book endeavors to cover a wide range of these approaches (*see* Fig. 1 and Table 1).

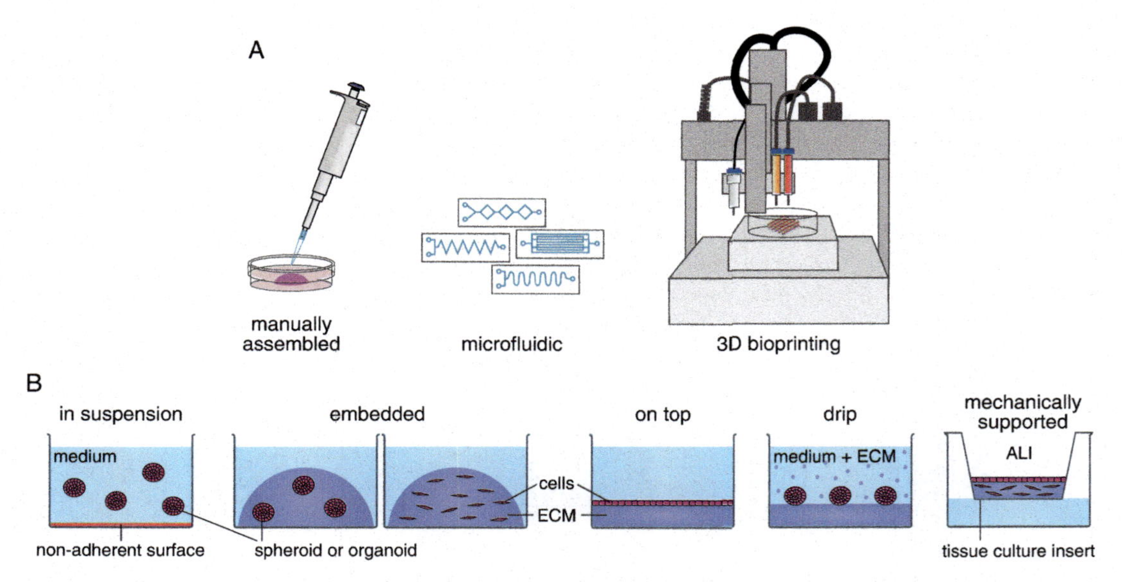

Fig. 1 3D cell culture techniques. (**a**) 3D cell cultures are assembled manually, using microfluidic devices or manufactured by 3D bioprinting. (**b**) For 3D cell culture, the cells can be cultured in suspension in dishes with non-adherent surfaces, embedded in ECM/hydrogel (as single cells, spheroids or organoids), cultured on top of the ECM or scaffold, on top of a layer of ECM with addition of ECM to the medium in a "drip" culture, or on tissue culture inserts for mechanical support as well as for the means of creating an air–liquid interface (ALI), if desired

The conceptually simplest and historically oldest type of 3D cell culture is ex vivo/in vitro culture of whole organs, tissue explants, or tissue slices, often mounted on mechanical supports (Chapter 16). Scientists have been culturing whole organs since late 1920s to study mechanisms of organ development [26, 27]. However, although this approach enables experimental observation and intervention within native tissue, tissue viability is limited and it is difficult to interpret results from these complex tissues.

Applying reductionist approach, organs/tissues/tumors can be deconstructed into their cellular subpopulations and ECM. These components can be selectively recombined in vitro to study specific questions. As the cellular component of 3D cell cultures, both primary cell and established cell lines can be used. Primary cells, either in the form of primary organoids (tissue fragments) or single-celled primary cells, are obtained by mechanical disruption and enzymatic digestion of donor tissue. Dissociated primary cells, including tumor cells, as well as established cell lines can be cultured randomly interspersed in ECM, grown in sheets or aggregated into spheroids (cell clusters) by a variety of methods. The most common methods to produce spheroids prevent cell surface adhesion and promote cell–cell attachment by culturing cells in non-adherent plates, such as agarose-coated flat-bottom plates (Chapters 8, 20, and 29), poly(2-hydroxyethyl methacrylate)-coated (Chapter 8) or ultra-low adhesion flat-, U-, or V-bottomed plates, micropatterned

Table 1
Overview of the 3D cell culture methods presented in this book

Chapter number	3D cell culture			Key method(s) described	Cell aggregation in 3D spheroids	Isolation of primary cells
	ECM/hydrogel	Format	Fluid dynamic			
2	Decellularized ECM	On top or embedded	Static	Tissue decellularization	No	No
3	Alginate with rBM	Embedded	Static	Production of interpenetrating networks of alginate and rBM	No	No
4	Biohybrid hydrogel	Embedded	Static	Production of biohybrid hydrogel (PEG with GAG and RGD)	No	No
5	Biohybrid hydrogel	Embedded	Static	Production of biohybrid hydrogel (TG–PEG with RGDSP)	No	No
6	CaP scaffold	On top	Static and dynamic	Manufacturing of CaP foam scaffold	No	No
7	rBM (Matrigel)	Embedded	Static	3D organoid culture	No	Mouse and human intestinal crypt cells
8	rBM (Matrigel)	Embedded	Static	3D coculture setup by micromanipulation	polyHEMA-treated plate or agarose-coated plate	Mouse mammary epithelial organoids and fibroblasts
9	Collagen	Embedded, floating	Static	3D culture in floating collagen gel	No	No
10	Collagen-based hydrogel	Embedded, floating	Static	3D culture in floating collagen gel	No	Human mammary epithelial organoids
11	rBM (Matrigel)	Embedded	Static	3D spheroid culture in non-adherent (polyHEMA-treated) plates	No	Mouse lung epithelial organoids and cells

12	N/A	Suspension	Static	3D spheroid culture in hanging drops	Hanging drop plate	No
13	N/A	Suspension	Dynamic	3D fluid-dynamic culture	No	No
14	Collagen	Embedded and ALI culture, mechanically supported	Static	3D cell culture in transwells	No	No
15	Collagen with Matrigel	Embedded and on top, mechanically supported	Static	Organotypic invasion assay	No	No
16	rBM (Cultrex)	Embedded, mechanically supported	Static	Organotypic tissue slice co-culture	No	No
17	Puramatrix or GMHA	Embedded	Static	Hydrogel micropatterning by photomask technique	No	No
18	Biohybrid hydrogel	Embedded	Static	Production of biohybrid hydrogels (SPELA, PEGDA and gelMA), micropatterning by photomask, micro-needle and soft lithography	No	No
19	SZ2080 scaffold	On top	Static	Micropatterning using two-photon polymerization	No	Bone marrow mesenchymal stem cells
20	Collagen	Embedded	Static	Microfuidic generation of collagen-embedded spheroids	Agarose-coated plate	No

(continued)

Table 1 (continued)

Chapter number	3D cell culture			Key method(s) described	Cell aggregation in 3D spheroids	Isolation of primary cells
	ECM/hydrogel	Format	Fluid dynamic			
21	N/A	Suspension	Static	Fabrication of PDMS microfluidic device	Microfluidic device (polyHEMA-treated)	No
22	N/A	Suspension	Dynamic	Fabrication of PDMS microfluidic device	Microfluidic device (Pluronic F127-treated)	No
23	N/A	Suspension	Static	Fabrication of PDMS microfluidic device	Microfluidic device	No
24	Laminin and fibrin	Embedded and on top, microfluidic	Dynamic	Fabrication of PDMS microfluidic device	No	No
25	Collagen	On top, Microfluidic	Dynamic	Fabrication of PDMS microfluidic device	No	Blood leukocytes
26	Alginate with gelM	Embedded	Static	PMMA microfluidic chip production by milling; Extrusion bioprinting	No	No
27	Decellularized ECM	Embedded	Static	Tissue decellularization, extrusion bioprinting	No	Human adipose-derived mesenchymal stem cells
28	PEGDA	Embedded	Static	Converting inkjet to 3D bioprinter, inkjet bioprinting	No	No

29	Collagen	Embedded	Static	3D real-time imaging by confocal microscopy	Agarose-coated plate	No
30	rBM (Matrigel)	Embedded	Static	Serial block-face imaging scanning electron microscopy, image processing and 3D visualization of ultrastructures	No	No
31	rBM (Matrigel)	Drip	Static	Confocal microscopy, quantitative phenotypic image analysis	No	No

ALI air–liquid interface, *CaP* calcium phosphate, *GAG* glycosaminoglycans, *gelM* gelatin methacrylate, *gelMA* gelatin methacrylic anhydride, *GMHA* glycidyl methacrylate hyaluronic acid, *PDMS* poly(dimethyl siloxane), *PEG* poly(ethylene glycol), *PEGDA* poly(ethylene glycol) diacrylate, *PMMA* poly(methyl methacrylate), *polyHEMA* poly(2-hydroxyethyl methacrylate), *rBM* reconstituted basement membrane, *RGD* and *RGDSP* adhesion receptor ligand peptides, *SPELA* star acrylate-functionalized lactide-chain-extended polyethylene glycol, *TG–PEG* transglutaminase–polyethylene glycol

plates and hanging drops (Chapter 12) (*see* Table 1). Alternatively, spheroids are generated by spontaneous aggregation or using spinner flasks and rotary cell culture systems [28].

For 3D cell culture per se, the cellular components are cultured: (1) in suspension on non-adherent surfaces (Chapter 11) or in bioreactors (Chapter 13); (2) embedded in hydrogels, such as collagen (Chapters 9 and 20), reconstituted basement membrane matrix (Chapters 7, 8, and 11), alginate, and their combinations (Chapter 3, 10, and 15), biohybrid hydrogels (Chapters 4, 5, and 18), and decellularized ECM (Chapters 2 and 27); (3) on mechanical supports, such as tissue culture inserts for air–liquid interface (Chapters 11, 14, and 16); or (4) on scaffolds (Chapter 6) and microcarriers (*see* Fig. 1 and Table 1). Conventionally, 3D cell cultures are assembled manually in cell culture dishes. Typically cells or organoids are suspended in hydrogel and plated in cell culture dish. The mechanical properties of the hydrogel can be controlled by adjusting its composition, such as concentration of collagen in collagen gels or concentration of calcium in alginate gels (Chapter 3) or by detaching an attached gel to create a floating-gel culture (Chapters 9 and 10). Furthermore, using several techniques, such as by photomask (Chapters 17 and 18), micro-needle (Chapter 18), two-photon polymerization (Chapter 19) or soft lithography (Chapter 18) techniques, the gel can be micropatterned to create specific patterns, such as of alternating low-stiffness and high-stiffness, or cell-permissive and cell-restrictive areas.

Recent technological advances have contributed microfluidic devices and 3D bioprinting to the 3D cell culture spectrum (Fig. 1). Microfluidic devices are used for high-throughput spheroid production (Chapters 21, 22, and 23), embedding into ECM (Chapter 20) and experimental testing (Chapter 23), or to create biomimetic models of organs, organs-on-chip (Chapters 24 and 25). Microfluidic organs-on-chip contain continuously perfused microchannels lined by living cells arranged to synthesize minimal functional units that simulate tissue- and organ-level functions [29]. By recapitulating the multicellular architectures, tissue–tissue interfaces, physiochemical microenvironments and vascular perfusion of the body, organs-on-chip enable to study organ level responses while allowing high-precision manipulation of the cell culture microenvironment and implementation through automation [29, 30]. Microfluidic devices are made by several techniques. Soft lithography, or replica molding on etched silicone chips for fabrication of elastomeric devices from poly(dimethyl siloxane), is the most commonly used technique and is employed to produce most of the microfluidic chips in this book (Chapters 21, 22, 23, 24, and 25). Other techniques for microchip fabrication include injection molding, micromolding, microetching, laser etching, micromilling (Chapter 26), photopolymerization and others [29].

3D bioprinting is a technique of additive manufacturing by which biological materials, biochemicals and living cells are layer-by-layer precisely positioned to fabricate 3D biomimetic structures

[31]. The main strategies for 3D bioprinting include inkjet, extrusion, laser-assisted and stereolithography bioprinting [31, 32], of which inkjet and extrusion bioprinting methods are described in this book for production of 3D tissue constructs (Chapters 26, 27, and 28) (*see* Table 1).

Besides the detailed protocols for 3D cell culture assembly, this book also provides useful protocols for 3D cell culture analysis. In comparison to 2D cell culture models, the analysis of 3D cell culture models requires more complex procedures. For biochemical and molecular biology analysis of samples, it is often necessary to remove the ECM, in which cells are embedded, prior to cell lysate preparation. For immunostaining of samples, longer times are required for cell culture fixation, antibody incubation as well as wash steps due to the presence of ECM that slows down solution penetration. Also, microscopy techniques that function well for thin, optically transparent cultures, are poorly suited for imaging rather thick and highly scattering 3D cell cultures. For non-destructive high-resolution imaging of 3D cell cultures, confocal microscopy (Chapters 29 and 31), multi-photon microscopy and optical coherence tomography can be used [33]. For ultrastructural analysis of 3D cell cultures, serial block-face imaging scanning electron microscopy (Chapter 30) is suitable.

3 3D Cell Culture: The Significance and Future Directions

3D cell culture techniques have become indispensable for in vitro modelling of fundamental developmental processes as well as human diseases. Due to realistic modelling of complex environment of a living tissue, 3D cell cultures have accelerated our understanding of mammalian organogenesis and cancerogenesis and have led to more advanced approaches in drug testing and personalized medicine. Yet several challenges remain in making fully physiologically relevant 3D cell culture systems, including incorporation of all the different cell types of a complex tissue, realistic modelling of regional differences in ECM composition, stiffness, and organization, or building of robust vascular networks that would enable the generation of larger tissue constructs.

This book provides standard, basic protocols for all the steps of 3D cell culture setup, from isolation of primary cells through preparation of decellularized ECM or biohybrid hydrogels to 3D cell culture assembly itself using several techniques, including manual assembly of 3D culture in cell culture dishes, microfluidic chips and 3D bioprinting. We believe that these protocols will be of use not only as a beginner's guide to the particular 3D cell culture techniques described but also as an inspiration and a starting point for development of new protocols and experimental designs as required according to specific scientific needs.

Acknowledgments

This work was supported by the project "Employment of Best Young Scientists for International Cooperation Empowerment" (CZ.1.07/2.3.00/30.0037) cofinanced from European Social Fund and the state budget of the Czech Republic, by the grant "Junior investigator 2015" (Faculty of Medicine, Masaryk University), and by the grant 16-20031Y from Czech Science Foundation (GACR).

References

1. Bissell MJ, Radisky D (2001) Putting tumours in context. Nat Rev Cancer 1(1):46–54
2. Wiseman BS, Werb Z (2002) Stromal effects on mammary gland development and breast cancer. Science 296(5570):1046–1049
3. Bissell MJ (1981) The differentiated state of normal and malignant cells or how to define a "normal" cell in culture. Int Rev Cytol 70:27–100
4. Petersen OW, Rønnov-Jessen L, Howlett AR et al (1992) Interaction with basement membrane serves to rapidly distinguish growth and differentiation pattern of normal and malignant human breast epithelial cells. Proc Natl Acad Sci U S A 89(19):9064–9068
5. Birgersdotter A, Baumforth KR, Porwit A et al (2007) Three-dimensional culturing of the Hodgkin lymphoma cell-line L1236 induces a HL tissue-like gene expression pattern. Leuk Lymphoma 48(10):2042–2053
6. Weaver VM, Petersen OW, Wang F et al (1997) Reversion of the malignant phenotype of human breast cells in three-dimensional culture and in vivo by integrin blocking antibodies. J Cell Biol 137(1):231–245
7. dit Faute MA, Laurent L, Ploton D et al (2002) Distinctive alterations of invasiveness, drug resistance and cell-cell organization in 3D-cultures of MCF-7, a human breast cancer cell line, and its multidrug resistant variant. Clin Exp Metastasis 19(2):161–168
8. Ghosh S, Spagnoli GC, Martin I et al (2005) Three-dimensional culture of melanoma cells profoundly affects gene expression profile: a high density oligonucleotide array study. J Cell Physiol 204(2):522–531
9. Li C, Kato M, Shiue L et al (2006) Cell type and culture condition-dependent alternative splicing in human breast cancer cells revealed by splicing-sensitive microarrays. Cancer Res 66(4):1990–1999
10. Ernst A, Hofmann S, Ahmadi R et al (2009) Genomic and expression profiling of glioblastoma stem cell-like spheroid cultures identifies novel tumor-relevant genes associated with survival. Clin Cancer Res 15(21): 6541–6550
11. De Witt Hamer PC, Van Tilborg AA, Eijk PP et al (2008) The genomic profile of human malignant glioma is altered early in primary cell culture and preserved in spheroids. Oncogene 27(14):2091–2096
12. Fischbach C, Chen R, Matsumoto T et al (2007) Engineering tumors with 3D scaffolds. Nat Methods 4(10):855–860
13. Friedrich J, Seidel C, Ebner R et al (2009) Spheroid-based drug screen: considerations and practical approach. Nat Protoc 4(3):309–324
14. Stoker AW, Streuli CH, Martins-Green M et al (1990) Designer microenvironments for the analysis of cell and tissue function. Curr Opin Cell Biol 2(5):864–874
15. Mueller-Klieser W (1997) Three-dimensional cell cultures: from molecular mechanisms to clinical applications. Am J Physiol 273(4 Pt 1):C1109–C1123
16. Friedl P (2004) Prespecification and plasticity: shifting mechanisms of cell migration. Curr Opin Cell Biol 16(1):14–23
17. Barcellos-Hoff MH, Aggeler J, Ram TG et al (1989) Functional differentiation and alveolar morphogenesis of primary mammary cultures on reconstituted basement membrane. Development 105(2):223–235
18. Cukierman E, Pankov R, Yamada KM (2002) Cell interactions with three-dimensional matrices. Curr Opin Cell Biol 14(5):633–639
19. Grinnell F (2003) Fibroblast biology in three-dimensional collagen matrices. Trends Cell Biol 13(5):264–269
20. Discher DE, Janmey P, Wang YL (2005) Tissue cells feel and respond to the stiffness of

their substrate. Science 310(5751): 1139–1143

21. Paszek MJ, Zahir N, Johnson KR et al (2005) Tensional homeostasis and the malignant phenotype. Cancer Cell 8(3):241–254
22. Chaudhuri O, Koshy ST, Branco da Cunha C et al (2014) Extracellular matrix stiffness and composition jointly regulate the induction of malignant phenotypes in mammary epithelium. Nat Mater 13(10):970–978
23. Linnemann JR, Miura H, Meixner LK et al (2015) Quantification of regenerative potential in primary human mammary epithelial cells. Development 142(18):3239–3251
24. Mroue R, Bissell MJ (2013) Three-dimensional cultures of mouse mammary epithelial cells. Methods Mol Biol 945:221–250
25. Shamir ER, Ewald AJ (2014) Three-dimensional organotypic culture: experimental models of mammalian biology and disease. Nat Rev Mol Cell Biol 15(10):647–664
26. Fell HB, Robinson R (1929) The growth, development and phosphatase activity of embryonic avian femora and limb-buds cultivated in vitro. Biochem J 23:767–784
27. Rak-Raszewska A, Hauser PV, Vainio S (2015) Organ in vitro culture: what have we learned about early kidney development? Stem Cells Int 2015:959807
28. Kim JB (2005) Three-dimensional tissue culture models in cancer biology. Semin Cancer Biol 15(5):365–377
29. Bhatia SN, Ingber DE (2014) Microfluidic organs-on-chips. Nat Biotechnol 32(8): 760–772
30. Luni C, Giulitti S, Serena E et al (2016) High-efficiency cellular reprogramming with microfluidics. Nat Methods 13(5):446–452
31. Murphy SV, Atala A (2014) 3D bioprinting of tissues and organs. Nat Biotechnol 32(8): 773–785
32. Mandrycky C, Wang Z, Kim K et al (2016) 3D bioprinting for engineering complex tissues. Biotechnol Adv 34(4):422–434
33. Graf BW, Boppart SA (2010) Imaging and analysis of three-dimensional cell culture models. In: Papkovsky DB (ed) Live cell imaging: methods and protocols, Methods in molecular biology, vol 591. Springer, New York, pp 211–227

Part I

Hydrogels and Scaffolds for 3D Cell Culture

Chapter 2

Preparation of Decellularized Biological Scaffolds for 3D Cell Culture

Bryan N. Brown, Michael J. Buckenmeyer, and Travis A. Prest

Abstract

The biggest challenge of designing and implementing an in vitro study is developing a microenvironment that most closely represents the interactions observed in vivo. Decellularization of tissues and organs has been shown to be an effective method for the removal of potentially immunogenic constituents while preserving essential growth factors and extracellular matrix (ECM) proteins necessary for proper cell function. Enzymatic digestion of decellularized tissues allows these tissue-specific components to be reconstituted into bioactive hydrogels through a physical crosslinking of collagen. In the following protocol, we describe unique decellularization methods for both dermis and urinary bladder matrix (UBM) derived from porcine tissues. We then provide details for hydrogel formation and subsequent three-dimensional (3D) culture of two cell types: NIH 3T3 fibroblasts and C2C12 myoblasts.

Key words Hydrogel, Extracellular matrix, Scaffolds, Cell culture, Decellularization

1 Introduction

It has been long understood that the extracellular matrix (ECM) not only provides structural support but also regulates cell growth [1, 2], survival, maturation, differentiation [3, 4], and development [5] of resident cells [6]. While many components of the ECM are conserved across several tissue types, each tissue is believed to possess a unique composition [7, 8]. Recently, scaffolds have been derived from ECM sourced from a variety of tissues including skin, fat, pericardium, heart, skeletal muscle, and liver for both in vivo and in vitro experiments [6, 9–12]. These scaffolds have been observed to assist with constructive remodeling or the formation of site-appropriate tissue when used as a biomaterial in vivo [13] and in some cases these ECM scaffolds have tissue-specific effects on cellular behavior [10].

Despite this unique interplay between the ECM and resident cells, in vitro studies often assess cell behavior on coatings consisting of either a single purified protein or directly on polystyrene

Zuzana Koledova (ed.), *3D Cell Culture: Methods and Protocols*, Methods in Molecular Biology, vol. 1612,
DOI 10.1007/978-1-4939-7021-6_2, © Springer Science+Business Media LLC 2017

tissue culture dishes [14]. These methods do not accurately mimic the complexity of the extracellular microenvironment and may significantly alter outcomes observed in vitro, making it difficult for studies to translate appropriately in vivo [6].

The use of ECM matrices during in vitro experiments is becoming increasingly common as these matrices are believed to better mimic the native cellular environments [6, 14, 15]. Here we describe the methods involved in decellularizing porcine dermis and UBM as well as the preparation of ECM hydrogels for subsequent three-dimensional (3D) culture of two cell types: NIH 3T3 fibroblasts and C2C12 myoblasts. Both porcine dermis and UBM are decellularized through mechanical delamination, followed by chemical and enzymatic washes. The resultant ECM scaffolds are solubilized using a hydrochloric acid and pepsin solution then reconstituted as hydrogels and seeded with fibroblasts or myoblasts. While the methods described below are designed for specific tissues and cell types, the basic principles can be applied for the decellularization of multiple tissue types and the subsequent formation of 3D cell culture models.

2 Materials

2.1 Reagents for Decellularization of Porcine Dermis and Urinary Bladder

1. Tissues: Porcine full thickness skin and urinary bladder acquired from a local abattoir.
2. Distilled H_2O (diH_2O).
3. Peracetic Acid (PAA) solution: 0.1% PAA and 4% ethanol. Mix diH_2O, 100% ethanol, and 15% PAA in the ratios shown in Table 1 based on tissue weight being decellularized.
4. 0.25% trypsin solution: Add 20 mL of 2.5% trypsin stock solution (10× solution in Hank's balanced salt solution) to 180 mL of diH_2O.
5. 3% H_2O_2: Add 20 mL of 30% H_2O_2 to 180 mL of diH_2O.
6. 70% ethanol.

Table 1
Volumes to prepare PAA solution based upon tissue weight

Tissue weight (g)	Volume diH_2O (mL)	Volume ethanol (200 proof, mL)	Total liquid volume (mL)	Volume of PAA (15%, mL)
10	192	8	200	1.33
20	384	16	400	2.67
30	576	24	600	4.00

7. Triton X-100 solution: 1% Triton X-100 in 0.26% EDTA and 0.69% Tris. Weigh out 56 mg of EDTA and 138 mg of Tris, combine in a graduated cylinder and then fill with diH_2O up to 200 mL. Transfer 198 mL of this EDTA and Tris solution to a flask and add 2 mL of Triton X-100 to this flask while a magnetic stir bar mixes the solution.
8. Phosphate-buffered saline (PBS, 1×): 137 mM NaCl, 2.7 mM KCl, 10 mM Na_2HPO_4, 2 mM KH_2PO_4 in diH_2O.

2.2 Reagents and Equipment for Enzymatic Digestion, and Hydrogel Formation

1. Porcine pepsin.
2. 0.1 N HCl.
3. 0.1 N NaOH.
4. 10× PBS.
5. 1× PBS.
6. Lyophilizer.
7. Mill with a size 40 mesh screen.
8. Non-humidified incubator.

2.3 Reagents for 3D Cell Culture

1. Cell lines: NIH 3T3 fibroblasts (CRL-1658, ATCC) and C2C12 myoblasts (CRL-1772, ATCC).
2. Cell culture medium: 10% fetal bovine serum (FBS), 100 U/mL penicillin, and 100 μg/mL streptomycin in Dulbecco's Modified Eagle Media (DMEM). In a cell culture hood (under sterile conditions), add 50 mL of FBS and 5 mL of 100× penicillin/streptomycin to 445 mL of DMEM and then sterile filter into an autoclaved bottle. Store at 4 °C.
3. Trypsin–EDTA solution: 2.5% (w/v) trypsin and 0.53 mM EDTA solution in diH_2O.
4. Sterilized stainless steel ring with an inner diameter 1.38 cm (*see* **Note 1**).

3 Methods

3.1 Decellularization of Porcine Dermis

1. Collect full thickness sections of skin from market weight, adult pigs. In particular, skin sections were collected from the dorsolateral flank.
2. Cut the skin into 35 cm × 50 cm rectangular sections and use a dermatome to separate the subcutaneous fat and connective tissue layers from the dermal layer. Alternatively, pre-delaminated skin may be sourced directly from your local abattoir (*see* **Note 2**).
3. The now isolated dermal layer can be stored until decellularization in a −80 °C freezer and thawed to room temperature before use.

Table 2
Approximate wash volumes used based on amount of tissue being decellularized

Weight (g)	5	10	15	20	25	30	35	40	45	50	55	60
Volume (mL)	100	200	300	400	500	600	700	800	900	1000	1100	1200

4. Place the dermis into the 0.25% trypsin solution and agitate within an Erlenmeyer flask using an orbital shaker at 300 rpm for 6 h. Approximate wash volumes based upon the amount of tissue being decellularized are shown in Table 2.
5. After 6 h, drain the trypsin solution and wash the dermal tissue three times in diH_2O for 15 min each while agitating at 300 rpm.
6. After the third diH_2O wash, agitate the tissue with a 70% ethanol solution for 10 h at 300 rpm.
7. Remove the 70% ethanol and replace with a 3% H_2O_2 solution and agitate for 15 min at 300 rpm.
8. After the completion of the H_2O_2 solution wash, immerse the dermis twice in diH_2O for 15 min each at 300 rpm.
9. Wash the dermis with Triton X-100 solution and shake for 6 h at 300 rpm. Drain and refill with fresh solution then agitate for an additional 16 h.
10. Follow with an additional three washes with diH_2O for 15 min each at 300 rpm.
11. Place the dermis in PAA solution and shake at 300 rpm for 2 h.
12. Wash the dermis in 1× PBS twice for 15 min at 300 rpm followed by two final washes in diH_2O again for 15 min at 300 rpm.
13. Store the decellularized dermis in the freezer at −80 °C.

3.2 Decellularization of Porcine Urinary Bladder

1. Collect urinary bladders from market weight, adult pigs.
2. Cut the apex and neck of the bladder off. Then cut along the connective tissue middle line forming a rectangular sheet that can be laid flat with the lumen side facing down.
3. Using a hard plastic scraper, stretch the tissue to approximately twice the previous area.
4. Then using a pair of scissors, lightly placed on the surface of the tissue, make partial thickness cuts down the centerline of the bladder.
5. Use forceps to carefully peel back the muscle and mucosal layers then discard. This should leave only the basement membrane and tunica propria of the urinary bladder remaining (*see* **Note 3**).

Fig. 1 Urinary bladder matrix (UBM) decellularization and hydrogel formation. Processing of tissues occurs in four general stages: (1) decellularization of tissues to remove genetic material while preserving ECM proteins, (2) lyophilization and milling of ECM into fine powder, (3) digestion of tissues using pepsin and HCl, and (4) hydrogel formation by the balancing of pH and salt concentrations and physical cross-linking of collagen at 37 °C

6. Rinse the remaining tissue in diH_2O to remove any extraneous materials.
7. Place the tissue in an Erlenmeyer flask containing PAA solution for 2 h on an orbital shaker at 300 rpm. The volume of this solution is relative to the amount of tissue being decellularized (Table 2).
8. Rinse the UBM twice with 1× PBS for 15 min each at 300 rpm.
9. Perform two final washes for 15 min each in diH_2O at 300 rpm.
10. Store the UBM in the freezer at −80 °C.

3.3 Lyophilization, Enzymatic Digestion, and Hydrogel Formation

This section describes the process of turning the solid ECM scaffold into a thermally responsive ECM hydrogel (Fig. 1).

1. Remove the frozen dermis and/or UBM from the −80 °C freezer and immediately place in the lyophilizer (*see* **Note 4**).
2. Ensure that the tissues are completely free of water by cutting through the thickest section (*see* **Note 5**).
3. After lyophilization is confirmed, grind the decellularized and lyophilized tissues into fine particles using a mill with a size 40 mesh screen (*see* **Note 6**).
4. Prepare 10 mL of stock solution of ECM digest at a concentration of 10 mg/mL by combining 100 mg of lyophilized and ground ECM with 10 mg porcine pepsin (final concentration 1 mg/mL pepsin) and 10 mL of 0.1 N HCl in a 25 mL Erlenmeyer flask. Adjust the necessary amount of pepsin and HCl according to the yield of ECM. Thoroughly mix the components using a magnetic stir bar for 48–72 h at room temperature until fully digested. Keep a piece of Parafilm on top of the flask to prevent evaporation (*see* **Notes 7** and **8**).
5. After the tissue has been fully digested, aliquot 1 mL of ECM into ten 1.5 mL microcentrifuge tubes and freeze at −80 °C until use (*see* **Note 9**).

Table 3
Calculated volumes of the necessary components for four concentrations for ECM hydrogel formation. All hydrogels are prepared using a starting concentration of 10 mg/mL and make a final gel volume of 0.5 mL

Starting concentration (mg/mL)	10	10	10	10
Final concentration (mg/mL)	2	4	6	8
Final volume (mL)	0.5	0.5	0.5	0.5
Volume ECM digest (μL)	100	200	300	400
Volume 0.1 N NaOH (μL)	10	20	30	40
Volume 10× PBS (μL)	11.11	22.22	33.33	44.44
Volume 1× PBS (μL)	378.89	257.78	136.67	15.56

3.4 Three-Dimensional Cell Culture of NIH 3T3 Fibroblasts and C2C12 Myoblasts In Vitro

This section describes a cell culture technique for preparation of single-celled suspension of cells and two techniques of 3D cell culture using the ECM hydrogel. All procedures should be performed using stringent cell culture techniques and all items should be sterilized prior to working with cells to minimize the potential for contamination.

1. Thaw the ECM digest on ice or at RT.
2. To prepare ECM hydrogels (*see* **Notes 10–13**), neutralize the pH of the ECM digest by adding 0.1 N NaOH (one-tenth digest volume), adjust the salt concentration by adding 10× PBS (one-ninth digest volume) and dilute the sample to the desired final ECM concentration using 1× PBS (*see* Table 3).

   While preparing ECM hydrogels, keep ECM digest and buffers on ice until immediate use. Mix the components in a new 1.5 mL microcentrifuge tube by gently pipetting up and down until the ECM digest goes into solution. Avoid formation of bubbles.
3. Place a sterilized stainless steel ring into a 6-well plate and pipette 0.5 mL of newly prepared hydrogel solution into it while avoiding bubble formation.
4. Carefully place the plate in a non-humidified incubator at 37 °C for 1 h. Remove all bubbles prior to moving the plate into the incubator (*see* **Notes 14** and **15**).
5. Remove the hydrogel from the incubator and ensure that the gel has solidified (Fig. 2).

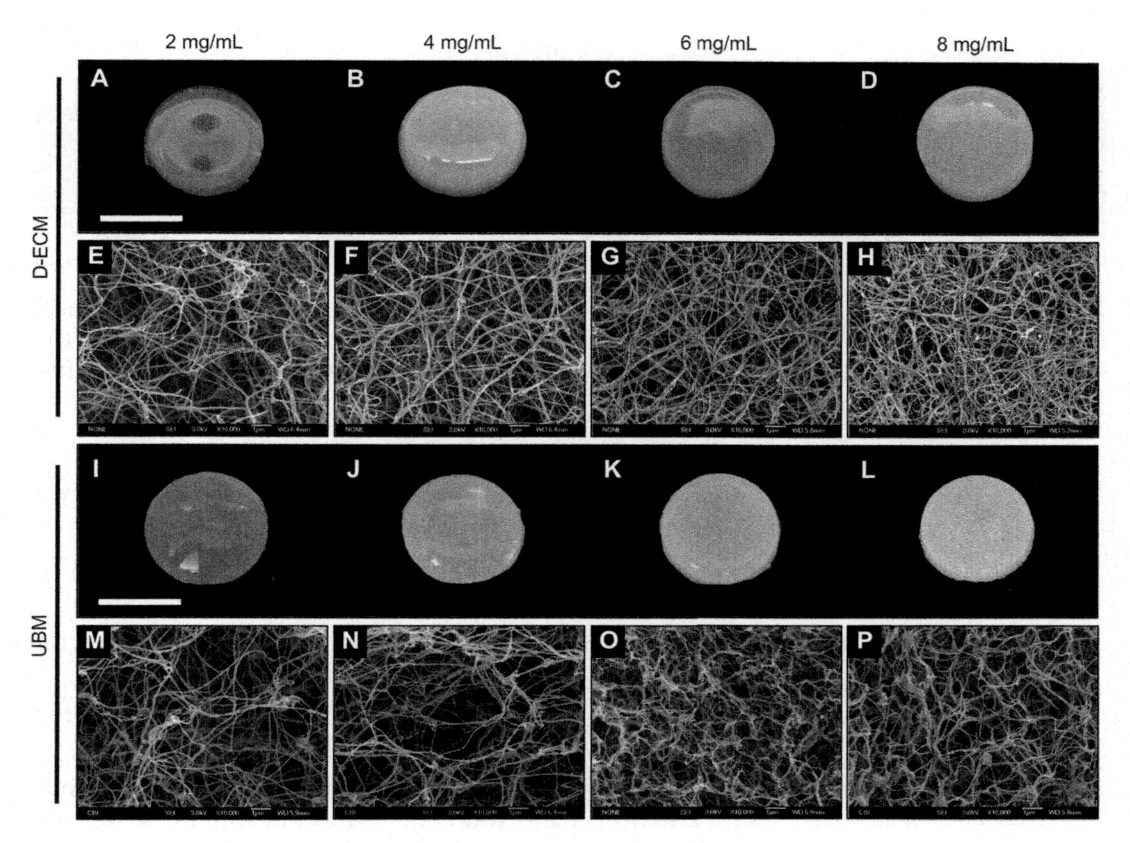

Fig. 2 Porcine dermis (D-ECM) (**A–D**) and UBM (**I–L**) hydrogels prepared at various concentrations: 2, 4, 6, 8 mg/mL. Hydrogels appear more opaque as ECM digest concentrations increase. Scanning electron microscopy provides detailed images of the ECM ultrastructure. (**E–H** and **M–P**) The fiber networks are significantly more advanced and complex at higher concentrations leading to increasing mechanical stiffness of the hydrogel. Reproduced from Wolf et al., 2012 [15] with permission from Elsevier B.V.

3.4.1 Preparation of Single-Celled Suspension and Sub-Culture of NIH 3T3 Fibroblasts and C2C12 Myoblast

1. To split cells (*see* **Note 16**), remove cell culture flask from the incubator and aspirate the medium without disturbing the cells.
2. Add 2–3 mL of warmed trypsin–EDTA solution to the flask and place in the incubator for 5–15 min.
3. After cells have detached, add 6–8 mL of fresh cell culture medium to the flask.
4. Transfer the cell suspension into a centrifuge tube and take an aliquot for counting e.g., using hemocytometer.
5. For subculturing, pipet the cell suspension at a seeding density of 2.25×10^5 to 3.75×10^5 cells/75 cm^2 for NIH 3T3 fibroblasts and 1.5×10^5 to 1×10^6 cells/75 cm^2 for C2C12 myoblasts into new flasks with appropriate amount of fresh cell culture medium and place in the cell culture incubator at 37 °C.

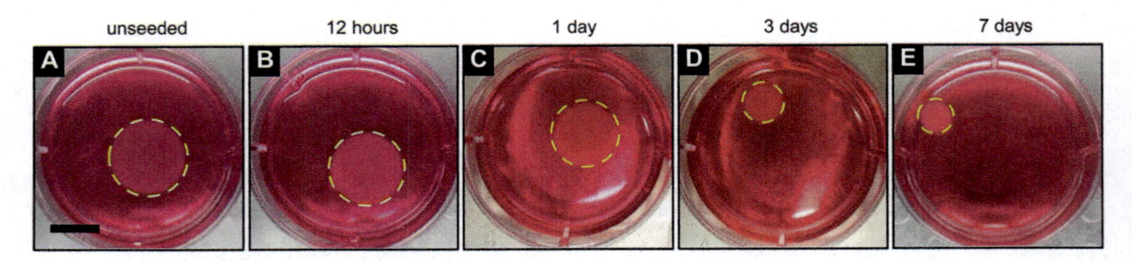

Fig. 3 In vitro cell culture of NIH 3T3 fibroblasts inside of UBM hydrogels at 6 mg/mL. Time lapse of the cell-hydrogel construct immersed in medium for 7 days shows a decrease in size due to cell contractility and migration. (**A**) unseeded gel, (**B**) 12 h, (**C**) 1 day, (**D**) 3 days, and (**E**) 7 days. Reproduced from Wolf et al., 2012 [15] with permission from Elsevier B.V.

3.4.2 3D Surface Culture

1. Under sterile conditions, prepare 500 μL of hydrogel at two concentrations: 6 and 8 mg/mL (*see* **Notes 17** and **18**).
2. Select a cell line for seeding onto the hydrogel: NIH 3T3 fibroblasts or C2C12 myoblasts (*see* **Note 19**).
3. Add 1 mL of cell suspension to the surface of the hydrogels for a final seeding density of 5.0×10^5 cells/cm^2.
4. Allow the cells about 16 h for adequate cell attachment to the hydrogel surface.
5. Once the cells have attached, remove the metal ring surrounding the hydrogel and add enough fresh medium to cover the hydrogel (Fig. 3) (*see* **Note 20**).
6. Replace the medium every 3 days to replenish essential nutrients for cell viability (Fig. 4) (*see* **Notes 21** and **22**).

3.4.3 3D Within-Gel Culture

1. Prepare single-celled suspension of cells (*see* Subheading 3.4.1).
2. Count the cells (e.g., using a hemocytometer) and adjust the cell suspension to a final concentration of 1.0×10^6 cells/mL of hydrogel. The cells for the "within-gel" culture replace the volume of the 1× PBS component with cell-containing medium. For example: In a 8 mg/mL hydrogel there should be 5.0×10^5 cells resuspended in 15.56 μL of cell media.
3. Pipet the pre-gel solution thoroughly to ensure uniform distribution of cells, then pipet the total mixture into the stainless steel ring.
4. Place the plate containing the cells and pre-gel mixture in a dry heat incubator at 37 °C for 45 min.
5. Once the gel has formed, remove the metal ring surrounding the hydrogel and add enough fresh medium to cover the hydrogel.
6. Replace the medium every 3 days to replenish essential nutrients for cell viability (*see* **Notes 21** and **22**).

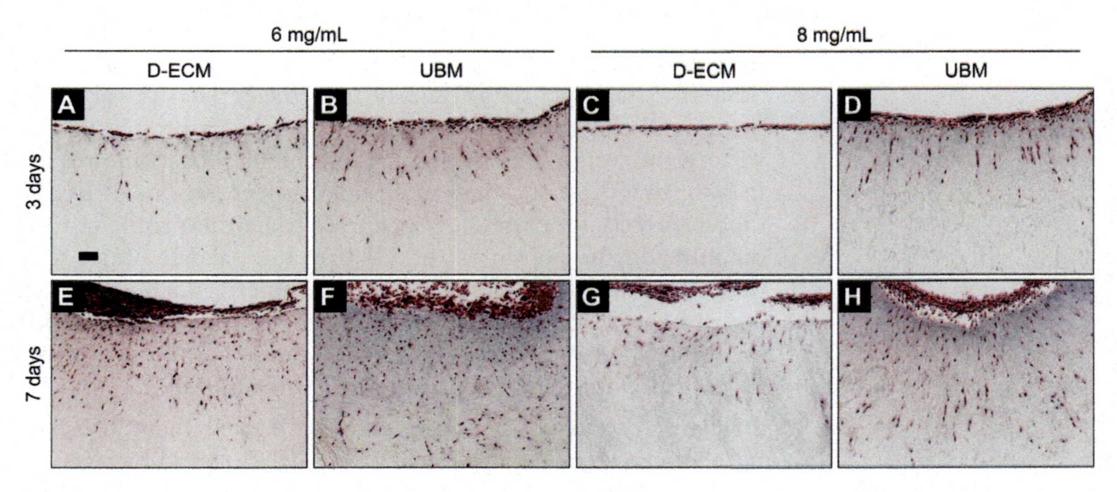

Fig. 4 NIH 3T3 fibroblasts seeded on top of both dermis and UBM hydrogels at two concentrations: 6 mg/mL and 8 mg/mL. Masson's Trichrome stained cross sections shows cell infiltration into ECM hydrogels after 3 (**a–d**) and 7 (**e–h**) days. Reproduced from Wolf et al., 2012 [15] with permission from Elsevier B.V.

4 Notes

1. Seeding rings are austenitic stainless steel with a 2.0 cm outer diameter and 1.38 cm inner diameter. Each ring has a height of 1.3 cm.
2. Isolating the dermal layer from full thickness skin without a dermatome, while not impossible, is very difficult. However, some abattoirs may sell the already isolated dermal tissue.
3. Effectiveness of the UBM decellularization protocol can be diminished if too many or too few tissue layers are removed during mechanical delamination. The basement membrane and tunica propria are approximately 100–150 μm in thickness. The tissue should appear almost but not completely transparent after delamination.
4. To freeze and lyophilize the UBM, spread the thin scaffold out on nonstick aluminum foil to increase surface area and hasten the drying process.
5. A non-lyophilized sample will make it difficult to collect the ECM from the mill. The moisture will prevent the tissue from freely flowing through the filter and into the collection chamber.
6. Using a 40 μm mesh screen with the mill provides small enough ECM particles to be easily digested. A larger mesh may lead to a longer incubation period for complete digestion of the ECM.

7. Pepsin can be difficult to weigh in small quantities due to its hygroscopic properties. It is very important to measure the desired mass of pepsin as quickly and efficiently as possible. A portion of the pepsin could be lost as it has a propensity to stick to the weigh boat during measurement. This could result in poor digestion and inaccurate final concentrations of ECM digest. A technique that has been used to fix this problem is to directly add the pepsin to the HCl on the balance. Once the desired volume of HCl is determined, pipette the acid into a weigh boat. Place the HCl and weigh boat onto the scale and "zero" the balance. Carefully add the desired amount of pepsin to the solution and add to the ECM for digestion.
8. Digesting the tissues in pepsin and HCl longer than 72 h has been shown to negatively impact both gelation and bioactivity of the hydrogels. Digestion must be monitored carefully to ensure that essential components of the ECM are conserved.
9. ECM digest should be kept to a minimal number of freeze–thaw cycles. Numerous freeze–thaw cycles are detrimental to the bioactivity of the hydrogels.
10. While preparing ECM hydrogels keep digest and buffers on ice until immediate use.
11. When creating the final concentrations for hydrogel formation, combine all buffers before adding the digest. For the "within-gel" culture, take care to add the 0.1 N NaOH and 10× PBS to the ECM digest prior to adding cells to ensure the cells are not damaged by either a basic or acidic environment. Be sure to mix the solution well by pipetting the mixture up and down.
12. The ECM digest and pre-gel solution is quite viscous and as much as 10% may be lost using normal air displacement pipettes. If available, a positive displacement pipette can reduce this volume loss and increase the precision of concentration measurements.
13. Basic mechanical characteristics of the gel can be determined using a parallel plate dynamic rheometer. The most basic characterization would be an amplitude sweep, while adding time and strain sweeps would be more encompassing.
14. Using a non-humidified chamber will assist with hydrogel gelation time. A humidified incubator adds moisture that limits the ability of the natural components in the gel to form. It is thus strongly recommended to use dry heat to prepare the hydrogels.
15. When mixing hydrogel components it can be very difficult to prevent bubble formation. However, there are ways to prevent bubbles and also to remove them. The best technique to prevent bubble formation is to slowly pipette the solution, while

keeping the pipette tip in the liquid. Once there is air in the tip bubbles will develop and can be difficult to remove. The other technique that has been successful is applying air pressure with a high volume micropipette. After the hydrogel solution has been prepared and pipetted into the metal ring, bubbles can be removed by directing large volumes of air directly toward the bubbles. The increased surface tension created by the air pressure should eliminate any residual bubbles. To create a uniform and consistent hydrogel architecture it is important to remove any and all bubbles from the gel mixture prior to gelation in the incubator.

16. Suggested volumes are based on cultures in a 75 cm^2 flask. The cells should be split at least twice per week at 80% confluency or less. It is important to not let the cells become confluent because this may result in contact inhibition. Media should be changed every 2–3 days. Cells may be frozen down for storage in complete growth medium supplemented with 5% (v/v) DMSO and stored in liquid nitrogen.
17. ECM derived from different tissues can be used in similar 3D cell culture experiments. However, ECM derived hydrogels from various tissue types including small intestinal submucosa, peripheral nerve, cardiac, and ovarian tissue can have varying properties. Therefore, the protocol described above may need to be adapted to work with ECM hydrogels derived from different tissues, particularly those with lower relative collagen I composition.
18. Selection of these concentrations was due to the differing mechanical and structural properties observed between these concentrations.
19. C2C12 myoblasts may be cultured both in growth and fusion medium. C2C12 myoblasts may be differentiated into myotubes using fusion medium containing DMEM with 2% horse serum, 100 U/mL penicillin and 100 μg/mL streptomycin for an additional 2 days after the 7 days of culture in growth medium. In a previous study, C2C12 myoblasts were used solely at a hydrogel concentration of 6 mg/mL.
20. For cell culture applications, it is essential that the hydrogel remains hydrated with medium to prevent the gel from drying out. A dehydrated hydrogel will result in altered mechanical properties leading to a detrimental cell response.
21. Cell viability of 3T3 fibroblasts and C2C12 myoblasts atop the ECM hydrogels can be analyzed using a Live/Dead assay kit.
22. Visualizing the cells within and atop the gels can be achieved by fixing the hydrogels in 10% neutral buffered formalin, splitting in half to expose a cross-section and embedding in paraffin. The paraffin block is then sectioned and stained with Masson's Trichrome (Fig. 5).

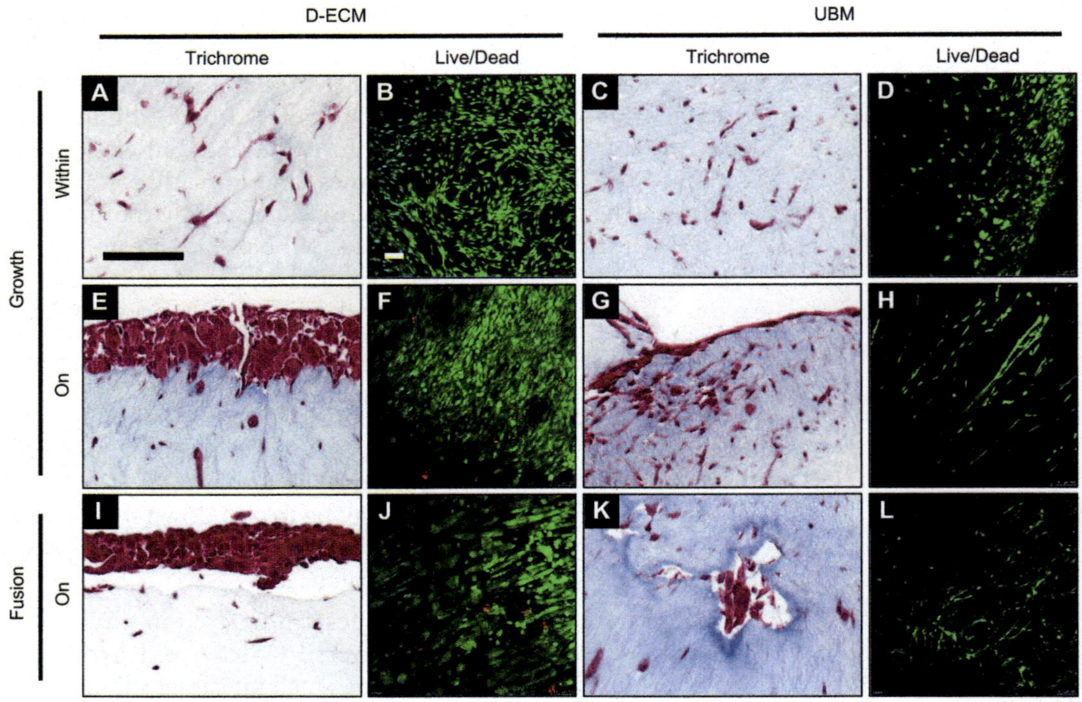

Fig. 5 C2C12 myoblasts cultured on the surface and within D-ECM (*left panel*) and UBM (*right panel*) hydrogels in both growth (**A–D** - *within gel*, **E–H** - *on top of gel*) and fusion (**I–L** - *on top of gel*) conditions (*see* **Note 19**). Cells were visualized and analyzed using histology and immunofluorescence staining methods. Masson's Trichrome stain was used on cross sections of the hydrogel while a Live/Dead staining of the hydrogel surface was used to visualize viable cells (*green*) and dead cell nuclei (*red*) (*see* **Notes 21–22**). Reproduced from Wolf et al., 2012 [15] with permission from Elsevier B.V.

Acknowledgments

A special thanks to Matt Wolf, Janet Reing, and Scott Johnson of the McGowan Institute for Regenerative Medicine for their assistance with technical details for the methods section.

References

1. Koochekpour S, Merzak A, Pilkington GJ (1995) Extracellular matrix proteins inhibit proliferation, upregulate migration and induce morphological changes in human glioma cell lines. Eur J Cancer 31A(3):375–380
2. Williams CM, Engler AJ, Slone RD et al (2008) Fibronectin expression modulates mammary epithelial cell proliferation during acinar differentiation. Cancer Res 68(9):3185–3192
3. Bosnakovski D, Mizuno M, Kim G et al (2006) Chondrogenic differentiation of bovine bone marrow mesenchymal stem cells (MSCs) in different hydrogels: influence of collagen type II extracellular matrix on MSC chondrogenesis. Biotechnol Bioeng 93(6):1152–1163
4. Chastain SR, Kundu AK, Dhar S et al (2006) Adhesion of mesenchymal stem cells to polymer scaffolds occurs via distinct ECM ligands and controls their osteogenic differentiation. J Biomed Mater Res A 78a(1):73–85
5. Simon-Assmann P, Kedinger M, De Arcangelis A et al (1995) Extracellular matrix components in intestinal development. Experientia 51(9–10):883–900

6. DeQuach JA, Mezzano V, Miglani A et al (2010) Simple and high yielding method for preparing tissue specific extracellular matrix coatings for cell culture. PLoS One 5(9):e13039
7. Lutolf MP, Hubbell JA (2005) Synthetic biomaterials as instructive extracellular microenvironments for morphogenesis in tissue engineering. Nat Biotechnol 23(1):47–55
8. Uriel S, Labay E, Francis-Sedlak M et al (2009) Extraction and assembly of tissue-derived gels for cell culture and tissue engineering. Tissue Eng Part C Methods 15(3):309–321
9. Stern MM, Myers RL, Hammam N et al (2009) The influence of extracellular matrix derived from skeletal muscle tissue on the proliferation and differentiation of myogenic progenitor cells ex vivo. Biomaterials 30(12):2393–2399
10. Zhang Y, He Y, Bharadwaj S et al (2009) Tissue-specific extracellular matrix coatings for the promotion of cell proliferation and maintenance of cell phenotype. Biomaterials 30(23–24):4021–4028
11. Young DA, Ibrahim DO, Hu D et al (2011) Injectable hydrogel scaffold from decellularized human lipoaspirate. Acta Biomater 7(3):1040–1049
12. Seif-Naraghi SB, Salvatore MA, Schup-Magoffin PJ et al (2010) Design and characterization of an injectable pericardial matrix gel: a potentially autologous scaffold for cardiac tissue engineering. Tissue Eng Part A 16(6):2017–2027
13. Freytes DO, Martin J, Velankar SS et al (2008) Preparation and rheological characterization of a gel form of the porcine urinary bladder matrix. Biomaterials 29(11):1630–1637
14. DeQuach JA, Yuan SH, Goldstein LS et al (2011) Decellularized porcine brain matrix for cell culture and tissue engineering scaffolds. Tissue Eng Part A 17(21–22): 2583–2592
15. Wolf MT, Daly KA, Brennan-Pierce EP et al (2012) A hydrogel derived from decellularized dermal extracellular matrix. Biomaterials 33(29):7028–7038

Chapter 3

3D Cell Culture in Interpenetrating Networks of Alginate and rBM Matrix

Katrina Wisdom and Ovijit Chaudhuri

Abstract

Altered tissue mechanical properties have been implicated in many key physiological and pathological processes. Hydrogel-based materials systems, made with native extracellular matrix (ECM) proteins, nonnative biopolymers, or synthetic polymers are often used to study these processes in vitro in 3D cell culture experiments. However, each of these materials systems present major limitations when used in mechanobiological studies. While native ECM-based hydrogels may enable good recapitulation of physiological behavior, the mechanics of these hydrogels are often manipulated by increasing or decreasing the protein concentration. This manipulation changes cell adhesion ligand density, thereby altering cell signaling. Alternatively, synthetic polymer-based hydrogels and nonnative biopolymer-based hydrogels can be mechanically tuned and engineered to present cell adhesion peptide motifs, but still may not fully promote physiologically relevant behavior. Here, we combine the advantages of native ECM proteins and nonnative biopolymers in interpenetrating network (IPN) hydrogels consisting of rBM matrix, which contains ligands native to epithelial basement membrane, and alginate, an inert biopolymer derived from seaweed. The following protocol details the generation of IPNs for mechanical testing or for 3D cell culture. This biomaterial system offers the ability to tune the stiffness of the 3D microenvironment without altering cell adhesion ligand concentration or pore size.

Key words Alginate, Reconstituted basement membrane, Interpenetrating networks, Hydrogels, ECM stiffness, 3D culture, 3D materials systems, Biomaterials

1 Introduction

Mechanical properties of the microenvironment have long been recognized to play key roles in development, tissue remodeling, and disease progression [1]. Within the last two decades, new technologies have started to enable measurement of mechanical forces on cellular and molecular scales, as well as new biomaterials to enable 3D cell culture studies [2, 3]. Due to these tools, compelling evidence continues to emerge showing that cells sense and respond to microenvironmental mechanics in 3D environments differently than they do on 2D substrates [2, 4]. This has

Zuzana Koledova (ed.), *3D Cell Culture: Methods and Protocols*, Methods in Molecular Biology, vol. 1612,
DOI 10.1007/978-1-4939-7021-6_3,

increasingly motivated the use of hydrogel-based, 3D materials systems in mechanobiological research [3]. Because native extracellular matrix (ECM)-based hydrogels like collagen and reconstituted Basement Membrane (rBM) matrix provide biologically relevant proteins, they are often used to recapitulate physiological behavior. However, the mechanics of these native ECM-based hydrogels are often manipulated by increasing or decreasing cell ligand concentration. Recent work has demonstrated that altered ligand density can change the binding of integrins, transmembrane receptors that facilitate cell-ECM interactions, thereby activating cell signaling that can obscure the effects of altered microenvironmental stiffness [5]. The mechanics of these native ECM-based hydrogels have also been altered directly through the addition of cross-linking sites within the matrix [6]. However, this approach may alter the pore size and ligand accessibility within the hydrogel [7]. Alternatively, synthetic polymer-based hydrogels (i.e. polyethylene-glycol and polyacrylamide) and nonnative biopolymer-based hydrogels (i.e. alginate and agarose) can be mechanically tuned and separately engineered to present cell adhesion peptide motifs, but these peptide motifs may not fully recapitulate the physiological behavior. To capture physiological behavior and to control stiffness and ligand density separately, interpenetrating networks (IPNs) consisting of rBM and alginate were designed and implemented in 3D cell culture experiments.

rBM, which was discovered in the 1970s and has been widely used for cell culture, is derived from the mouse Engelbreth-Holm-Swarm (EHS) tumor and contains key proteins normally found in the basement membrane, including laminin-111 and collagen IV [8]. rBM forms a matrix at physiological temperatures due to bonding interactions between constituent proteins [9].

The tunable alginate network serves to mechanically reinforce the rBM matrix. Alginate, a polysaccharide derived from seaweed, has been used extensively in cell culture and tissue engineering applications [10]. Because alginate does not present cell adhesion ligands and is not susceptible to degradation by mammalian enzymes, it is useful as the inert, tunable, mechanical framework of this 3D culture system [10]. Alginate is a non-repeating copolymer of β-D mannuronic acid (M) and α-L guluronic acid (G), which are arranged in continuous blocks of M-residues (M-blocks), blocks of G-residues (G-blocks), or alternating MG-residues (MG-blocks) [11]. To form a 3D hydrogel, the G-blocks along the alginate polymer chains can be ionically cross-linked by divalent cations, such as calcium (Ca^{2+}) [10]. To alter the mechanical properties of the network, the concentration of calcium used to cross-link the network can be modulated, with increased cross-linking leading to a stiffer hydrogel.

By mixing rBM and alginate, and by modulating the initial concentration of calcium used to cross-link the alginate, the overall stiffness (G' ~20 to ~300 Pa, corresponding to E ~60 Pa to ~1 kPa) of these IPNs can be tuned while holding the concentration of rBM, and therefore ligand density, constant. The pore structure and ligand accessibility are not altered with increased calcium concentration, as has been observed in alginate gels [5]. This is likely because the G-blocks become aligned at low concentrations of cross-linking, and further cross-linking serves only to fill in the G-blocks [12, 13]. While the presence of calcium could influence cells, altered cross-linking over a range of 0–20 mM was not found to impact epithelial morphogenesis [5]. Here, we detail the design and generation of IPNs, as well as how to encapsulate cells within them, for broad use in 3D cell culture studies.

2 Materials

2.1 IPN Components

1. Alginate (FMC Biopolymer, Protanal® LF 20/40). Alginate must be purified and lyophilized, as previously described [10].
2. Corning® Matrigel® Growth Factor Reduced (GFR) Basement Membrane Matrix (rBM) (Corning Life Sciences, 354230) (*see* **Note 1**).
3. 1.22 M Sterile Calcium Slurry: Dissolve calcium sulfate dihydrate (Sigma, 255548) in deionized water (*see* **Note 2**).
4. BD™ Luer-Lok™ Disposable Syringes without Needles (Becton Dickinson, 309628 or 309657) (*see* **Note 3**).
5. Female Luer Thread Style Coupler, White Nylon (Nordson Medical, FTLLC-1). Autoclaved and kept sterile.
6. Micro stir bars (VWR International).
7. Kimble 20 mL Glass Screw-Thread Scintillation Vials (Fisher Scientific, 22-043-725).

2.2 Cell Culture Components

1. Flat bottomed, tissue culture treated, 24- or 12-well plates.
2. Corning® Costar® Microcentrifuge 1.7 mL Tubes with Snap Cap.
3. 15 mL conical centrifuge tubes.
4. Container of ice.
5. Aluminum foil.
6. 40 μm cell strainer.
7. Trypsin-EDTA, 0.05%.
8. Serum-free medium.
9. Complete growth medium.

10. Phosphate buffered saline (PBS).
11. Cell counting instrument (i.e. Vi-Cell™ or hemocytometer).
12. Optional: 12 mm or 6.5 mm Transwell® 3.0 μm Pore Polycarbonate Membrane Inserts, Sterile (Corning Life Sciences/Thermo Fisher Scientific, 3402 or 3415).
13. Optional: Razor blade.

3 Methods

The procedure for encapsulating cells into IPNs is broken up into three sections: *Days Prior to Encapsulation, Hours Prior to Encapsulation, and Encapsulation.* Handling of reagents prior to and during the encapsulation procedure should be carried out in a sterile environment suitable for cell culture (i.e. tissue culture hood). If IPNs will be utilized for mechanical testing rather than for 3D cell culture, *see* **Note 4**.

3.1 Days Prior to Encapsulation

1. Design IPNs. Determine (a) total volume, (b) alginate concentration, (c) rBM concentration, and (d) calcium concentration (based on desired stiffness) of each IPN in the system (*see* Fig. 1 and **Note 5**).
2. Using sterile technique, reconstitute alginate at the desired concentration in serum-free medium within a scintillation vial. Stir for several hours, or at most overnight, at room temperature with a sterile micro stir bar until alginate fully dissolves (*see* **Note 6**).
3. Using a pipette or syringe, aliquot desired volume of alginate for each IPN into 1.7 mL centrifuge tubes (*see* **Note 7**).
4. Thaw rBM overnight at 4 °C on ice.

3.2 Hours Prior to Encapsulation

1. Dilute the Sterile Calcium Slurry with serum-free medium to make desired calcium working solutions, or Calcium Stocks

	IPN Design Parameters					Materials				Cells and Polymers Syringe				Crosslinking Syringe	
Type	Total Vol (μL)	IPN Alg Conc. (mg/mL)	IPN rBM Conc. (mg/mL)	IPN Cell Conc. (cells/mL)	Ca in IPN (mM)	Alg (mg/mL)	rBM Batch Conc. (mg/mL)	Cell conc. (cells/mL)	Ca in Stock (mM)	Vol rBM Needed + 10% extra (μL)	Alginate (μL)	Extra Medium (μL)	Cells (μL)	Ca Stock (μL)	Extra Medium for Ca (μL)
STIFF (E~1 kPa)	500	5	4.4	4.39E+05	20	25	10.3	5.40E+06	488	353.1	150	68	61.0	20	80
SOFT (E~60 Pa)	500	5	4.4	4.39E+05	0	25	10.3	5.40E+06	488	353.1	150	68	61.0	0	100

Black-shaded parameters can be altered to change IPN stiffness while gray-shaded parameters are constant across IPNs in the system. IPN Design Parameters and Materials determine the volumes of (1) rBM, Alginate, Extra Medium, and Cell Suspension in the *Cells and Polymers* syringe, and (2) Calcium Stock and Extra Medium in the *Crosslinking* syringe. Account for rBM sticking to pipette tips by withdrawing 10% more rBM than is needed for the *Cells and Polymers* solution. For instance, in this example, assume for calculations that 321 μL is delivered into the *Cells and Polymers* solution, even though 353.1 μL of rBM was withdrawn from the rBM stock

Fig. 1 Example Interpenetrating Network (IPN) formulations for stiff and soft hydrogels

(between 10% and 40% Sterile Calcium Slurry, i.e. between 122 and 488 mM calcium) in 15 mL Falcon tubes (*see* **Note 8**).

2. Make the *Cross-linking* syringes for all IPNs. Label syringes and chill on foil-covered ice. Deposit the appropriate volume of Calcium Stock, and volume of serum-free medium, in a syringe such that (a) the final concentration of calcium in the IPN will be as desired, and (b) the total volume in the *Cross-linking* syringe is 1/5 of the total volume of the final IPN.
3. Open the other syringes that will be used in the experiment (for the *Cells and Polymers* syringes). Leave these, also, to cool on foil-covered ice.
4. Coat the bottom of the well plates that will be used in the encapsulation with a small volume (~30–50 μL) of rBM. This will help prevent cells from contacting the tissue culture plastic on the bottom of the well plates and growing as a 2D monolayer.

3.3 Encapsulation

1. Trypsinize cells and resuspend them in a small volume (~1–3 mL) of serum-free medium, as per cell type-specific cell culture recommendations. Keep cells on ice during the encapsulation procedure.
2. Use a 40 μm cell strainer to obtain a single-cell suspension. Count the number of viable cells. To obtain the desired cell concentration in the IPNs, dilute cell suspension as needed (*see* **Note 9**). The extra volume of serum-free medium that has been included in the *Cells and Polymers* syringe solution can be increased or decreased in order to accommodate the volume of cell solution needed to achieve the desired cell concentration in the final IPNs.
3. *Make each IPN separately*, using the following procedure (*see* Fig. 2).

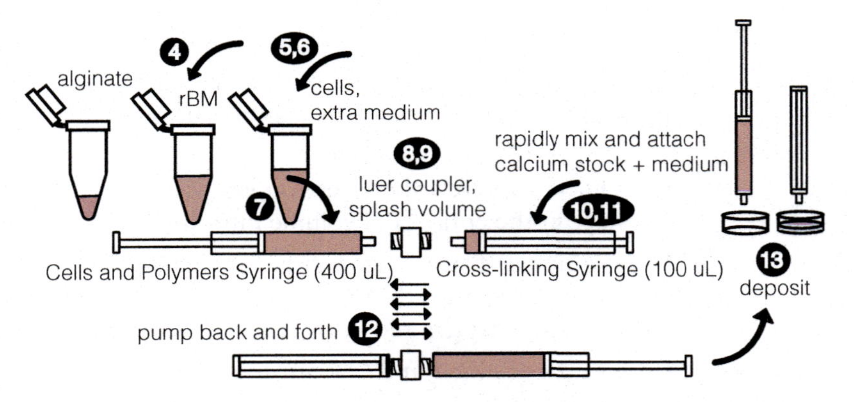

Fig. 2 Schematic depicting the steps for making the Interpenetrating Network (IPN) hydrogel with alginate and rBM for use in 3D cell culture studies

4. Mix rBM stock 5–10 times by pipetting. Withdraw the volume required for one IPN, plus an extra 10% to compensate for some of the rBM sticking to the pipette tip. Deposit this volume into a pre-aliquotted 1.7 mL centrifuge tube of reconstituted alginate. Mix by pipetting 20–30 times, being careful not to create bubbles (*see* **Note 10**).
5. Next, deposit the desired volume of single cell suspension into the alginate-rBM mixture (*see* **Note 11**). Mix the *Cells and Polymers* solution by pipetting and/or stirring.
6. Lastly, deposit the appropriate amount of serum-free medium to reach the desired volume of the *Cells and Polymers* solution, which is 4/5 of the final IPN volume, plus 50% extra (*see* **Note 12**).
7. Using a P1000, withdraw the *Cells and Polymers* solution (*see* **Note 13**). To transfer this solution into an empty, cooled syringe, insert the pipette tip into the syringe nozzle and then pull the syringe plunger. While holding the syringe upright, gently tap the syringe and cycle the plunger so bubbles rise to the top of the syringe. Expel bubbles and excess solution until the final volume in this syringe is 4/5 of the final IPN volume (*see* **Note 14**).
8. Secure a Luer coupler onto the top of the *Cells and Polymers* syringe, and push the plunger up such that the solution is ~2 mm from the top of the Luer coupler.
9. Deposit 40 μL of serum-free medium into the top of the Luer coupler. This will serve as the "splash volume" that will be lost when the second syringe is connected to the coupler. Place syringe back on foiled ice.
10. The next steps should be done very rapidly as hydrogel gelation occurs quickly. First, agitate the *Cross-linking* syringe using small and rapid motions with the plunger.
11. Before the calcium can settle, twist the *Cross-linking* syringe into the other side of the Luer coupler.
12. Quickly pump both plungers back and forth, six times total, depressing the entire IPN solution into one plunger with the last pump.
13. Quickly unscrew the Luer coupler with the other syringe, and deposit the entire IPN volume into the rBM-coated well plate.
14. (Optional) Place Transwell inserts on top of the IPNs. These will help maintain the structural integrity of the hydrogel during medium changes and keep gels from floating.
15. Let the IPN gel in 37 °C incubator with CO_2 for 30–50 min (*see* **Note 15**).

16. Drip 2× the IPN volume's worth of growth medium down the side of the well. Change medium as needed for 3D culture experiment (*see* **Note 16**).

4 Notes

1. Matrigel® concentration varies by lot. Due to volume constraints from mixing with other components for IPNs, we recommend purchasing rBM that is a concentration of at least 9.0 mg/mL.
2. Add a magnetic stir bar to this solution and then autoclave. Stir overnight after autoclaving to get a uniform dispersion of calcium sulfate slurry in water.
3. We suggest using 1 mL syringes for 500 μL IPNs and 3 mL syringes for 1 mL IPNs, as free space in the syringe is needed to generate enough force for thorough mixing.
4. For IPN mechanical testing rather than cell encapsulation, follow Subheadings 3.1 (all) and 3.2 (**steps 1–3**). In Subheading 3.3, proceed directly to **step 3** through **13**. Cell suspension volumes can be replaced with equal volume of serum-free medium. Because there may be batch-to-batch variability with both rBM and alginate, we recommend testing the mechanical properties of the IPNs, and adjusting IPN design parameters accordingly, before using IPNs for 3D culture.
5. The maximum concentration of rBM possible in the final IPNs will depend on the lot-specific concentration of rBM used. Because we have seen that normal mammary epithelial cells (MCF10A) recapitulate their canonical acinar structure in IPNs containing 4.4 mg/mL rBM [5], we recommend using at least this concentration of rBM in the final IPNs. With regard to alginate concentration, we have used IPNs with final concentrations ranging from 5.0 to 10.0 mg/mL of alginate. For imaging and biochemical studies, 500 μL IPNs are typically used and placed in one well of a 24-well plate.
6. Reconstitute the alginate to between 15 and 35 mg/mL, depending on the desired concentration of alginate in the final IPNs, within 2–3 weeks of encapsulating and store at 4 °C.
7. Depending on the concentration at which the alginate was reconstituted, it may be too viscous to pipette. In this case, we recommend using a syringe or cut pipette tip to aliquot. For cut tip, cut 5–10 mm off the tip of a P1000 pipette tip using a clean razor blade.
8. Because the calcium settles quickly, the calcium slurry is hard to work with in small volumes. Vortex, pipette, and stir the Sterile Calcium Slurry vigorously when making Calcium Stocks.

9. The concentration of cells needed in the single-cell suspension, given a desired cell concentration in the final IPNs, can be calculated as follows:

$$C_{cells} = C_{cells\,IPN} = \left(\frac{TV_{cellpolymer}}{TV_{cells}}\right) \times \left(\frac{FV_{IPN}}{FV_{cellpolymer}}\right)$$

C_{cellsIPN} = Desired concentration of cells in final IPN.

C_{cells} = Concentration of single cell suspension (cells/mL).

$TV_{\text{cellpolymer}}$ = Total volume made of Cells and Polymers Solution (μL).

TV_{cells} = Total volume of cells used in making Cells and Polymers Solution (μL).

FV_{IPN} = Final volume of IPN (μL).

$FV_{\text{cellpolymer}}$ = Final volume of Cells and Polymers Solution used in IPN (μL).

10. For highly viscous solutions, cut the pipette tip with a razor blade to pipette, or mix the alginate-rBM solution by thoroughly stirring with a pipette tip.
11. Make sure to mix by pipetting the cell suspension 15–20 times before withdrawing the solution, as the cells will settle over time.
12. It is recommended to make 50% extra of the *Cells and Polymers* solution in order to account for losses due to pipette transferring and bubbles. Any extra amount can be expelled once the solution is transferred to the syringe.
13. If this solution is highly viscous, set the P1000 to its lowest volume and submerge it in the solution. While simultaneously stirring the solution, slowly turn the knob of the P1000 to gradually pipette up the viscous solution.
14. Do not forget to account for the volume in the barrel of the syringe, which is about 40–60 μL. For example, for a 500 μL IPN, the *Cells and Polymers* syringe should contain 400 μL (i.e. 4/5 total volume of IPN). Therefore the *Cells and Polymers* syringe should be filled to the 360 μL demarcation on the syringe. Extra solution can be expelled at this point.
15. Each IPN takes about 4–5 min to make. If an experiment requires many gels to be made in sequence, we suggest using separate well plates and adding media at different times, accordingly.
16. Be careful not to pipette directly onto the gel during medium changes. Instead, carefully withdraw or drip medium from sides of the wells.

Acknowledgements

This work was supported by the National Science Foundation Graduate Research Fellowship to Katrina Wisdom.

References

1. Eyckmans J, Boudou T, Yu X et al (2011) A hitchhiker's guide to mechanobiology. Dev Cell 21:35–47
2. Pedersen JA, Swartz MA (2005) Mechanobiology in the third dimension. Ann Biomed Eng 33:1469–1490
3. Tibbitt MW, Anseth KS (2009) Hydrogels as extracellular matrix mimics for 3D cell culture. Biotechnol Bioeng 103:655–663
4. Cukierman E, Pankov R, Stevens DR et al (2001) Taking cell-matrix adhesions to the third dimension. Science 294:1708–1712
5. Chaudhuri O, Koshy S, Branco da Cunha C et al (2014) Extracellular matrix stiffness and composition jointly regulate the induction of malignant phenotypes in mammary epithelium. Nat Mater 13:970–978
6. Mammoto A, Connor KM, Mammoto T et al (2009) A mechanosensitive transcriptional mechanism that controls angiogenesis. Nature 457:1103–1108
7. Boyd-White J, Williams JC (1996) Effect of cross-linking on matrix permeability: a model for AGE-modified basement membranes. Diabetes 45:348–353
8. Kleinman HK, Martin GR (2005) Matrigel: basement membrane matrix with biological activity. Semin Cancer Biol 15:378–386
9. Yurchenco PD, Schittny JC (1990) Molecular architecture of basement membranes. FASEB J 4:1577–1590
10. Rowley JA, Madlambayan G, Mooney DJ (1999) Alginate hydrogels as synthetic extracellular matrix materials. Biomaterials 20:45–53
11. Rehm BHA (2009) Alginates: biology and applications. In: Steinbüchel A (ed) Microbiology monographs, vol 13. Springer, Heidelberg
12. Huebsch N, Arany PR, Mao AS et al (2010) Harnessing traction-mediated manipulation of the cell/matrix interface to control stem-cell fate. Nat Mater 9:518–526
13. Kong HJ, Kim CJ, Huebsch N et al (2007) Noninvasive probing of the spatial organization of polymer chains in hydrogels using fluorescence resonance energy transfer (FRET). J Am Chem Soc 129:4518–4519

Chapter 4

Hydrogel-Based In Vitro Models of Tumor Angiogenesis

Laura J. Bray, Marcus Binner, Uwe Freudenberg, and Carsten Werner

Abstract

In situ forming hydrogels prepared from multi-armed poly(ethylene glycol) (PEG), glycosaminoglycans (GAG) and various peptides enable the development of advanced three dimensional (3D) culture models. Herein, we report methods for the PEG-GAG gel-based 3D co-cultivation of human umbilical vein endothelial cells, mesenchymal stromal cells, and different cancer cell lines. The resulting constructs allow for the exploration of interactions between solid tumors with 3D vascular networks in vitro to study the mechanistic aspects of cancer development and to perform drug testing.

Key words Hydrogel, Tumor, Angiogenesis, 3D model

1 Introduction

Improvements in modeling the tumor microenvironment are essential in producing more effective research outcomes in the future [1]. Various materials are available for the culture of cells within a 3D platform [1–7]. Our laboratory focuses on the use of a set of biohybrid hydrogels prepared from multi-armed, end-functionalized poly(ethylene glycol) (PEG), glycosaminoglycans (GAG), and functional peptides [8–10].

Several characteristics of PEG-heparin hydrogels suggest that this material could provide a useful alternative to Matrigel™ for supporting 3D cell cultures. Michael-type addition reaction schemes are utilized for the in situ formation of the polymer networks from soluble precursors to embed cells in 3D. Moreover, conjugation of matrix metalloprotease (MMP)-sensitive sequences to PEG allows for localized, cell-mediated degradation of the matrix to enable migration and proliferation [9]. By altering the ratio of PEG to GAG within the system, loading of various GAG-binding growth factors and conjugation of adhesion receptor ligand peptides (such as RGD), the hydrogels can be customized to adjust the biophysical characteristics, morphogen presentation, and adhesiveness to the tissue microenvironment of choice. Using these options, the

Zuzana Koledova (ed.), *3D Cell Culture: Methods and Protocols*, Methods in Molecular Biology, vol. 1612,
DOI 10.1007/978-1-4939-7021-6_4,

hydrogels have been utilized in the culture of human umbilical vein endothelial cells (HUVECs) and mesenchymal stromal cells (MSCs) for creating mature capillary networks [11]. The obtained HUVECs network structures were validated by testing routinely applied antiangiogenic drugs [2]. PEG-GAG hydrogels have also been used for the culture of tumors in order to investigate the effects of matrix stiffness on tumor development in vitro [2]. In situ forming PEG-GAG hydrogel materials are advantageous compared to other 3D matrices including collagen or Matrigel™ due to the far-reaching, well-defined, and independent tunability of exogenous, cell-instructive signals and excellent reproducibility.

Herein, we report methods used in our laboratory to construct PEG-GAG hydrogel based 3D culture models to mimic tumor angiogenesis. The components polymerize, using Michael-type addition, to form transparent hydrogels (Fig. 1) that serve as a scaffold for the co-culture of tumor cells with HUVECs and MSCs. Six key steps are described in detail; (1) synthesis of RGDSP peptide, (2) synthesis of the PEG-peptide conjugate PEG-$(MMP)_4$, (3) synthesis of heparin-maleimide 6, (4) preparation of tumor

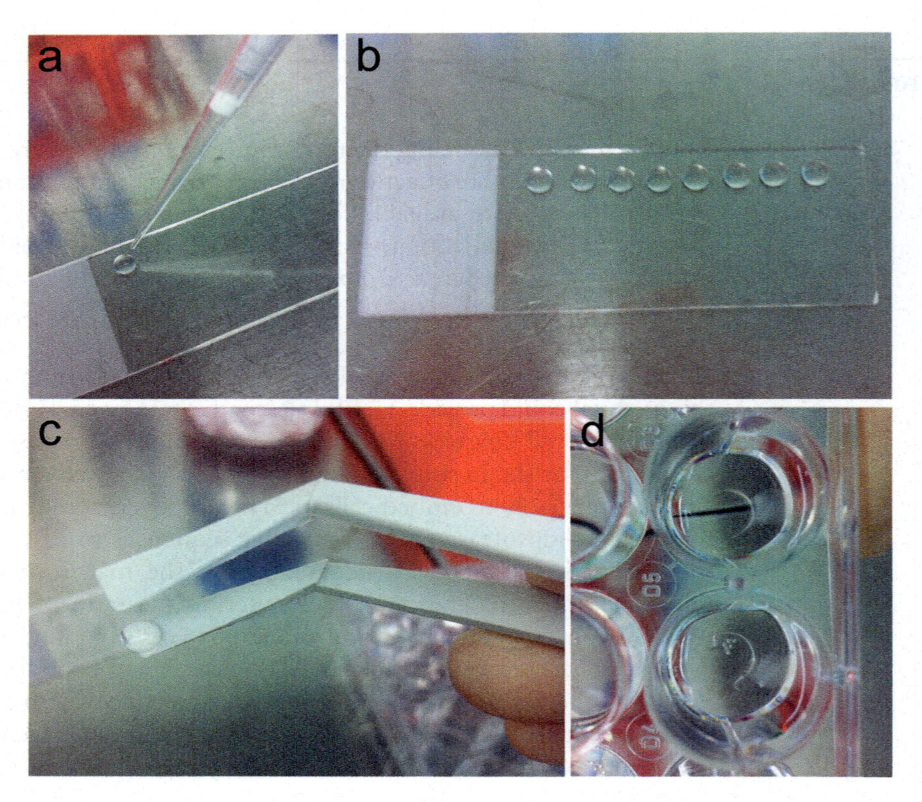

Fig. 1 Illustration of hydrogel preparation and culture. (**a**) Pipetting of hydrogel onto Sigmacote® slide. (**b**) Example of hydrogels cast together on a single microscope slide. (**c**) Collection of hydrogel from microscope slide using plastic forceps. (**d**) Hydrogel submerged within ECGM ready for cell culture

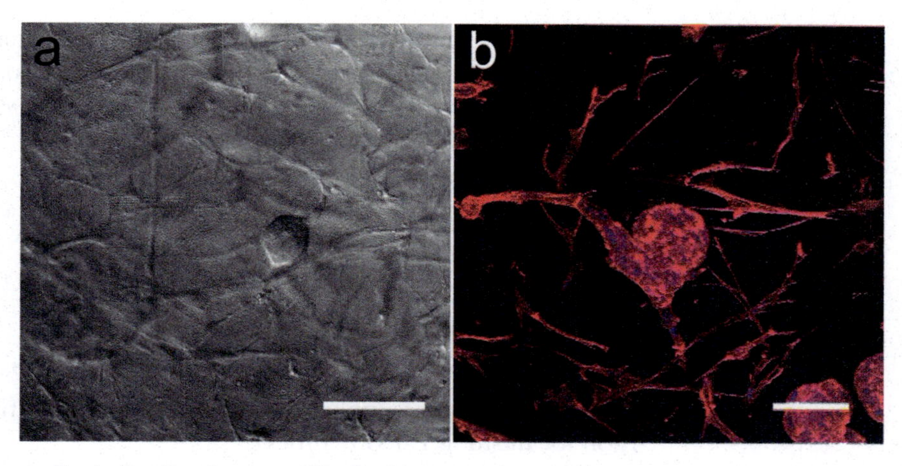

Fig. 2 Images illustrating the structure of the final tumor angiogenesis construct. (**a**) MCF-7 tumor formed from single cells co-cultured with HUVECs and MSCs visualized by light microscopy. (**b**) Confocal image of LNCaP tumor formed from single cells cocultured with HUVECs and MSCs, stained with Phalloidin (*red*) and Hoechst (*blue*). Scale bars, 200 μm and 80 μm, respectively

only constructs, (5) preparation of tumor angiogenesis constructs, and (6) cultivation of the final construct. The final product (Fig. 2) serves as a platform for further investigations into drug efficacy and tumor biology.

2 Materials

2.1 Peptide Synthesis

1. Fume hood.
2. Semi-automated peptide synthesizer Activotec® P-14 or comparable, including 150 mL glass reactor (*see* **Note 1**) and pressure-resistant glass reagent bottles (1 L capacity).
3. Amber glass bottles (2 × 500 mL or 2 × 250 mL capacity) with cap.
4. Fmoc-Rink amide aminomethyl-polystyrene resin (200–400 mesh).
5. Piperidine (Pip) solution: 20% (v/v) Pip (≥99%) in *N,N*-Dimethylformamide (DMF, ≥99%) (approximately 110 mL per amino acid (AA) needed).
6. HOBt/TBTU solution: 0.5 M 1-hydroxybenzotriazole hydrate (HOBt, ≥97%) and 0.5 M 2-(1H-benzotriazol-1-yl)-1,1,3,3-tetramethyluronium tetrafluoroborate (TBTU, ≥97%) in DMF (28 mL volume per AA, plus extra volume; *see* **Note 2**).
7. DiPEA solution: 1 M *N,N*-diisopropylethylamine (DiPEA, ≥99%) in DMF (28 mL volume per AA, plus extra volume; *see* **Note 3**).

8. *N*-(Fmoc)-protected AA's (Fmoc = 9-fluorenylmethyloxycarbonyl) with orthogonal side chain protecting groups, dissolved in 28 mL HOBt/TBTU solution and 28 mL DiPEA solution (*see* **Note 4**).
9. Two 5000 μL pipettes with appropriate pipette tips (keep separated; use one *only* for HOBt/TBTU and one *only* for DiPEA).
10. Acetic anhydride (Ac_2O, ≥99%).
11. Dichloromethane (DCM, ≥99%).
12. Falcon™ polypropylene centrifuge tubes (175 mL capacity).
13. Nitrogen and argon supply (nitrogen for peptide synthesizer, argon for flushing Falcon® tubes).
14. Silicone tubing (>80 cm in length, inner diameter 1–1.5 cm) attached to inert gas hose connector.
15. Ultrasonic water bath.

2.2 Peptide Cleavage

1. Fume hood.
2. Glass Erlenmeyer flask (500 mL capacity) with cap (*see* **Note 5**).
3. Magnetic stirrer.
4. Cylindrical PTFE-covered stir bar (25 × 6 mm in size).
5. Graduated glass measuring cylinder (250 mL capacity).
6. Resin with peptide to cleave.
7. Cleavage cocktail: 85% (v/v) trifluoroacetic acid (TFA, ≥99.9%), 5% (v/v) distilled water, 5% (v/v) phenol (unstabilized, ≥99%), 2.5% (v/v) triisopropylsilane (TIS, ≥98%) and 2.5% (v/v) dithiothreitol (DTT, racemic, ≥99%, *see* **Note 6**). Prepare fresh for each experiment.
8. DCM.
9. Nitrogen and argon supply.
10. POR 2 and POR 4 glass filter (50 mL capacity each).
11. Evaporating flask, pear-shaped (2 L capacity) with cap.
12. Rotary evaporator (with heating bath) attached to gas washing bottle (filled with 5 M aqueous NaOH solution) and water-jet vacuum pump.
13. Diethylether (Et_2O, ≥99.9%).
14. Explosion proof refrigerator.
15. Vacuum pump (*see* **Note 7**).
16. Spatula with flat end.
17. Büchner flask (500 mL capacity).
18. Silicone tubing connecting vacuum pump and Büchner flask.
19. Silicone tubing (>80 cm in length) with glass cone, right angle connector, attached to inert gas hose connector.

20. Rubber filtration cones to seal connection between glass cone and glass filter.
21. Glass vial (25 mL capacity) with cap.
22. Microcentrifuge tubes (0.5 mL capacity).

2.3 Poly(Ethylene Glycol) (PEG)-Peptide Conjugate Synthesis

1. Ultra-high purity (UHP) water (prepared to 18.2 MΩ cm at 25 °C).
2. Acetonitrile (AcN, ≥99.9%).
3. 621 mg (0.0621 mmol) maleimide-functionalized four-arm PEG, MW 10,000 g/mol (PEG-Mal; JenKem Technology USA) in a 100 mL glass Erlenmeyer flask with stir bar, dissolved in 10–20 mL 50% AcN–50% UHP water (*see* **Note 8**).
4. 381.5 mg (0.261 mmol, 4.2 equivalents) of purified MMP peptide (produced in Subheading 3.2) in a 25 mL glass vial, dissolved in 5 mL 50% AcN–50% UHP water (*see* **Note 9**).
5. 20% AcN–80% UHP water.
6. Phosphate buffered saline (PBS), pH 7.4.
7. 0.5 M NaOH in UHP water.
8. 1 M HCl in UHP water, only if necessary.
9. 374 mg tris(2-carboxyethyl)phosphine HCl salt (TCEP; 1.3 mmol, 5 equivalents per peptide) in aqueous 5 M NaOH (~230 μL per 100 mg), adjust pH to 7–8 if necessary (*see* **Note 10**).
10. 100, 1000, and 5000 μL pipettes with appropriate pipette tips.
11. Glass Erlenmeyer flask (2 × 100 mL capacity) with cap and ground joint (14/23).
12. Glass vial (25 mL capacity) with cap.
13. Magnetic stirrer.
14. Cylindrical PTFE-covered stir bar (25 × 6 mm in size).
15. Conical rubber septum, big enough to fit glass Erlenmeyer flask.
16. Syringe needles (20 G).
17. Nitrogen or argon supply.
18. Silicone tubing (>80 cm in length) with tube-needle connector Luer-Lock male, attached to inert gas hose connector.
19. Fume hood.
20. Vacuum pump (*see* **Note 7**).
21. Glass straight vacuum take-off adapter with 14/23 ground joint attached to Erlenmeyer flask.
22. POR 4 glass filter (50 mL capacity).

23. Rubber filtration cone between vacuum adapter and glass filter.
24. Silicone tubing connecting vacuum pump and vacuum adapter.
25. 50% AcN–50% UHP water.

2.4 Heparin-Maleimide Synthesis

1. Glass vial (10 mL capacity) with cap.
2. Microcentrifuge tubes (0.5 and 1.5 mL capacity).
3. Ice bath.
4. Magnetic stirrer.
5. Cylindrical PTFE-covered stir bar (15 × 4.5 mm in size).
6. 1000 μL pipette with appropriate pipette tips.
7. UHP water.
8. 500 mg (0.0357 mmol) heparin, sodium salt, porcine intestinal mucosa (14,000 g/mol from Merck Millipore, Darmstadt, Germany), in 10 mL glass vial with stir bar, placed in ice bath and dissolved in 2.7 mL ice-cold UHP water (*see* **Note 11**).
9. 73.3 mg (0.337 mmol) *N*-hydroxysulfosuccinimide sodium salt (sulfo-NHS, ≥98%) in 1.5 mL microcentrifuge tube, in 400 μL ice-cold UHP water (*see* **Note 12**).
10. 129.4 mg (0.675 mmol) *N*-(3-dimethylaminopropyl)-*N′*-ethylcarbodiimide hydrochloride (EDC, ≥98%) in 0.5 mL microcentrifuge tube, dissolved in 200 μL ice-cold UHP water.
11. 57.2 mg (0.225 mmol) *N*-(2-aminoethyl)maleimide trifluoroacetate salt (maleimide amine ≥95%) in 0.5 mL microcentrifuge tube, dissolved in 200 μL ice-cold UHP water.
12. Glass beaker (1 L capacity) filled with 0.8 L of 1 M NaCl. Dissolve 46.7 g NaCl (≥99%) in 700 mL UHP water and fill up to 800 mL with UHP water.
13. Cylindrical PTFE-covered stir bar (40 × 8 mm in size).
14. Dialysis membrane (flat width 40 mm, molecular weight cutoff 5000).
15. Two clamps to close dialysis bag.
16. Round-bottom flask (500 mL capacity).
17. Freeze-drying unit.

2.5 Analytical High Performance Liquid Chromatography (HPLC) and Electrospray Ionization Mass Spectrometry (ESI-MS)

1. Analytical HPLC unit Agilent 1100 series or comparable, equipped with degasser, binary pump, autosampler, and UV/Vis diode array detector (DAD).
2. ESI-MS unit Agilent 6200 series or comparable, coupled to HPLC unit (DAD outlet).
3. Analytical C18 column (250 × 3 mm in size, 5 μm particle size, 100 Å pore size).

4. Analytical C18 column protection guard/pre-filter unit (4 × 3 mm in size) attached to column.
5. Mobile phase A: 5% AcN–95% UHP water with 0.1% formic acid (≥98%) in a solvent bottle (1 L capacity).
6. Mobile phase B: 95% AcN–5% UHP water with 0.1% formic acid in a solvent bottle (1 L capacity).
7. Autosampler glass vials with bonded caps.
8. Microvolume inserts for autosampler vials (200 μL capacity).
9. 10 μL and 100 μL pipette with appropriate pipette tips.

2.6 Preparative HPLC

1. Preparative HPLC unit Agilent 1200 series or comparable, equipped with 20 mL sample loop, two preparative scale pumps, UV/VIS diode array detector and preparative scale fraction collector.
2. Preparative C18 column (250 × 30 mm in size, 10 μm particle size, 100 Å pore size).
3. Preparative C18 column protection guard/pre-filter unit (15 × 30 mm) attached to column.
4. Glass sample vials for fraction collector (25 × 100 mm, 50 mL capacity).
5. Mobile phase A: 5% AcN–95% UHP water with 0.1% TFA in a solvent bottle (10 L capacity).
6. Mobile phase B: AcN with 0.1% TFA in a solvent bottle (10 L capacity).
7. Falcon™ polypropylene centrifuge tubes (15 mL capacity).
8. 50% AcN–50% UHP water.
9. 10% AcN–90% UHP water.
10. Ultrasonic water bath.
11. Syringe (25 mL capacity, *see* **Note 13**) with long syringe needle (21G × 4 ¾").
12. Mixed cellulose ester (MCE) syringe filters 0.8 μm pore size.
13. Evaporating flask, pear-shaped (1 L capacity).
14. 1 M HCl in UHP water.
15. Rotary evaporator with heating bath.
16. Round-bottom flask (500 mL capacity).
17. Freeze-drying unit.
18. Glass vials (15 mL capacity) with cap.
19. Microcentrifuge tubes (0.5 mL capacity).

2.7 Matrix-Assisted Laser Desorption/Ionization Time-of-Flight (MALDI-TOF) Mass Spectrometry

1. MALDI-TOF instrument Bruker autoflex speed TOF/TOF or comparable, equipped with Nd:YAG laser emitting at 355 nm.
2. Ground steel target plate attached to target holder.
3. 30% AcN–70% UHP water with 0.1% TFA (TA-30).
4. Saturated solution of α-cyano-4-hydroxycinnamic acid (HCCA, suitable for MALDI-TOF, ≥99%) in TA-30 (≥4.4 mg/mL).
5. Microcentrifuge tubes (0.5 mL capacity).

2.8 Analytical High Performance Size Exclusion Chromatography (HPSEC)

1. Analytical HPLC unit (*see* Subheading 2.5).
2. Analytical hydrophilic silica column (300 × 7.8 mm in size, 5 μm particle size, 145 Å pore size).
3. Hydrophilic silica guard column (35 × 7.8 mm in size, 5 μm particle size, 145 Å pore size) attached to column.
4. Solvent bottle (1 L capacity) with mobile phase PBS.
5. Microcentrifuge tubes (2 mL capacity).
6. PBS.
7. Autosampler glass vials with bonded caps.
8. Microvolume inserts for autosampler vials (e.g., 200 μL capacity).
9. 10 μL and 100 μL pipette with appropriate pipette tips.
10. RGDSP peptide for quantitative analysis (produced in Subheading 3.1).

2.9 Preparation of Hydrogels

1. Sigmacote® reagent (Sigma Aldrich, SL2-100ML).
2. Microscope slides.
3. 0.5 mL and 1.5 mL microcentrifuge tubes.
4. Flat plastic forceps.
5. RGDSP peptide (produced in Subheading 3.1).
6. PEG-MMP peptide conjugate (produced in Subheading 3.3).
7. Heparin-maleimide conjugate (produced in Subheading 3.4).
8. PBS.
9. 100 μL, 200 μL, and 1000 μL pipettes with appropriate pipette tips.
10. Ultrasonic water bath.
11. Vortex mixer.
12. Mini centrifuge.

2.10 Cell Culture

1. MCF-7 breast adenocarcinoma cell line.
2. MDA-MB-231 breast adenocarcinoma cell line.
3. LNCaP, lymph node-derived prostate carcinoma cell line.
4. PC3, bone-derived prostate adenocarcinoma cell line.

5. HUVECs.
6. MSCs.
7. MCF-7 medium: 10% fetal bovine serum (FBS), 100 U/mL penicillin and 100 μg/mL streptomycin, 1× MEM nonessential amino acids, 1 mM sodium pyruvate, and 0.1% human insulin in Roswell Park Memorial Institute Medium (RPMI).
8. MDA-MB-231 medium: 10% FBS, 100 U/mL penicillin, and 100 μg/mL streptomycin in Dulbecco's Modified Eagle's Medium (DMEM).
9. LNCaP and PC3 medium: 10% FBS, 100 U/mL penicillin, and 100 μg/mL streptomycin in RPMI medium.
10. Endothelial Cell Growth Medium (ECGM): 2% FBS, 0.4% Endothelial Cell Growth Supplement, 0.1 ng/mL epidermal growth factor, 1 ng/mL basic fibroblast growth factor, 90 μg/mL heparin, and 1 μg/mL hydrocortisone (all part of Endothelial Cell Growth Medium Supplement Mix; Promocell, C-39215) in Endothelial Cell Basal Medium (Promocell, C-22210).
11. MSC medium: 10% FBS, 100 U/mL penicillin, and 100 μg/mL streptomycin in DMEM.
12. 24-well tissue culture plate.
13. Microscope slides.
14. Collagenase solution: 2.5 mg/mL of collagenase (0.45 U/mL; SERVA Electrophoresis GmbH, 17465) in PBS with calcium and magnesium.
15. MACS Quant Running Buffer (Miltenyi Biotec, 130-092-747). Other buffers containing minimum 0.5 mM ethylenediamine-tetraacetic acid (EDTA) could be substituted.
16. Serological pipettes (5 mL and 10 mL).
17. 0.25% trypsin–EDTA.
18. PBS.
19. 37 °C cell culture incubator.
20. 4% paraformaldehyde (PFA) in PBS.

3 Methods

3.1 Synthesis, Cleavage, Analysis, and Purification of RGDSP Peptide for Hydrogel Functionalization

3.1.1 Semi-automated Synthesis

1. Weigh 4 mmol of resin in 150 mL reactor, attach to synthesizer.
2. Weigh 14 mmol of *N*-(Fmoc)-protected AA's in 175 mL Falcon™ tubes, sequence is H_2N-GCWGGRGDSP-$CONH_2$.
3. Set up peptide synthesizer to perform resin swelling twice in 60 mL DMF for 15 min.
4. Peptide synthesizer should perform the following actions per coupling step (all steps under nitrogen atmosphere):

Deprotection twice (55 mL Pip solution each) for 15 min and 5 min, respectively; resin wash five times with 55 mL DMF for 1 min each; AA coupling after manual addition of activated AA, resin wash four times with 55 mL DMF for 1 min each.

5. Add activated AA to reactor for coupling (*see* **Notes 4** and **14**).
6. Repeat **steps 4–5** for every AA coupling.
7. Set up the peptide synthesizer to perform final deprotection (under nitrogen atmosphere) twice with 55 mL Pip solution for 15 min and 5 min, respectively.
8. Wash resin three times with 60 mL DCM for 1 min mixing under nitrogen atmosphere.
9. Dry resin well.

3.1.2 Cleavage

1. Transfer dried resin to a 500 mL glass Erlenmeyer flask, add stir bar.
2. Carefully add cleavage cocktail to the resin (*see* **Note 15**).
3. While stirring, add cleavage cocktail carefully to the resin.
4. Flush the reactor flask three times with DCM to remove residual resin and add to the cleavage mixture (*see* **Note 16**).
5. Stir for 4 h at room temperature (21 °C) under argon atmosphere.
6. During the last 10–15 min of cleavage, attach rotary evaporator to gas washing bottle and water-jet vacuum pump, heat water bath to 40 °C.
7. Filter cleavage mixture through POR 2 glass filter into a 2 L pear-shaped evaporating flask.
8. At 40 °C, evaporate DCM and most TFA from the mixture (*see* **Note 17**).
9. Add a few drops of Et_2O to the mixture and dissolve occurring precipitate.
10. Repeat **step 8** until the precipitate is not dissolving anymore.
11. Add a few drops of DCM to dissolve residual precipitate.
12. Quickly add 1.5–2 L Et_2O, mix thoroughly and close the flask.
13. Store solution for at least 4 h or overnight refrigerated at 4 °C (*see* **Note 18**).
14. Warm up the flask to room temperature (21 °C) for approximately 1–2 h.
15. Filtrate mixture through POR 4 glass filter (*see* **Note 19**). Use small membrane vacuum pump at the end of filtration process.
16. Wash peptide precipitate twice with Et_2O.
17. Dry under nitrogen gas flow while vacuum pump is running (place right angle connector glass cone with rubber filtration cones above filter) and comminute to powder with spatula.

18. Wash once with DCM and once with Et_2O.
19. Repeat **step 16**.
20. Transfer peptide to 25 mL glass vial.
21. Dry the crude peptide in vacuum for 1–2 h.
22. Take a small amount (~0.2–0.5 mg) for analytical purposes, store peptide well sealed at −20 °C until purification.

3.1.3 Analysis with Analytical HPLC and ESI-MS

1. Dissolve peptide in mobile phase A.
2. Adjust peptide concentration to 0.02 M with UHP water.
3. Inject 1 μL and start the gradient program (*see* **Note 20**) using mobile phase A and mobile phase B.
4. Follow UV signal at 278 nm and corresponding MS spectrum at peptide peak (*see* **Note 21**).

3.1.4 Purification with Preparative HPLC

1. In a 15 mL Falcon™ tube, dissolve 150–300 mg of crude peptide in 2–4 mL 50% AcN–50% UHP water (*see* **Note 22**) and soak solution into syringe.
2. Wash tube with 4 mL 10% AcN–90% UHP water and soak solution into syringe.
3. Wash tube with 4 mL UHP water and soak solution into syringe.
4. Load the peptide solution through a 0.8 μm MCE filter to the sample loop.
5. Start the run, follow the UV signal at 278 nm and collect only the main part of the peptide peak, discard peak front and the peak tail (*see* **Note 23**).
6. Transfer the collected fractions to 1 L pear-shaped evaporating flask.
7. For every 10 mL of collected fraction, add 13.06 μL of 1 M HCl to the solution (*see* **Note 24**).
8. Evaporate most solvent from the fraction in vacuum at 40 °C (*see* **Note 25**).
9. Repeat **steps 2–9** until the total amount of crude peptide is purified.
10. When purification is finished, transfer collected fractions to a 500 mL round-bottom flask, wash evaporating flask three times with UHP water and freeze-dry the product for at least 24 h. Repeat freeze-drying with sufficient amounts of UHP water 2–3 times if necessary, until product is generated in a powder form.
11. Take a small amount (~0.2–0.5 mg) for analytical purposes and prepare 100–150 mg aliquots in 15 mL glass vials, seal well and store at −20 °C until use (*see* **Note 26**).
12. Analyze peptide with HPLC ESI-MS (*see* **steps 1–4** in Subheading 3.1.3).

3.2 Synthesis, Cleavage, Analysis, and Purification of MMP Peptide for PEG Conjugation

1. For semi-automated synthesis, follow **steps 1–7** from Subheading 3.1, sequence is Ac-GC(StBu) GGPQGIWGQGGCG-$CONH_2$ (*see* **Note 27**).
2. After final deprotection, perform manual acetylation with 50 mL Ac_2O for 15 min.
3. Repeat **step 2**.
4. Follow **steps 8–9** from to Subheading 3.1.
5. For cleavage, follow steps from Subheading 3.1.2 (*see* **Note 28**). Do not add DTT to the cleavage cocktail.
6. For analysis, follow steps from Subheading 3.1.3 (*see* **Note 29**).
7. For purification with preparative HPLC, follow **steps 1–4** from Subheading 3.1.4.
8. Start the run, follow the UV signal at 278 nm and collect peptide peak (*see* **Note 30**).
9. Follow **steps 6–12** from Subheading 3.1.4 (*see* **Note 31**).

3.3 Synthesis, Analysis, and Purification of PEG-Peptide Conjugate PEG-$(MMP)_4$

3.3.1 Synthesis

1. Add 5 mL of fresh PBS to PEG-Mal solution.
2. Add fresh MMP peptide (produced in Subheading 3.2) solution to PEG-Mal solution.
3. Wash glass vial three times with 2–4 mL 20% AcN–80% UHP water, once with 4 mL UHP water and add to PEG-Mal solution.
4. While stirring, carefully adjust pH to 7.5–8 with aqueous 0.5 M NaOH.
5. After 20 min under argon atmosphere, reaction is controlled by analytical HPLC (*see* **Note 32**).
6. Add TCEP solution to reaction mixture.
7. Attach rubber septum with 5–6 20G syringe needles to Erlenmeyer flask; connect one needle to nitrogen supply/ Luer-Lock connector.
8. Stir solution under a strong flow of nitrogen until there is no specific smell of tBu-SH protection group (overnight), add UHP water if necessary (*see* **Note 33**).
9. Filtrate the reaction mixture through POR 4 glass filter using gentle vacuum avoiding bubbles/foam formation by keeping the filter wet at any time.
10. Wash the Erlenmeyer flask twice with 4–5 mL 50% AcN–50% UHP water and twice with 4–5 mL UHP water and filtrate washing solution through POR 4 glass filter.
11. Keep the reaction mixture cooled at 4 °C during purification with preparative HPLC. If not purified directly after filtration, solution is stored well-sealed at −20 °C.

3.3.2 Purification with Preparative HPLC

1. Load the filtrated reaction mixture to the sample loop, use of 0.8 μm MCE filter is not necessary (*see* **Note 34**).
2. Start the run, follow the UV signal at 278 nm and collect the conjugate peak (*see* **Note 35**).
3. Follow **steps 6–11** from Subheading 3.1.4.
4. Analyze conjugate with HPLC and MALDI (*see* Subheadings 3.3.3 and 3.3.4).

3.3.3 Analysis with Analytical HPLC

1. Dissolve PEG-$(MMP)_4$ in mobile phase A.
2. If necessary, adjust concentration to 0.03 mg/μL with UHP water.
3. Inject 1 μL (*see* **Note 20**).
4. Follow the UV signal at 278 nm (*see* **Note 36**).

3.3.4 Analysis with MALDI

1. Apply 1 μL of matrix solution (saturated HCCA in TA-30, *see* Subheading 2.7) on target and let air-dry at room temperature (21 °C, *see* **Note 37**).
2. Dissolve PEG-peptide conjugate in TA-30, adjust to a concentration of 0.1 mg/mL.
3. Mix 1 μL of conjugate solution with 1 μL of matrix solution, apply 1 μL onto dried matrix spot, and let air-dry at room temperature (21 °C, *see* **Notes 37** and **38**).
4. Measure spectrum using linear mode TOF (*see* **Note 39**).

3.4 Synthesis, Purification, and Analysis of Heparin-Maleimide 6 (HM6) Conjugate

3.4.1 Synthesis

1. While stirring in ice bath, add sulfo-NHS solution to the heparin solution.
2. Add EDC solution to the heparin/sulfo-NHS solution.
3. Stir for 20 min under ice bath cooling (*see* **Note 40**).
4. When activation time is finished, add the maleimide amine solution to the Heparin/EDC/sulfo-NHS solution.
5. After 5 min remove ice bath, stir overnight.

3.4.2 Purification with Dialysis

1. Add stir bar to the 1 M NaCl solution.
2. Wash dialysis tubing well with UHP water, close one side with clamp (= bottom clamp, *see* **Note 41**).
3. Transfer reaction mixture to dialysis tubing, wash vial 4–5 times with UHP water and close tubing with (top) clamp.
4. Dialyze against 800 mL of 1 M NaCl for 1 h with medium to low stir speed.
5. Dialyze four times against 800 mL of UHP water for 1 h each (*see* **Note 42**).
6. When dialysis is finished, transfer solution to a 500 mL round-bottom flask and freeze-dry the product for at least 24 h (*see* **Note 43**).

7. Take a small amount (two times ~1–2 mg) for analytical purposes and prepare 100–150 mg aliquots in 15 mL glass vials, seal well and store at −20 °C until use (*see* **Note 26**).

3.4.3 Qualitative Analysis with Analytical HPSEC

1. Dissolve HM6 conjugate in PBS to concentration of 2 mg/mL.
2. Inject 50 μL (*see* **Note 44**).
3. Follow UV signals at 210 nm and 300 nm (*see* **Note 45**).

3.4.4 Quantitative Analysis with Analytical HPSEC

1. Prepare HM6 solution at 4 mg/mL in UHP water and RGDSP peptide (produced in Subheading 3.1) solution at 1.76 mg/mL in UHP water (*see* **Note 46**).
2. Mix 0.5 mL of HM6 solution with 0.5 mL of peptide solution (*see* **Note 47**).
3. Inject 50 μL (*see* **Note 44**).
4. Follow the UV signal at 278 nm (*see* **Note 48**).

3.5 Preparation of Tumor only Constructs (See Note 49)

1. Weigh the synthesized PEG-MMP and HM6 (produced in Subheadings 3.3 and 3.4) into smaller aliquots in 0.5 mL or 1.5 mL microcentrifuge tubes (*see* **Note 50**).
2. Pre-coat the microscope slides with a film of Sigmacote®:
   (a) Dip the microscope slide into a container containing 100% Sigmacote® solution for approximately 10 s.
   (b) Remove the microscope slide and allow the solvent to evaporate.
   (c) Wash microscope slide by dipping into a container containing UHP water.
   (d) Let air-dry at room temperature (21 °C).
3. Spray the slides with ethanol to sterilize before cell culture use. Allow to dry under sterile conditions (in a cell culture hood).
4. Remove supernatant from a flask of cancer cells. Wash once with PBS.
5. Add 2 mL of 0.25% trypsin–EDTA solution and incubate the flask for 2–3 min at 37 °C.
6. Add 8 mL of cell culture medium (respective to cells utilized; *see* Subheading 2.10) to wash, take a sample for counting and centrifuge the cells for 5 min at 300 × *g*.
7. Count the cells using a hemocytometer or similar.
8. Remove the supernatant from the pelleted cells and resuspend cells at a concentration of 3×10^6 per mL in PBS. This concentration will be equivalent to 15,000 cells per 20 μL hydrogel.
9. Calculate the volume of heparin solution required to cast an appropriate number of 20 μL gels based upon the desired size of experiment.

10. To make 20 gels, dissolve 9.85 mg of HM6 in 110 μL of PBS. Vortex until completely dissolved and centrifuge briefly.
11. Dissolve 10.2 mg of PEG-MMP in 220 μL of PBS (*see* **Note 51**). Vortex solution, then sonicate in water bath for 1–2 min. Vortex and centrifuge briefly.
12. Add 110 μL of the 3×10^6 per mL cell suspension to the HM6 tube. Vortex briefly.
13. Using a pipette, transfer 11 μL of HM6-cell solution to individual 0.5 mL microcentrifuge tubes (1 tube required per gel cast).
14. Draw 11 μL of PEG-MMP solution into the pipette, and then change the pipette volume to 20 μL. Mix quickly with HM6-cell solution in a 0.5 mL tube several times before casting the gel onto the Sigmacote® surface (Fig. 1a) (*see* **Notes 52** and **53**).
15. After allowing the gels to polymerize (*see* **Note 54**), use flat plastic forceps to gently remove the hydrogel from the Sigmacote® slide (Fig. 1c).
16. Transfer the hydrogel to a 24-well plate with 1.5 mL of appropriate cell culture medium, depending on the cell line (Fig. 1d).
17. Culture the tumor only constructs using the respective cancer cell medium. Use 1.5 mL of medium per well in a 24-well plate format.
18. After initial gel formation, change medium after 24 h. Thereafter, change medium as necessary (2–3 times per week) (*see* **Note 55**). Cultures can be maintained for up to 3 weeks to achieve adequate tumor size required for the subsequent assay.

3.6 Preparation of Tumor Angiogenesis Constructs

Using cancer cell lines with HUVECs and MSCs, a tumor angiogenesis model can be generated by seeding these cells within the four-arm PEG-heparin hydrogels (*see* **Note 56**). If pre-grown tumors are not used, the protocol can be started at **step 7**.

1. After 1–3 weeks of pre-growth, depending on required tumor size, combine up to six tumor constructs per well in a 24-well plate. Remove any medium.
2. Add 500 μL of collagenase solution to the well, and incubate for 20–30 min at 37 °C.
3. Check sample under a light microscope. When the tumors have been removed from the gel and are floating freely within the solution, add 500 μL of a buffer containing EDTA. We utilize MACS Quant Running Buffer for this purpose.
4. Add 4 mL of medium to wash the tumors, take a sample for counting, and centrifuge at $70 \times g$ for 5 min (*see* **Note 57**).
5. Check tumor density by pipetting a known volume of the sample onto a microscope slide or similar.

6. Resuspend at an appropriate density for the required experiments (high tumor numbers or low tumor numbers).
7. Remove supernatant from a flask of cancer cells, HUVECs, and MSCs as required. Wash once with PBS.
8. Add 2 mL of trypsin–EDTA solution per a flask and incubate the flasks for 2–3 min at 37 °C.
9. Add 8 mL of cell culture medium (respective to cells utilized; *see* Subheading 2.10) to each tube to wash, take a sample for counting and centrifuge the cells for 5 min at 300 × *g*.
10. Count the cells using a hemocytometer or similar.
11. Remove the supernatant from the pelleted cells and resuspend HUVECs at a concentration of 4.8×10^7 per mL and the MSCs at a concentration of 4.8×10^6 per mL. This concentration will be equivalent to 1.2×10^5 HUVECs and 1.2×10^4 MSC per 20 μL hydrogel. If single cancer cells are to be utilized, resuspend these at a concentration of 9×10^6 per mL (1×10^4 cells per 20 μL hydrogel). If pre-grown tumors are to be utilized, add an appropriate concentration of tumors in 24.6 μL.
12. Calculate the volume of heparin solution required to cast an appropriate number of 20 μL gels based upon the desired size of experiment.
13. For 20 gels, dissolve 9.85 mg of HM6 (produced in Subheading 3.4) in 55 μL of PBS. Vortex until completely dissolved and centrifuge briefly.
14. Add 1.3 mg of RGDSP (produced in Subheading 3.1) in 26 μL to the side of the microcentrifuge tube without touching HM6 solution, and vortex for at least 10 s to mix (*see* **Note 58**).
15. Add 2.2 μg of VEGF, SDF-1 and FGF-2 in 4.4 μL each for a final concentration of 5 μg of growth factor per mL of hydrogel (*see* **Note 59**). Vortex.
16. Dissolve 7.25 mg of PEG-MMP (produced in Subheading 3.3) in 220 μL of PBS (*see* **Note 60**). Sonicate solution in a water bath for 1–2 min.
17. Add 55 μL of HUVECs and 55 μL of MSCs to the HM6-RGDSP-growth factor solution (*see* **Note 61**).
18. Add 24.6 μL of single cancer cells or tumors to the same solution.
19. Using a pipette, transfer 11 μL of HM6-cell solution to individual 0.5 mL microcentrifuge tubes (1 tube required per gel cast).
20. Draw 11 μL of PEG-MMP solution into the pipette, and then change the pipette volume to 20 μL. Mix quickly with HM6-cell solution in a 0.5 mL tube several times before casting the gel onto the Sigmacote® surface (Fig. 1a) (*see* **Note 53**).

21. After allowing the gels to polymerize (*see* **Note 54**), use flat plastic forceps to gently remove the hydrogel from the Sigmacote® slide (Fig. 1c).
22. Transfer the hydrogel to a 24-well plate with 1.5 mL of ECGM (Fig. 1d).
23. Once tri-cultures have been formed, maintain the tumor angiogenesis constructs for 7 days in ECGM (*see* **Note 55**).
24. The resulting constructs can be fixed for 30 min in 4% PFA and processed for immunostaining and confocal imaging (Fig. 2).

4 Notes

1. For 4 mmol scale, reactor with 150 mL capacity is needed.
2. Prepare solution in an amber glass bottle and store under argon atmosphere, discard if yellow color occurs. 2-(1H-benzotriazol-1-yl)-1,1,3,3-tetramethyluronium hexafluorophosphate (HBTU, ≥97%) can be used as an alternative to TBTU.
3. Prepare solution in an amber glass bottle and store under argon atmosphere.
4. Amino acid building blocks for RGDSP peptide are (all *L*-stereoisomers): Fmoc-Arg(Pbf)-OH, Fmoc-Asp(OtBu)-OH, Fmoc-Cys(Trt)-OH, Fmoc-Gly-OH, Fmoc-Pro-OH, Fmoc-Ser(tBu)-OH, Fmoc-Trp(Boc)-OH. Amino acid building blocks for MMP peptide are (all *L*-stereoisomers): Fmoc-Cys(StBu)-OH, Fmoc-Cys(Trt)-OH, Fmoc-Gln(Trt)-OH, Fmoc-Gly-OH, Fmoc-Ile-OH, Fmoc-Pro-OH, Fmoc-Trp(Boc)-OH. Very important: start dissolving the AA powder during the third or fourth washing step after deprotection, flush Falcon™ tube with argon and use ultrasonic water bath if necessary. After complete dissolution of AA, add 28 mL of DiPEA solution (during the last washing step after deprotection), flush Falcon™ tube gas space with argon and mix briefly. *Always* prepare fresh solutions to ensure maximum reactivity and less side reactions during coupling.
5. For 4 mmol scale, Erlenmeyer flask with 500 mL capacity is needed, since the total volume of the cleavage cocktail is approximately 200–300 mL.
6. Very important: for Cys(StBu)-containing MMP peptide, *no DTT* is added to the cleavage cocktail in order not to reduce the S-StBu-bond, substitute the 2.5% DTT fraction with TFA.
7. A small, oil-free membrane vacuum pump is sufficient enough for the filtration purpose.
8. Quality of PEG products vary significantly by choice of supplier. Solubility and degree of functionalization was best for the

polymer from JenKem Technology USA. The given amounts are for synthesis of 1 g conjugate. Upscaling is possible up to 5 g. PEG needs to be completely dissolved before adding the peptide, check the solution carefully and treat with ultrasonic water bath if necessary.

9. A peptide excess of 5% per arm is necessary for complete functionalization of the four-arm PEG to compensate the TFA content in the peptide, which is generated as TFA salt after preparative HPLC. Dissolve the peptide only after the PEG is completely dissolved, always use the solution prepared as fresh as possible.
10. Keep TCEP cooled (4 °C) until use (after first reaction control at the earliest).
11. Leave maleimide amine frozen at −20 °C until the end of heparin EDC/sulfo-NHS activation. Prepare all solutions as fresh as possible. Especially EDC should not be kept for too long time in solution prior use.
12. Excess of maleimide amine is 1.05× per modification. EDC is 3× excess and sulfo-NHS 1.5× excess per maleimide, respectively. Solubility of sulfo-NHS in water approximately 0.37 mg/μL.
13. Syringe volume depending on volume of the injector sample loop (20 mL version used for this work). When loading, be aware not to exceed the sample loop volume.
14. Coupling times: (1) P 25 min, (2) S 45 min, (3) D 25 min, (4) G 25 min, (5) R 30 min, (6) G 35 min, (7) G 20 min, (8) W 30 min, (9) C 40 min, (10) G 25 min. If synthesis is interrupted overnight, store resin swollen under DMF (best: in third wash solution after coupling).
15. For 4 mmol scale of the 10mer peptide RGDSP, the total volume of cleavage cocktail is 180 mL; adjust for lower scales. As the first steps, add water and TIS to the measuring cylinder and then add 30 mL of TFA. Phenol and DTT are solid at room temperature. Add as much to raise the meniscus about the volume fraction needed for cleavage cocktail. Fill the cylinder with TFA to 80–90% of the total volume and make sure to completely dissolve TIS. After addition to the resin, use residual 10–20% TFA to flush the cylinder twice.
16. For 150 mL reactor and 4 mmol scale, use 8–10 mL DCM for each washing step. DCM as a low volume fraction (5–15%) in the cleavage mixture ensures good resin swelling and solubility of hydrophobic cleavage intermediates. If necessary, add additional DCM to the cleavage cocktail if precipitate occurs.
17. Remove the flask from rotary evaporator only after cooling down to room temperature under vacuum.

18. Storing the solution for few hours up to overnight refrigerated yields to better precipitate quality which is easier to filtrate later on. In that time, peptide precipitate sediments at the flasks' bottom. Move the flask carefully not to disperse the precipitate back through solution.
19. For 4 mmol scale, sequentially wash and dry the peptide precipitate in 5–6 aliquots. Do not overload the glass filter since for great amounts, peptide moisturizing becomes problematic.
20. Gradient program for analytical HPLC with the given equipment is provided in Table 1.
21. Use UV signal to calculate crude purity of peptide. MS signal at 990.4 *m/z* corresponds to the $[M+H]^{+}$-ion of RGDSP peptide.
22. The peptide amount for one run depends on the crude purity of the peptide. Start with low amounts and increase if separation is good. Use ultrasonic water bath for dissolving peptide if necessary. Do not forget to wash the flask from peptide precipitation and filter from peptide filtration.
23. Gradient program for preparative HPLC for RGDSP peptide with the given equipment is provided in Table 2.

Table 1
Gradient program for analytical HPLC

Time [min]	%A	%B	Flow [mL/min]
0	100	0	0.5
30	40	60	0.5
32	20	80	0.5
33	2	98	0.5
35	100	0	0.5

Table 2
Gradient program for preparative HPLC for RGDSP peptide

Time [min]	%A	%B	Flow [mL/min]
0	100	0	30
5	92	8	30
20	70	30	30
21	35	65	30
23	5	95	30
24	100	0	30

24. Adding HCl is necessary to remove cell-toxic TFA from the product during freeze-drying. The amount is calculated from the volume fraction of TFA in the mobile phase. With a density of 1.49 g/mL and M = 114.02 g/mol, the molar concentration of 0.1% (v/v) TFA is 13.06 mM. Adding 13.06 μL of 1 M HCl to every 10 mL of collected fraction is equal to 10% of moles TFA in the solution. Since HCl is a ~10× stronger acid then TFA, this is sufficient enough to protonate TFA to the more volatile acid form. The total volume of collected fraction is calculated from the flow rate and duration of collecting.
25. Remove AcN at 70–90 mbar, then water at 15–30 mbar. Never let the solution run dry during evaporation. Very important: Before adding new fractions to the evaporating flask, remove the flask from heating bath while being still under vacuum. Due to evaporation enthalpy, the solution will cool down first. During venting, stop evaporation flask rotation.
26. For storage greater than 4 weeks, it is strongly recommended to seal the vials in aluminum bags with 2 silica desiccants to ensure compound reactivity. Store at −20 °C and wait for at least 30 min at room temperature (21 °C) before open the bag/vial for preparing small aliquots.
27. Coupling times: (1) G 20 min, (2) C(Trt) 20 min, (3) G 25 min, (4) G 20 min, (5) Q 30 min, (6) G 30 min, (7) W 30 min, (8) I 40 min, (9) G 30 min, (10) Q 25 min, (11) P 30 min, (12) G 45 min, (13) G 25 min, (14) C(StBu) 25 min (15) G 30 min. If synthesis is interrupted overnight, store resin swollen under DMF (best: in third wash solution after coupling).
28. For 4 mmol scale of the 15mer peptide MMP, total volume of cleavage cocktail is 240 mL; adjust for lower scales. Follow the steps as in **Note 15**. Very important: *no DTT* is added to the cleavage cocktail in order not to reduce the S-StBu-bond, substitute the 2.5% DTT fraction with TFA.
29. Use UV signal to calculate crude purity of peptide. MS signal at 1462.6 *m/z* corresponds to the $[M+H]^+$-ion of MMP peptide.
30. Gradient program for preparative HPLC for MMP peptide with the given equipment is provided in Table 3.
31. After purification, it is not necessary to add 1 M HCl to the collected fractions since the peptide is later used for conjugation reaction and not directly used for cell culture.
32. Use gradient program provided in Table 1. Follow the UV signal at 278 nm. Additionally to the conjugate peak (~25 min), the peptide peak (~16 min) needs to be visible. Very important: if no excess peptide is detected, add another 3–5% peptide

Table 3
Gradient program for preparative HPLC for MMP peptide

Time [min]	%A	%B	Flow [mL/min]
0	100	0	30
0.5	100	0	30
4	73	27	30
12	65	35	45
13	5	95	45
14.5	30	70	45
15.5	100	0	45
16	100	0	30

Table 4
Gradient program for preparative HPLC for PEG-$(MMP)_4$ conjugate

Time [min]	%A	%B	Flow [mL/min]
0	100	0	30
4	70	30	25
14	58	42	20
15	5	95	20
17	30	70	20
18	100	0	20

per PEG functionality (few mg), adjust pH and run for 20 min under argon atmosphere. Check again for completion of reaction with analytical HPLC. If necessary, repeat these steps until completion of reaction.

33. Reaction mixture should not run dry when stirring overnight.
34. After stirring overnight under nitrogen gas flow and filtration, overall solid content of the conjugation reaction mixture is 20–25 mg/mL. For using a 250 × 30 mm preparative column, purification of 300 mg with one injection is possible, meaning 12–15 mL injections. Adjust the injection volume if necessary.
35. Gradient program for preparative HPLC for PEG-$(MMP)_4$ conjugate with the given equipment is provided in Table 4.
36. With ESI-MS it is not possible to gain information about the molecular weight of PEG-$(MMP)_4$. MALDI is used for that purpose.

37. Avoid scratching the pipette tip at any surface while in contact with HCCA solution. This will avoid formation of big crystals on the target, improving MALDI spectrum resolution.
38. Mix conjugate solution with matrix solution in the cap of a 0.5 mL microcentrifuge tube.
39. PEG-peptide conjugates are not stable in reflectron mode, most probably due to fragmentation during the flight time. Therefore it is necessary to use linear mode to generate a mass spectrum.
40. After adding EDC to the heparin/sulfo-NHS solution, warm up maleimide amine (closed vial) to room temperature (21 °C), weigh after approximately 15 min.
41. Length of dialysis tubing should be adjusted to 70–80% of the fill level of the glass beaker filled with 800 mL of 1 M NaCl.
42. For every dialysis solution exchange, wash dialysis tubing well with UHP water from the outside to remove residues from the previous dialysis solution.
43. Wash dialysis tubing well with UHP water from the outside after completion of dialysis. Turn the dialysis tubing around and open the bottom clamp to access dialyzed solution and transfer to round-bottom flask. Wash the inner part of the tubing four times with UHP water and transfer into to the round-bottom flask.
44. With the given equipment, HPSEC run is performed at 0.5 mL/min for 50 min.
45. With the given equipment, HM6 conjugate is eluting at 10–15 min and free maleimide amine at 20–23 min. After manual baseline-correction and integration, at 210 nm the area between 20 and 23 min must not exceed 1% of the area between 10 and 15 min. Furthermore, at 300 nm, no signal should be observable between 20 and 23 min. If both points are fulfilled, the conjugate is ready to use. Otherwise, the conjugate is not useable for cell experiments due to free maleimide amine residues that are cell-toxic. A repeated dialysis would result in an increased hydrolysis of the maleimide groups, so that a new synthesis with fresh reagents should be set up again.
46. The molecular weight (Mw) of RGDSP peptide (H_2N-GCWGGRGDSP-$CONH_2$) used for quantitative analysis is 990.05 g/mol. Mw for HM6: M = 14,000 + 6 × 139.13 = 14,8 35 g/mol. For a solution of 2 mg/mL, molar concentration of HM6 is 1.35×10^{-4} mM and 8.1×10^{-4} mM maleimide-units (6 equivalents). The peptide needs to be added as 1.1 equivalents per maleimide to compensate the TFA content in the peptide, which is generated as TFA salt after preparative HPLC. In total, 6.6 equivalents relative to HM6 need to be added, which is 0.88 mg per mL HM6 solution with 2 mg/mL.

47. With the given concentrations and mixing volumes, the final HM6 concentration is adjusted to 2 mg/mL and final peptide concentration to 0.88 mg per mL HM6 solution with 2 mg/mL.
48. With the given equipment, HM6-RGDSP conjugate elutes at 10–15 min and excess peptide at 25–30 min. After manual baseline-correction and integration, at 278 nm the area between 25 and 30 min must not exceed 5% of the area between 10 and 15 min, which proves the functionality of 6 maleimide groups per heparin molecule.
49. This step can be omitted if the single cell method is preferred. We have cultivated tumors both seeded as single cells into the tumor angiogenesis culture, and those pre-grown in hydrogels before seeding into the tumor angiogenesis cultures.
50. We try to limit the amount of freeze–thaw cycles that the PEG-MMP and HM6 materials undertake. Our routine is to aliquot small experiment-sized amounts of material into 0.5 mL or 1.5 mL microcentrifuge tubes and freeze for use over several months.
51. After HM6 powder has been dissolved in PBS, you can in principle keep the solution for up to 4 h at room temperature. However, the dissolved PEG-MMP solution must be utilized within 30 min of PBS being added. We recommend that everything else for the experiment is ready before dissolving the PEG and casting of the hydrogels.
52. For tumor only constructs, we routinely use hydrogels at a cross-linking degree of gamma 1 (gamma = molar ratio PEG/HM6), which result in a stiffness of approximately 1.5 kPa storage modulus. It is necessary to use gels which are relatively stiff in mechanical properties for pre-growing the tumors, due to the fast degradation of the hydrogels by tumor cells.
53. To avoid bubbles forming within the cast hydrogels, when the PEG-containing pipette is changed to 20 μL, slowly press on the pipette to remove the extra 10 μL of space before starting to mix with the HM6-cell solution. We find that mixing at least 6–8 times produces the most consistent 3D cell culture results. How many gels you can make before transferring them into medium depends upon how fast you cast the gels.
54. It is important that the hydrogels do not over-dry post-casting, as this can dehydrate the cells inside. To achieve an optimum polymerization time, manipulate the gelation time, by increasing or decreasing the pH of the PEG-MMP solution (by 1 μL stepwise addition of 1 M NaOH, or 1 M HCl), to within 30 s. Afterwards, allow the gels to form for a further 2–3 min on the Sigmacote® slide. Test if the gels are completely ready with your forceps, and then immediately transfer to the 24-well plate with medium.

55. Great care must be taken during feeding of cultures and afterwards during the immunostaining process. We do not recommend using vacuum devices or serological pipettes to change medium or buffers, this is due to the hydrogels often being destroyed by these devices. Instead, use a 1000-μL pipette to manually remove medium or buffers for this purpose to ensure the safety of your hydrogels.
56. We have developed these tumor angiogenesis cultures using the breast cancer cell lines, MCF-7 and MDA-MB-231, and prostate cancer cell lines, LNCaP and PC3 [2].
57. In order to protect the tumors once formed, slow centrifugation allows for greater tumor yield. Moreover, for post-tumor harvest, 1 mL pipette tips should be used to transfer samples and 200 μL pipette tips should be cut to ensure tumor stability.
58. Our RGDSP, synthesized in our laboratory, is dissolved at a concentration of 50 mg/mL in MilliQ water, adjusted to pH 5 and then aliquoted at different amounts from 10 to 50 μL. One aliquot is taken for the experiment and then discarded afterwards. The RGDSP binds quickly to the HM6, therefore to ensure homogenous distribution we recommend that the RGDSP is not mixed by pipette into the HM6 solution but rather by vortex as described.
59. We routinely keep all of our growth factors at a consistent concentration of 500 ng/μL stock solution. This stock solution is frozen into small workable aliquots.
60. We routinely make hydrogels that are between 200 and 400 Pa in stiffness proved by rheometry. In the case of hydrogels dissolving too fast or HUVECs not spreading, the stiffness can be altered by increasing or decreasing the cross-linking degree via adjusting the molar ratio of PEG to heparin.
61. The maleimide functionalization of the heparin is known to cause cell toxicity. It is important that cells are not in contact with the maleimide for longer than required. We recommend that cells are added to the solution just before the aliquots are made and hydrogels are cast.

Acknowledgments

L. J. B. was supported by the Endeavour Awards as part of the Prime Minister's Australia Awards. Financial support was provided by the German Research Foundation (Deutsche Forschungsgemeinschaft) through grant numbers: SFB-TR 67, WE 2539-7 and FOR/EXC999, by the Leibniz Association (SAW-2011-IPF-2 68) and by the European Union through the

Integrated Project ANGIOSCAFF (Seventh Framework Program). Finally, we wish to acknowledge the advice and assistance received from Dr. Karolina Chwalek (Harvard University), Prof. Dietmar Hutmacher (Queensland University of Technology), Ms. Milauscha Grimmer (Leibniz Institute of Polymer Research Dresden), Mr. Ulrich Bonda (Leibniz Institute of Polymer Research Dresden), and Dr. Mikhail Tsurkan (Leibniz Institute of Polymer Research Dresden).

References

1. Hutmacher DW (2010) Biomaterials offer cancer research the third dimension. Nat Mater 9(2):90–93
2. Bray LJ, Binner M, Holzheu A et al (2015) Multi-parametric hydrogels support 3D in vitro bioengineered microenvironment models of tumour angiogenesis. Biomaterials 53:609–620
3. Chwalek K, Bray LJ, Werner C (2014) Tissue-engineered 3D tumor angiogenesis models: potential technologies for anti-cancer drug discovery. Adv Drug Deliv Rev 79-80:30–39
4. Fischbach C, Chen R, Matsumoto T et al (2007) Engineering tumors with 3D scaffolds. Nat Methods 4(10):855–860
5. Lee GY, Kenny PA, Lee EH et al (2007) Three-dimensional culture models of normal and malignant breast epithelial cells. Nat Methods 4(4):359–365
6. Loessner D, Stok KS, Lutolf MP et al (2010) Bioengineered 3D platform to explore cell-ECM interactions and drug resistance of epithelial ovarian cancer cells. Biomaterials 31(32):8494–8506
7. Sieh S, Taubenberger AV, Rizzi SC et al (2012) Phenotypic characterization of prostate cancer LNCaP cells cultured within a bioengineered microenvironment. PLoS One 7(9):e40217
8. Freudenberg U, Hermann A, Welzel PB et al (2009) A star-PEG-heparin hydrogel platform to aid cell replacement therapies for neurodegenerative diseases. Biomaterials 30(28):5049–5060
9. Tsurkan MV, Chwalek K, Prokoph S et al (2013) Defined polymer-peptide conjugates to form cell-instructive starPEG-heparin matrices in situ. Adv Mater 25(18):2606–2610
10. Zieris A, Dockhorn R, Rohrich A et al (2014) Biohybrid networks of selectively desulfated glycosaminoglycans for tunable growth factor delivery. Biomacromolecules 15(12):4439–4446
11. Chwalek K, Tsurkan MV, Freudenberg U et al (2014) Glycosaminoglycan-based hydrogels to modulate heterocellular communication in in vitro angiogenesis models. Sci Rep 4:4414

Chapter 5

Generation of Induced Pluripotent Stem Cells in Defined Three-Dimensional Hydrogels

Massimiliano Caiazzo, Yoji Tabata, and Matthias Lutolf

Abstract

Since the groundbreaking discovery of induced pluripotent stem cells (iPSCs) many research groups have attempted to improve the efficiency of the classical cell reprogramming process. Surprisingly, the contribution of the three-dimensional (3D) microenvironment to iPSC generation has been largely overlooked. Here we describe a protocol for the generation of iPSCs in defined poly(ethylene glycol) (PEG)-based hydrogels that, besides allowing higher reprogramming efficiency, are also a powerful tool to study the influence of biophysical parameters on iPSC generation.

Key words Cell reprogramming, iPSC, 3D culture, PEG hydrogels, Synthetic microenvironment, Bioengineering

1 Introduction

Since the first publication of iPSC generation [1], many groups have focused their attention on the optimization of the reprogramming protocol, employing different transcription factors and/or additional factors in the medium such as cytokines or chromatin modifiers [2]. All these methods are based on conventional two-dimensional (2D) culture systems that are all sharing the drawback of a limited control of key parameters of a cell's microenvironment. Downing and colleagues [3] provided the first demonstration that cell reprogramming could be influenced by a modulation of the physical characteristics of the microenvironment, revealing a synergic effect of reprogramming transcription factors with cell confinement in microgrooves.

More recently, we reported the use of synthetic 3D matrices to improve reprogramming efficiency [4]. Using a transglutaminase (TG)-PEG-hydrogel system, it is possible to tune the stiffness and degradability of the matrix by modulating, respectively, the concentration and the proteolytically active sites of the PEG precursors [5, 6]. Moreover, these defined microenvironments can be further

Zuzana Koledova (ed.), *3D Cell Culture: Methods and Protocols*, Methods in Molecular Biology, vol. 1612,
DOI 10.1007/978-1-4939-7021-6_5, © Springer Science+Business Media LLC 2017

enriched by tethering cell–cell contact or extracellular matrix (ECM) proteins [7]. As previously shown, our 3D reprogramming system bears several advantages such as an increased efficiency and speed both in the generation of mouse and human iPSCs. A key factor of the 3D reprogramming protocol is the possibility to keep a constant proliferation rate throughout the reprogramming process, thus avoiding an initial lag phase that is caused by cell confluence as typically found in protocols based on 2D cell culture. We found that these improvements are linked to an accelerated mesenchymal–epithelial transition and to an increased epigenetic plasticity exerted by the 3D microenvironment.

Here we describe the protocol to get the highest reprogramming efficiency in TG-PEG hydrogels with mouse and human cells.

2 Materials

All the solutions are prepared with autoclaved Milli-Q H_2O at room temperature (RT).

2.1 TG-PEG Gel Preparation

1. 10× Tris buffered saline (TBS): 500 mM Tris base, 100 mM $CaCl_2$, 1 M NaCl, pH 7.6. Add 60.57 mg of Tris, 11.10 mg of $CaCl_2$, and 5.84 mg of NaCl in 1 L of water. Store at RT.
2. PEG polymer: 8-arm PEG-vinylsulphone (PEG-VS) 40 kDa (NOF Europe, SUNBRIGHT® HGEO-400VS). Store at −80 °C.
3. Peptides for PEG functionalization, store both at −80 °C:
   (a) Glutamine peptide (H-NQEQVSPLERCG-NH2, Bachem).
   (b) MMP-sensitive lysine peptide (Ac-FKGGGPQGIWGQ ERCG-NH2, Bachem).
4. 0.3 M triethanolamine (TEOA), pH 8. Store at RT.
5. SnakeSkin dialysis tubing 3.7 mL/cm, MWCO 10 KDa (Pierce).
6. Dialysis clips (Pierce) and floater (Sagex).
7. Fibrogammin P1250 (CSL Behring), 200 U/mL in 2.5 mM $CaCl_2$. Store at −80 °C.
8. Thrombin (Sigma-Aldrich, T6884), 20 U/mL in 2.5 mM $CaCl_2$. Store at −80 °C.
9. Sigmacote (Sigma-Aldrich, SL2). Store at 4 °C.
10. Adhesion peptide RGDSP (Ac-FKGGRGDSPG-NH2, GL Biochem). 10× stock 0.5 M solution can be stored at −20 °C.
11. 100 μg/mL Fc-tagged Epcam (R&D systems, 960-EP-050) in sterile PBS. Keep aliquots at −80 °C.

12. 1 mg/mL Laminin ultra-pure (BD Biosciences, 354239) in TBS. Store aliquots at −80 °C.
13. 50 mg/mL Protein A (Biovision) in water. Aliquot and store at −80 °C.
14. NHS-PEG-maleimide, 3.5 kDa (JenKem Technology). Make small aliquots (1–3 mg) and store as powder at −20 °C.
15. KimWipes.
16. Liquid nitrogen.
17. Lyophilizer.
18. Sterile distilled deionized water.
19. Microscope glass slides.
20. Plasma cleaner.
21. Rheometer, e.g., Bohlin CV 120 rheometer (Bohlin Instruments).

2.2 Virus Production

1. 293T cells (ATCC, CRL11268).
2. Plasmids:
   (a) FUW-M2rtTA (Addgene, clone 20342).
   (b) TetO-FUW-OSKM (Addgene, clone 20321).
   (c) pMD2.G (Addgene, clone 12259).
   (d) pMDLg/pRRE (Addgene, clone 12251).
   (e) pRSV-Rev (Addgene, clone 12253).
3. GenElute HP Endotoxin-Free Plasmid Maxiprep Kit (Sigma-Aldrich).
4. CalPhos Mammalian Transfection kit (Clontech, 631312).
5. 15 cm cell culture dishes.
6. 0.22 μm Stericup filter units (Millipore).
7. Utracentrifuge Tubes, Thinwall, Ultra-Clear™, 38.5 mL, 25 × 89 mm (Beckman Coulter, 344058).
8. 293T cell medium: 2 mM GlutaMAX, 10% (v/v) fetal bovine serum (FBS), 100 U/mL penicillin, and 100 μg/mL streptomycin in Dulbecco's modified Eagle's medium (DMEM). Store up to 4 weeks at 4 °C.

2.3 Cell Reprogramming and 3D-iPSC Culture

1. Fibroblast medium: 2 mM GlutaMAX, 10% (v/v) FBS, 2 mM L-glutamine, 100 U/mL penicillin and 100 μg/mL streptomycin, 1× non-essential amino acids (NEEA), 1 mM sodium pyruvate in DMEM. Store up to 4 weeks at 4 °C.
2. Mouse embryonic stem cell (mESC) medium: 2 mM GlutaMAX, 15% (v/v) ESC-qualified FBS, 1 mM sodium pyruvate, 1× NEAA, 100 U/mL penicillin and 100 μg/mL streptomycin, 0.1 mM 2-mercaptoethanol, and 103 U/mL

mouse leukemia inhibitory factor (Millipore) in DMEM. Store up to 1 week at 4 °C.

3. mTeSR™1 medium (STEMCELL Technologies). Store with added mTeSR™1 supplement at 4 °C for 1 month.
4. 2 mg/mL doxycycline hyclate (Sigma-Aldrich), stock 1000×, filter-sterilized (0.22 μm). Store at −20 °C (*see* **Note 1**).
5. 10 mM CHIR99021, GSK-3β inhibitor (Stemgent), stock 1000×. Store at −20 °C.
6. 12-well plates.
7. 10 cm cell culture dishes.
8. TrypLE™ Express Enzyme, 1× (Gibco, 12605028).
9. TrypLE™ Select Enzyme, 1× (Gibco, 12563011).
10. 100 mm stainless steel Chattaway spatula micro (Nickel-Electro Ltd., 2350).
11. 35 mm glass-bottom dishes (Nunc, 150680).
12. Matrigel® Growth Factor Reduced Basement Membrane Matrix (Corning, 356231). Thaw on ice overnight (ON) and store aliquots at −20 °C. 1% (v/v) working solution in mESC medium can be stored for 2 weeks at 4 °C (*see* **Note 2**).
13. Matrigel® hESC-Qualified Matrix (Corning, 354277). Thaw on ice ON and store aliquots at −20 °C. 1% (v/v) working solution in mTeSR™1 can be stored for 2 weeks at 4 °C (*see* **Note 2**).

2.4 Immunofluorescence Staining of 3D-iPSCs

1. 4% paraformaldehyde (PFA) in PBS. Prepare aliquots and store at −20 °C.
2. 1 mg/mL DAPI (Thermo Fisher Scientific), stock 1000×. Prepare aliquots and store at −20 °C protected from light. Aliquots in use can be stored at 4 °C protected from light for 1 month.
3. 1% Triton X-100 in PBS.
4. Blocking solution: 10% (v/v) serum (the animal of origin should be the same as the one of the secondary antibody) in PBS.
5. Primary staining solution: 10% (v/v) serum (the animal of origin should be the same as the one of the secondary antibody) and 0.5% (v/v) Triton X-100 in PBS.
6. Secondary staining solution: 10% (v/v) serum (the animal of origin should be the same as the one of the secondary antibody) and 0.1% (v/v) Triton X-100 in PBS.
7. Primary and secondary antibodies as desired.
8. 35 mm glass-bottom dish.

3 Methods

All the procedures are performed at RT unless specified otherwise.

3.1 Activation of FXIII

1. Mix 900 μL of 200 U/mL fibrogammin P1250 (FXIII) with 100 μL of 20 U/mL thrombin.
2. Incubate for 30 min at 37 °C. Every 10 min shake the mix.
3. Immediately put the activated FXIII (FXIIIa) on ice.
4. Make aliquots of FXIIIa according to the exact calculated amount for the PEG-hydrogel that you need to prepare and store at −80 °C [8] (*see* **Note 3**).

3.2 Preparation of Functionalized PEG Polymers

1. Take the PEG-VS and peptide powders out from the −20 °C freezer and wait until they reach RT.
2. Prepare two suspensions of 1 g of PEG-VS in 50 mL of TEOA each for both the glutamine and the MMP-sensitive lysine peptides.
3. To prepare PEG-Gln and PEG-MMP-Lys solutions, add peptide powders to the PEG-VS solutions in a 1.2 molar excess over functional VS groups. Functionality as well as purity of PEG-VS and peptide batches must be taken into account for the calculations (*see* **Note 4**).
4. Incubate for 2 h at 37 °C and mix time to time (*see* **Note 5**).
5. Dialyze the obtained PEG-Gln and PEG-MMP-Lys solutions.
   (a) Fill a big bucket (5 L) with distilled deionized water for both the PEG-peptides, put them at 4 °C gently agitated on a magnetic stirring plate.
   (b) Cut about 22 cm of SnakeSkin for both the PEG-peptide solutions. Fold the bottom part of SnakeSkin 3 times and clip with a heavy clip.
   (c) Carefully pour the PEG-peptide solution into the SnakeSkin. Fold two times the top part of the SnakeSkin and clip with a lighter clip. Equip the filled SnakeSkin with a Sagex floater (Fig. 1) and place it inside the 5 L water bucket.
   (d) Incubate the setup at 4 °C for 4 days while changing the water 3 times a day. Hence, unreacted peptide groups are diluted about 1024 times.
6. Filter the dialyzed PEG-peptide solutions with hydrophilic 0.22 μm filter and split the content in two 50 mL falcon tubes.
7. Place a tube holder in a Sagex container filled with liquid nitrogen to about 5 cm from the bottom.

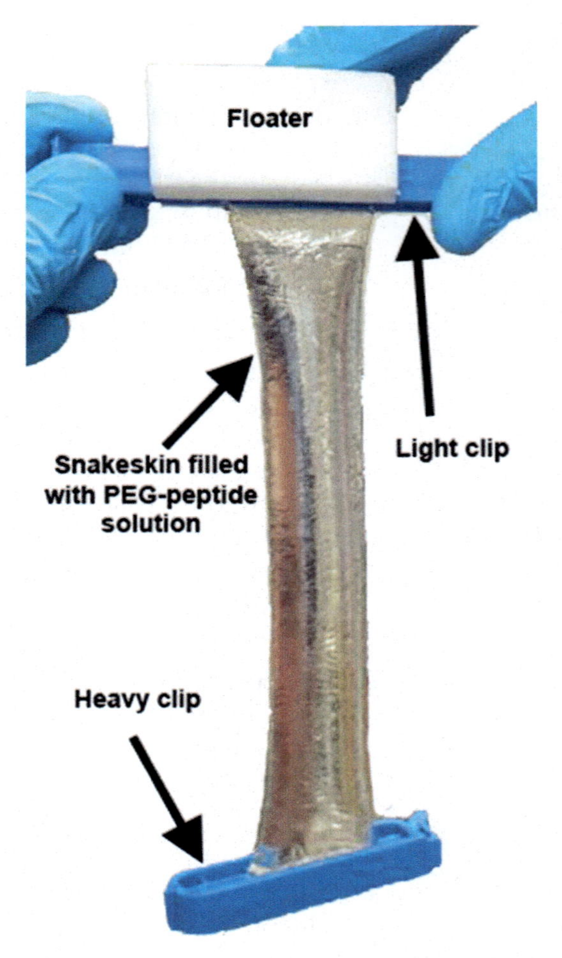

Fig. 1 Preparation of the dialysis sausage used to purify the PEG-Gln or PEG-MMP-Lys peptides from the peptide excess

8. To increase gas permeability, replace the lid of each falcon tube with a KimWipes wiper fixed with rubber bands (*see* **Note 6**).
9. Place and instantly freeze the falcon tubes in liquid nitrogen. Keep the samples in liquid nitrogen until lyophilization.
10. Put the falcon tubes in a lyophilizer and keep them for 2–3 days under constant vacuum.
11. Weigh the product and resuspend the powder in autoclaved distilled deionized water at 13.33% (w/v). Make aliquots and store at −20 °C. PEG-peptide aliquots can be thawed and frozen 2–3 times.
12. Mix the PEG-Gln and PEG-MMP-Lys solutions in a stochiometrically balanced fashion ([Gln]/[Lys] = 1) and then adjust the concentration of the total weight of PEG-peptide mix to 10% (w/v) with sterile distilled deionized water.

Functionality of each PEG population must be taken into account (*see* **Note 7**).

13. Gelation of the functionalized PEG precursor solution will occur within few minutes upon addition of TBS and FXIIIa (*see* **Note 8**). For example, to make 100 μL of 2% (w/v) PEG gel, mix 20 μL of PEG precursor solution, 10 μL of 10× TBS, 65 μL of water, and 5 μL of FXIIIa. After mixing, the sample is placed at 37 °C in a humidified environment for about 30 min, and then soaked in the appropriate medium.

3.3 Characterization of Hydrogel Mechanical Properties

1. Treat glass slides with O_2 plasma using a plasma cleaner.
2. Coat the plasma-treated side of the glass slide with Sigmacote and let it dry for 2 h.
3. Prepare 100 μL of 1.5% and 2%, 2.5%, and 3% precursor solutions in 1.5 mL eppendorf tubes. For 1.5% gels, mix in a tube 15 μL PEG-peptide mix, 10 μL 10× TBS buffer, 70 μL H_2O.
4. Add dedicated amount of FXIIIa to gel precursor solution. Quickly mix with pipette and cast three drops of 30 μL onto the Sigmacote-glass slide. Place also 1 mm spacer on the slide and gently overlay another Sigmacote-glass slide to sandwich the gel precursor drops to make a disk-like shape (Fig. 2).

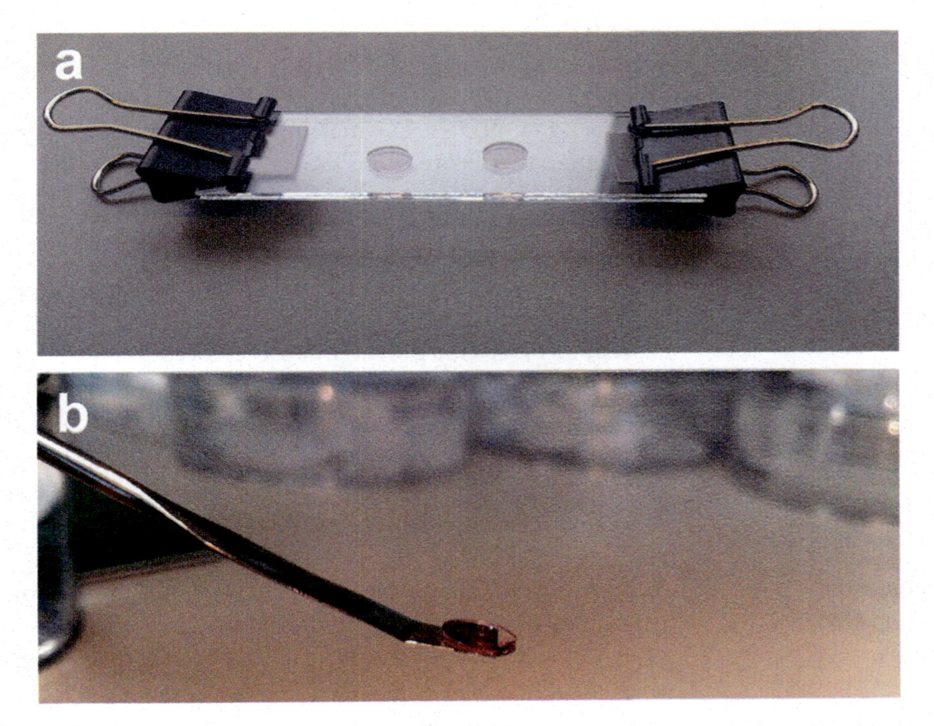

Fig. 2 Preparation of PEG-hydrogel test disc. (**a**) PEG-hydrogel discs cast in a sandwich of Sigmacote-treated glass slides. (**b**) Swollen PEG-hydrogel disc decast with a micro-spatula and ready to be tested by rheological analysis

5. Incubate for 30 min at 37 °C in a humidified environment to completely polymerize the gel.
6. Put the sandwiched slides into a 10 cm dish and cover them with 1× TBS, then carefully remove the first slide and gently detach the gel discs with a micro-spatula.
7. Incubate the resulting gel discs in culture medium at RT overnight to let it swell.
8. Put the swollen hydrogel discs of 1 mm thickness between the two plates of a rheometer and compress them up to 80% of their original thickness.
9. Perform measurements in constant strain (5%) mode. Record shear stress over the frequency range of 0.1–1 Hz and average storage moduli (G') over the frequency range.
10. Plot storage modulus (G′) as a function of PEG content in order to correlate PEG concentration with stiffness. Generally, hydrogels between 1.5% and 2.5% of PEG content correspond to a range of 300–1500 Pa (G').

3.4 Tethering Proteins in PEG-Hydrogels

ECM proteins such as fibronectin, laminins, vitronectin, and collagens are natural substrates of FXIIIa and they will be automatically conjugated in the PEG-hydrogel network upon mixing them with PEG-peptides before adding the transglutaminase enzyme (*see* **Note 9**).

Small proteins (<150 kDa) such as the cell–cell interaction protein EpCAM can be tethered in PEG-hydrogel as Fc-tagged versions using a NHS-PEG-maleimide heterobifunctional PEG linker to generate a Protein A-PEG-Gln peptide polymer.

1. Mix Protein A with NHS-PEG-maleimide in tenfold molar excess ratio for 1 h at RT (*see* **Note 10**).
2. Mix the resulting Protein A-PEG-maleimide with Gln peptide (resuspended in H_2O) in 1:1 ratio ON at 4 °C. The resulting Protein A-PEG-Gln polymer can be stored at −20 °C in 30% glycerol.
3. Mix the Fc-tagged protein of interest with Protein A-PEG-Gln in 1.66 fold molar excess ratio for 30 min at RT to finally obtain a Protein-Fc-Protein A-PEG-Gln polymer. Make aliquots and store at −20 °C in 30% glycerol.

3.5 Preparation of OSKM and M2rtTA Lentiviruses

1. Prepare plasmid maxi-prep of TetO-FUW-OSKM construct according to manufacturer's instructions and resupend the pellet in sterile Milli-Q water in order to have a DNA concentration between 1 and 2 μg/μL.
2. The day before the transfection seed 7.5 × 10^6 293T cells on 15 cm dishes and incubate ON (*see* **Note 11**).

3. 2 h before the transfection change the medium for 22.5 mL of fresh 293T cell medium in order to induce cell cycling.
4. Prepare in 15 mL falcon tube the following mix for each transfected dish (*see* **Note 12**):
   (a) 32 μg of TetO-FUW-OSKM, 6.25 μg of pRSV-Rev, 9 μg of pMD.2, 12.5 μg of pMDLg/pRRE, in 1.1 mL of sterile Milli-Q water.
   (b) For the M2rtTA virus add 32 μg of FUW-M2rtTA to the mix instead of TetO-FUW-OSKM.
5. Add 156 μL of 2 M $CaCl_2$ (CalPhos Mammalian Transfection kit), mix by pipetting 3–4 times and wait 5 min.
6. Add 1.25 mL of HEPES-buffered saline (CalPhos Mammalian Transfection kit) drop by drop while vortexing at full speed and wait 15 min. At this point the solution should appear slightly cloudy.
7. Add the 2.5 mL mix solution to a 15 cm dish (*see* **Note 13**).
8. 16 h after the transfection, change the medium for 16 mL of fresh 293T cell medium.
9. The day after (~48 h after the transfection) collect the supernatants for each type of viral preparation and filter them with 0.22 μm Stericups filter units. Put up to 30 mL of viral preparation in an Ultra-Clear tube and ultracentrifuge it at 20,000 × *g* for 2 h at 20 °C.
10. Carefully discard the supernatant and turn the tubes upside-down on absorbent paper and let dry the virus pellet for ~10 min.
11. Add 80 μL of PBS in each tube and carefully pipette several times to resuspend the pellet (*see* **Note 14**). Make aliquots that can be used in a single experiment and store them at −80 °C (*see* **Note 15**).

3.6 3D Reprogramming of Fibroblasts into iPSCs

1. Seed 1 × 10^6 fibroblasts in 10 cm dish and incubate ON.
2. Infect the cells with both TetOFUW-OSKM and M2rtTA viruses in 4 mL of fibroblast medium (15 μL/virus should be sufficient to have 100% efficiency of co-infected cells) and incubate ON.
3. Wash the cells with PBS, trypsinize them, and prepare a fibroblast suspension at 5 × 10^6 cells/mL (*see* **Note 16**).
4. Thaw and keep in ice PEG-peptides, RGDSP peptides, and FXIIIa.
5. Prepare a working mix of PEG-Gln and PEG-MMP-Lys, 1:1 (*see* **Note 17**).
6. Prepare the following mix in a 1.5 mL eppendorf tube to obtain 100 μL of 600 Pa PEG-hydrogel (generally it

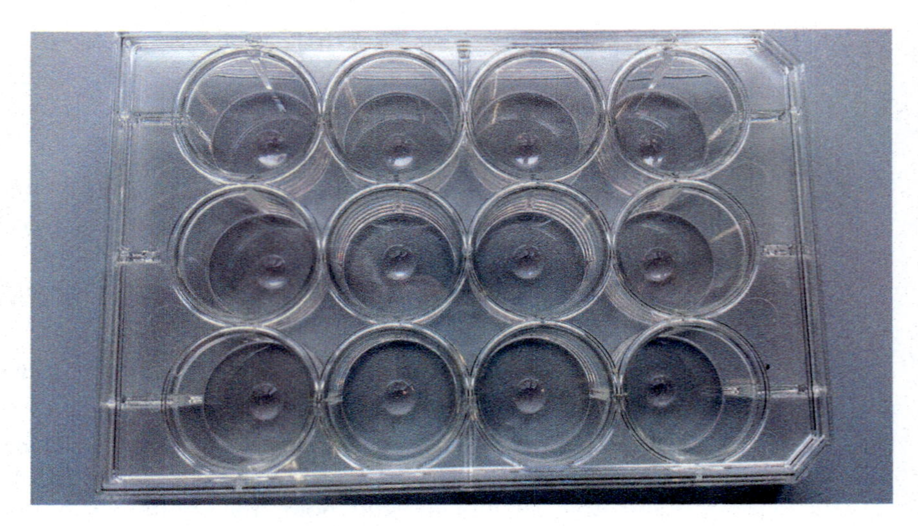

Fig. 3 PEG-hydrogel drops of 30 μL plated in a 12-well dish

corresponds to ~1.8% PEG): 51.74 μL of H_2O, 10 μL of 10× TBS buffer, 13.26 μL of PEG-peptide mix, 10 μL of 10× RGDSP, 10 μL of fibroblast suspension, 5 μL of FXIIIa (add the components in the specified order). Mix by pipetting 3–4 times (*see* **Note 18**).

7. Wait 30 s and pipette single drops of 30 μL in the middle of a 12-well plate well (Fig. 3). Keep 10 μL leftover of the mix to follow the polymerization of the hydrogel using a 200 μL pipette tip.
8. When the hydrogel mix becomes viscous, turn the plate upside-down and put it in a cell culture incubator at 37 °C for 30 min in order to allow complete crosslinking of the hydrogels.
9. Flip the plate and add 2 mL of fibroblasts medium per well.
10. The day after seeding/encapsulation shift cells to mouse ESC medium containing 2 μg/mL doxycycline and 10 μM CHIR. Change medium every 48 h during the entire reprogramming process. For human fibroblast reprogramming shift cells to mTeSR™1 medium with 2 μg/mL doxycycline and 10 μM CHIR and change the medium daily.
11. Within 5 days from doxycycline administration, OCT4-positive colonies should appear. 14 days of doxycycline exposure are sufficient to generate iPSCs (Fig. 4). A further week of culture without doxycycline will be needed to keep only the stably reprogrammed colonies. Human iPSC colonies will appear within 2 weeks and take ~6 weeks to get completely reprogrammed.

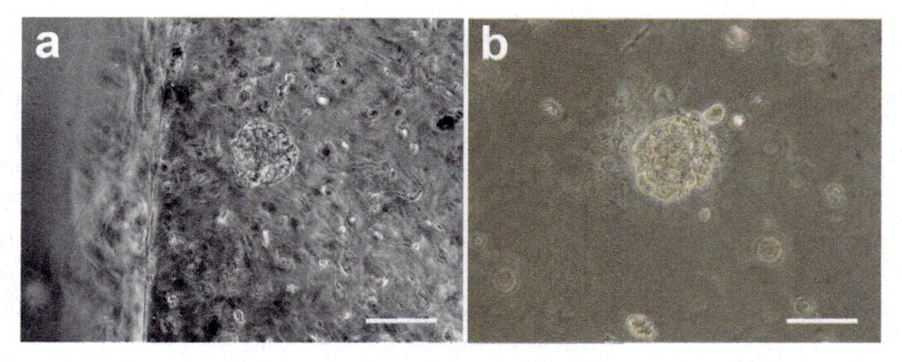

Fig. 4 Bright-field images showing 3D-iPSC generation in PEG-hydrogels. (**a**) Mouse iPSC colony after 7 days of OSKM overexpression. (**b**) Human iPSC colony after 14 days of reprogramming process. Scale bars, 50 μm (**a**) and 30 μm (**b**)

3.7 Adaptation of 3D-iPSCs to 2D Culture

1. In order to release 3D-iPSCs from PEG hydrogels discard the medium from the wells and wash with PBS three times for 5 min each, keeping the cells in the incubator at 37 °C in the meantime.
2. Discard most of the PBS, leaving in the wells a thin film of PBS in order to avoid hydrogels getting dry. Use the 45° angle end of a micro-spatula to carefully detach PEG-hydrogels from the wells and put them in a 15 mL falcon tube with 1 mL of TrypLE Express Enzyme (for mouse 3D-iPSCs) or TrypLE Select Enzyme (for human 3D-iPSCs) per gel (*see* **Note 19**).
3. Incubate in a water bath at 37 °C for 10 min shaking them continuously.
4. Add 1 mL of mouse ESC medium per PEG-hydrogel to inactivate TrypLE Express Enzyme or 1 mL of PBS per PEG-hydrogel to dilute TrypLE Select Enzyme.
5. Pellet cells at 200 × *g* for 5 min.
6. Plate 1 × 10^6 3D-iPSCs on 10 cm dish coated with Matrigel (use hESC-qualified Matrigel for human 3D-iPSCs) at 37 °C for 30 min.
7. Add mouse ESC medium or mTeSR1 medium for mouse and human 3D-iPSCs, respectively. Thereafter 3D-iPSCs can be cultured using normal feeder-free 2D culture protocols.

3.8 Immunocytochemical Characterization of 3D-iPSCs

1. To directly characterize 3D-iPSCs within PEG-hydrogels by immunostaining, fix the cell-PEG-hydrogel constructs for 30 min in 4% PFA.
2. Wash for 30 min in PBS three times at RT (this step can be performed directly in the 12-well dish). Fixed PEG-hydrogels can be stored in PBS for several weeks at 4 °C.
3. Permeabilize PEG-hydrogels with 1% Triton X-100 in PBS for 30 min at RT.

4. Block the samples with blocking solution for 2 h at RT.
5. Detach PEG-hydrogels using a micro-spatula and put them in a 48- or 96-well plate in primary staining solution (150 μL per well) with primary antibody (diluted according to data sheet indications). Agitate very gently (~30 rpm) the samples on a plate shaker ON at 4 °C.
6. Transfer PEG-hydrogels in a 12-well plate and wash them in PBS three times for 1 h each at RT.
7. Transfer PEG-hydrogels back in the 48- or 96-well plate and incubate in secondary staining solution (150 μL per well) with 1× DAPI and secondary antibody (diluted according to data sheet indications) ON at 4 °C on a plate shaker at ~100 rpm.
8. Transfer PEG-hydrogels into a 12-well plate and wash in PBS three times for 1 h each at RT.
9. Transfer PEG-hydrogels in a 35 mm glass-bottom dish. Place the PEG-hydrogel in the center on a thin film of PBS to perform epifluorescence or confocal imaging (*see* **Note 20**).

4 Notes

1. Protect doxycycline from light by switching off hood light and using aluminum foil on the aliquots. Used aliquots can be kept at 4 °C for few days, otherwise keep them at −20 °C and reuse them only once.
2. In order to avoid the generation of nanofibers, Matrigel should be always thawed on ice. To this aim, it is useful to prepare 200 μL aliquots that require only ~30 min to thaw on ice.
3. Preparation of aliquots should be performed quickly while keeping FXIIIa on ice. Do not reuse thawed aliquots.
4. For a good functionalization of PEG macromers the weight of the peptide can be slightly more than the calculated amount but not less. The excess would be washed out by dialysis anyway.
5. Peptides covalently bind to PEG macromers via Michael-type addition reaction through the thiol group of the cysteine side chain [9].
6. A simple way to obtain rubber bands fitting 50 mL falcon tubes is to cut 1 cm sections from nitrile disposable glove fingers.
7. The estimated molecular weight of PEG-peptide can be derived by summing up the molecular weight of the PEG macromer and dedicated number of peptides that one macromere can carry (number of arm corrected with functionality of the macromer). In the actual study, the estimated molecular weight of

PEG-Gln and PEG-MMP-Lys is 50,735.64 Da and 53,415.61 Da, respectively. The mass ratio to achieve stoichiometric balance is 0.4871:0.5129. For making 1 mL of 10% PEG gel precursor solution, 365.35 μL of 13.33% PEG-Gln, 384.65 μL of 13.33% PEG-MMP-Lys, and 250 μL of water should be mixed.

8. Gelation proceeds very slowly (hours) until FXIIIa is added to the mix. PEG-hydrogel gelation time is proportionally related to both PEG concentration and to temperature. A 1.5% PEG-hydrogel at RT starts to gel in 2 min. In order to follow the gelation, prepare a few extra microliters of PEG-hydrogel solution and use a 200 μL tip to check its stickiness in the eppendorf tube.
9. Most of the proteins with a molecular weight >150 kDa do not need any further functionalization to be tethered in PEG-hydrogels because they will be trapped in the PEG mash independently of their biochemical composition.
10. Resuspend NHS powder in PBS at ~10 mg/mL just before mixing with Protein A in order to avoid spontaneous hydrolysis occurring in aqueous buffer. Discard the resuspended NHS leftover.
11. Plated 293T cells should be ~30% confluence. Virus production lasts 48 h and higher starting confluence will result in high risk of cell detachment.
12. It is suggested to prepare a plasmid DNA mix for each type of virus. Then split the content of the mix in the number of 15 mL falcon tubes equal to the number of 15 cm dishes to be transfected.
13. Small dark precipitates should be visible by optic microscope soon after adding the transfection mix to the cells.
14. Avoid generation of bubbles during virus resuspension. If bubbles are present, spin down the virus solution at 100 × *g* for 5 min before freezing it.
15. Viral titer can be checked by ELISA assay for p24 viral capsid protein [10] (Clontech, 632200). A good virus preparation should be close to 10^9 viral particles/mL. In order to have a brief estimation of the amount of virus to be used for a reprogramming experiment, plate 5×10^4 cells in a well of 24-well plate and infect them with 0.2, 0.5, 1, 2, or 5 μL of virus in 250 μL of medium. Then, after 48 h of doxycycline treatment check by immunostaining the amount of cells expressing one of the transgenes of interest. The amount of virus used should have ~100% of transduction efficiency.
16. It is very important to discard any residual amount of TrypLE Express Enzyme in order to avoid following PEG-hydrogel

degradation. Therefore, it is highly recommended to resuspend the cell in at least 10 mL of fibroblast medium to inactivate the TrypLE Express Enzyme, and then to pellet again the cells to resuspend them in a smaller volume.

17. PEG–peptide mix can be kept at −20 °C and thawed 3–4 times.
18. ECM or PEGylated proteins should always be added before cells and FXIIIa.
19. TrypLE Select Enzyme does not need serum to be inactivated, and therefore it is enough to dilute it with PBS.
20. PEG-hydrogels can be imaged also in a multi-well plate at 5× or 10× magnification. 20× imaging could be possible with long distance objectives.

Acknowledgments

We thank Y. Okawa for setting-up the 3D reprogramming procedure in the lab, N. Brandenberg for helping with hydrogel precursor preparation, and D. Blondel for taking the picture shown in Fig. 1. This work was supported by the EU framework 7 HEALTH research programme PluriMes (http://www.plurimes.eu), the SystemsX.ch RTD project StoNets, an ERC grant (StG_311422), a Swiss National Science Foundation Singergia grant (CRSII3_147684), and FIRB 2013 grant from the Italian Ministry of University and Research (to M.C.).

References

1. Takahashi K, Yamanaka S (2006) Induction of pluripotent stem cells from mouse embryonic and adult fibroblast cultures by defined factors. Cell 126:663–676
2. Malik N, Rao MS (2013) A review of the methods for human iPSC derivation. Methods Mol Biol 997:23–33
3. Downing TL, Soto J, Morez C et al (2013) Biophysical regulation of epigenetic state and cell reprogramming. Nat Mater 12:1154–1162
4. Caiazzo M, Okawa Y, Ranga A et al (2016) Defined three-dimensional microenvironments boost induction of pluripotency. Nat Mater 15:344–352
5. Ehrbar M, Rizzi SC, Hlushchuk R et al (2007) Enzymatic formation of modular cell-instructive fibrin analogs for tissue engineering. Biomaterials 28:3856–3866
6. Ehrbar M, Rizzi SC, Schoenmakers RG et al (2007) Biomolecular hydrogels formed and degraded via site-specific enzymatic reactions. Biomacromolecules 8:3000–3007
7. Ranga A, Gobaa S, Okawa Y et al (2014) 3D niche microarrays for systems-level analyses of cell fate. Nat Commun 5:4324
8. Ehrbar M, Sala A, Lienemann P et al (2011) Elucidating the role of matrix stiffness in 3D cell migration and remodeling. Biophys J 100:284–293
9. Lutolf MP, Tirelli N, Cerritelli S et al (2001) Systematic modulation of Michael-type reactivity of thiols through the use of charged amino acids. Bioconjug Chem 12:1051–1056
10. Tiscornia G, Singer O, Verma IM (2006) Production and purification of lentiviral vectors. Nat Protoc 1:241–245

Chapter 6

Calcium Phosphate Foams: Potential Scaffolds for Bone Tissue Modeling in Three Dimensions

Edgar B. Montufar, Lucy Vojtova, Ladislav Celko, and Maria-Pau Ginebra

Abstract

The present method describes the procedure to fabricate calcium phosphate foams with suitable open porosity, pore size, and composition to perform three-dimensional (3D) cell cultures with the objective to simulate the bone tissue microenvironment in vitro. Foams with two compositions but equivalent porosity can be fabricated. On the one hand, hydroxyapatite foams obtained by hydrolysis at 37 °C, with microstructure that mimics the small crystal size of the mineral component of bones, and on the other hand, beta tricalcium phosphate foams with polygonal grains obtained by sintering at 1100 °C. In the first part of the chapter the calcium phosphate foams are briefly described. Afterwards, the foaming process is described in detail, including alternatives to overcome processing problems than can arise. Finally, insights are provided on how to perform 3D cell cultures using the calcium phosphate foams as substrates.

Key words Calcium phosphate foam, Hydroxyapatite, Tricalcium phosphate, Bone model, Cell culture, Three-dimensional, Osteoblast, Tissue engineering, Scaffold

1 Introduction

Calcium phosphate foams (CPFs) were originally designed as bone filling materials to support new bone formation in vivo [1]. These solid foams consist of a continuous trabecular structure from calcium phosphate having an interconnected network of spherical pores (Fig. 1). The diameter of spherical pores ranges from 100 to 500 μm, occupying between 45 and 55% of total volume of the foam. These pores are connected from the surface to the interior of the foam through circular apertures with diameters between 10 and 350 μm (Fig. 1). The connected pore network allows ingrowth of new bone in vivo [2] and the foam colonization by cells migrating from the surface in vitro [3]. The penetration of cells in culture is limited by the diffusion of nutrients rather than the pore connectivity. However, the diffusion of nutrients could be improved using a bioreactor such as perfusion system [1]. The calcium

Zuzana Koledova (ed.), *3D Cell Culture: Methods and Protocols*, Methods in Molecular Biology, vol. 1612,
DOI 10.1007/978-1-4939-7021-6_6, © Springer Science+Business Media LLC 2017

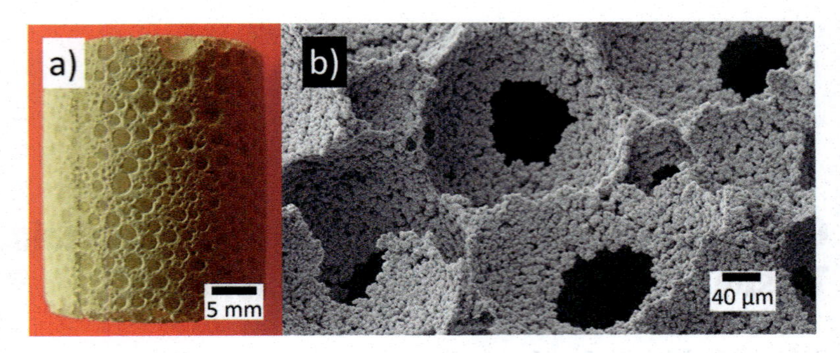

Fig. 1 Images of the CPFs. (**a**) Representative optical microscope image of the general aspect of a CPF, showing the open porosity on its surface. (**b**) Detail image of a CPF obtained with scanning electron microscope, showing the fractured surface. Note that the spherical pores are connected by at least one circular aperture, with suitable size for cell migration. Furthermore, smaller pores at walls of the foam can be observed

phosphate that forms the trabeculae in the CPFs can be either hydroxyapatite [3–7], brushite [8] or beta tricalcium phosphate (β-TCP) [9]. All these calcium phosphate ceramics are successfully used as synthetic bone grafting materials or as scaffolds for bone tissue engineering [10, 11]. Their success in bone regeneration is due to the fact that the mineral component of bone tissue also belongs to the family of calcium phosphates (hydroxyapatite is the most similar calcium phosphate to the bone mineral phase). The microstructure of the trabeculae of the CPFs is also porous but in a smaller scale. The trabeculae contain a network of micropores with sizes below 2–5 µm (Fig. 2). Although the size of these pores is unsuitable for cell colonization, they are important for protein adsorption [12], cell adhesion (providing larger surface area) and if necessary for the loading of grow factors or drugs that stimulate or inhibit cell responses [13]. Therefore, due to their porous structure at different scales and biomimetic chemical composition, CPFs are promising candidates to perform three-dimensional (3D) cell cultures with the particular aim of mimicking in vitro the bone tissue niche.

The first stage for preparing the CPF consists of the production of a stable liquid foam. This can be achieved by mechanical stirring of a water solution containing a biocompatible surface active molecule (surfactant). The second stage comprises gently mixing the liquid foam with alpha tricalcium phosphate (α-TCP) powder [3]. Upon mixing, the liquid foam acts as a template for the formation of spherical pores and α-TCP undergoes a chemical reaction at constant temperature that allows the consolidation of the foam [3]. More specifically, α-TCP hydrolyses and hardens according to Eq. 1 resulting in calcium deficient hydroxyapatite (CDHA).

$$3\alpha - Ca_3\left(PO_4\right)_2 + \mathrm{H_2O} \rightarrow Ca_9\left(HPO_4\right)\left(PO_4\right)_5 OH \qquad (1)$$

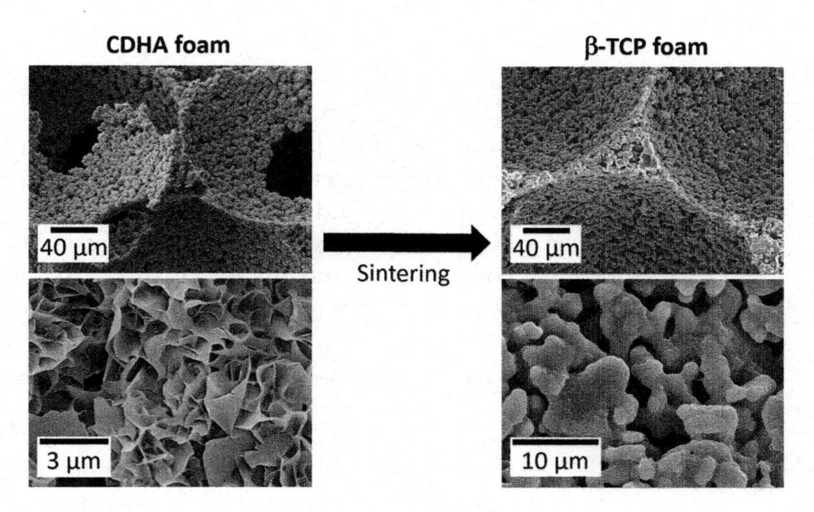

Fig. 2 Representative scanning electron microscope images of the microstructure of calcium deficient hydroxyapatite (CDHA) foam (*left*) and beta-tricalcium phosphate (β-TCP) foam (*right*). While the CDHA foam is composed of plate-like crystals, the β-TCP foam is composed of smooth polygonal grains. The polygonal grains are formed by the CDHA plate-like crystals fusion during sintering [9]. Note that after sintering, the morphology of the spherical pores (*top images*) is maintained

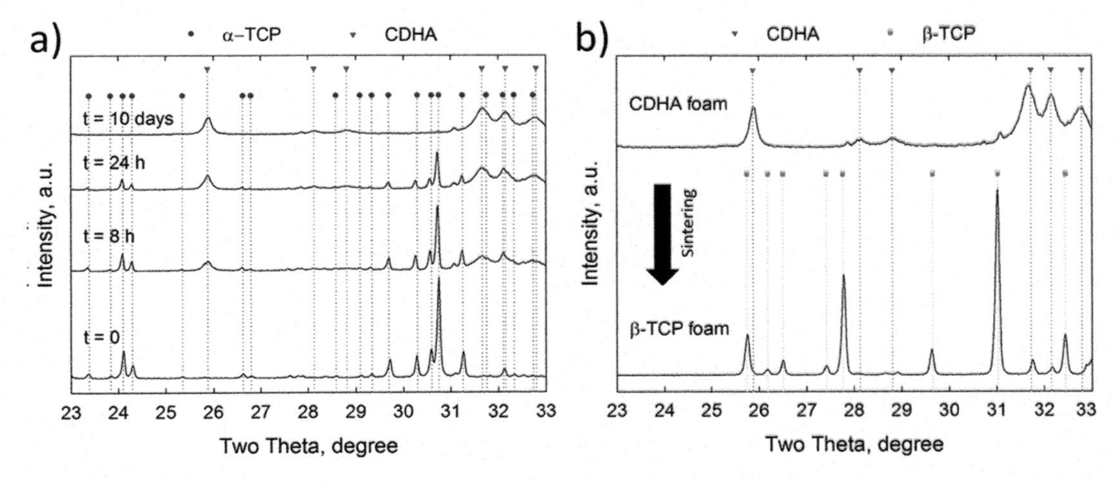

Fig. 3 X-ray diffraction patterns. (**a**) The α-TCP powder hydrolysis to CDHA crystals by dissolution and precipitation reaction. In this case the reaction was performed at 37 °C by immersing samples in water. At day 10, the peaks of α-TCP disappear and the pattern corresponds to hydroxyapatite with low crystalline order due to the small size of formed crystals (*see* Fig. 2). (**b**) After sintering, the CDHA foam transforms into β-TCP, with high intensity peaks due to crystal growth

As Fig. 3a shows, the progress of the reaction can be followed by X-ray diffraction. The total time needed to complete the reaction depends on several conditions, for example, at 37 °C, 100 kPa and 100% humidity the maximum strength of the foam is reached after 10 days. A particular advantage of this process is that the foam is moldable for a few minutes after mixing. Therefore, the mixture can be easily shaped to obtain almost any geometry using a mold.

To obtain the β-TCP foam an extra sintering step of the CDHA foam is required [9]. During sintering, the microstructure (*see* Fig. 2) and composition (*see* Fig. 3b) of the foam are changed, while the morphology and the porosity are maintained. Therefore, the process described herein offers the possibility to select two different compositions and microstructures with a broad range of similar open porosities to study the bone biology in 3D cell cultures.

2 Materials

Prepare all solutions at room temperature (RT) using ultrapure water and analytical grade reagents. Store all reagents and products at RT (unless indicated otherwise). Diligently follow all waste disposal regulations when disposing waste materials.

2.1 α-TCP Powder

1. Calcium carbonate ($CaCO_3$) powder: ≥99%.
2. Calcium hydrogen phosphate ($CaHPO_4$) powder: ≥98%.
3. Magnetic stirrer.
4. Glass beaker.
5. Ethanol (≥95%).
6. Oven for drying.
7. Platinum crucible or a crucible that supports 1400 °C and quenching.
8. High temperature chamber furnace (furnace): up to 1400 °C.
9. Appropriate protective gear for high temperatures, i.e., gloves, apron, and glasses.
10. Steel drawer (40 × 40 × 20 cm, approximately).
11. Household fan.
12. Steel hammer.
13. X-ray diffractometer (optional).
14. Planetary ball mill (mill) (Fritsch Pulverisette 6): with agate jar and 10 agate balls of 30 mm in diameter (*see* **Note 1**).
15. Desiccator.

2.2 Liquid Foam Template

1. Polysorbate 80 solution: 1% (w/w) polysorbate 80 (also known as Tween 80®) and 2.5% (w/w) Na_2HPO_4 in water. Weight 0.50 g of liquid polysorbate 80, add water to a volume of 20 mL and dissolve. Weight 1.25 g of Na_2HPO_4, add water to a volume of 20 mL and dissolve. Transfer the two solutions to a volumetric flask and make up to 50 mL with water. Use the solution within 2–3 days.

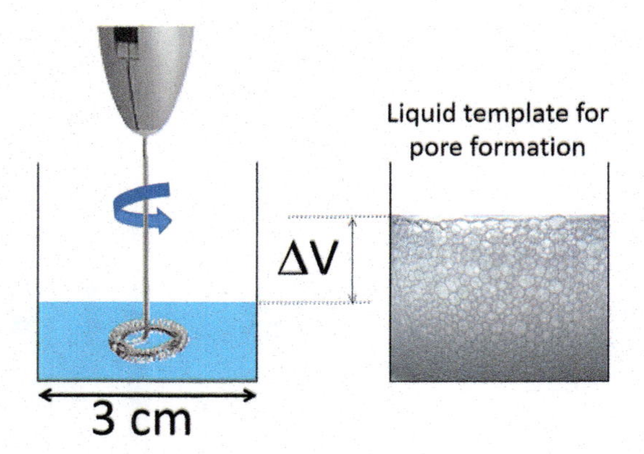

Fig. 4 Illustration of the foaming process. The left side shows the shape and position of the steel blade used to foam the polysorbate 80 solution. The right side shows the aspect of the obtained foam observed by optical microscope. Note the increment in volume (ΔV) due to the foaming process. The polysorbate 80 foam (*right side*) acts as a template for spherical pores formation

2. Plastic container for foaming: approximately 3 cm in diameter with flat bottom.
3. Foaming tool: stainless steel blade adapted to a rotatory tool (*see* Fig. 4 and **Note 2**).
4. Chronometer.
5. Spatula and scalpel.
6. Polytetrafluoroethylene mold(s).
7. Food grade silicone release agent in spray.
8. Oven (37 °C).
9. Plastic containers for the molds.
10. Gamma irradiator (optional).
11. 70% ethanol in water (optional).

2.3 3D Cell Seeding and Culture in Perfusion Bioreactor System

1. Rat multipotent stromal cells (rMSC), extracted from bone marrow of long bones of Lewis rats. For 3D cell culture, use passage 3–5 (*see* **Note 3**).
2. Cell culture medium (Adv-DMEM): 10% fetal bovine serum, 2 mM L-glutamine, 50 U/mL penicillin, 50 μg/mL streptomycin and 20 mM HEPES in Advanced Dulbecco's modified Eagle's medium.
3. Two CDHA foams (6 mm of diameter × 10 mm of height), sterilized by gamma radiation at 25 kGy.
4. Perfusion bioreactor system with two culture chambers connected in series, three-way Luer connector for cell loading, medium reservoir for cell culture and a peristaltic pump (Fig. 5) [14].

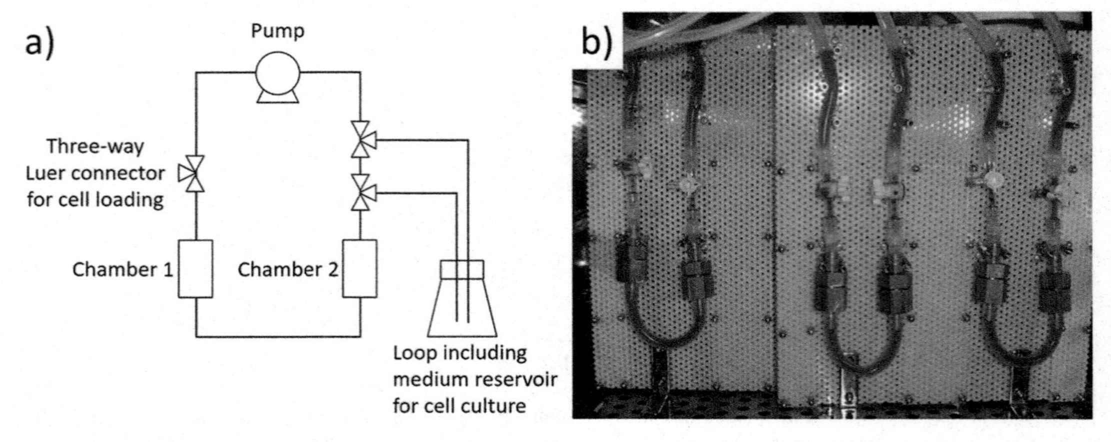

Fig. 5 Example of a perfusion bioreactor system used for dynamic seeding and culture of rMSC on CDHA foams. (**a**) Diagram of the main components of the bioreactor. Note that the two three-way Luer connectors on the right side of the bioreactor are used to open/close the loop for the reservoir of cell culture medium. The loop is closed during seeding to allow cell suspension oscillation and is open during culture to constantly supply fresh culture medium [14]. (**b**) Photograph of three bioreactors inside the incubator after injection of 10 mL of cell suspension. Each bioreactor has two metallic chambers with one CDHA foam

5. 6-well cell culture plates (optional for static 3D cell culture).
6. Laminar flow hood.
7. Cell culture incubator (37 °C and 5% CO_2).
8. Phosphate buffer solution (PBS).
9. Lysis buffer for protein extraction, e.g., Mammalian Protein Extraction Reagent (M-PER).
10. Lactate dehydrogenase (LDH) assay kit (optional).
11. Microplate spectrophotometer (optional).

3 Methods

3.1 Synthesis and Milling of α-TCP

1. Mix 1 mol of $CaCO_3$ powder with 2 mol of $CaHPO_4$ powder (*see* **Note 4**).
2. Introduce the dry mixture in the crucible, place the crucible containing the mixture in the furnace, and perform the thermal treatment as indicated in Table 1 (*see* **Note 5**).
3. Introduce the agate balls in the jar of the mill.
4. Add 145 g of the quenched α-TCP and mill the material for 15 min at 450 rpm in dry conditions (*see* **Notes 1** and **6**).
5. Store the powder in dry atmosphere.

3.2 Foaming Process

1. Add 3 mL of polysorbate 80 solution in the plastic container (Fig. 4).
2. Foam the solution with the foaming tool for 1 min at 12,000 rpm (*see* **Note 7**).

Table 1
Stages of the thermal treatment to obtain α-TCP

Stage	Description	Temperature [°C]	Heating rate [°C/min]	Time [h]	Reason
1	Heating	From RT to 300	2.5	–	
2	Dwelling	300	–	2	$2CaHPO_4 \rightarrow Ca_2P_2O_7 + H_2O$
3	Heating	From 300 to 1100	2.5	–	$CaCO_3 \rightarrow CaO + CO_2$
4	Dwelling	1100	–	2	$CaO + Ca_2P_2O_7 \rightarrow \beta\text{-}Ca_3(PO_4)_2$
5	Heating	From 1100 to 1400	2.5	–	
6	Dwelling	1400	–	2	$\beta\text{-}Ca_3(PO_4)_2 \rightarrow \alpha\text{-}Ca_3(PO_4)_2$
7	Air quenching	From 1400 to RT			Retain the alpha phase

3. Immediately, add 5.454 g of α-TCP powder to the foam (*see* **Note 8**) and gently mix the powder with the foam preventing bubble disruption (*see* **Notes 9** and **10**).
4. Cast the paste manually into the molds to shape the foam to the required dimensions (*see* **Note 11**). Avoid thumping to fill the mold.
5. Keep molds with foams for 12 h at 37 °C and within 100% of relative humidity (*see* **Note 12**).
6. Afterwards, immerse the molds with foams in water at 37 °C for 9 days for total consolidation and conversion into CDHA.
7. Extract the CDHA foams from the molds (*see* **Note 13**), rinse them with water to remove solid debris and dry at 90 °C for 24 h.
8. Introduce the CDHA foams into the furnace and perform the following thermal treatment to obtain the β-TCP foams. Heating from RT to 1100 °C at 2.5 °C/min, dwelling during 9 h and turn off the power of the furnace to cool down the β-TCP foams. Remove the foams from the furnace at a temperature below 100 °C (*see* **Note 14**).
9. Sterilize foams using one of the following recommended methods: Gamma irradiation at 25 kGy or immersion in 70% ethanol for 30 min.

3.3 Dynamic Seeding and Culture of rMSC on CDHA Foams: 3D Cell Culture

1. Place the sterile CDHA foams inside culture chambers of the perfusion bioreactor system (one foam per chamber, *see* Fig. 5) and precondition them with cell culture medium for 24 h in cell culture incubator (*see* **Note 15**). Use the three-way Luer connector to insert and remove the cell culture medium. Work in laminar flow hood to guaranty the sterility.

2. Prepare suspension of rMSCs in cell culture medium (10 mL), count the cells and adjust cell concentration as needed (e.g., 5×10^5 cells per CDHA foam).
3. For dynamic cell seeding (*see* **Note 16**), inject 10 mL of the cell suspension inside the perfusion bioreactor system using a three-way Luer connector and use a peristaltic pump to oscillate the levels of the cell suspension, e.g., at the speed 10 mm/s (*see* **Note 17**). Continue pushing the suspension through the two CDHA foams repeatedly for 3 h in cell culture incubator to allow homogenous cell distribution and wait 30 min to promote cell attachment.
4. After seeding, extract the foams from the bioreactor and individually rinse them with PBS.
5. Lyse the cells, e.g., using M-PER (*see* **Note 18**), and quantify the cell number (*see* **Note 19**), e.g., quantifying LDH content. Example results are presented in Fig. 6 as seeding efficiency per condition. Experiments suggest that higher velocity and longer seeding time increase cell seeding efficiency. However, under the studied conditions, double the initial cell number does not significantly affect seeding efficiency [15].
6. There are at least two options to continue the 3D cell culture (*see* **Note 20**).
   (a) For dynamic 3D cell culture remove the cell suspension, fill the bioreactor with fresh cell culture medium and keep the foams loaded with cells inside the bioreactor (*see* **Note 21**).

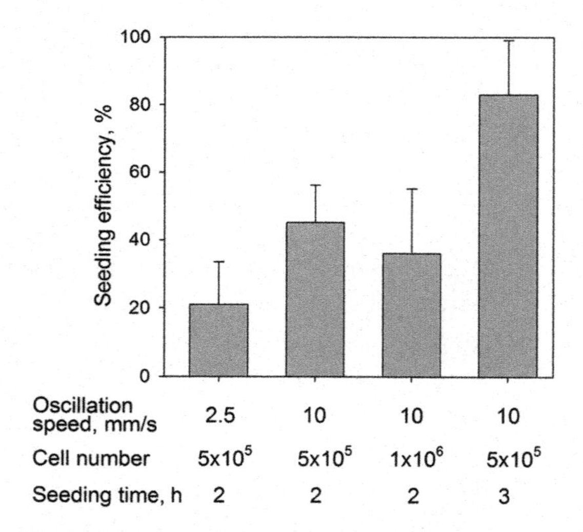

Fig. 6 The plot shows cell seeding efficiency for different seeding conditions. The parameters studied were: speed of the cell suspension during oscillation (2.5 or 10 mm/s), cell number per CDHA foam (5×10^5 or 1×10^6 cells), and time allowed for cell attachment (seeding time; 2 or 3 h)

During culture apply a constant unidirectional medium flow at a rate of 1 mm/s.

(b) For static cell culture, extract the foams loaded with cells from the bioreactor and culture them in 6-well plates. Use enough cell culture medium to completely cover the foams.

4 Notes

1. Other mill types and milling conditions can be used to prepare the α-TCP powder achieving similar results. However, the process described herein has been optimized using the quoted planetary ball mill. Notice that it is highly recommended to avoid the use of metallic milling media to prevent the contamination of the powder.
2. The stainless steel blade can be obtained from any milk frother (e.g., from IKEA®) and can be adapted to any rotatory tool system that supplies a constant rotation speed of 11,000–15,000 rpm (e.g., Dremel 4000, Robert Bosch Tool Corporation). Alternatively, the milk frother can be used directly for foaming. However, periodic renewal of batteries is required.
3. Other cell types in different passages can be seeded and cultured on CPFs depending on the particular experiment to be performed. However, as an example, in this chapter we describe a 3D cell culture performed with rMSC. Other cells cultured on CPFs are SAOS-2 and MG-63 osteosarcoma cell lines [3, 6, 7].
4. To obtain 150 g of α-TCP, mix 48.40 g of $CaCO_3$ with 131.59 g of $CaHPO_4$. To guarantee a homogenous mixture place the powders inside a glass beaker, incorporate a magnetic stir bar and 250 mL of ethanol (≥95%). Stir the powders during 3 h at RT. Afterwards, decant the liquid and dry the powder mixture at 50 °C during 3 h.
5. To quench the sintered block prepare the steel drawer, circulate air near the drawer (e.g., using a fan) and prepare a steel hammer to break the block. Stop the heating of the furnace, remove as fast as possible the crucible using proper protective gear, flip the crucible to allow the block to fall and break the block with the hammer in smaller pieces (from millimeters to several centimeters of size; Fig. 7). It is highly recommended to perform the quenching process in two people. X-ray powder diffraction can be used to determine the purity of the α-TCP obtained. α-TCP has monoclinic crystalline structure (space group P21/a) and its inorganic crystal structure database (ICSD) is the number 923.

Fig. 7 Steps followed for quenching of the α-TCP. (**a**) Flip the crucible to extract the block, (**b**) the second person closes the furnace door and the operator breaks the block with a metallic hammer. (**c**) Detail of how the α-TCP particles are broken with the hammer impact

6. Alternatively, α-TCP powder can be commercially acquired. Some commercial providers of α-TCP powders are Ensail Beijing Co. Ltd., Sigma Aldrich Co., Zimmer Dental GmbH and Wako Pure Chemical Industries Ltd. For more details about commercially available α-TCP powders the following excellent review is recommended [16]. Since the fabrication process of the foams has been optimized for specific powder particle size distribution, the use of commercial α-TCP powders or different milling conditions than those described herein (*see* **Note 1**) can require adjusting some processing parameters. As an indication, the optimal particle size distribution (determined by laser diffraction) of the α-TCP powder for the fabrication the CPFs is as follows: average particle size of 7.34 μm with accumulated particle sizes at 25, 50, and 75% of 1.52, 4.95, and 10.82 μm, respectively [3, 17].
7. Immerse the stainless steel blade in the polysorbate 80 solution but avoid touching the wall or the bottom of the container with the blade. Avoid circular or up and down movements of the blade. An external support can be used to hold the foaming tool and prevent variations due to the operator. To prove if the foaming was properly performed, i.e., total conversion of the liquid into the foam, flip the plastic container, no liquid should flow and the white foam should stay at the bottom of the container. If necessary discard the foam and repeat the foaming process.
8. It is recommended to break the powder agglomerates prior to mixing with the liquid foam. To this end, expand the powder in a flat surface and cut the powder with a sharp-edged instrument, i.e., flat spatula or scalpel. The liquid to powder ratio of 0.55 mL/g (volume refers to the liquid before foaming) has been optimized to obtain a workable and moldable paste that

after hardening retains the bubbles as spherical macropores interconnected by circular windows. The use of milling conditions different from those described herein or the use of commercial α-TCP powders could require a readjustment of the liquid to powder ratio. In such case, it is recommended to keep constant the volume of polysorbate 80 solution and vary the amount of powder to be incorporated.

9. After adding the powder, it will go to the bottom of the container; use a spatula to slowly move the powder allowing it get wet with the foam. At the beginning, it will be a difficult task, however, after a few seconds (about 20–30 s), when the powder is completely wetted, the workability of the paste considerably increases. From this moment, mix more thoroughly and slightly faster to homogenize the paste. Use circular movements and avoid changes of direction. After additional 30 s (around a total of 1 min, since the powder has contacted the foam) the paste should be ready to use.
10. Alternatively, the foaming and mixing steps can be performed simultaneously. In such case, the foaming tool should have enough power to keep the velocity constant during the process. Add 3 mL of polysorbate 80 solution and 5.454 g of α-TCP powder in the plastic container (*see* **Note 8**) and immediately foam the mixture with the foaming tool during 30 s at 7000 rpm, following instruction as indicated in **Note 7** (note that the proof of foaming indicated in **Note 7** is not applied in case of simultaneous foaming and mixing). Afterwards, the paste should be ready to use. It is recommended to get used to the two steps method (foaming liquid + mixing) before trying the one step method.
11. It is strongly recommended that the paste is completely casted within the first 5 min after its preparation; otherwise the spherical pore morphology can be compromised. The molds can have any shape providing the paste can flow, fill the cavity and prevent the entrapment of air bubbles. To promote the formation of open pores at the surface of the CPF, it is recommended to impregnate the mold with a hydrophobic liquid (i.e., food grade silicone release agent; a spray system makes the task easier) prior to casting. Leave a small paste meniscus over the surface of the mold to compensate a small contraction due to slight bubble collapse before hardening.
12. One alternative to get 100% of relative humidity at 37 °C without special equipment is using two containers, one inside the other. Place the mold(s) with the foam in the internal container and keep it open. Put water in the external container (approximately up to one-fourths of the height) and close it hermetically. The internal container avoids direct contact of

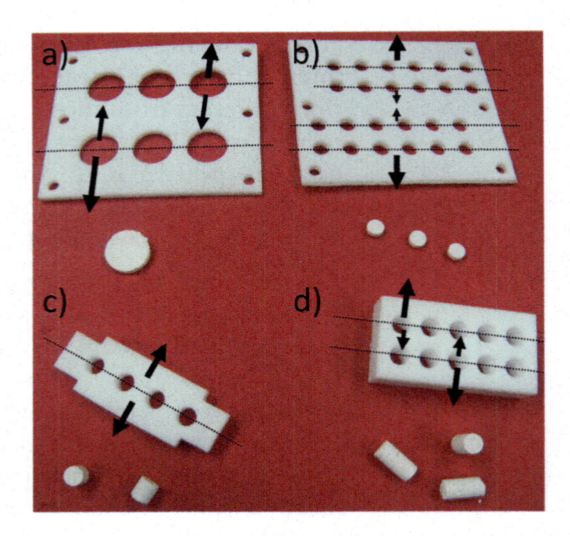

Fig. 8 Examples of polytetrafluoroethylene molds used to fabricate discs of 12 (**a**) or 6 (**b**) mm in diameter, respectively, both with 2 mm of thickness, and (**c**) and (**d**) cylinders with 6 or 12 mm in height, respectively, both with 6 mm of diameter. While the *dashed lines* indicate the join between the different parts of the molds that allow the easiest sample removal, the *arrows* show the sliding direction to open the molds

liquid water with the foams. The whole set can be placed in an oven or incubator at 37 °C.

13. This is the most difficult step. Probably many attempts have to be performed to extract the foams without fracturing them. However, with practice this will become easier, just keep calm. The mold design is of paramount relevance to extracting the foams without problems. The shape of the mold must prevent that the sample remains trapped. As Fig. 8 shows, for the most common shapes, i.e., discs or cylinders, molds with at least two parts that can be separated by hand are the best alternative. To make sample removal easier it is recommended to fabricate the part of the mold in contact with the foams from polytetrafluoroethylene (Teflon®). Furthermore, to prevent contaminations with metallic ions, avoid the use of metallic parts, such as screws.

14. As shown in Figs. 2 and 3, the sintering step modifies the microstructure and the composition of the wall of the foams. Besides, sintering produces a contraction of the overall sample. Therefore, the initial CDHA foam should be overdimensioned to end with the desired β-TCP foam size. For the sintering process described herein the linear contraction of the sample corresponded to 11.5%. However, it is recommended to run a preliminary test to guarantee the accuracy in the final β-TCP foam size.

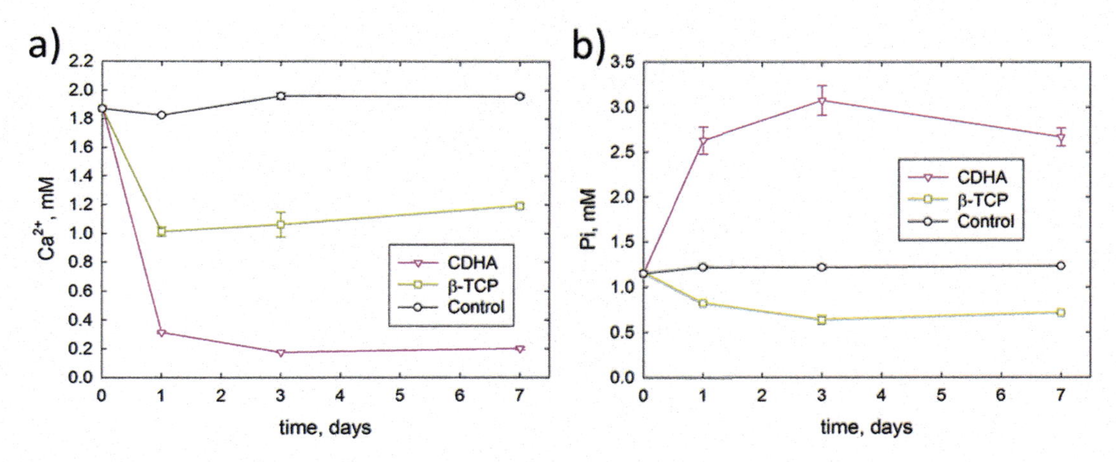

Fig. 9 Variation in (**a**) calcium (Ca^{2+}) and (**b**) phosphate (Pi) concentration of the cell culture medium in contact with CDHA, β-TCP or the cell culture plastic used as control. The ionic concentrations were determined by inductively coupled plasma-optical emission spectroscopy in cell culture medium samples that have been in contact with the materials during 24 h. The total volume of medium (1 mL) was renewed every 24 h. Samples after 1, 3, and 7 days of experiment were analyzed in triplicate

15. The preconditioning allows the proteins of the cell culture medium to adsorb on the sample surface improving cell attachment. Furthermore, it should be considered that the foams interact with the cell culture medium, resulting in ionic exchanges, which in turn could have some effects on cell behavior. As an example (Fig. 9), in contact with Adv-DMEM, CDHA foams tend to reduce the calcium (Ca^{2+}) content and increase phosphate (Pi) content regarding the basal concentration of Adv-DMEM. For this reason, to minimize ionic fluctuations, it is recommended to use a small ratio between material mass and cell culture medium volume, i.e., to use small samples. In contrast, β-TCP foams reduce both, Ca^{2+} and Pi, but to smaller extent when compared with CDHA foams.
16. There is not a standard method for cell seeding in 3D porous materials. Cell seeding efficiency will depend on several parameters such as cell type, cell number, volume of used medium, sample size, sample geometry, static or dynamic seeding conditions, time allowed for cell attachment, among others. In general, it is well accepted that dynamic seeding conditions improve the penetration of the cells, allowing more homogeneous cell distribution. However, devices for dynamic cell seeding, such as perfusion bioreactors, are not commonly available. In case of having access to dynamic cell seeding methods, we recommend to follow the seeding conditions according to your own expertise. For the alternative of static cell seeding, according to our experience, we recommend to remove the medium used to precondition the samples and

seed the cells in the exact volume of medium needed to wet the sample, without spilling it outside. The process can be performed in normal cell culture wells under common cell culture conditions. To determine the time needed for initial cell attachment you can seed cells in parallel on plastic and follow their morphology by optical microscopy. After proper incubation time (at least the same time as needed for cell attachment on plastic), add enough cell culture medium to cover the whole sample.

17. During the oscillation the cell suspension will be pushed through the CPFs ensuring a more homogenous cell distribution. Make sure that during the oscillation the cell culture medium always wets the foams.

18. In case of cell lysis, it should be considered that calcium phosphates and in particular hydroxyapatite have large affinity to proteins and nucleic acids. Therefore, the lysate should be removed as fast as possible from the CPFs to prevent loss. In particular, avoid storing the lysate together with the CFPs. Besides, due to the 3D structure, the lysis methodology should allow to remove the cells from the interior of the CPFs and recover the maximum amount of lysate. One alternative is to break the CPF prior to lysis of the cells.

19. Common procedures and protocols can be used to evaluate the outcome of the cell culture on CPFs. However, due to the high porosity content with pore sizes below 5 μm, some precautions should be tacked to stain the cells, because staining agents can be easily adsorbed in the CPFs, increasing the background of the signal during microscope observation. It is recommended to minimize the concentration of the dye and the incubation time to prevent excessive background. In addition, histological preparation of the samples can be performed to analyze the cell culture inside the CPFs.

20. The exact conditions for cell culture depend on the type of experiment that will be performed. General aspects to be considered are the following:

    (a) As shown in Fig. 9, although the cell culture medium is renewed every day, the ionic exchange continues for several days (*see* **Note 15**). Therefore, to decrease the effects of the ionic exchanges on cell responses, more frequent renewal of cell culture medium or use of larger volume of medium is recommended.

    (b) Although that static cell culture conditions allow cell proliferation, cell survival will be limited by the diffusion of oxygen inside the CPF. The use of thinner samples or dynamic culture conditions can improve cell survival and cell proliferation in the whole volume of the CPF. Our results showed twice number of cells after 3 days of culture in dynamic conditions [15], revealing the advantage of dynamic cell culture over static conditions.

(c) Cell response depends on the microstructure and composition of the material. For example, some works have shown that apatites with plate-like or needle-like crystals (such as in the CDHA foam; Fig. 2) promote the differentiation of mesenchymal stem cells into osteoblasts [18] and promote the osteoblastic phenotype [19, 20]. The more osteoinductive effect is associated with a microstructure and composition more similar to the mineral phase of bone tissue (nanocrystals of non-stoichiometry and ion substituted apatite). In contrast, conventional sintered pure tricalcium phosphate with polygonal grains, such as β-TCP foams (Fig. 2), do not induce osteoblastic differentiation in vitro [21]. Furthermore, it was observed that cell proliferation was slower on calcium phosphate ceramics than on cell culture plastic, and this effect was more obvious in the case of apatites with plate-like crystals than for sintered β-TCP [21].

21. The tubing of the perfusion bioreactor should be switched to perform the dynamic cell culture. Basically a reservoir with 150 mL cell culture medium has to be incorporated using the three-way Luer connectors (Fig. 5a).

Acknowledgments

This project has received funding from the European Union's Horizon 2020 research and innovation program under the Marie Skłodowska-Curie and it is cofinanced by the South Moravian Region under grant agreement no. 665860. We also thank the Spanish Government for financial support through the Project MAT2015-65601-R, co-funded by the EU through European Regional Development Funds. Support for the research of MPG was received through the "ICREA Academia" award for excellence in research, funded by the Generalitat de Catalunya. The project CEITEC 2020 (LQ1601) was supported from the Ministry of Education, Youth and Sports of the Czech Republic under the National Sustainability Programme II.

References

1. Ginebra MP, Montufar EB (2014) Injectable biomedical foams for bone regeneration. In: Netti P (ed) Biomedical foams for tissue engineering applications, 1st edn. Woodhead Publishing Ltd, UK
2. Kovtun A, Goeckelmann MJ, Niclas AA et al (2015) In vivo performance of novel soybean/gelatin-based bioactive and injectable hydroxyapatite foams. Acta Biomater 12:242–249

3. Montufar EB, Traykova T, Gil C et al (2010) Foamed surfactant solution as a template for self-setting injectable hydroxyapatite scaffolds for bone regeneration. Acta Biomater 6:876–885
4. Ginebra MP, Delgado JA, Harr I et al (2007) Factors affecting the structure and properties of an injectable self-setting calcium phosphate foam. J Biomed Mater Res 80A:351–361
5. Montufar EB, Traykova T, Schacht E et al (2010) Self-hardening calcium deficient hydroxyapatite/gelatine foams for bone regeneration. J Mater Sci Mater Med 21:863–869
6. Montufar EB, Traykova T, Planell JA et al (2011) Comparison of a low molecular weight and a macromolecular surfactant as foamingagents for injectable self setting hydroxyapatite foams: polysorbate 80 versus gelatin. Mater Sci Eng C 31:1498–1504
7. Perut F, Montufar EB, Ciapetti G et al (2011) Novel soybean/gelatine-based bioactive and injectable hydroxyapatite foam: material properties and cell response. Acta Biomater 7: 1780–1787
8. Engstrand J, Montufar EB, Engqvist H et al (2016) Brushite foams - the effect of tween® 80 and Pluronic® F-127 on foam porosity and mechanical properties. J Biomed Mater Res Part B 104B:67–77
9. Montufar EB, Gil C, Traykova T et al (2008) Foamed beta-tricalcium phosphate scaffolds. Key Eng Mater 361–363:323–326
10. LeGeros RZ, LeGeros JP (2003) Calcium phosphate bioceramics: past, present and future. Key Eng Mater 240-242:3–10
11. Bohner M (2000) Calcium orthophosphates in medicine: from ceramics to calcium phosphate cements. Injury 31:D37–D47
12. Espanol M, Perez RA, Montufar EB et al (2009) Intrinsic porosity of calcium phosphate cements and its significance for drug delivery and tissue engineering applications. Acta Biomater 5:2752–2762
13. Ginebra MP, Canal C, Espanol M et al (2012) Calcium phosphate cements as drug delivery materials. Adv Drug Deliv Rev 64:1090–1110
14. Koch MA, Vrij EJ, Engel E et al (2010) Perfusion cell seeding on large porous PLA/calcium phosphate composite scaffolds in a perfusion bioreactor system under varying perfusion parameters. J Biomed Mater Res A 95A:1011–1018
15. Gómez-Martínez M (2011) Cultivo de células mesenquimales de rata sobre espumas de hidroxiapatita mediante bioreactor de perfusion. Dissertation, Technical University of Catalonia
16. Carrodeguas RG, De Aza S (2011) α-tricalcium phosphate: synthesis, properties and biomedical applications. Acta Biomater 7:3536–3546
17. Montufar EB, Maazouz Y, Ginebra MP (2013) Relevance of the setting reaction to the injectability of tricalcium phosphate pastes. Acta Biomater 9:6188–6198
18. Hesaraki S, Nazarian H, Pourbaghi-Masouleh M et al (2014) Comparative study of mesenchymal stem cells osteogenic differentiation on low-temperature biomineralized nanocrystalline carbonated hydroxyapatite and sintered hydroxyapatite. J Biomed Mater Res B Appl Biomater 102:108–118
19. Engel E, Del Valle S, Aparicio C et al (2008) Discerning the role of topography and ion exchange in cell response of bioactive tissue engineering scaffolds. Tissue Eng Part A 4: 1341–1351
20. Chou YF, Huang W, Dunn JCY et al (2005) The effect of biomimetic apatite structure on osteoblast viability, proliferation, and gene expression. Biomaterials 26:285–295
21. Sadowska JM, Guillem-Marti J, Montufar EB, et al (2017) Biomimetic versus sintered calcium phosphates: the in vitro behavior of osteoblasts and mesenchymal stem cells. Tissue Engineering Part A, ahead of print. doi:10.1089/ten.TEA.2016.0406.

Part II

3D Organoid and Organotypic Cultures

Chapter 7

Establishment of 3D Intestinal Organoid Cultures from Intestinal Stem Cells

Shinya Sugimoto and Toshiro Sato

Abstract

The intestinal epithelium is the most rapidly renewed tissue in adult mammals, and its renewal is strictly controlled by intestinal stem cells. Extensive studies using genetic models of intestinal epithelium have revealed the mechanisms underlying the self-renewal of intestinal stem cells. Exploiting this knowledge, we developed a novel 3D culture system that enables the outgrowth of intestinal Lgr5⁺ stem cells derived from mouse and human tissues into ever-expanding crypt–villus mini-guts, known as intestinal epithelial organoids. These organoids are maintained by the self-renewal of stem cells and give rise to all differentiated cell types of the intestinal epithelium. Once established, organoids can be cryopreserved and thawed when needed. This culture system has been widely used for studying stem cell behavior and gene function and for disease modeling.

Key words Intestinal stem cells, Lgr5, Crypt isolation, Organoid culture, Mini-gut, Wnt signal, Small intestine, Colon, Mouse, Human

1 Introduction

The intestinal epithelium is the most rapidly renewed tissue in adult mammals. The renewal of intestinal epithelium is strictly controlled by intestinal stem cell (ISC) maintenance and self-renewal. The epithelium of the small intestine is histologically compartmentalized into two units: the crypts and the villi. Crypts are mainly composed of undifferentiated progenitor cells, such as transit amplifying cells and ISCs, whereas the villi consist of mature epithelial cells, which contribute to major intestinal functions such as digestion and absorption of luminal nutrients. ISCs continuously self-renew throughout an organism's lifespan and give rise to all types of differentiated lineages in the tissue. Studies using genetic lineage tracing systems have demonstrated that the crypt base columnar cells, marked by the expression of *Leucine-rich repeat-containing G protein-coupled receptor 5* (*Lgr5*), serve as ISCs [1]. Although Lgr5⁺ ISCs robustly propagate in native tissue environments, they do not survive in conventional

Zuzana Koledova (ed.), *3D Cell Culture: Methods and Protocols*, Methods in Molecular Biology, vol. 1612,
DOI 10.1007/978-1-4939-7021-6_7, © Springer Science+Business Media LLC 2017

monolayer 2D culture systems. By recapitulating the in vivo extracellular niche microenvironments of ISCs, we developed a 3D culture method, called organoid culture, in which $Lgr5^{+}$ ISCs efficiently expand and form stereotypical structures resembling the intestinal crypts. The culture conditions were first optimized for mouse small ISCs. These conditions involved an extracellular matrix that mimicked the basement membrane (Matrigel) and addition of three niche factors: epidermal growth factor (EGF), Noggin (bone morphogenic protein inhibitor), and R-spondin1 (Wnt agonist) [2]. Subsequently, the organoid culture conditions were optimized for human ISCs by including Wnt-3a, an activin-like kinase 4/5/7 inhibitor (A83-01), a p38 mitogen-activated protein kinase inhibitor (SB202190), and gastrin [3]. When culturing single ISCs, their propagation is hampered by detachment-induced cell death, known as anoikis. Supplementation with a ROCK inhibitor (Y-27632) suppresses anoikis and enables robust recovery from single ISCs [2, 4]. Intestinal organoids differentiate into all cell types of the intestinal epithelium, including the absorptive enterocytes and secretory cells, such as Paneth, goblet, tuft, and enteroendocrine cells [2]. These organoid cultures are phenotypically and genetically stable and have allowed disease modeling with organoids established from diseased tissue from patients. Surgical specimens or even small biopsy specimens are sufficient to generate large numbers of human organoids [3]. Once established, organoids can be cryopreserved and thawed when needed. The establishment of 3D intestinal organoid cultures has allowed the development of new models for use in physiological and biochemical/molecular studies of stem cell behavior, gene function, and disease modeling [5]. In this protocol, we provide details regarding the isolation of crypts, the culturing of organoids, the cryopreservation of cultures, and the thawing of organoids harvested from mouse and human intestinal tissue.

2 Materials

2.1 Intestinal Crypt Isolation

1. Dulbecco's phosphate-buffered saline without Ca^{2+} and Mg^{2+} (DPBS).
2. Crypt isolation buffer: 2.5 mM ethylenediaminetetraacetic acid (EDTA) in DPBS (*see* **Note 1**).
3. Basal culture medium: 10 mM HEPES, 2 mM GlutaMAX, 100 U/mL penicillin, and 100 μg/mL streptomycin in advanced DMEM/F12 (*see* **Note 2**).
4. 70 μm cell strainer.
5. 50 mL tube coated with 1% (w/v) bovine serum albumin in PBS (BSA/PBS) (*see* **Note 3**).

6. 10 mL disposable serological pipette coated with 1% BSA/PBS (*see* **Note 3**).
7. Intestinal samples (*see* **Note 4**).

2.2 Intestinal Organoid Culture and Thawing

1. Matrigel, basement membrane matrix, growth factor reduced (GFR), phenol red-free (Corning) (*see* **Note 5**).
2. Y-27632 (1000× stock; 10 mM in 0.1% BSA/PBS).
3. Mouse ENR medium: 50 ng/mL recombinant mouse EGF, 100 ng/mL recombinant mouse Noggin, 1 μg/mL recombinant human R-spondin1 (*see* **Note 6**), 500 nM A83-01, 10 μM SB202190, 1× B27 supplement, 1× N2 supplement, and 1 mM *N*-acetyl-L-cysteine in basal culture medium (*see* **Notes 7** and **8**).
4. Mouse WENR medium: 50% (v/v) Wnt-3a-conditioned medium (*see* **Note 9**), 50 ng/mL recombinant mouse EGF, 100 ng/mL recombinant mouse Noggin, 1 μg/mL recombinant human R-spondin1, 1× B27 supplement, 1× N2 supplement, and 1 mM *N*-acetyl-L-cysteine in basal culture medium (*see* **Note 8**).
5. Human WENRAS medium: 50% (v/v) Wnt-3a-conditioned medium, 50 ng/mL recombinant mouse EGF, 100 ng/mL recombinant mouse Noggin, 1 μg/mL recombinant human R-spondin1, 500 nM A83-01, 10 μM SB202190, 1× B27 supplement, 1 mM *N*-Acetyl-L-cysteine, and 10 nM [Leu^{15}]-Gastrin I in basal culture medium (*see* **Note 8**).
6. Tissue culture plates, 48-well flat bottom.
7. TrypLE Express with phenol red (Life Technologies).

2.3 Cryopreservation of Organoids

1. Recovery Cell Culture Freezing Medium (Life Technologies).
2. CoolCell Cell Freezing Container (BioCision).
3. 1 mL cryo tubes.

3 Methods

3.1 Intestinal Crypt Isolation

1. Before isolation, thaw aliquots of Matrigel on ice and keep them cold. Pre-warm a 48-well plate by placing it in a CO_2 incubator (5% CO_2, 37 °C).
2. Wash the intestinal samples with ice-cold DPBS in a petri dish until nearly all visible luminal contents are removed (*see* **Note 10**).
3. Cut the intestinal sample into pieces of ~5-mm in size with fine scissors on a petri dish, and transfer the pieces into a 15 mL centrifuge tube.

4. Add 10 mL of ice-cold DPBS and wash the fragments. Gently pipette the fragments up and down at least ten times with a 10 mL disposable pipette coated with 1% BSA/PBS, and discard the supernatant after allowing the fragments to settle for 1 min or less. Repeat this step five to ten times until the supernatant is almost clear.
5. Add 10 mL of crypt isolation buffer and place the tube on a rocking shaker. Rock the tube gently at 4 °C for 30 min.
6. Allow the fragments to settle, then discard the supernatant.
7. Add 10 mL of ice-cold DPBS and pipette up and down at least ten times with a 10 mL pipette. After allowing the fragments to settle, examine one drop of the supernatant under a microscope to determine whether it contains enriched crypt fractions. Collect the supernatants containing the crypts and filter them through a 70 μm cell strainer into a BSA-coated 50 mL tube to remove the villus fraction and debris. Repeat this procedure several times until most of the crypts are collected (*see* **Note 11**).
8. To separate crypts from single cells, centrifuge the crypt fractions at 300 × *g* for 3 min at 4 °C. Discard the supernatant carefully without disturbing the cell pellet.
9. Resuspend the pellets in 1 mL of ice-cold DPBS and transfer into a 15 mL centrifuge tube placed on ice.
10. Place 20 μL of the crypt suspension in a petri dish and count the crypts under a microscope. Calculate the total number of crypts (counted crypts × 50).

3.2 Intestinal Organoid Culture

1. Add 10 mL of ice-cold DPBS to the tube (*see* Subheading 3.1, **step 9**) and centrifuge the crypt suspension at 400 × *g* for 3 min at 4 °C. Discard the supernatant without disturbing the cell pellet (*see* **Note 12**).
2. Using a 200 μL pipette and prechilled sterile pipette tips, resuspend the crypt pellet in Matrigel (25 μL of Matrigel for 50–200 crypts), being careful to avoid bubbling.
3. Apply 25 μL of the crypt suspension in Matrigel (containing 50–200 crypts) to the center of each well of a pre-warmed 48-well plate (Fig. 1).
4. Place the plate in a CO_2 incubator (5% CO_2, 37 °C) for 10 min to allow complete polymerization of the Matrigel.
5. Add 250 μL of optimized medium to each well and incubate at 37 °C. Use mouse ENR medium for mouse small intestine, mouse WENR medium for mouse colon, and human WENRAS medium for human intestine (*see* Table 1). To avoid anoikis, supplement the culture medium with 10 μM Y-27632 for the first 2 days.

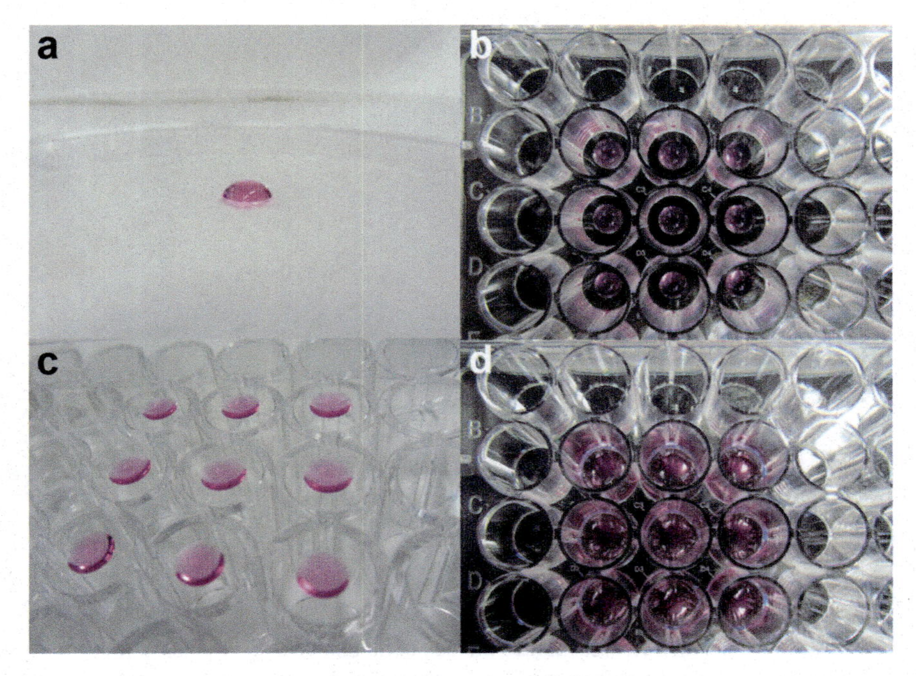

Fig. 1 Intestinal organoid culture. (**a**) Hemispherical 3D droplet of Matrigel. (**b**) Top view of the resuspended Matrigel (25 μL) placed at the center of each well of a pre-warmed 48-well plate. (**c**) Matrigel becomes solid after complete polymerization. Solidification of the Matrigel can be confirmed by inverting the plates. (**d**) Optimized medium is added to each well

Table 1
Optimized organoid culture media components

Species		**Mouse**		**Human**	
Organ		**Small intestine**	**Colon**	**Small intestine**	**Colon**
Reagents	**Final concentration**	**ENR**	**WENR**	**WENRAS**	**WENRAS**
Wnt-3a	50% (v/v)	–	+	+	+
EGF	50 ng/mL	+	+	+	+
Noggin	100 ng/mL	+	+	+	+
R-spondin1	1 μg/mL	+	+	+	+
A83-01	500 nM	–	–	+	+
SB202190	10 μM	–	–	+	+
Gastrin	10 nM	–	–	+	+
N2 supplement	1×	+	+	–	–
B27 supplement	1×	+	+	+	+
N-acetyl-L-cysteine	1 mM	+	+	+	+

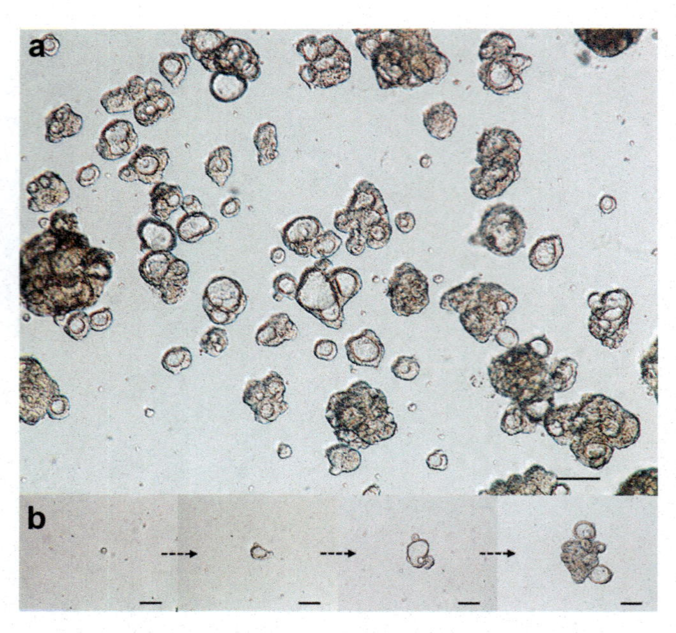

Fig. 2 Representative images of human colonic organoids. (**a**) Organoids can be observed in a variety of sizes without single-cell dissociation or use of a cell strainer. (**b**) Representative time course within 7 days. Scale bars, 100 μm

6. Every 2–3 days after plating the crypts replace the medium with 250 μL of the optimized medium for each crypt type.
7. Examine the cultures daily and passage the organoids upon outgrowth (Fig. 2). Mouse and human organoids are generally passaged at a 1:5–6 split ratio every 5 and 7 days, respectively.
8. Before passaging, thaw aliquots of Matrigel on ice and keep them cold. Pre-warm a 48-well plate in a CO_2 incubator (5% CO_2, 37 °C).
9. Remove the culture medium and add 500 μL of TrypLE Express to each well with a 1000 μL pipette.
10. Scrape and suspend the crypt cultures in TrypLE Express and transfer into a 15 mL centrifuge tube.
11. Incubate the tube at 37 °C in a water bath for 5 min (*see* **Note 13**).
12. Pipette up and down gently 10–15 times with a 1000 μL pipette.
13. Add 10 mL of basal culture medium to the tube and centrifuge the crypts at 400 × *g* for 3 min at 4 °C.
14. Discard the supernatant carefully without disturbing the cell pellet.
15. Resuspend the cell pellet in Matrigel (50–200 crypts/25 μL) after cell counting or visual estimation of the number of organoids in the pellet.
16. Repeat **steps 3–5** to plate and culture organoids in a new 48-well plate.

3.3 Cryopreservation of Organoids

Organoids can be frozen 2–3 days after passage.

1. Remove the culture medium and add 500 μL of Recovery™ Cell Culture Freezing Medium to each well with a 1000 μL pipette.
2. Scrape and suspend the Matrigel™ containing the organoid culture in the freezing solution and transfer the suspension into a 1 mL cryotube.
3. Place the cryotubes in a CoolCell® cell freezing container and store at −80 °C.
4. After overnight freezing, transfer the tubes into liquid N_2. Samples can be stored in liquid N_2 for at least several years.

3.4 Thawing Cryopreserved Organoids

1. Before thawing the cryopreserved organoids, thaw aliquots of Matrigel on ice and keep them cold. Pre-warm a 48-well plate in a CO_2 incubator (5% CO_2, 37 °C).
2. Quickly thaw the frozen tube containing the organoids at 37 °C in a water bath.
3. Transfer the suspension of organoids into a 15 mL centrifuge tube, quickly add 10 mL of basal culture medium to the tube, and centrifuge at 400 × *g* for 3 min at 4 °C.
4. Discard the supernatant carefully without disturbing the cell pellet.
5. Using a 200 μL pipette and prechilled sterile pipette tips, resuspend the crypt pellet in Matrigel, being careful to avoid bubbling.
6. Plate the organoids in Matrigel and culture in a new 48-well plate, as described in Subheading 3.2, **steps 3–5** (*see* **Note 14**).

4 Notes

1. Add 50 μL of 0.5 M EDTA, pH 8.0 to 10 mL of DPBS.
2. Reconstituted basal culture medium can be stored at 4 °C for at least 1 month.
3. To minimize cell adhesion to the sides of the tubes and pipettes, fill the tubes with 1% BSA/PBS, and then aspirate the solution by pipettes.
4. Samples should be collected only after informed consent from patients and approval from the appropriate ethics committee is obtained. Keep the samples in ice-cold DPBS until crypt isolation. The samples can be stored overnight in ice-cold DPBS.
5. A 10 mL vial of Matrigel must be thawed on ice and aliquoted into precooled, sterile cryogenic tubes and stored at −20 °C. The 1 mL aliquots must be thawed on ice before use and can be refrozen and thawed several times without substantial loss of culture efficiency.

6. Recombinant Noggin or R-spondin1 can be substituted with conditioned medium from mouse R-spondin1-Fc- [6] or Noggin-Fc- [7] producing HEK293T cell lines. Divide into 2 mL aliquots in 2 mL cryotubes and store at −20 °C for up to 6 months. Supplement the basal culture medium with 10% (v/v) R-spondin1-conditioned medium and 10% (v/v) Noggin-conditioned medium. When using a new batch of medium, batch-to-batch differences should be tested by growing organoids with the earlier batch and the new batch at various concentrations.
7. The basal culture medium becomes yellowish after the addition of B27 supplement and *N*-Acetyl-L-cysteine, owing to their acidic nature.
8. Complete culture medium can be stored at 4 °C for up to 7 days.
9. Wnt-3a-conditioned medium is prepared using the cell line L Wnt-3a (ATCC®, CRL-2647™) as described elsewhere [8]. The TOPflash assay can be used to test the transcriptional activity of Wnt [9]. Prepare 50 mL aliquots of freshly collected Wnt-3a-conditioned medium in 50 mL conical tubes and store at −20 °C for up to 6 months. Thawed aliquots can be stored for at least 1 month at 4 °C without loss of activity.
10. The number of washes can be reduced if the villi of the small intestine are scraped off with a coverslip. Conversely, the scraping procedure can be replaced with extensive DPBS washing. In the case of human intestinal samples, surgically resected intestinal specimens or endoscopic biopsy samples are stripped of the underlying muscle layers and connective tissues with fine scissors. A tissue sample of 5 mm size is sufficient for crypt isolation.
11. In isolation of small intestinal crypts, the first and second supernatants usually contain mostly villous fractions. We always inspect fractions under a microscope and discard the supernatants in which crypts are rare.
12. Discard as much supernatant as possible to avoid dilution of Matrigel. Diluted Matrigel breaks easily, thus resulting in loss of cells.
13. Organoids are passaged either as cell colonies or as single cells. For single-cell passage, incubate the tube at 37 °C in a water bath for 10 min and dissociate the colonies into single cells by pipetting. Because single-cell dissociation results in poor recovery of the organoids, it is essential to supplement the cell culture medium with Y-27632 to avoid anoikis [2].
14. Typically, we use frozen organoids from one well to seed cultures in five wells.

Acknowledgments

We thank the past and present members of the Sato laboratory for their continued support. S.S. was supported by the Japan Society for the Promotion of Science Research Fellowships for Young Scientists. This work was supported in part by JSPS KAKENHI Grant Number 15J00981.

References

1. Barker N, van Es JH, Kuipers J et al (2007) Identification of stem cells in small intestine and colon by marker gene Lgr5. Nature 449:1003–1007
2. Sato T, Vries RG, Snippert HJ et al (2009) Single Lgr5 stem cells build crypt-villus structures in vitro without a mesenchymal niche. Nature 459:262–265
3. Sato T, Stange DE, Ferrante M et al (2011) Long-term expansion of epithelial organoids from human colon, adenoma, adenocarcinoma, and Barrett's epithelium. Gastroenterology 141:1762–1772
4. Watanabe K, Ueno M, Kamiya D et al (2007) A ROCK inhibitor permits survival of dissociated human embryonic stem cells. Nat Biotechnol 25:681–686
5. Sato T, Clevers H (2013) Growing self-organizing mini-guts from a single intestinal stem cell: mechanism and applications. Science 340:1190–1194
6. Ootani A, Li X, Sangiorgi E et al (2009) Sustained in vitro intestinal epithelial culture within a Wnt-dependent stem cell niche. Nat Med 15:701–706
7. Farin HF, Van Es JH, Clevers H (2012) Redundant sources of Wnt regulate intestinal stem cells and promote formation of Paneth cells. Gastroenterology 143:1518–1529
8. Willert K, Brown JD, Danenberg E et al (2003) Wnt proteins are lipid-modified and can act as stem cell growth factors. Nature 423:448–452
9. Korinek V, Barker N, Morin PJ et al (1997) Constitutive transcriptional activation by a beta-catenin-Tcf complex in APC−/− colon carcinoma. Science 275:1784–1787

Chapter 8

3D Coculture of Mammary Organoids with Fibrospheres: A Model for Studying Epithelial–Stromal Interactions During Mammary Branching Morphogenesis

Zuzana Koledova

Abstract

Mammary gland is composed of branched epithelial structure embedded within a complex stroma formed by several stromal cell types, including fibroblasts, and extracellular matrix (ECM). Development of mammary gland is tightly regulated by bidirectional epithelial–stromal interactions that include paracrine signaling, ECM remodeling and mechanosensing. Importantly, these interactions play crucial role in mammary gland homeostasis and when deregulated they contribute to tumorigenesis. Therefore, understanding the mechanisms underlying epithelial–stromal interactions is critical for elucidating regulation of normal mammary gland development and homeostasis and revealing novel strategies for breast cancer therapy. To this end, several three-dimensional (3D) cell culture models have been developed to study these interactions in vitro. In this chapter, a novel 3D organoid–fibrosphere coculture model of mammary gland is described with the capacity for studying not only the qualitative and quantitative aspects of interactions between mammary fibroblasts and epithelial organoids but also their radius and directionality.

Key words 3D culture, Branching morphogenesis, Collagen, Extracellular matrix, Fibroblasts, Mammary gland, Matrigel, Organoids, Paracrine signaling, Spheroid

1 Introduction

Mammary gland consists of two major compartments: epithelium and stroma. The epithelial compartment forms a two-layer, tree-like branched structure that is embedded within stromal environment, composed of extracellular matrix (ECM) and multiple stromal cell types, including adipocytes, fibroblasts, endothelial cells, and infiltrating leukocytes. While the major function of mammary epithelium is production of milk for progeny, stroma has a pivotal role in controlling tissue structure and epithelial cell function [1]. Epithelial–stromal interactions via chemical and mechanical signals play a major regulatory role during normal mammary gland development and mammary tumorigenesis [1–4].

Zuzana Koledova (ed.), *3D Cell Culture: Methods and Protocols*, Methods in Molecular Biology, vol. 1612,
DOI 10.1007/978-1-4939-7021-6_8, © Springer Science+Business Media LLC 2017

For investigation of epithelial–stromal cross talk in vitro, development of novel three-dimensional (3D) cultures has been necessary. Classical two-dimensional cultures are not suitable because they are unable to recapitulate the in vivo structure and function of mammary epithelium. For accurate models of mammary epithelial morphogenesis, recapitulation of 3D architecture of mammary epithelium in appropriate ECM matrices is critical [5].

To study the role of stromal cells in mammary epithelial morphogenesis, mammary stromal cells need to be introduced into the 3D cultures of mammary epithelial cells. Although transwell cultures are useful tools for certain investigations, placing stromal cells in intimate association with epithelium has important implications for mammary gland modeling [6]. To this end, we developed a 3D fibroblast–epithelium coculture model in which primary mammary organoids are embedded within Matrigel matrix containing mammary fibroblasts [7].

This model enables to study in vitro the qualitative and quantitative effects of mammary fibroblasts on mammary epithelial branching morphogenesis. As a modification of this assay, a novel 3D coculture model was further developed that has a unique capacity to study also directionality and radius of epithelial–stromal interactions. This is achieved by aggregation of fibroblasts into spheroids, the so-called fibrospheres, and their subsequent precise localization near mammary epithelial organoid (Fig. 1). Thus, in the 3D organoid–fibrosphere coculture, the fibrosphere provides an amplified, yet localized source of paracrine signals and ECM remodeling activities for the epithelial organoid and, vice versa, the epithelial organoid serves as a localized source of signals to fibrosphere [8]. Further improvement in regard to the previous model is employment of physiologically more accurate ECM; a mixture of Matrigel with collagen I pre-assembled into fibrils [9].

In this chapter, a detailed protocol for isolation of primary mouse mammary organoids and fibroblasts, aggregation of

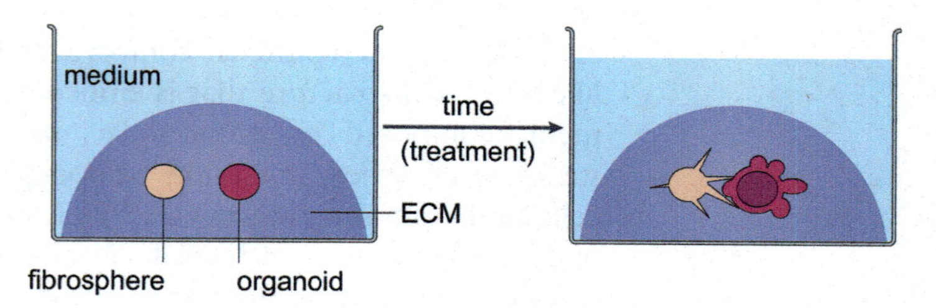

Fig. 1 Diagram of the 3D organoid–fibrosphere coculture model. Fibroblasts aggregated into a fibrosphere are positioned near an epithelial organoid in 3D extracellular matrix (ECM) of pre-assembled collagen I and Matrigel and overlaid with cell culture medium. Over time, the spheroids interact, leading to changes in spheroid morphology, including fibroblast migration and epithelial branching or cyst formation

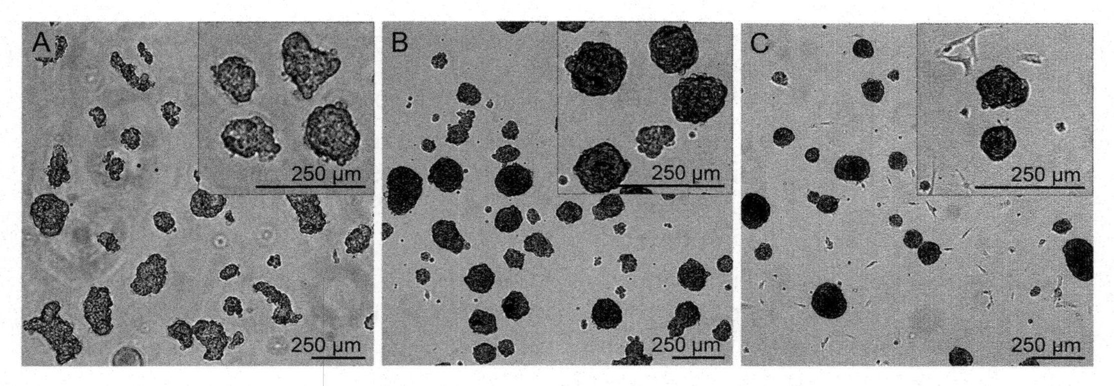

Fig. 2 Photographs of mammary spheroids. (**a**) Freshly isolated primary mammary organoids, and (**b**, **c**) fibrospheres generated by aggregation of primary mammary fibroblasts on polyHEMA (**b**) and non-treated dishes (**c**)

fibroblasts into fibrospheres (Fig. 2), preparation of pre-assembled collagen I matrix and 3D organoid–fibrosphere coculture setup is provided.

2 Materials

All reagents for tissue culture should be sterilized by autoclaving or filtration through a 0.22 μm filter.

2.1 Isolation of Mouse Mammary Epithelial Organoids and Fibroblasts

1. Mouse mammary glands, from pubertal females (6–10 weeks of age) (*see* **Notes 1** and **2**), freshly dissected or short-term stored in PBS on ice.
2. Phosphate buffered saline (PBS) without Ca^{2+}, Mg^{2+}.
3. Dulbecco's Modified Eagle Medium: Nutrient Mixture F-12 (DMEM/F12).
4. Digestion solution: 2 mg/mL collagenase A, 2 mg/mL trypsin, 5% (v/v) fetal bovine serum (FBS), 5 μg/mL insulin, 50 μg/mL gentamicin in DMEM/F12. For pooled mammary glands from one pubertal mouse, prepare 10 mL of digestion solution by combining 8.7 mL of DMEM/F12, 0.5 mL of FBS, 10 μL of 50 mg/mL gentamicin, 50 μL of 1 mg/mL insulin, 400 μL of 50 mg/mL trypsin, and 400 μL of 50 mg/mL collagenase (*see* **Notes 3–5**). Filter-sterilize and warm up to 37 °C in a water bath before use. Prepare freshly for each primary cell preparation.
5. Bovine serum albumin (BSA) solution: 2.5% BSA in PBS. Store at 4 °C and warm up before use.
6. Red blood cell (RBC) lysis buffer: 155 mM NH_4Cl, 12 mM $NaHCO_3$, 0.1 mM EDTA in distilled H_2O, pH 7.4, sterile. Store at room temperature (RT).

7. Deoxyribonuclease I (DNase I) solution: 20 U/mL DNase I in DMEM/F12. Add 40 μL of 2 U/μL DNase I to 4 mL of DMEM/F12 (*see* **Note 6**). Prepare fresh for each experiment.
8. Basal organoid coculture medium: 1× insulin–transferrin–selenium (ITS), 100 U/mL of penicillin, and 100 μg/mL of streptomycin in DMEM/F12. Store at 4 °C and use within a month.
9. Fibroblast medium: 10% FBS, 1× ITS, 100 U/mL of penicillin, and 100 μg/mL of streptomycin in DMEM. Store at 4 °C and use within a month.
10. Forceps, sterile.
11. Polystyrene petri dishes, non-treated, 100 mm, sterile (*see* **Note 7**).
12. Disposable scalpel blades, number 20 or 21, sterile.
13. Scalpel handles, sterile.
14. Benchtop orbital shaker, incubated.
15. 100 mm cell culture dishes.

2.2 Primary Fibroblast Culture and Preparation of Fibroblast Spheroids (Fibrospheres)

1. PBS.
2. Trypsin–EDTA: 0.05% trypsin and 0.2% EDTA in PBS.
3. Trypsin neutralizing (TN) medium: 10% FBS in PBS.
4. 100 mm cell culture dishes.
5. Polystyrene petri dishes, non-treated, 60 mm, sterile (*see* **Note 7**).
6. PolyHEMA [poly(2-hydroxyethyl methacrylate), Sigma] solution: In a sterile bottle, dissolve 12 g of polyHEMA in 1 L of 95% ethanol on a heated plate. Store at RT.
7. Agarose coating solution: 1.5% agarose in PBS, sterile. Add 0.75 g of agarose in 50 mL of PBS and autoclave. Use immediately after autoclaving (while hot) or microwave to dissolve before each use.
8. 6-well and 96-well cell culture dishes, flat bottom.
9. Hemocytometer.

2.3 3D Coculture of Mammary Organoids and Fibrospheres

1. μ-Dish 35 mm, high (ibidi).
2. PBS.
3. Basal organoid coculture medium.
4. Collagen I from rat tail (Corning). Adjust concentration to 4.59 mg/mL with 0.02 N acetic acid (*see* **Note 8**).
5. 5× collagen gel solution (5× CGS): Mix 2.5 g of MEM powder, 5 mL of 1 M HEPES (pH 7.5), 1 g of $NaHCO_3$ and top up to 50 mL with water. Mix well and filter-sterilize (0.22 μm).

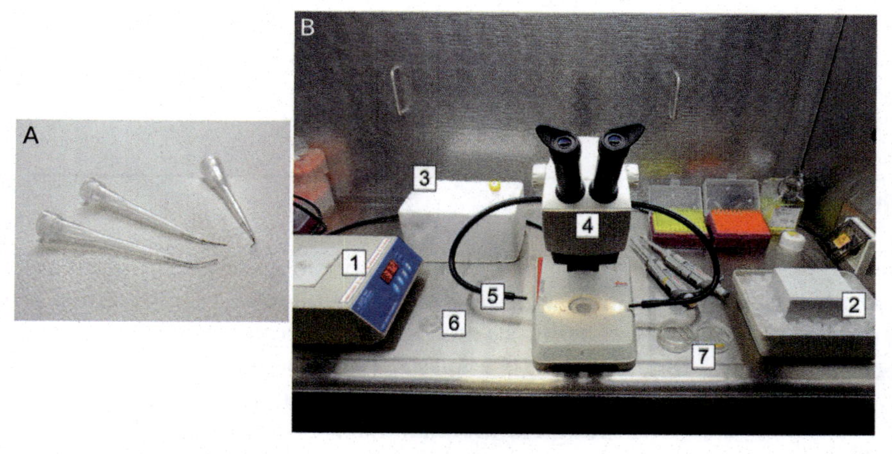

Fig. 3 Equipment for setting up 3D spheroid coculture. (**a**) Spheroid positioning tools. (**b**) Horizontal flow hood with heat block at 37 °C (*1*), cold block (*2*), ice bucket (for dishes with organoids or fibrospheres and sterile PBS in a tube; (*3*), stereoscope (*4*) with two-armed fiber light guides (*5*), ibidi dish (*6*), and sterilized spheroid positioning tools stored in dishes (*7*)

Aliquot and store at −20 °C. Working aliquot can be stored at 4 °C for several weeks.

6. 0.22 N NaOH.
7. DMEM.
8. Matrigel® Basement Membrane Matrix Growth Factor Reduced (Corning, 354230) (*see* **Note 9**).
9. Stereoscope equipped with cold light source with two-armed fiber light guides, set up in a horizontal flow hood (Fig. 3).
10. Heat block.
11. Ice bucket.
12. Stainless steel (or tungsten) wire, 100–150 μm in diameter.
13. Pliers and wire cutters.
14. Forceps (optional).
15. Super-fast glue.
16. 70% ethanol.
17. Polystyrene petri dishes, non-treated, 60 mm, sterile.
18. Round-bottom 2 mL eppendorf tubes.
19. Fixation solution: 4% paraformaldehyde (PFA) in PBS.
20. (Optional) CellTracker™ dyes, such as CellTracker™ Blue CMAC, CellTracker™ Green BODIPY® etc.

2.4 General Instruments, Equipment, and Other Materials

1. Centrifuge with swing-bucket rotor.
2. Cell culture incubator, 37 °C and 5% CO_2.
3. Centrifuge tubes, 15 mL and 50 mL.
4. Disposable plastic serological pipettes, 5 mL, 10 mL, 25 mL.

5. Electronic pipettor.
6. Automatic pipettes and pipette tips.
7. Inverted microscope.

3 Methods

3.1 Isolation of Primary Mammary Epithelial Organoids and Fibroblasts

Primary mammary epithelial organoids and fibroblasts are obtained by mammary tissue digestion, followed by differential centrifugation to obtain organoids and subsequent selective adhesion procedure to obtain fibroblasts. The procedure presented in this section is a modification of our recently published protocol [7].

Although both mammary organoids and fibroblasts can be obtained from a single mammary prep, for the spheroid coculture experiment described in this chapter two separate mammary preparations (on different days) are required. Mammary organoids need to be prepared freshly on the day of the coculture experiment setup, while fibroblasts need to be cultured and aggregated into fibrospheres beforehand, and therefore isolated approximately a week before the spheroid coculture experiment setup.

1. Using sterile forceps transfer the freshly dissected mammary gland tissue into a non-treated petri dish. Remove any excess of PBS or medium.
2. Using two or three scalpels at the same time, chop the mammary gland tissue finely to a homogenous mince of approximately 1 mm^3 pieces.
3. Using one of the scalpels used for chopping, carefully transfer the minced tissue to a 50 mL tube with digestion solution.
4. Incubate the digestion suspension in an incubated shaker at 37 °C for 30 min while shaking at 100 rpm.
5. After digestion, centrifuge the digestion suspension at 450 × *g* for 10 min at RT.
6. For all subsequent steps, use BSA-coated pipettes and tubes to avoid cell sticking to the plastic, thereby preventing loss of cells during the procedure (*see* **Note 10**).
7. Centrifugation yields three layers of the digestion suspension: a top white fatty layer, a middle aqueous phase (with the largest volume) and cell pellet at the bottom of the tube. Using a BSA-coated pipette, transfer the fatty layer into a 15 mL tube. Add 5 mL of DMEM/F12, mix well with the pipette and centrifuge at 450 × *g* for 10 min (*see* **Note 11**).
8. In the meantime, carefully aspirate the remaining supernatant from the 50 mL tube (*see* **Note 12**) and resuspend the pellet in 5 mL of DMEM/F12.

9. After centrifugation, aspirate the supernatant (both the top fatty and middle aqueous layers) from the 15 mL tube and to the pellet add the cells suspension from the 50 mL tube.
10. Using 5 mL of DMEM/F12, wash out any remaining cells from the 50 mL tube, combine the resulting suspension with the suspension in the 15 mL tube and mix well.
11. Centrifuge the 15 mL tube at 450 × *g* for 10 min.
12. Discard the supernatant, resuspend the pellet in 1–1.5 mL of RBC lysis buffer and incubate at RT for 1 min without shaking.
13. Add 8 mL of DMEM/F12, mix well by pipetting and centrifuge at 450 × *g* for 10 min.
14. Discard the supernatant, resuspend the pellet in 4 mL of DNase I solution and shake by hand or on an orbital shaker (50 rpm) for 3–5 min.
15. Add 6 mL of DMEM/F12, mix well and centrifuge at 450 × *g* for 10 min.
16. Discard the supernatant and perform differential centrifugation (*see* **Note 13**):
    (a) Resupend the pellet in 8–9 mL of DMEM/F12.
    (b) Centrifuge the tube at 450 × *g* for 10 s at RT (*see* **Note 14**).
    (c) Transfer the supernatant into a 50 mL tube.
17. Repeat previous **step 4** more times (five times in total), while collecting the supernatants in one 50 mL tube.
18. After performing five rounds of differential centrifugation, resuspend the pellet containing epithelial organoids in a small volume (0.5–2 mL) of basal organoid medium and place in ice.
19. Centrifuge the 50 mL tube containing the collected supernatant at 600 × *g* for 3 min.
20. Aspirate the supernatant, resuspend the cell pellet in 10 mL of warm fibroblast medium and transfer the suspension onto a 100 mm cell culture dish (*see* **Note 15**).
21. Incubate the dish in cell culture incubator at 37 °C for 30 min (*see* **Note 16**).
22. After 30 min, check under the microscope that fibroblasts have attached to the dish while most other stromal cells as well as epithelial organoids have remained in the suspension (selective adhesion).
23. Remove the unattached cells by aspirating the medium from the dish, wash the cell culture dish twice with PBS (5–10 mL) to get rid of any leftover unattached cells and add 10 mL of fresh fibroblast medium.

24. Incubate the plate with fibroblasts in a cell culture incubator (37 °C, 5% CO_2) for 24 h. This is passage 1 (p1).
25. Next day, change the medium for fresh fibroblast medium.
26. Maintain the fibroblast culture in a cell culture incubator until the cells reach about 80% confluence while changing the medium every 2–3 days.

3.2 Culture of Primary Mammary Fibroblasts

After isolation of primary mammary fibroblasts, a small contamination by epithelial cells will be unavoidably present. These cells will not efficiently proliferate in fibroblast culture conditions and, consequently, they will be diluted from the population during prolonged culture. Also, second round of selective adhesion procedure is used during the first passaging (i.e., from passage 1 to 2) to "purify" the fibroblast population.

1. To trypsinize the cells when they reach about 80% confluence, aspirate the medium from the cell culture dish and wash the dish twice with 5 mL of PBS.
2. Add 1.5 mL of trypsin–EDTA to the dish, spread it all over the cell culture area and incubate the dish at 37 °C (in cell culture incubator) for 5 min.
3. Check the cell dissociation process under microscope. Gently and repeatedly tap on side of the dish to help releasing cells.
4. After cell detachment add 5 mL of TN medium (or fibroblast medium) to the dish and pipette the suspension several times to detach remaining cells from the bottom and to resuspend cells.
5. Transfer the cell suspension to a 15 mL tube and centrifuge at 600 × *g* for 2 min.
6. Aspirate the supernatant and resuspend the pellet in fibroblast medium and split (1:3–1:6 according to the cell culture area) into new dishes with fresh fibroblast medium.
7. To perform selective adhesion procedure during first passaging, place the dishes in the cell culture incubator and incubate for 20 min (*see* **Note 16**).
8. After 20 min check under the microscope that fibroblasts have attached to the dish.
9. Aspirate the medium from the dishes, wash the dishes twice with PBS (5 mL) and add fresh fibroblast medium.
10. Maintain the cells in a cell culture incubator. Change the medium every 2–3 days. When the cells reach about 80% confluence, subculture them into new dishes, use for formation of fibrospheres and/or freeze for later use.

3.3 Preparation of Fibroblast Spheroids (Fibrospheres)

Fibrospheres are formed by spontaneous aggregation of fibroblasts during culture in non-adherent conditions. To this end, several techniques can be used, including non-adherent dishes (such as non-treated dishes, ultra-low attachment plates, polyHEMA- or agarose-coated plates) and hanging drop assay. In this section, we describe preparation of spheroids using non-adherent dishes. The technique using non-treated or polyHEMA dishes is less laborious and yields high numbers of spheroids of various sizes. On the other hand, the more laborious method using agarose-coated plates enables formation of spheroids of more uniform size.

For preparation of fibrospheres use fibroblasts from passage 2 to 4 (*see* **Note 17**).

3.3.1 Preparation of polyHEMA Dishes

1. In a cell culture hood, pipette 1 mL of polyHEMA solution in each well of a 6-well plate (*see* **Note 18**).
2. Make sure that polyHEMA solution is spread over the whole bottom of each well. Gently shake or swirl the plate to spread the solution if needed (*see* **Notes 19** and **20**).
3. Transfer the plate in an oven and incubate at 40 °C until completely dry (about 2 days).
4. Store dried polyHEMA plate at room temperature until use (*see* **Note 21**).

3.3.2 Preparation of Fibrospheres Using Non-treated or polyHEMA Dishes

1. Trypsinize fibroblasts (at 80% confluence) from cell culture dishes and prepare single-celled suspension by repetitive pipetting as described in Subheading 3.2.
2. Centrifuge at 600 × *g* to pellet the cells.
3. Resuspend the cell pellet in fibroblast medium.
4. Determine cell concentration using a hemocytometer or similar device.
5. Seed the cells in non-treated ("bacterial") or polyHEMA dish:
   (a) For one 60 mm non-treated dish, seed 0.8×10^6–1.4×10^6 cells in 4 mL of fibroblast medium (*see* **Note 22**).
   (b) For one well of polyHEMA 6-well plate, seed 2×10^5–4.5×10^5 cells in 3 mL of fibroblast medium (*see* **Note 23**).
6. Incubate in cell culture incubator for at least 16 h.
7. Next day (i.e., on the day of 3D coculture setup) check using an inverted microscope that the spheroids have formed (Fig. 2) (*see* **Note 24**).
8. Collect the fibrospheres from non-adherent plates in a 15 mL tube and centrifuge at 450 × *g* for 1 min.
9. Discard the supernatant and perform differential centrifugation (*see* **Notes 25** and **26**):
   (a) Resuspend the pellet in 5–8 mL of PBS.
   (b) Centrifuge at 450 × *g* for 5 s.

(c) Aspirate supernatant.

(d) Repeat two more times with PBS, followed by one time with DMEM/F12.

10. Resuspend the fibrosphere pellet in basal organoid coculture medium (1–2 mL) and proceed with CellTracker labeling (when fluorescently labeled fibrospheres are required; *see* Subheading 3.5.2) or put the suspension in ice until 3D coculture setup.

3.3.3 Preparation of Fibrospheres Using Agarose-Coated Dishes

1. To prepare an agarose-coated 96-well plate for spheroid formation, autoclave or microwave the agarose coating solution until the agarose is fully dissolved and immediately dispense 100 μL of the solution in each well of a 96-well plate.
2. Incubate the plate at RT for 15–20 min to allow the agarose to cool down and harden.
3. Trypsinize fibroblasts and prepare a single-celled suspension by repetitive pipetting as described in Subheading 3.2.
4. Count the cells, e.g., using a hemocytometer, and resuspend fibroblasts in fibroblast medium at a concentration of 2.5×10^4 cells/mL.
5. Add 200 μL of the cell suspension to each well so that 5000 cells are seeded per well.
6. Incubate in cell culture incubator until next day (at least 16 h).
7. Next day check using an inverted microscope that the spheroids have formed (Fig. 2) and collect the fibrospheres as described in Subheading 3.3.2, **steps 8–10**.

3.4 Preparation of Spheroid Positioning Tool

For the positioning of spheroids within 3D gels, we use simple house-made tools consisting of a 10 μL pipette tip and stainless steel wire glued to it. The tools can be used repeatedly ensuring thorough cleaning and sterilization between experiments.

1. Cut the stainless steel (or tungsten) wire in approximately 1–1.5 cm long pieces.
2. Using pliers, bend some of the wires to create various circular curvatures.
3. Glue the cut wire to the 10 μL pipette tip so that a part of the wire is inside the pipette tip. Allow the glue to completely dry.
4. Wash the positioning tools thoroughly with 70% ethanol and let them dry.
5. Sterilize by UV irradiation (*see* **Notes 27** and **28**).

3.5 3D Coculture of Mammary Organoids and Fibrospheres

For 3D coculture, use freshly isolated mammary organoids and fibrospheres prepared from mammary fibroblasts of passage 2 to 4. In the method described here, the spheroids are embedded in a mixture of Matrigel and pre-assembled collagen I that closely mimics the ECM of mammary gland tissue [9]. This basic protocol can be easily modified according to the experimental needs. Fibrospheres and organoids can be fluorescently labeled, such as using CellTracker dyes, prior to embedding into the ECM (*see* Subheading 3.5.2). Moreover, the spheroids can be embedded in pure Matrigel, pure collagen I or any Matrigel- and collagen-based ECM mixtures as desired, including fluorescently labeled ECM [10].

3.5.1 Preparation of Pre-assembled Collagen I

1. Place all components required for collagen I neutralization in ice, including collagen I (4.59 mg/mL), 5× CGS, 0.22 N NaOH, and DMEM. Place a round-bottom 2 mL eppendorf tube in ice to chill.
2. In the chilled eppendorf tube, pipette 80 μL of 5× CGS, 32 μL of NaOH, 100 μL of DMEM, and 400 μL of collagen I (*see* **Note 29**). Mix well to create a homogenous mixture while avoiding air introduction and bubble formation.
3. Incubate neutralized collagen in ice or at 4 °C for 1 h. During this time, collagen pre-assembles into fibrils.
4. In the meantime (i.e., during the collagen incubation time) label organoids and fibrospheres if required and set up all equipment for spheroid positioning in a horizontal flow hood (Fig. 3).

3.5.2 Labeling of Organoids and Fibrospheres (Optional)

1. Transfer the organoid suspension in basal organoid medium (e.g., 1.5 mL) into a well of polyHEMA-treated 6-well plate. Similarly, transfer the fibrosphere suspension into another polyHEMA-treated well.
2. In two separate eppendorf tubes, dilute CellTracker™ dyes in basal organoid medium to such a concentration, so that when combined with the organoid or fibrosphere suspension in the polyHEMA plate, the final concentration of the dye will be 10 μM. For example, add 0.5 mL of 40 μM CellTracker Blue to 1.5 mL of organoid suspension, and 0.5 mL of 40 μM CellTracker Green to 1.5 mL of fibrosphere suspension (*see* **Note 30**).
3. From now on, avoid any unnecessary exposure of the labeled cells to the light. Cover the plate with aluminum foil and incubate for 30–45 min in cell culture incubator (37 °C, 5% CO_2).
4. Collect the organoids and fibrospheres into 15 mL centrifuge tubes and centrifuge at 450 × *g* for 1 min.
5. Aspirate the supernatants, wash once with basal organoid medium (about 2 mL) and centrifuge again.
6. Resuspend the organoid and fibrosphere pellets in basal organoid medium (1–2 mL) and put the tubes in ice.

3.5.3 3D Spheroid Coculture Set Up

1. Transfer a part of the fibrosphere suspension to a non-treated 60 mm dish with 2–3 mL of basal organoid coculture medium and place on ice.
2. Take a part of the organoid suspension (prepared in Subheading 3.1) and transfer it to another non-treated 60 mm dish with 2–3 mL of basal organoid coculture medium. Place it on ice.
3. After 1 h of collagen incubation in ice, when the collagen is pre-assembled, add 262 μL of Matrigel to the mixture (the ratio is 7 parts of collagen I to 3 parts of Matrigel) (*see* **Note 31**). Mix the ECM well while being careful not to introduce any bubbles.
4. Place ibidi dish on a cold plate. Add 20 μL of Matrigel to the center of the dish, creating a small patch about 5 mm in diameter.
5. Move the ibidi dish on a heated plate (37 °C) and allow the Matrigel to gel for 5 min.
6. Move the ibidi dish back on the cold plate and add 15 μL of the ECM on top of the Matrigel patch.
7. Under a stereoscope, select a fibrosphere from the 60 mm dish with fibrosphere suspension and take it up using a pipette with 10 μL pipette tip with as little medium as possible (*see* **Notes 32** and **33**).
8. Using the same pipette tip, select an organoid from the 60 mm dish with organoid suspension and take it up to the fibrosphere with as little medium as possible.
9. Place the organoid and fibrosphere in the middle of the ECM patch.
10. Add 20 μL of the ECM on top of the ECM patch with organoid and fibrosphere.
11. Move the ibidi dish under the stereoscope and using the spheroid positioning tool, place the organoid and fibrosphere about 150 μm from each other (*see* **Note 34**).
12. Incubate the dish at 37 °C for 20 min in cell culture incubator (5% CO_2) to solidify the ECM.
13. Add 3 mL of pre-warmed (37 °C) basal organoid coculture medium to the dish (*see* **Notes 35** and **36**).
14. Incubate the cocultures in the cell incubator, changing medium every 2–3 days. Alternatively, place the coculture in a live cell imaging microscope to monitor cellular/spheroid phenotypes using time-lapse microscopy (Fig. 4) (*see* **Note 37**).

3.6 Fixation of the 3D Cocultures

When the experiment is finished, the cocultures can be fixed and used for subsequent analysis, e.g., immunofluorescence staining.

1. Carefully aspirate the culture medium and wash the coculture with PBS (*see* **Note 38**).

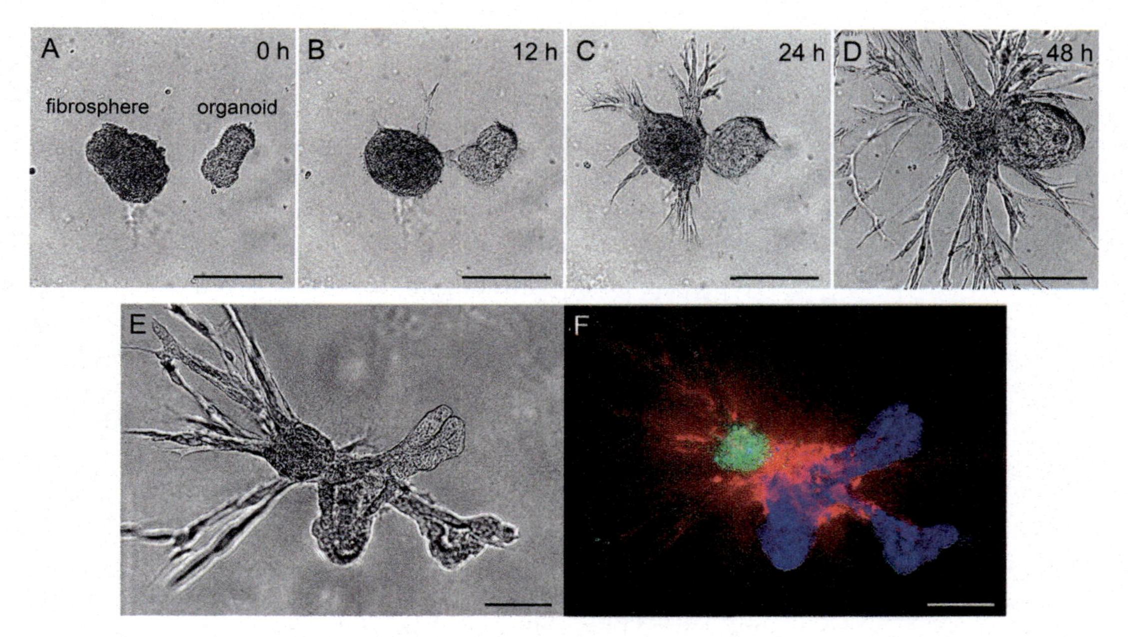

Fig. 4 3D organoid–fibrosphere coculture as a model of mammary morphogenesis. (**a–d**) Time-lapse microscopy images of an organoid–fibrosphere coculture in 3D collagen I/Matrigel. Scale bar, 200 μm. (**e**, **f**) 3D mammary organoid–fibrosphere coculture on day 4. Bright-field (**e**) and maximum projection (**f**) images from an inverted confocal microscope. Fibrosphere was labeled with CellTracker Green, organoid was labeled with CellTracker Blue, collagen I was labeled with TAMRA (*red*). Scale bar, 100 μm

2. Fix the coculture with 4% PFA for 15–30 min at RT.
3. Carefully aspirate PFA and wash the coculture two to three times with PBS.
4. Fixed sample can be stored in PBS at 4 °C for weeks.

4 Notes

1. To obtain mammary organoids and fibroblasts, female mice of any strain or age can be used. However, the developmental stage and parity of donor mice have a major impact on mammary epithelial morphogenesis and therefore they need to be carefully considered.
2. The donor mice should be euthanized by a procedure approved by committee for ethical work with laboratory animals (such as by cervical dislocation), according to institutional regulations and under a valid project license, and immediately used for tissue harvesting. The dissection should be carried out in a laminar flow hood using sterile instruments. For the mammary gland harvest procedure, *see* Koledova and Lu, 2017 [7].

3. To prepare 1 mg/mL insulin, dissolve 50 mg insulin in 50 mL of 0.1 M HCl, pH 2–3. Filter-sterilize, aliquot (100–250 μL) and store at −20 °C.
4. To prepare 50 mg/mL collagenase, dissolve 1 g of collagenase from Clostridium histolyticum (Sigma, C9891) in 20 mL of DMEM/F12. The solution will be cloudy. Aliqout (400 and 800 μL) and store at −20 °C.
5. To prepare 50 mg/mL trypsin, dissolve 1 g of trypsin (1:250, powder; ThermoFisher Scientific, 27250018) in 20 mL of DMEM/F12. Aliqout (400 and 800 μL) and store at −20 °C.
6. To prepare 2 U/μL DNase I, dissolve 2000 U of DNase I (Sigma) in 1 mL of sterile 0.15 M NaCl. Aliquot (40 μL) and store at −20 °C.
7. The designation "non-treated" culture dishes refers to dishes that have not been treated for cell attachment. These dishes are manufactured from virgin polystyrene and have hydrophobic surface. They are used in applications where cell attachment is not desired or required, such as for bacterial cultures on agar.
8. 0.02 N acetic acid is the solvent in which collagen I is formulated and supplied. Keep collagen I in ice or at 4 °C at all times. Do not freeze.
9. Thaw 10 mL vial of Matrigel in ice in refrigerator overnight. Avoid exposing Matrigel to temperatures over 10 °C because it would lead to its premature gelation. Using chilled pipette tips, aliquot Matrigel into prechilled eppendorf tubes on ice and store the aliquots at −20 °C.
10. To coat a pipette, aspirate the BSA solution into the pipette until its whole working surface is covered, then return the used BSA solution back into the stock tube/bottle containing the rest of the BSA solution. To coat a tube, carefully pipette the BSA solution on the walls of the tube or fill the tube with the BSA solution, then aspirate the BSA solution with a pipette and return it back into the stock tube/bottle. The BSA solution can be reused many times as long as it is kept sterile. Preferentially, filter-sterilize the solution after each experiment.
11. This step is optional. It serves to obtain maximal cell yields from the digested tissue because during the procedure some cells become trapped in the fat. These cells can be released by resuspending the fatty layer in fresh DMEM/F12 and collected by centrifugation. If you wish, this step can be skipped and the fatty layer discarded.
12. To aspirate the supernatant, use a pipette attached to a pipettor rather than vacuum aspiration system.
13. Differential centrifugation separates epithelial organoids from the stromal cells based on differences in their density. Epithelial

organoids are composed of hundreds of cells, and therefore they are larger and heavier and sink to the bottom of the tube faster than single-celled stromal cells. Stromal cells will remain in the supernatant. During the five rounds of differential centrifugation, the pellet becomes successively devoid of single cells. However, this technique is not absolute and some organoids (smaller fraction) will not sink to the bottom and will remain in the supernatant with stromal cells.

14. The centrifugation time of 10 s applies to the actual time of centrifugation at exactly 450 × *g*. Because most centrifuges count the total time of centrifugation and not the time of centrifugation at the set speed only, we recommend setting the time of centrifugation to 1 min and watching carefully when the rotor reaches 450 × *g*. From this moment on, measure 10 s and then stop the centrifuge manually by pressing the stop button.
15. The suggested volume of fibroblast medium and one 100 mm cell culture dish are suitable for cell yield from mammary glands of two pubertal mice. If more mammary gland tissue is processed, the cell yield will be higher. Adjust the volume of fibroblast medium and split the cell suspension into more cell culture dishes proportionally.
16. If you have more than one dish, place all dishes in direct contact with the warm shelves of incubator (i.e., not on top of each other). During this incubation time, do not move the dishes and avoid any vibrations, such as by harsh opening and closing of cell culture incubator.
17. Primary fibroblasts have limited replication capacity and after several population doublings they will undergo senescence or crisis and associated changes in cell physiology and signaling. When cultured in atmospheric (20%) oxygen concentration, commonly provided in cell culture incubators, mouse fibroblasts will senesce or enter crisis within ~14 population doublings [11], approximately corresponding to passage 5. Prolonged culture might lead to cell culture adaptation and other changes in cell behavior.
18. For coating of smaller wells, adjust the volume of polyHEMA solution accordingly. I.e., use 0.75 mL and 0.5 mL per one well of a 12-well and 24-well plate, respectively. We recommend using Corning/Costar multiwell plates for polyHEMA coating.
19. Uneven coating of a well will result in patches accessible for cell attachment.
20. To prevent accidental opening of the plate lid outside cell culture hood and thus breaching sterility of the plate (e.g., during the transfer to the oven or during storage), tape the lid to the bottom of the dish using a masking or autoclave tape.

21. The polyHEMA plates can be stored for several weeks, and therefore it is convenient to prepare several polyHEMA dishes at once.
22. This number of cells can be obtained from one to two 100 mm cell culture dishes of fibroblasts with 80% confluence and will yield approximately 7000 to 9000 fibrospheres, respectively. The number of cells seeded influences both fibrosphere number and size; with more cells seeded, larger fibrospheres will form.
23. This number of cells can be obtained from one to two 60 mm cell culture dishes of fibroblasts with 80% confluence and will yield approximately 1500 to 2100 fibrospheres, respectively. The number of cells seeded influences both fibrosphere number and size; with more cells seeded, larger fibrospheres will form.
24. While polyHEMA completely prevents fibroblast attachment and on polyHEMA plates all spheroids will be in suspension, the non-treated plate is somewhat permissive to fibroblast attachment and a small fraction of cells and spheroids will be loosely attached to the bottom. Gently tap on a side of the dish to release cells and spheroids from the plate.
25. Differential centrifugation is performed to get rid of single-celled fibroblasts and small fibrospheres.
26. This step also serves to wash the fibrospheres and to get rid of any FBS left from the fibroblast medium.
27. For UV sterilization, UV lamps built in the cell culture hood can be used. Place the tools inside a cell culture dish, remove the dish lid and irradiate with UV for 30 min. After irradiation, close the cell culture dish with the lid. Store the tools in the dish.
28. If available, gamma irradiation can be used to sterilize the tools as well.
29. Before pipetting collagen I stock solution, chill the pipette tip in ice-cold 0.02 N acetic acid solution (this is the solvent in which collagen I is formulated) or use pipette tips prechilled in the freezer (−20 °C) for at least 30 min before use. The collagen I stock solution is quite viscous, take your time while pipetting it to ensure taking and transferring the correct volume. Keep collagen I and its mixtures in ice at all times.
30. Protect the CellTracker dye stock solutions, labeling solutions as well as the labeled cells from light.
31. Keep Matrigel (and its mixtures) in ice at all times. Use ice-cold sterile PBS to cool pipette tips before using them to pipette Matrigel or use pipette tips prechilled in the freezer (−20 °C).

32. Set the pipette to 10 μL, press the operating button and immerse into the spheroid suspension. To take up a spheroid, release the operating button only partially, until the spheroid with minimum volume of medium is aspirated into the tip. Hold the pipette operating button in this position until you find next spheroid for aspiration. Then again release the operating button only partially until the desired spheroid is aspirated into the tip. Hold the pipette operating button in this position until dispensing the spheroids on the ibidi dish with ECM.
33. Several fibrospheres and organoids can be picked up at once and assembled into a few pairs in one ECM gel.
34. When positioning the spheroids, work quickly because as the ECM warms up, it soon begins to gel. Alternatively, use a cold plate for dish incubation during spheroid positioning. Avoid reaching too deep into the bottom layer of ECM and scratching the bottom of the dish with the positioning tool. Avoid placing the organoid–fibrosphere pair in the edge region of the ECM gel; preferably place it in the middle part of the ECM gel. If you are positioning several pairs in one ECM dish, make sure to place the pairs at least 400 μm apart from each other, otherwise they might interact.
35. Temperatures of medium and solidified ECM must be the same to avoid formation of bubbles in ECM. Bubbles negatively interfere with assessment of the cocultures under a microscope.
36. It is possible to add growth factors, inhibitors or other agents in the basal organoid coculture medium, as required for the experiment.
37. For an excellent guide to time-lapse imaging of 3D culture models of mammary morphogenesis, useful tips and troubleshooting, *see* Ewald, 2013 [12].
38. Take care not to disturb the gel construct with a pipette tip and avoid strong aspiration forces near the gel. Especially during prolonged culture, the gel becomes fragile due to ECM remodeling activities of the embedded cells.

Acknowledgments

This work was supported by the project “Employment of Best Young Scientists for International Cooperation Empowerment“ (CZ.1.07/2.3.00/30.0037) cofinanced from European Social Fund and the state budget of the Czech Republic, and by the grants “Junior investigator 2015” (Faculty of Medicine, Masaryk University) and 16-20031Y (Czech Science Foundation, GACR).

References

1. Polyak K, Kalluri R (2010) The role of the microenvironment in mammary gland development and cancer. Cold Spring Harb Perspect Biol 2(11):a003244
2. Muschler J, Streuli CH (2010) Cell-matrix interactions in mammary gland development and breast cancer. Cold Spring Harb Perspect Biol 2(10):a003202
3. Schedin P, Keely PJ (2011) Mammary gland ECM remodeling, stiffness, and mechanosignaling in normal development and tumor progression. Cold Spring Harb Perspect Biol 3(1):a003228
4. Bonnans C, Chou J, Werb Z (2014) Remodelling the extracellular matrix in development and disease. Nat Rev Mol Cell Biol 15(12):786–801
5. Bissell MJ, Rizki A, Mian IS (2003) Tissue architecture: the ultimate regulator of breast epithelial function. Curr Opin Cell Biol 15(6): 753–762
6. Campbell JJ, Watson CJ (2009) Three-dimensional culture models of mammary gland. Organogenesis 5(2):43–49
7. Koledova Z, Lu P (2017) A 3D fibroblast-epithelium co-culture model for understanding micro-environmental role in branching morphogenesis of the mammary gland. In: Martin F, Stein T, Howlin J (eds) Mammary gland development: methods and protocols, Methods in molecular biology, vol 1501. Springer, New York, pp 217–231
8. Koledova Z, Zhang X, Streuli C et al (2016) SPRY1 regulates mammary epithelial morphogenesis by modulating EGFR-dependent stromal paracrine signaling and ECM remodeling. Proc Natl Acad Sci U S A 113(39): E5731–E5740
9. Nguyen-Ngoc KV, Ewald AJ (2013) Mammary ductal elongation and myoepithelial migration are regulated by the composition of the extracellular matrix. J Microsc 251(3):212–223
10. Geraldo S, Simon A, Vignjevic DM (2013) Revealing the cytoskeletal organization of invasive cancer cells in 3D. J Vis Exp 80:e50763
11. Parrinello S, Samper E, Krtolica A et al (2003) Oxygen sensitivity severely limits the replicative lifespan of murine fibroblasts. Nat Cell Biol 5(8):741–747
12. Ewald AJ (2013) Practical considerations for long-term time-lapse imaging of epithelial morphogenesis in three-dimensional organotypic cultures. In: Sharpe J, Wong RO (eds) Imaging in developmental biology: a laboratory manual. Cold Spring Harbor Laboratory Press, Woodbury, pp 100–177

Chapter 9

An Organotypic 3D Assay for Primary Human Mammary Epithelial Cells that Recapitulates Branching Morphogenesis

Jelena R. Linnemann, Lisa K. Meixner, Haruko Miura, and Christina H. Scheel

Abstract

We have developed a three-dimensional organotypic culture system for primary human mammary epithelial cells (HMECs) in which the cells are cultured in free floating collagen type I gels. In this assay, luminal cells predominantly form multicellular spheres, while basal/myoepithelial cells form complex branched structures resembling terminal ductal lobular units (TDLUs), the functional units of the human mammary gland in situ. The TDLU-like organoids can be cultured for at least 3 weeks and can then be passaged multiple times. Subsequently, collagen gels can be stained with carmine or by immunofluorescence to allow for the analysis of morphology, protein expression and polarization, and to facilitate quantification of structures. In addition, structures can be isolated for gene expression analysis. In summary, this technique is suitable for studying branching morphogenesis, regeneration, and differentiation of HMECs as well as their dependence on the physical environment.

Key words Primary human mammary epithelial cells, Branching morphogenesis, Organoid, 3D assay, Regenerative potential, Collagen gel, Mechanotransduction, Forskolin, Rho-associated protein kinase

1 Introduction

The mammary gland (MG) is a unique characteristic of mammals. Its function is to produce milk following pregnancy for nursing offspring. The mature human MG consists of a network of branched ducts that terminate in alveoli-like clusters, the terminal ductal lobular units (TDLUs) [1, 2]. In the human MG, the TDLUs are surrounded by collagen-rich stroma that contains fibroblasts and immune cells. The epithelial ducts and alveoli contain two major cell lineages, a basal layer of myoepithelial cells that contract and thereby excrete the milk, and an inner layer of cells surrounding the lumen of the ducts, the luminal cells, which produce milk. A unique characteristic of the MG is that most of its development takes place postnatally: under the influence of steroid hormones during puberty,

Zuzana Koledova (ed.), *3D Cell Culture: Methods and Protocols*, Methods in Molecular Biology, vol. 1612, DOI 10.1007/978-1-4939-7021-6_9, © Springer Science+Business Media LLC 2017

a rudimentary ductal tree grows out into the concomitantly developing fat pad, undergoing branching morphogenesis through ductal elongation and side-branching [3]. Taken together, the function of the MG is derived directly from its branched morphology: differentiation of both myoepithelial and luminal cells depend on their respective in situ localization. Moreover, an increasing loss of glandular, branched morphology is a hallmark of tumor progression. Therefore, we reasoned that organotypic assays to model pubertal MG development need to recapitulate branching morphogenesis.

Consequently, it is apparent that culture of human mammary epithelial cells (HMECs) in standard two-dimensional conditions does not allow for the analysis of morphological aspects and limits assessment of functional features. However, in contrast to mouse mammary epithelial cells, HMECs do not recapitulate branching morphogenesis in the commonly used three-dimensional Matrigel in vitro assay.

We have developed an organotypic assay for HMECs that recapitulates branching morphogenesis of the breast (Figs. 1 and 2). This assay is based on a protocol for free-floating collagen gels described to enable breast cancer cell lines to generate tube-like structures [4].

In our protocol, the addition of Forskolin is critical to enable the generation of TDLU-like organoids [5]. Forskolin is a direct agonist of adenylyl cyclase, thereby increasing intracellular cAMP levels [6]. Moreover, we add the Rho-associated protein kinase (ROCK) inhibitor Y-27632, for the initial plating of cells, which promotes survival.

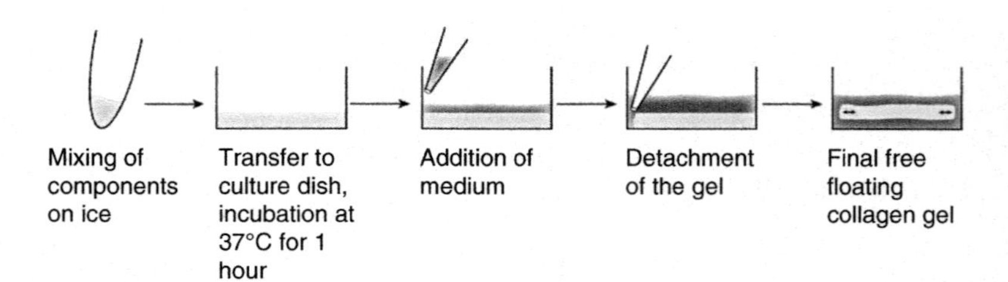

Fig. 1 Workflow for the generation of floating collagen type I gels. First, cell suspension, neutralizing solution, and collagen type I are added to a 15 mL conical tube on ice. Mind the correct order and mix by thoroughly pipetting up and down after every step. Next, the mix is transferred to a silicon-coated culture dish and incubated at 37 °C. After 1 h, medium is added and gels are detached by encircling the gel with a pipet tip and shaking the plate carefully

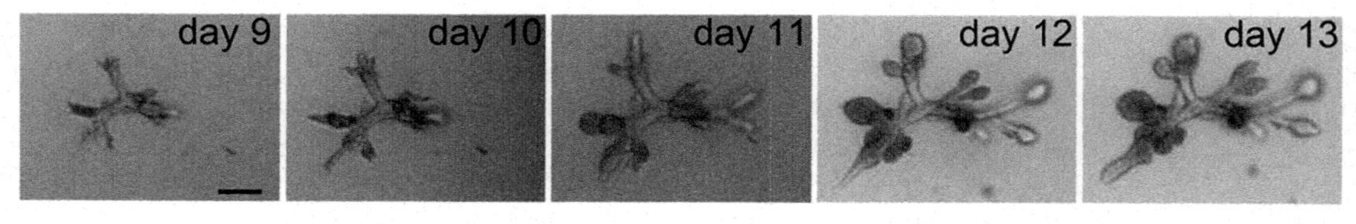

Fig. 2 Development of a TDLU-like branched structure. Bright-field images were taken starting at day 9 of culture and then every 24 h. Scale bar, 200 μm

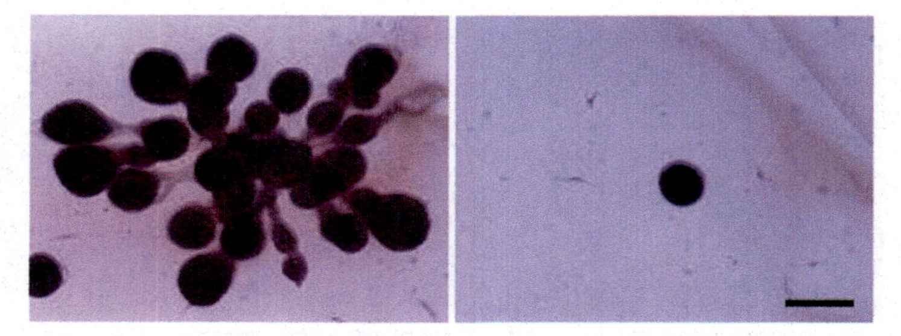

Fig. 3 Typical luminal cell- (*left*) and basal cell- (*right*) derived structures stained with carmine. Scale bar, 200 μm

Taken together, our protocol enables single primary HMECS to generate both spheres and complex branched structures. Specifically, by using published cell surface markers for fluorescent-activated cells sorting (FACS), we determined that luminal cells predominantly generate spheres while basal/myoepithelial cells generate complex branched structures (Fig. 3). This is in line with the finding that in transplantation assays using mouse mammary epithelial cells, regenerative capacity is enriched in myoepithelial cells [7–9].

These basal/myoepithelial cell-derived structures resemble the TDLUs of the MG: they contain p63-positive cells (p63, a p53 homologue, is a selective nuclear marker of myoepithelial cells in the human breast) at a basal position [10] that secrete laminin-1 to generate a basement membrane-like extracellular structure. On top of these p63-positive cells are one or more layers of p63-negative cells, many of which express the luminal markers GATA-3 and ZO-1 [11] (Fig. 4). The degree of lumen formation varies between donors and also depends on time in culture. 10–12 days after initial plating of single cells, the p63-positive, myoepithelial cells within the TDLU-like structures spontaneously contract the collagen gel in which they are embedded. This can be observed by shrinking of the gels circumference. Thereby, the contractile functional features of basal/myoepithelial cell in vivo are recapitulated. Importantly, free-floating collagen gels represent a compliant matrix that closely resembles the mechanical properties of the human MG extracellular matrix [12]. These mechanical properties are crucial to obtain TDLU-like organoids. If the collagen gels are left attached to the bottom and sides of the well, the myoepithelial cells within the branched structures are unable to contract the gels. As a consequence, both luminal differentiation, as gauged by the appearance of GATA-3 positive cells at a luminal position, and the formation of alveoli-like structures at the tip of the ducts are inhibited.

The organoids obtained in this assay can be cultured for at least 3 weeks and passaged multiple times. The structures within the gel can be visualized by carmine staining for morphological analysis and quantification. Immunofluorescence staining allows for the analysis of protein expression and polarization. In addition, cells can be isolated for gene expression analysis.

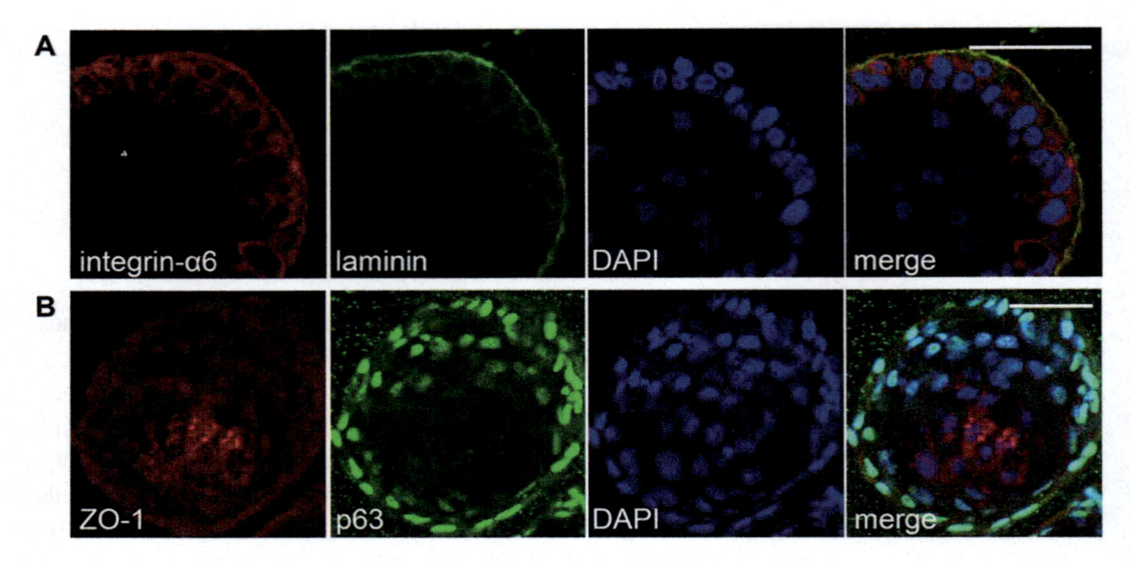

Fig. 4 Confocal microscopy of branched structures stained by immunofluorescent antibodies. (**a**) The presence of the basement membrane component laminin and expression of its receptor integrin-α6 indicate deposition of a basement membrane. (**b**) Branched structures show expression of luminal cell marker ZO-1 and basal cell marker p63 at correct positions. Scale bars, 50 μm

In summary, the assay allows for the analysis of branching morphogenesis, the quantification and characterization of regenerative cells that give rise to branched TDLU-like structures, and the influence of the physical environment on cell function and differentiation.

2 Materials

2.1 Coating Components

1. Silicone coating reagent: 2.5% siloxane in n-heptane. Add 2.5 mL of 1,7-dichloro-1,1,3,3,5,5,7,7-octamethyltetrasiloxane to 1 L of n-heptane. Swirl to mix. Close bottle tightly and store in a dry, well-ventilated space at room temperature.
2. Phosphate-buffered saline (PBS).
3. Ultrapure water (Milli-Q = MQ).
4. 24-well polystyrene tissue culture plate.

2.2 Collagen Gel Preparation, Culture, and Passaging Components

1. Cells: Cryopreserved primary HMECs. The cells are commercially available from different providers or can be isolated from reduction mammoplasties [5, 13].
2. Neutralizing solution: 550 mM HEPES in 11× PBS. Weigh 5.25 g of PBS powder and 6.55 g of HEPES into a beaker. Dissolve in approximately 40 mL of MQ. Adjust pH to 7.4 using NaOH. Add MQ to 50 mL. Filter sterilize, then aliquot into Eppendorf tubes and store at −20 °C. Aliquots in use can be stored at 4 °C.

3. MECGM basic medium: 100 U/mL penicillin and 100 μg/mL streptomycin in Mammary Epithelial Cell Growth Medium (MECGM, PromoCell).
4. 10× initial plating medium: 5% FCS, 100 μM forskolin, and 30 μM Y-27632 in MECGM basic medium (*see* **Notes 1** and **2**).
5. 10× passaging medium: 100 μM forskolin and 30 μM Y-27632 in MECGM basic medium (*see* **Note 3**).
6. Maintenance medium: 10 μM forskolin in MECGM basic medium.
7. Collagen type I from rat tail (BD Biosciences).
8. Digestion medium: 300 U/mL collagenase in MECGM basic medium (*see* **Note 4**).
9. 40 μm cell strainer.
10. 0.15% trypsin, pre-warm in water bath at 37 °C before use.
11. Trypsin-neutralizing solution.

2.3 Fixation, Immunofluorescence, and Carmine Staining Components

1. Permeabilization solution: 0.2% Triton X-100 in PBS. Dilute 100 μL Triton X-100 in 50 mL PBS (*see* **Note 5**).
2. PFA fixation solution: 4% PFA in PBS.
3. 0.15 M glycine in PBS.
4. Antibody buffer solution: 0.1% BSA in PBS. For a 1% BSA stock solution, dissolve 0.5 g BSA fraction V in 50 mL of PBS, sterile filter, and store at −20 °C. For a 0.1% BSA solution, dilute the stock 1:10 in PBS.
5. Blocking solution: 10% goat or donkey serum and 0.1% BSA in PBS. Add 1 mL of goat or donkey serum and 1 mL of 1% BSA stock to 8 mL of PBS (*see* **Note 6**).
6. Primary antibody solution: Prepare 250 μL of primary antibody solution per gel by adding the desired volume of primary antibody to antibody buffer solution (*see* **Note 7**).
7. Secondary antibody solution: Prepare 250 μL of secondary antibody solution per gel by adding the desired volume of secondary antibody to antibody buffer solution (*see* **Note 8**).
8. DAPI staining solution: 167 ng/mL 4′,6-diamidino-2-phenylindole (DAPI) in PBS (*see* **Note 9**).
9. Mounting medium suitable for immunofluorescence.
10. Clear nail polish.
11. Carmine alum staining solution: Weigh 1 g of carmine and 2.5 g of aluminum potassium sulfate and place in a beaker. Add 500 mL of distilled water. Boil for 20 min while mixing with a magnetic stirrer. Adjust final volume to 500 mL with distilled water. Filter-sterilize, add a crystal of thymol for preservation. Store at 4 °C.
12. Tweezers.

3 Methods

The methods described below are optimized for 24-well cell culture plates. We have also successfully used other formats like 48-well and 6-well cell culture plates by simply adjusting the used volumes. All centrifugation steps are performed at 500 × *g* at 4 °C for 5 min. Cells are incubated at 37 °C in the presence of 5% CO_2 and 3% O_2.

3.1 Silicone Coating of Cell Culture Plates

1. In a sterile flow cabinet, transfer 1 mL of fresh coating solution to the first well of a 24-well cell culture plate using a Pasteur pipette. Wait for 10 s, then transfer coating solution to the second well, wait for 10 s and proceed in this manner until the whole plate is coated (*see* **Notes 10** and **11**).
2. Remove the remaining coating solution with the Pasteur pipet and transfer to a designated waste container (*see* **Note 11**).
3. Rinse each well with 1 mL of PBS.
4. Rinse each well with 1 mL of MQ.
5. Open the lid of the plate and allow the wells to dry. Plates can be used immediately or sealed with Parafilm, wrapped in cling wrap and stored at 4 °C until usage (*see* **Note 12**).

3.2 Determination of Final Volumes of Collagen Solution, Neutralizing Solution, and Cell Suspension

1. Based on the concentration of the collagen I stock solution, determine the volume of collagen I stock solution needed to make a 400 μL collagen gel with a final collagen concentration of 1.3 mg/mL.
2. Calculate the amount of neutralizing solution which is one-tenth of the volume of collagen stock solution.
3. The remaining volume consists of cell suspension with the desired amount of cells in MECGM basic medium (*see* Table 1).
4. Multiply determined volumes with the number of gels you wish to produce, add volume for one extra gel to account for fluid loss due to the viscosity of collagen.

3.3 Plating Cells into Floating Collagen Gels

Until polymerization of collagen, all steps are carried out at 4 °C or on wet ice unless otherwise described.

1. Place collagen stock solution on ice and bring neutralizing solution to room temperature to allow for complete dissolution of salts. If necessary vortex until all crystals have resolved.
2. Quickly thaw primary HMECs in a water bath at 37 °C.
3. Take up cell suspension in MECGM basic medium.
4. Filter cell suspension using a 40 μm strainer, to remove residual fragments and cell aggregates.
5. Spin down cell suspension, remove supernatant and resuspend the cells in the desired volume of MECGM basic medium

Table 1
Example for volumes used to make one or five collagen gels, assuming a collagen type I stock concentration of 4.04 mg/mL. Figures in μL

	One collagen gel	Five collagen gels
Collagen I stock solution	128.71	643.55
Neutralizing solution	12.87	64.36
Cell suspension	258.42	1292.10
Total volume	400	2000

(determined in Subheading 3.2). For bulk HMECs, we recommend cell densities of 2000–3000 cells per 400 μL of collagen gel to obtain 2–6 TDLU-like organoids (*see* **Note 13**).

6. The following steps should be performed quickly in order to prevent premature polymerization of collagen: Add the desired volume of neutralizing solution (determined in Subheading 3.2) to the cell suspension and mix well by pipetting.
7. Add according amount of collagen stock solution (determined in Subheading 3.2) and mix well by pipetting without introducing any bubbles.
8. Immediately transfer 400 μL of the mixture to one well of the silicone-coated 24-well cell culture plate. Repeat to plate desired number of gels.
9. Carefully tilt the plate to ensure even coverage of the wells. Place the plate in an incubator at 37 °C for 1 h to allow for polymerization of collagen.
10. Add 500 μL of MECGM basic medium and 100 μL of 10× initial plating medium to each well (total volume per well including 400 μL of collagen gel volume: 1 mL).
11. Encircle the gels using a small pipet tip and carefully shake the plate to detach the gels from the bottom of the culture dish. Place the plate back in incubator.
12. Structures start to form at about 7 days of culture. Structures grow and expand in size until about 14 days of culture. After this, the size of structures does not increase but organoids continue to differentiate, form alveoli-like buds and contract the gels. Feeding and passaging of cells is described in Subheadings 3.4 and 3.5. For analysis of organoids *see* Subheadings 3.6–3.8.

3.4 Feeding of Cells in Floating Collagen Gels

After 5 days of culture the initial plating medium is replaced by maintenance medium. After this, fresh maintenance medium is added every 2–4 days, depending on the growth rate of primary mammary epithelial cells and on the number of cells seeded per well.

1. To remove medium, carefully hold back the floating collagen gel to one side of the well using a small pipet tip.
2. Take a 1000 μL pipet in the other hand and remove the medium from the opposite side of the well.
3. Carefully add 600 μL of fresh maintenance medium to the well. Make sure not to pipet the medium on the gel directly, to avoid damaging it.

3.5 Passaging of Cells in Floating Collagen Gels

1. Preheat water bath to 37 °C. Prepare a tube containing digestion medium (1 mL digestion medium per collagen gel).
2. Use sterile tweezers to transfer the collagen gels from the cell culture plate into the digestion medium.
3. Put the tube into the preheated water bath for around 30 min up to 1 h. Occasionally check dissolution of collagen gels and vortex briefly to ensure even digestion of collagen gels.
4. Spin down cells and remove supernatant.
5. Resuspend the pellet in 0.15% pre-warmed trypsin (use 1.5 mL trypsin for a pellet of five digested collagen gels) and mix well by pipetting for about 3 min.
6. Add trypsin-neutralizing solution (for the required volume of trypsin neutralizing solution, refer to your manufacturer's manual).
7. Filter cell suspension through a 40 μm mesh to get rid of any residual clumped cells. If desired, cells can now be counted.
8. Spin down cell suspension, remove supernatant and resuspend the cell pellet in desired volume of maintenance medium.
9. Cells can now be plated into floating collagen gels as described in Subheading 3.3. Use 10× passaging medium instead of 10× initial plating medium and switch to maintenance medium after 5 days of culture.

3.6 Immunofluorescent Staining of Breast Epithelial Cells Cultured in Floating Collagen Gels

All steps are carried out at room temperature unless otherwise described. Between all incubation and washing steps, carefully remove the solutions with a 1000 μL pipet while holding back the gel with a 200 μL pipet as described above. Perform all fixation, permeabilization, blocking, washing, and staining steps on an orbital shaker (*see* **Note 14**).

1. To pre-digest collagen, pre-warm collagen digestion solution at 37 °C in water bath and prepare a container with wet ice.
2. Remove the medium from the collagen gels, add 1 mL of digestion medium to each collagen gel and place the plate back into the incubator for 5 min.
3. Put the plate on ice to stop collagenase digestion immediately.

4. Carefully remove the digestion medium and wash with 1 mL of PBS (*see* **Note 15**).
5. To fix collagen gels, add 1 mL of PFA fixation solution to each well and incubate for 15 min.
6. Remove PFA fixation solution and wash the gels in 1 mL of PBS for 10 min.
7. Add 1 mL of 0.15 M glycine to each gel and incubate for 10 min to quench the paraformaldehyde.
8. Wash again with 1 mL of PBS for 10 min. Collagen gels can now be stained directly or stored in PBS at 4 °C for several months (*see* **Note 16**).
9. To permeabilize the gels, add 1 mL of permeabilization solution to each well and incubate for 10 min.
10. Wash the gels in 1 mL of PBS for 10 min.
11. Next, add 1 mL of blocking solution and incubate at 4 °C overnight.
12. Remove blocking solution from collagen gels and wash with 1 mL of PBS for 10 min (*see* **Note 17**).
13. Add 250 μL of primary antibody solution (*see* **Notes 7** and **18**) to each collagen gel and incubate at 4 °C overnight (*see* **Note 19**).
14. Wash collagen gels three times with 1 mL PBS for 10 min to remove unbound primary antibodies.
15. Add 250 μL of secondary antibody solution (*see* **Notes 8** and **20**) to each collagen gel and incubate in the dark for around 3 h. From this point on, exposure to light should be avoided (e.g., by wrapping the plate in aluminum foil).
16. Wash collagen gels two times with 1 mL of PBS for 10 min.
17. Add 250 μL of DAPI staining solution to each collagen gel and incubate for 2 min.
18. Wash three times with 1 mL of PBS for 10 min and two times with MQ for 5 min.
19. For mounting, transfer the gel onto a microscope glass slide using tweezers, spread the gel out with a pipet tip, and carefully soak up excess water with a tissue.
20. Add approximately two drops of mounting medium on top of the gel (*see* **Note 21**) and lay a coverslip on top without introducing any bubbles (*see* **Note 22**).
21. Let the mounting medium dry at room temperature overnight (*see* **Note 23**).
22. Seal the rims of the coverslip with colorless nail polish as soon as the mounting medium has dried.

3.7 Carmine Staining of Mammary Epithelial Cells Cultured in Floating Collagen Gels

Carmine alum solution stains cellular structures red and thereby allows better visualization.

All steps are carried out at room temperature unless otherwise described. Between all incubation and washing steps, carefully remove the solutions with a 1000 μL pipet while holding back the gel with a 200 μL pipet. Perform all fixation, permeabilization, blocking, washing, and staining steps on a shaker (*see* **Note 14**).

1. Fix structures as described in Subheading 3.6, **steps 5–8** (*see* **Note 24**).
2. Add 1 mL carmine alum staining solution to each gel and incubate plate on a shaker overnight at room temperature.
3. The next day, use tweezers to transfer the gels to a glass microscopy slide (*see* **Note 25**). Using the tweezers, make sure that the gels lay flat and in the middle of the microscopy slide. Remove excess fluid with a tissue.
4. Mount as described in Subheading 3.6, **steps 19–22**.

3.8 Cell Lysis for RNA Extraction

1. Digest collagen gels as described in Subheading 3.5, **steps 1–4**.
2. After the gels have dissolved, fill up the tube with PBS and centrifuge.
3. Wash the pellet with PBS and centrifuge again.
4. Use the pellet for your regular RNA isolation protocol (*see* **Note 26**).

4 Notes

1. 10× initial plating medium is prepared for easier handling. The resulting concentration in cell culture (1× initial plating medium) is 0.5% FCS, 10 μM Forskolin and 3 μM Y-27632 in MECGM basic medium.
2. We recommend making 10 mM Forskolin and 10 mM Y-27632 stock solutions in DMSO. These can be stored in small aliquots at −20 °C.
3. 10× passaging medium is prepared for easier handling. The resulting concentration in cell culture (1× passaging medium) is 10 μM Forskolin and 3 μM Y-27632 in MECGM basic medium.
4. Preparation of collagenase stock: We recommend preparing a 100× collagenase stock. For this, get the collagen digestion unit (CDU) of your collagenase from the data specification sheet and dilute the collagenase in PBS to a final stock concentration of 30,000 CDU/mL.
5. For pipetting viscous Triton X-100, cut off the end of pipet tip.
6. Use serum from the same species as the secondary antibody.

7. Use antibody concentrations recommended by the manufacturer or determine the optimal concentration empirically; 1:100 is often a good starting point.
8. Use antibody concentrations recommended by manufacturer or determine the optimal concentration empirically; 1:250 often is a good starting point.
9. Prepare 5 mg/mL DAPI stock solution in sterile H_2O, keep at 4 °C and protect from light. To get DAPI staining solution, add 0.33 μL of stock solution to 10 mL of PBS, keep at 4 °C and protect from light.
10. We suggest transferring 1 mL of coating solution from well to well, as this saves a lot of coating solution. Theoretically, more than one 24-well plate can be coated with the described volume. To save time, it is also possible to coat more than one well at a time. For instance, add 1 mL of coating solution to four wells each, wait for 10 s and then transfer the coating solution to the next four wells. As there is loss of coating solution with every well coated, one might have to add fresh coating solution for any additional plates.
11. The siloxane coating solution is flammable and caustic. Therefore, wear protective clothing when handling. Refer to safety data sheet for correct disposal.
12. The siloxane coating solution reacts with plastic and forms a hydrophobic surface. With this coating, collagen hydrogels are much easier to detach and are less likely to get damaged by detachment.
13. Importantly, this approach does not guarantee clonality. However, at densities of 100–500 cells, most organoids are clonal. The enrichment of myoepithelial cells with the capacity to generate TDLU-like organoids and limiting dilution assays to determine the frequency of these cells are described in Linnemann et al. 2015 [5].
14. Do not use a vacuum pump or an electric pipet aid to remove fluids as the gels might be aspirated.
15. After the collagenase predigest and before fixation the gels are very fragile and slippery and tear easily! Be very careful when removing collagenase solution and PBS. Also, in order to better visualize the pre-digested gels, this step can be done on the bench where one can look into the well from a closer distance. Visibility of the gel is also promoted by placing the wells on a dark surface.
16. For storage, plates should be sealed with Parafilm and cling wrap to avoid dehydration.

17. If desired, the collagen gels can now be cut into smaller pieces using a scalpel. This allows for more antibody combinations for one collagen gel.
18. Not all antibodies recommended for immunofluorescence work in 3D staining. The suitability of each antibody has to be determined by the experimenter. For a list of suitable antibodies, refer to Linnemann et al., 2015 [5].
19. If primary antibodies are directly conjugated to a fluorophore, the plate should be wrapped with aluminum foil to avoid bleaching and to ensure high quality of staining.
20. Phalloidin for visualization of F-actin can be added at this point.
21. When mounting multiple gels on one glass slide, we usually place an additional drop of mounting medium between the gels.
22. We do this by carefully putting one side of the coverslip on the microscope slide and then slowly lowering the coverslip. This prevents the formation of bubbles.
23. Alternatively, mounting medium can be dried in the fridge, which takes about 2–3 days. We have the impression that this longer process of drying reduces the formation of bubbles.
24. For the carmine alum staining, the gels do not need to be pre-digested.
25. Be careful not to break the gels. We use tweezers like a shovel in order not to squish the gels.
26. We usually take up the pellet in RLT lysis buffer (RNeasy mini kit, Qiagen) containing beta-mercaptoethanol, mix by pipetting ten times and vortex for 30 s. Next, the lysates are frozen at −80 °C for at least 1 h or until further usage. After thawing, the lysate is loaded on QIAshredder and centrifuged at maximum speed for 2 min. The flow-through is used for isolation of RNA using the RNeasy mini kit (Qiagen).

Acknowledgments

This work was supported by a Max Eder Grant of the German Cancer Aid Foundation.
(Deutsche Krebshilfe 110225 to C.H.S.).

References

1. Visvader JE, Stingl J (2014) Mammary stem cells and the differentiation hierarchy: current status and perspectives. Genes Dev 28(11): 1143–1158
2. Inman JL, Robertson C, Mott JD et al (2015) Mammary gland development: cell fate specification, stem cells and the microenvironment. Development 142(6):1028–1042
3. Nelson CM, Gleghorn JP (2012) Sculpting organs: mechanical regulation of tissue development. Annu Rev Biomed Eng 14: 129–154

4. Wozniak MA, Keely PJ (2005) Use of three-dimensional collagen gels to study mechano-transduction in t47d breast epithelial cells. Biol Proced Online 7:144–161
5. Linnemann JR, Miura H, Meixner LK et al (2015) Quantification of regenerative potential in primary human mammary epithelial cells. Development 142(18):3239–3251
6. Fradkin JE, Cook GH, Kilhoffer MC et al (1982) Forskolin stimulation of thyroid adenylate cyclase and cyclic 3',5'-adenosine monophosphate accumulation. Endocrinology 111(3): 849–856
7. Prater MD, Petit V, Alasdair Russell I et al (2014) Mammary stem cells have myoepithelial cell properties. Nat Cell Biol 16(10):942–950, 941–947
8. Shackleton M, Vaillant F, Simpson KJ et al (2006) Generation of a functional mammary gland from a single stem cell. Nature 439(7072): 84–88
9. Stingl J, Eirew P, Ricketson I et al (2006) Purification and unique properties of mammary epithelial stem cells. Nature 439(7079): 993–997
10. Barbareschi M, Pecciarini L, Cangi MG et al (2001) P63, a p53 homologue, is a selective nuclear marker of myoepithelial cells of the human breast. Am J Surg Pathol 25(8): 1054–1060
11. Kouros-Mehr H, Slorach EM, Sternlicht MD et al (2006) Gata-3 maintains the differentiation of the luminal cell fate in the mammary gland. Cell 127(5):1041–1055
12. Lui C, Lee K, Nelson CM (2012) Matrix compliance and rhoa direct the differentiation of mammary progenitor cells. Biomech Model Mechanobiol 11(8):1241–1249
13. Stingl J, Emerman JT, Eaves CJ (2005) Enzymatic dissociation and culture of normal human mammary tissue to detect progenitor activity. Methods Mol Biol 290:249–263

Chapter 10

3D Primary Culture Model to Study Human Mammary Development

Daniel H. Miller, Ethan S. Sokol, and Piyush B. Gupta

Abstract

We present a protocol for expanding human mammary tissues from primary patient-derived cells in three-dimensional (3D) cultures. The primary epithelial cells are seeded into 3D hydrogels with defined components, which include both proteins and carbohydrates present in mammary tissue. Over a span of 10–14 days, the seeded cells form mammary tissues with complex ductal-lobular topologies and include luminal and basal cells in the correct orientation, together with cells that stain positively for stem cell markers. In addition to recapitulating key architectural features of human mammary tissue, the expanded tissues also respond to lactogenic hormones including estrogen, progesterone, and prolactin. We anticipate that these cultures will prove useful for studies of mammary development and breast cancer.

Key words Hydrogel, 3D culture, Mammary gland, Primary culture, Ex vivo culture, Extracellular matrix, Immunofluorescence, Collagen, Human mammary epithelial cells

1 Introduction

Mammary gland development and maintenance is strongly shaped by the extracellular matrix (ECM), which regulates both cellular proliferation and differentiation [1, 2]. The ECM of the mammary gland consists of cross-linked collagen fibrils and other structural proteins—e.g., fibronectin and laminin—together with high molecular-weight glycosaminoglycans. At a structural level, the various protein components provide resistance to tensile forces and serve as a scaffold for cell–ECM interactions, while carbohydrates alter the elasticity and water content of tissues and provide resistance to compressive forces [3].

Previous studies have described three-dimensional (3D) culture systems for diverse epithelial tissues including kidney and intestine [4–7]. However, until recently 3D cultures for expanding primary human mammary tissues from patient-derived cells were not available. We have addressed this deficit by engineering hydrogels that incorporate key ECM proteins, carbohydrates, and

Zuzana Koledova (ed.), *3D Cell Culture: Methods and Protocols*, Methods in Molecular Biology, vol. 1612,
DOI 10.1007/978-1-4939-7021-6_10, © Springer Science+Business Media LLC 2017

growth signals present within human mammary tissue. These ECM hydrogels support the growth of patient-derived mammary epithelial tissue, allowing for cell proliferation and differentiation into mature mammary cell types. As described in our recent publication [8], the tissues grown in culture remain hormonally responsive and stain positively for markers of differentiation. These tissues can also be monitored in real-time using time-lapse imaging.

Here we describe protocols for patient tissue processing, ECM hydrogel fabrication, and downstream applications such as immunofluorescent staining, RNA collection, and protein analysis. These protocols serve as a foundation of our culture methods, but can be modified for a number of possible applications including treatment with small molecules, cell sorting, or gene perturbation. We anticipate that this technique will be useful for studies of human mammary tissue morphogenesis and biology.

2 Materials

2.1 Tissue Processing and Preparation

1. MEGM: Complete Mammary Epithelial Cell Growth Medium (Lonza, CC-3150).
2. Dissociation medium: 3 mg/mL Collagenase A, and 7 mg/mL hyaluronidase in MEGM.
3. Wash medium: 5% fetal bovine serum (FBS) in phosphate-buffered saline (PBS).
4. Freeze medium: 10% dimethyl sulfoxide in MEGM.
5. Fibroblast depletion medium (FDM): 10% FBS in Dulbecco's Modified Eagle Medium.
6. Scalpels, sterile.
7. Centrifuge tubes, 15 and 50 mL.
8. Cryotubes.
9. 10 cm tissue culture plate.

2.2 Seeding and Culturing Hydrogels

1. 25× extracellular matrix supplements (ES): 0.5 mg/mL laminin (Thermo Fisher, 23017015), 0.25 mg/mL hyaluronic acid (Millipore, 385908), and 0.5 mg/mL fibronectin (Thermo Fisher, 33016015) in PBS (*see* **Note 1**).
2. Rat tail collagen I (Corning) (*see* **Note 2**).
3. 4-chamber culture slides (BD Falcon).
4. 0.1 N NaOH.

2.3 Immunofluorescence

1. Wash buffer (PBST): 0.05% Tween 20 in PBS.
2. Fixative buffer: 4% paraformaldehyde in PBST.
3. Permeabilization buffer: 0.1% Triton X-100 in PBST.

4. Blocking buffer: 3% bovine serum albumin and 3% goat serum in PBST.
5. DAPI solution: 5 μg/mL DAPI in PBST.
6. Primary antibody solution: 1:300 rabbit anti-cytokeratin 14 antibody (Thermo Fisher, RB-9020) and 1:500 mouse anti-cytokeratin 8/18 antibody (Vector, VP-C407) in blocking buffer.
7. Secondary antibody solution: 1:1000 goat anti-rabbit IgG, Alexa Fluor 488 conjugate; 1:1000 goat anti-mouse IgG, Alexa Fluor 555 conjugate; and 1:100 Alexa Fluor 647 Phalloidin in blocking buffer.
8. Antifade reagent.

2.4 RNA and Protein Collection

1. Digestion buffer: 15 mg/mL collagenase A in PBS.
2. RNeasy Mini kit (Qiagen, 74106).
3. RIPA buffer: 50 mM Tris pH 7.5, 150 mM NaCl, 2 mM EDTA, 1% NP-40, 1% SDS, 1× Phosphatase Inhibitor Cocktail 2 (Sigma, P5726), 1× Phosphatase Inhibitor Cocktail 3 (Sigma, P0044), 1× Protease Inhibitor Cocktail (Sigma, P8340).

3 Methods

3.1 Tissue Processing and Preparation

1. In a sterile hood, use surgical scalpels to chop the tissue (*see* **Note 3**) into approximately 2–3 g pieces. Mince each piece into roughly 3–5 mm^3 fragments, and place the fragments into a 15 mL conical tube. To each conical tube, add 10 mL of dissociation medium.
2. Incubate tubes at 37 °C, on a rocker table, for 12–18 h (*see* **Note 4**).
3. Remove the tubes from the rocker and allow epithelial tissue to pellet by gravity (about 5 min). Decant and discard supernatant (*see* **Note 5**).
4. Resuspend the pellets in wash medium and spin them at 250 × *g* for 5 min. Decant and discard the supernatant, and repeat three additional times.
5. Resuspend the washed pellets in 1 mL of freezing medium per gram of unprocessed tissue and transfer 1 mL aliquots to cryotubes and freeze them at −80 °C (*see* **Notes 6** and **7**).

3.2 Recovery of Frozen Mammary Epithelial Organoids

1. Thaw the frozen tissue fragments in a 37 °C water bath. Resuspend them in 10 mL of FDM and spin at 250 × *g* for 5 min. Discard the supernatant.
2. Resuspend the pellet in 10 mL of FDM and plate on a 10 cm tissue culture plate. Incubate the plate in cell culture incubator

(37 °C, 5% CO_2) for 90 min. During this incubation fibroblasts still attached to the epithelial tissue will adhere to the plate.

3. Decant the medium from the plate into a 50 mL conical tube.
4. Wash the plate very gently with 10 mL of PBS, and decant into the same tube.
5. Spin at 250 × *g* for 5 min.
6. Discard the supernatant, and resuspend the pellet in 1 mL of MEGM (*see* **Note 8**).

3.3 Seeding and Culturing Hydrogels

1. Place collagen I, ES, 0.1 N NaOH, and a vial of MEGM on ice.
2. Add 436 μL of collagen I (*see* **Note 9**) to a microcentrifuge tube, and keep the tube on ice at all times.
3. Add 419.5 μL of prechilled MEGM to the tube with collagen I.
4. Add 54.5 μL of 0.1 N NaOH (*see* **Notes 10** and **11**) and quickly cap the tube and invert vigorously (*see* **Note 12**).
5. Add 40 μL of ES to the tube.
6. Using a wide-bore tip, add 50 μL of organoids (*see* **Note 13**), and mix by vigorously inverting the tube.
7. Using a wide-bore tip, pipette 200 μL of the solution into each chamber of a four-chamber slide, sliding the tip of the pipette across the surface of the chamber as you pipette out, so that you produce a thin, oval-shaped pad.
8. Incubate the slide at 37 °C for 60 min.
9. Add 1 mL of pre-warmed MEGM to each chamber of the slide.
10. Use a P200 tip to gently scrape each pad off of the surface of the slide, so that the pads float.
11. Twice per week, remove the medium and replace with fresh, pre-warmed MEGM (Fig. 1).

3.4 Immunofluorescence Staining

1. Using forceps, remove the hydrogels from their slides and place them in a 15 mL conical tube. Wash once with PBS for 5 min at room temperature (RT).
2. Decant and discard the PBS and add 2 mL of fixative buffer. Incubate at RT for 30 min on a rocker.
3. Decant and discard the fixative buffer and add 10 mL of wash buffer. Incubate on a rocker for 10 min at RT, and repeat twice more.
4. Decant and discard wash buffer and add 2 mL of permeabilization buffer. Incubate overnight at RT on a rocker.
5. Wash, as in **step 3**, three times.

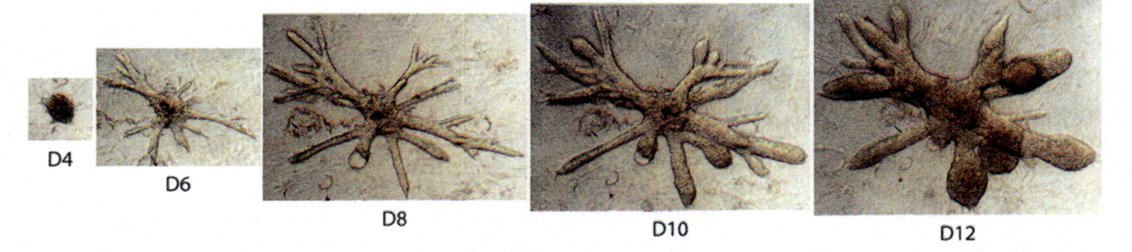

Fig. 1 Tissue grown in hydrogels undergoes ductal initiation, branching, and alveogenesis over the course of 12 days. By day 4, early ducts can be seen, which elongate and thicken, giving rise to lobules starting between day 8 and day 10, and expanding through day 12

6. Decant and discard wash buffer and add 2 mL of blocking buffer. Incubate at RT for at least 2 h on a rocker (*see* **Note 14**).
7. Decant and discard blocking buffer and transfer hydrogels to a microcentrifuge tube. Add 500 μL of the primary antibody solution to the microcentrifuge tube. Incubate at 4 °C on a tube rotator overnight.
8. Transfer hydrogels to a 15 mL conical tube and wash, as in **step 3**, three times.
9. Decant and discard wash buffer. Transfer hydrogels to a microcentrifuge tube and add 500 μL of secondary antibody staining solution to the microcentrifuge tube. Incubate overnight at 4 °C or for 2 h at RT on a tube rotator.
10. Transfer hydrogels to a 15 mL conical tube and wash, as in **step 3**, three times.
11. Decant and discard wash buffer and add 2 mL of DAPI solution. Incubate for 20 min at RT on a rocker.
12. Wash, as in **step 3**, three times.
13. Mount hydrogels on a glass microscopy slide, in antifade reagent (*see* **Note 15**) (Fig. 2). The stained structures are now ready for imaging, either using epifluorescence microscopy, or confocal microscopy (Fig. 3).

3.5 RNA and Protein Collection

1. Using forceps, remove the hydrogels from their slides and place them in a 15 mL conical tube. Wash once with PBS briefly at room temperature (RT). Decant and discard the PBS.
2. Add 2 mL of digestion buffer for up to 10 hydrogels. Scale the volumes linearly for more than ten hydrogels. Incubate at 37 °C for 20–30 min in a water bath, inverting every 3–5 min (*see* **Note 16**).

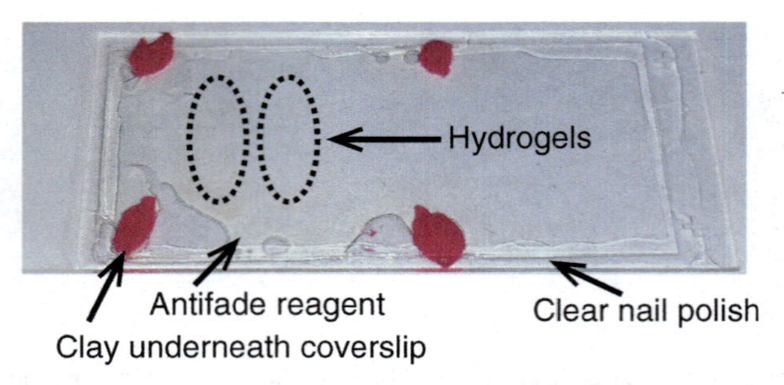

Fig. 2 Example of hydrogels mounted in antifade reagent on a microscope slide. Dots of clay are placed at the corners of the coverslip to raise the coverslip to the height of the hydrogels. The coverslip is then sealed onto the slide using clear nail polish. The translucent hydrogels are indicated by the *dotted lines*

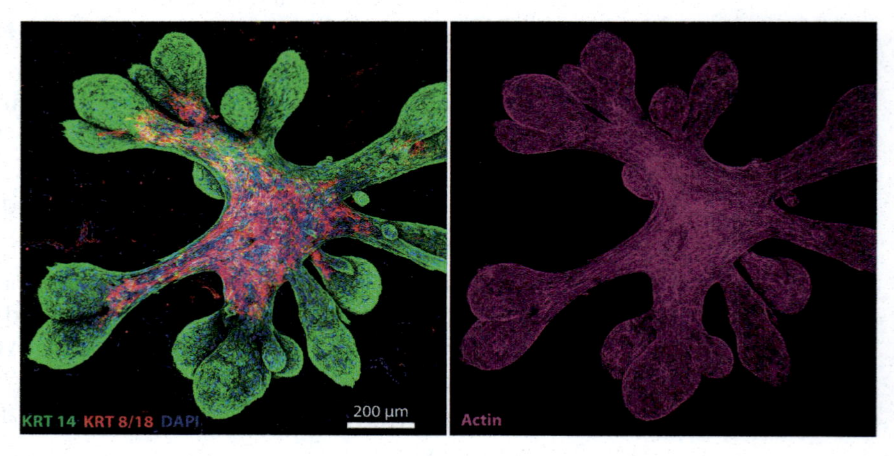

Fig. 3 Example of staining for basal (KRT14) and luminal (KRT8/18) cytokeratins. Keratin 14 is shown in *green*, keratin 8/18 is shown in *red*, DAPI is shown in *blue*, and actin is shown in *pink* (*right*). Scale bar, 200 μm. The tissue shown was cultured in a hydrogel, as described in this protocol, for 11 days. Adapted from [8]

3. Add MEGM to a total volume of 15 mL and spin at 500 × *g* for 3 min.
4. Discard the supernatant and wash with 10 mL of PBS.
5. For RNA collection, decant the PBS wash and proceed directly with cell lysis (350 μL) using the RNeasy Mini Kit.
6. For protein collection, wash the pellet of tissues an additional two times with PBS to remove any traces of residual collagen.
7. On the final wash, carefully remove any remaining supernatant with a P200 micropipette.
8. Resuspend the pellet in 60 μL of ice cold RIPA buffer and transfer to a microcentrifuge tube.
9. Incubate on ice for 15 min, then spin at 16,000 × *g* for 15 min.
10. Transfer the supernatant to a new tube. The protein lysate is ready for quantification and Western blot analysis (*see* **Note 17**).

4 Notes

1. Components are resuspended according to the manufacturers' recommendations. Laminin (Thermo Fisher, 23017015) is supplied in solution at 1.18 mg/mL. Hyaluronan (Millipore, 385908) is resuspended in sterile water at 1 mg/mL. Fibronectin (Thermo Fisher, 33016015) is resuspended in sterile water at 2 mg/mL. Aliquots of each individual component are stored long term at −80 °C. To make 2 mL of ES, mix 850 μL of laminin, 500 μL of hyaluronans, 500 μL of fibronectin, and 150 μL of PBS. Store at 4 °C for up to 1 month.
2. Be careful to always store stock solutions of collagen at 4 °C. Even short term increases in temperature can lead to partial polymerization of the collagen and impact subsequent experiments.
3. After surgery, tissue should be stored and delivered cold, on ice, but not frozen.
4. The digestion is complete when the chunks of dissociated tissue are no more than 1 mm^3 in size and the fat has completely separated from the epithelial tissue.
5. The supernatant can be collected and is highly enriched for fibroblasts. The supernatant can be processed in parallel with the epithelium.
6. Take care to pipette the freezing solution slowly; the organoids resist the fluid forces and will become unevenly distributed between the tubes.
7. For long term storage, place the frozen tubes (after minimum 24 h at −80 °C) in a liquid nitrogen freezer.
8. Only use wide bore tips when pipetting any solution containing tissue fragments. Narrow bore micropipette tips will sheer apart the tissue.
9. Volumes in this protocol are for 1 mL of hydrogel, using a collagen stock that is 3.9 mg/mL. The final concentration of collagen will be 1.7 mg/mL. Adjust volumes of collagen and MEGM accordingly for collagen stocks produced at different concentrations. Be sure to also adjust the volume of NaOH added, to 12.5% of the volume of collagen. Volumes can be scaled up linearly to make more than 1 mL of hydrogels.
10. The volume of 0.1 N NaOH added is calculated as 12.5% of the volume of collagen stock solution added.

11. Any time a new lot of collagen or a newly made 0.1 N NaOH stock is used, we recommend checking the pH of the final solution to ensure that the pH is roughly 7.2. If the pH is incorrect, adjust the volume of NaOH added accordingly.
12. Everything after this step is time-sensitive, as the collagen will start to polymerize as soon as it is pH neutralized. Make sure to keep the tube on ice whenever possible, to slow the collagen polymerization process.
13. The amount of tissue to add varies from patient to patient, and the ideal amount will need to be determined empirically, dependent on how much material is in each frozen tube, generally between 10 and 100 μL per slide. If too much tissue is seeded per hydrogel, the gel will rapidly condense during culturing, and will quickly become unusable.
14. The blocking step, and all downstream incubation and wash steps, can be extended to longer times. The longer these incubations and washes, the better. Additional washes, or longer washes, will reduce background during imaging. Longer incubations with antibody will increase signal.
15. Because the hydrogels are fairly thick, the corners of the coverslip need to be propped up with small dots of clay before the coverslip is sealed into the slide (*see* Fig. 2).
16. For protein collection, be sure the pads are completely dissolved. After about 10–15 min the pads will be thin and wispy. After 30 min there should be no visible evidence of the collagen pads. If there is remaining collagen after 30 min, allow to digest for an additional 15 min if necessary, to prevent collagen protein contamination.
17. Even with additional washes, there will be contaminating collagen from the hydrogel. Skipping the quantification and estimating protein concentration based on the number of digested pads can often times be more accurate than quantifying protein concentration using a Bradford Assay.

Acknowledgments

We gratefully acknowledge Wendy Salmon and the Keck Imaging Facility at the Whitehead Institute for microscopy services. This research was supported in part through the National Science Foundation Graduate Research Fellowship Program (award 1122374 to ESS) and by the Whitehead Institute.

References

1. Streuli CH, Bailey N, Bissell MJ (1991) Control of mammary epithelial differentiation: basement membrane induces tissue-specific gene expression in the absence of cell-cell interaction and morphological polarity. J Cell Biol 115:1383–1395
2. Schedin P, Keely PJ (2011) Mammary gland ECM remodeling, stiffness, and mechanosignaling in normal development and tumor progression. Cold Spring Harb Perspect Biol 3:1–22
3. Fraser JR, Laurent TC, Laurent UB (1997) Hyaluronan: its nature, distribution, functions and turnover. J Intern Med 242:27–33
4. McCracken KW, Catá EM, Crawford CM et al (2014) Modelling human development and disease in pluripotent stem-cell-derived gastric organoids. Nature 516:400–404
5. Sato T, Vries RG, Snippert HJ et al (2009) Single Lgr5 stem cells build crypt-villus structures in vitro without a mesenchymal niche. Nature 459:262–265
6. Takasato M, Er PX, Becroft M et al (2014) Directing human embryonic stem cell differentiation towards a renal lineage generates a self-organizing kidney. Nat Cell Biol 16:118–126
7. Lancaster MA, Renner M, Martin CA et al (2013) Cerebral organoids model human brain development and microcephaly. Nature 501:373–379
8. Sokol ES, Miller DH, Breggia A et al (2016) Growth of human breast tissues from patient cells in 3D hydrogel scaffolds. Breast Cancer Res 18:19

Chapter 11

Lungosphere Assay: 3D Culture of Lung Epithelial Stem/Progenitor Cells

Anas Rabata, Ales Hampl, and Zuzana Koledova

Abstract

Lung epithelium contains distinctive subpopulations of lung stem/progenitor cells (LSPCs) that are essential for lung epithelial maintenance and repair in vivo. Hence, LSPCs are in the center of interest of lung biology due to their promising therapeutic applications. To reach this goal, proper characterization of LSPCs, understanding of their proliferation and differentiation potentials and elucidation of mechanisms that control them are necessary. Therefore, development of reliable in vitro clonogenic assays has been needed. We established lungosphere assay, an in vitro sphere-forming 3D culture assay that enables to evaluate stem/progenitor cell activity, self-renewal and differentiation capacity of LSPCs and to conveniently test the effect of various treatments on LSPCs. Here we provide a detailed description of procedures for isolation of adult mouse lung epithelial cells, their culture in non-adherent conditions to form LSPC-derived spheroids (lungospheres) and for embedding of lungospheres into 3D extracellular matrix to model processes of lung tissue maintenance in a physiologically relevant microenvironment.

Key words Lung epithelial stem/progenitor cells, 3D culture, Spheroid culture, Lungosphere assay

1 Introduction

The ability of adult lung tissue to maintain itself, remodel, regenerate and repair after injury is dependent on the activity of resident adult lung stem/progenitor cells (LSPCs). In vivo lineage tracing studies and mouse lung injury models revealed that lung epithelium contains divergent stem/progenitor cell populations, including basal cells, pulmonary neuroendocrine cells, bronchioalveolar stem cells and type II alveolar cells that reside in distinct microenvironmental niches along the proximal–distal axis of the respiratory tree [1–9]. To dissect the mechanisms that regulate lung regenerative potential, including deciphering the niche components and the mechanisms regulating LSPC proliferation and differentiation, development of 3D in vitro culture systems has been necessary.

3D cultures of lung epithelial cells in extracellular matrix (ECM), lung organoids, were established by several groups by seeding lung

Zuzana Koledova (ed.), *3D Cell Culture: Methods and Protocols*, Methods in Molecular Biology, vol. 1612,
DOI 10.1007/978-1-4939-7021-6_11, © Springer Science+Business Media LLC 2017

epithelial cells (alone or in coculture with stromal cells) in 3D Matrigel and their culture in air–liquid interface (ALI) [10]. These lung organoid cultures revealed, for example, that distal LSPCs require coculture with fibroblasts [7, 11–14] or endothelial cells [15] to proliferate and form clonal spheres in 3D Matrigel while proximal LSPCs (basal cells of proximal airways) do not—they can self-renew and form spheres in Matrigel without stromal cells [3, 4, 14]. Furthermore, lung organoid cultures have provided insights into differentiation spectra of various lung epithelial cell populations [14, 15]. By manipulating stromal cell content and identity, ECM composition, epithelial cell localization within the ECM, culture medium composition etc., lung organoid cultures offer a versatile 3D culture system to study the effects of complex microenvironment on LSPC biology in vitro.

For quantification of LSPC activity and self-renewal, we developed a lungosphere assay. Same as neurosphere and mammosphere assays, this assay takes advantage of the unique abilities of stem cells to survive without attachment (resistance to anoikis) and to form spheres when clonally seeded in suspension [16, 17]. In the lungosphere assay, lung epithelial cells are seeded in suspension in defined serum-free medium on non-adherent plates at low densities and cultured to form primary lungospheres. The formation of primary lunospheres is a measure of LSPC activity. Next, primary lungospheres are counted, harvested, digested to single cells and seeded in secondary lungosphere assay for quantification of their self-renewal. Secondary lungospheres can be further passaged for tertiary lungosphere assay if desired. Alternatively, lungospheres can be embedded in 3D ECM, such as Matrigel, and cultured in submerged or ALI culture. The lungosphere assay provides an efficient and reliable method to test effects of soluble agents on LSPC activity and self-renewal and to grow LSPC-derived organoids that can be embedded in 3D ECM for further studies of LSPC proliferation and differentiation and epithelial morphogenesis in response to various factors, including soluble agents, coculture with other cell types or ECM.

This chapter provides detailed protocols for isolation of adult mouse lung epithelial cells (Fig. 1), establishment of primary lungosphere culture, passaging of lungospheres for secondary lungosphere assay, and embedding of lungospheres into 3D ECM (Fig. 2).

2 Materials

All tools, materials and solutions for mouse dissection, tissue dissociation, cell processing for fluorescence-activated cell sorting (FACS), and cell culture have to be sterile.

2.1 Mouse Dissection

1. Mice of any strain, age, and sex can be used as donors of lung tissue.
2. Dissection board and pins.

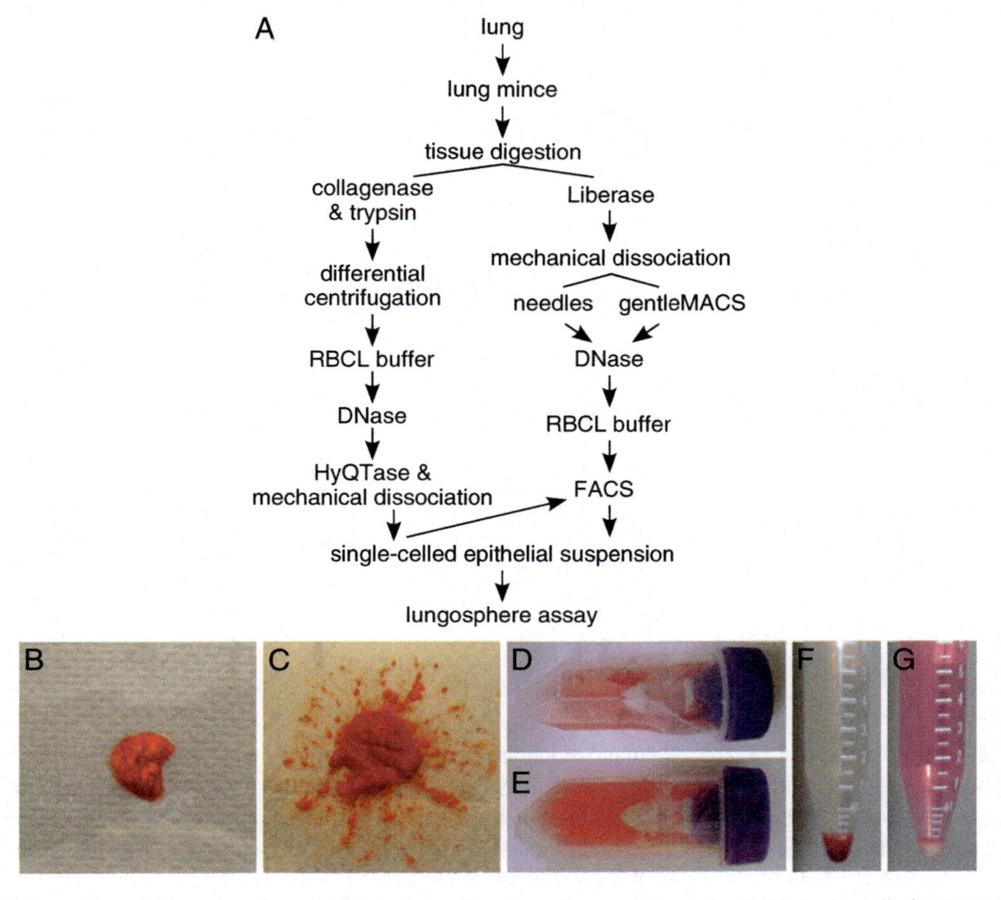

Fig. 1 Lung dissociation procedure. (**a**) The schematic of lung tissue dissociation into single-celled suspension for lungosphere assay. Freshly isolated lung (**b**) is cut by scalpel into pieces (**c**) and transferred into digestion solution (such as collagenase and trypsin solution or Liberase solution) and incubated in it. To promote tissue dissociation in Liberase solution, needles or gentleMACS dissociator are used. The photographs show lung tissue in C tube before (**d**) and after (**e**) gentleMACS-mediated dissociation. The resulting whole lung cell suspension is next treated by DNase to remove sticky DNA and by red blood cell lysis (RBCL) buffer to remove red blood cells. The photographs show cell pellet before (**f**) and after (**g**) RBCL treatment. Finally, to separate epithelial cells from stromal cells, FACS is used. Alternatively, tissue is digested only partially by collagenase and trypsin and in the next steps, after removal of sticky DNA and red blood cells by DNase and red blood cell lysis (RBCL) buffer, respectively, stromal cells are separated from epithelial organoids by differential centrifugation. Finally, the organoids are dissociated into single-celled epithelial suspension using HyQTase and mechanical dissociation

3. Standard forceps, 12–13 cm, straight, pointed.
4. Standard forceps, 9 cm, straight, pointed.
5. Operating scissors.
6. Iris/eye scissors, straight.
7. Sterile cotton buds (optional).
8. Polystyrene petri dish, 60 mm.
9. Phosphate buffered saline (PBS).

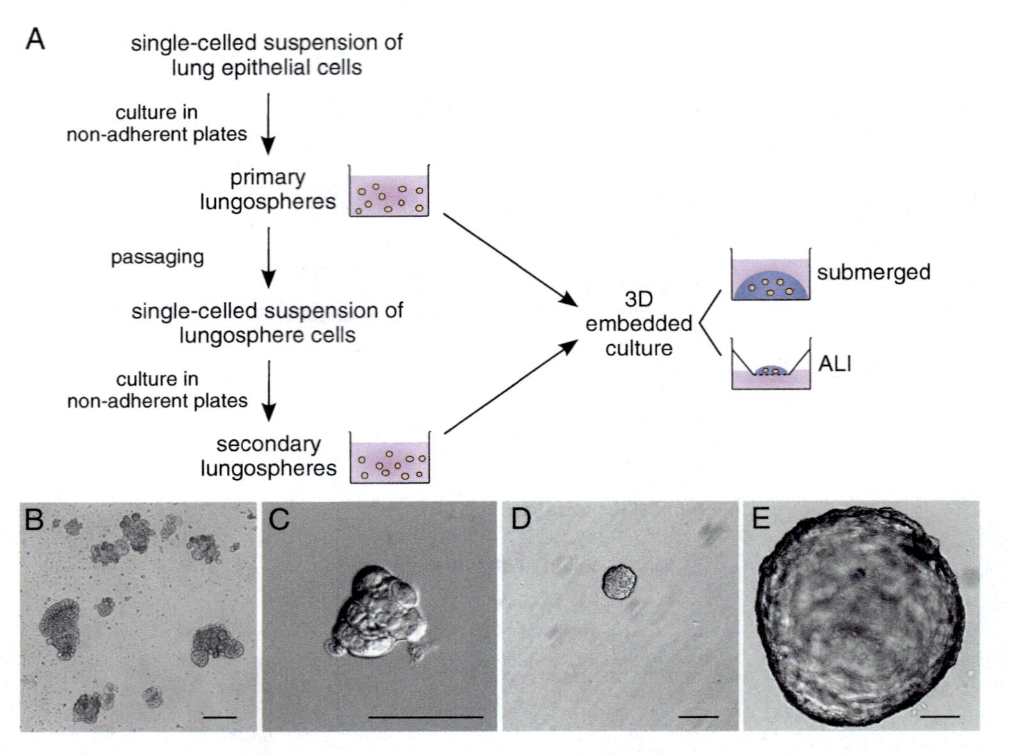

Fig. 2 Lungosphere assay. (**a**) The schematic of lungosphere assay. Lung epithelial cells are cultured at low density in suspension in non-adherent (polyHEMA) plates to form primary lungospheres. Primary lungospheres are dissociated and seeded in suspension in non-adherent plates to form secondary lungospheres. Primary or secondary lungospheres can be collected and embedded in Matrigel for 3D culture in air–liquid interface or submerged culture. (**b**–**e**) Photographs of lungospheres. Scale bars, 100 μm. (**b**, **c**) Lungospheres in suspension culture. (**d**, **e**) Lungosphere in 3D embedded submerged culture after 3 (**d**) or 20 (**e**) days of culture

2.2 Lung Tissue Dissociation

1. Polystyrene Petri dish, 100 mm.
2. Disposable scalpels blades, no. 20 or 21.
3. Scalpel blade holders.
4. Hank's balanced salt solution (HBSS).
5. Dulbecco's Modified Eagle Medium: Nutrient Mixture F-12 (DMEM/F12).
6. Collagenase digestion solution: 2 mg/mL collagenase A, 2 mg/mL trypsin, 5% (v/v) fetal bovine serum (FBS), 5 μg/mL insulin, 50 μg/mL gentamicin in DMEM/F12. For each lungs from one mouse, prepare 10 mL of digestion solution by combining 8.7 mL of DMEM/F-12, 0.5 mL of FBS, 10 μL of 50 mg/mL gentamicin, 50 μL of 1 mg/mL insulin, 400 μL of 50 mg/mL trypsin, and 400 μL of 50 mg/mL collagenase (*see* **Notes 1–3**). Filter-sterilize and warm up to 37 °C in a water bath before use. Prepare freshly for each primary cell preparation (*see* **Note 4**).

7. Coating solution: 2.5% (v/v) bovine serum albumin (BSA) in PBS.
8. Red blood cells lysis (RBCL) buffer: 155 mM NH_4Cl, 12 mM $NaHCO_3$, 0.1 mM EDTA in distilled H_2O, pH 7.4, filter sterilized.
9. Deoxyribonuclease I (DNase I) solution: 20 U/mL DNase I in DMEM/F12. Add 40 μL of 2 U/μL DNase I to 4 mL of DMEM/F12 (*see* **Note 5**). Prepare fresh for each experiment.
10. Liberase solution: For each mouse lungs, prepare 4 mL of the solution by adding 40 μL of the Liberase stock solution (4.8 mg/mL Liberase in HBSS) (*see* **Note 6**) to 4 mL of HBSS in a 50 mL tube. Preheat to 37 °C. Prepare fresh for each cell isolation experiment.
11. Liberase wash buffer: 5% FBS in HBSS.
12. HyQTase (GE Healthcare Life Sciences).
13. 6-well cell culture plate.
14. Syringe needles, 18 gauge (G), 21G and 24G.
15. Syringes, 5 and 20 mL.
16. Orbital shaker (*see* **Note 7**).
17. Cell strainers, 100 and 40 μm.
18. Hemocytometer.
19. (optional) gentleMACS™ Dissociator and gentleMACS™ C tubes (Miltenyi Biotec).

2.3 FACS

1. Blocking buffer: 1% BSA in HBSS.
2. FACS buffer: 0.2% BSA in HBSS.
3. Polypropylene round-bottom tubes for FACS, 5 and 15 mL.
4. 30 μm cell strainer.
5. Cell sorter, such as BD FACSAria II.
6. Selection antibody cocktail, such as as Alexa Fluor 647-conjugated anti-mouse CD104, FITC-conjugated anti-mouse EpCAM, eFluor 450-conjugated anti-mouse CD24, PE-conjugated anti-mouse CD49f, and PECy7-conjugated anti-mouse CD45 in blocking buffer (*see* **Note 8**).
7. Viability dye, e.g., 7-amino-actinomycin D (7-ADD).
8. Collection medium: 1× B-27 (without vitamin A), 100 U/mL penicillin, 100 μg/mL streptomycin in phenol red-free DMEM/F12.

2.4 Cell Culture

1. PolyHEMA [Poly(2-hydroxyethyl methacrylate), Sigma] solution: In a sterile bottle, dissolve 12 g of polyHEMA in 1 L of 95% ethanol on a heated plate. Store at room temperature (RT).

2. 6-well, 12-well, 24-well tissue culture plates.
3. Lungosphere medium: 1× B-27 (without vitamin A), 100 U/mL penicillin, 100 μg/mL streptomycin, 4 μg/mL heparin, 20 ng/mL epidermal growth factor (EGF), 10 ng/mL fibroblast growth factor 2 (FGF2), 10 μM Y-27632 (Rho-associated protein kinase inhibitor) in phenol red-free DMEM/F12.
4. Matrigel® Basement Membrane Matrix Growth Factor Reduced (Corning, 354230) (*see* **Note 9**).
5. Basal culture medium: 1× ITS (10 μg/mL insulin, 5.5 μg/mL transferrin, 6.7 ng/mL selenium), 100 U/mL penicillin, 100 μg/mL streptomycin in DMEM/F12.
6. EGF, FGF2, and/or other growth factors of interest.
7. (optional) 24-well cell culture plate with corresponding permeable cell culture inserts, 0.4 μm pore size.
8. (optional) 4% paraformaldehyde (PFA) in PBS.

2.5 General Materials and Equipment

1. 15 and 50 mL sterile centrifuge tubes.
2. Pipettes (5, 10, 50 mL) and micropipettes.
3. Centrifuge.
4. Cell culture incubator.

3 Methods

3.1 Preparation of PolyHEMA-Treated Dishes

1. Under sterile conditions (in a cell culture hood), pipette polyHEMA solution into the wells of 6-well, 12-well or 24-well plates. Use 1 mL, 0.75 mL, and 0.5 mL per one well of 6-well, 12-well and 24-well plate, respectively (*see* **Note 10**).
2. Make sure that polyHEMA solution is spread over the whole bottom of each well. Gently shake or swirl the plate to spread the solution if needed (*see* **Note 11**).
3. Transfer the plates into an oven (*see* **Note 12**) and incubate at 40 °C until completely dry (about 2 days).
4. Store dried polyHEMA-treated plates at RT until use (*see* **Note 13**).

3.2 Mouse Dissection

1. Cull the mice and immediately proceed to harvest lung tissue (*see* **Note 14**).
2. Generously spray a mouse with 70% ethanol and using pins affix it to the dissection board in a laminar hood (*see* **Note 15**).
3. Using operating scissors cut the skin medially from approximately the middle of the belly to the neck (while holding the skin with the bigger forceps). Take care to avoid unnecessary cutting through the peritoneum. Then cut the skin from the

medial incision towards the forelimbs. Separate the skin from the underlying tissues by gently pulling the skin with forceps while pushing back the underlying tissues with cotton buds. Pin the skin down to expose thoracic cage (*see* **Note 16**).

4. Using the operating scissors and the bigger forceps carefully cut through the ribs on both the left and the right side and through the diaphragm to open thoracic cage. Take care to avoid damage to the lungs. Remove any excessive blood or blood clots using cotton buds.
5. Using eye scissors and finer forceps, excise the lungs. Remove extra lobular airways and place the lung lobes in a petri dish containing 10 mL of sterile PBS (Fig. 1b).

3.3 Lung Dissociation Using Collagenase and Trypsin

This lung dissociation method enables to first separate epithelial and stromal cell fractions (using differential centrifugation) and then to get single-celled suspension enriched for epithelial cells (by further digestion of epithelial organoids with HyQTase) that can be sorted by FACS to obtain specific subpopulations or seeded directly in non-adherent conditions to grow lungospheres (Fig. 1a).

1. Using sterile forceps transfer the lungs into a new 100 mm Petri dish. Remove excessive PBS.
2. Chop up the lungs using scalpels into a mince of approximately 1 mm^3 pieces.
3. Transfer the minced tissue into a 50 mL tube with the pre-heated collagenase digestion solution (use 10 mL of the solution per lungs from one mouse).
4. Place the tube in the incubated shaker at 37 °C and shake at 100 rpm for 45 min (*see* **Notes 17** and **18**).
5. Centrifuge at 450 × *g* at 4 °C for 10 min.
6. Carefully, without disturbing the cell pellet, transfer the supernatant into a 15 mL tube and centrifuge at 450 × *g*, at 4 °C for 10 min (*see* **Note 19**). In the meantime, resuspend the pellet (remaining in the 50 mL tube) in 4–5 mL of DMEM/F12.
7. After centrifugation, discard the supernatant from the 15 mL tube and resuspend the pellet in 4–5 mL of DMEM/F12. Add the cell suspension from the 50 mL tube to it and centrifuge at 450 × *g* at 4 °C for 10 min.
8. Discard the supernatant and resuspend the pellet in 2–3 mL of RBCL buffer. Incubate at RT for 2 min. Then add 6 mL of DMEM/F12, mix using a pipette and centrifuge at 450 × *g* at 4 °C for 10 min.
9. Remove supernatant and repeat **step 8** one or two more times until the cell pellet becomes white.

10. Resuspend the pellet in 4 mL of DNase solution. Shake by hand for 3–5 min at RT. Then add 6 mL of DMEM/F12 and mix by pipetting. Centrifuge at 450 × *g* at 4 °C for 10 min.
11. Remove supernatant and perform differential centrifugation:
    (a) Resuspend the pellet in 8–10 mL of DMEM/F12.
    (b) Centrifuge at 450 × *g* at 4 °C for 10 s.
    (c) Transfer the supernatant to a 50 mL tube.
    (d) Repeat **steps 11a–11c** four more times, while collecting the supernatants in one 50 mL tube.
    (e) The resulting pellet in the 15 mL tube will contain mostly organoids, while the supernatant in the 50 mL tube will contain stromal cells (*see* **Note 20**).
12. Wash the organoids with 2 mL of HBSS, centrifuge at 450 × *g* at 4 °C for 3 min, remove supernatant.
13. Resuspend the organoids in 2 mL of HyQTase, transfer the suspension into a 6-well plate and incubate at 37 °C for 5 min.
14. Dissociate the organoids into a single-celled suspension:
    (a) Using a 1000 μL pipette, pipette the suspension repeatedly (about 60 times).
    (b) Check the progress of dissociation into single cells under a microscope.
    (c) If the organoids are not dissociated enough (clumps of cells are remaining), incubate the suspension for 5 more minutes at 37 °C and repeat **steps 14a** and **b** until the organoids become single cells (*see* **Note 21**).
15. Transfer the single-cell suspension into a 15 mL tube, add the same volume of wash buffer and mix.
16. Strain the suspension through a 40 μm cell strainer into a fresh 50 mL tube to eliminate debris and clumps and centrifuge at 450 × *g* at 4 °C for 3 min.
17. Remove the supernatant and wash the cell pellet with approximately 2 mL of HBSS or collection medium. Centrifuge at 450 × *g* at 4 °C for 3 min.
18. Resuspend the cell pellet in 1–2 mL of HBSS (if the cells are going to be stained for FACS) or collection medium (if the cells are to be seeded for lungospheres without sorting).
19. Take a small aliquot for cell counting (e.g., using hemocytometer) and calculate the cell concentration (*see* **Note 22**). If the cells are to be seeded for lungospheres directly, adjust the cell concentration to 1×10^5–5×10^6 cells/mL with collection medium.

3.4 Lung Dissociation Using Liberase

Using the Liberase dissociation method, adapted from McQualter and Bertoncello [18], the whole lung tissue is dissociated into a single-celled suspension that needs to be further sorted to obtain lung epithelial cells and/or their subpopulations. The tissue can be dissociated either manually (the "needle method") or using gentleMACS dissociator (Fig. 1a).

1. Using forceps transfer the lungs into a new 100 mm Petri dish. Remove excessive PBS.
2. Chop up the lungs using scalpels (Fig. 1c):
   (a) Into a mince of approximately 1 mm^3 pieces for the "needle method".
   (b) Into approximately 2 × 2 × 2 mm pieces for the gentleMACS method.
3. Dissociate the tissue using either the "needle method" (*see* **step 4**) or using gentleMACS dissociator (*see* **step 5**).
4. To dissociate the tissue using the "needle method":
   (a) Transfer the minced tissue into 50 mL tube with the preheated Liberase solution (use 4 mL of the solution per lungs from one mouse). Place the tube in the incubated shaker at 37 °C and shake at 100 rpm for 30 min.
   (b) Passage the sample through an 18 G needle attached to a 20 mL syringe until the tissue passes freely through the needle. Place the tube in the incubated shaker and shake at 37 °C for 15 min.
   (c) Passage the sample through a 21 G needle attached to a 20 mL syringe until the tissue passes freely through the needle.
   (d) Centrifuge at 450 × *g* at 4 °C for 10 min.
   (e) Remove the supernatant and resuspend the pellet in 4 mL of DNase solution. Shake by hand for 3–5 min at RT.
   (f) Add 6 mL of DMEM/F12, mix by pipetting and centrifuge at 450 × *g* at 4 °C for 10 min.
   (g) Remove the supernatant and resuspend the pellet in wash buffer.
5. To dissociate the tissue using gentleMACS (Fig. 1d, e):
   (a) Transfer the lung tissue pieces into a C tube containing the preheated Liberase solution (4 mL per lungs from one mouse) and place the tube in gentleMACS dissociator, choose the program "m_lung_01" and process the tissue for 10 s.
   (b) Incubate the C tube with lung tissue suspension for 60 min at 37 °C.

(c) Add 80 U of DNAse (40 μL of 2 U/μL of DNase stock) per each 4 mL of Liberase solution and incubate for 5 min at 37 °C.

(d) Process the tissue digest using the gentleMACS dissociator set at "m_lung_01" for 10 s.

6. Pass the tissue digest through a 100 μm cell strainer into a 50 mL tube to remove tissue debris and cell clumps. Add the same volume of Liberase wash buffer, mix by pipetting and centrifuge at 450 × *g* at 4 °C for 5 min.
7. Remove the supernatant, resuspend the pellet in Liberase wash buffer, and centrifuge at 450 × *g* at 4 °C for 5 min.
8. Resuspend the cell pellet in 2–3 mL of RBCL buffer and incubate for 2 min at RT. Then add 6 mL of DMEM/F12, mix by pipetting and centrifuge at 450 × *g* at 4 °C for 5 min.
9. Remove the supernatant and repeat **step 8** one or two more times until the cell pellet becomes white (Fig. 1g).
10. Resuspend the cell pellet in 2–3 mL of HBSS and pass the suspension through a 40 μm cell strainer into a 50 mL tube to get a single-celled suspension.
11. Take a small aliquot for cell counting and calculate the cell number.

3.5 Cell Staining for FACS

1. Centrifuge the freshly isolated lung cell suspension at 450 × *g* at 4 °C for 5 min, remove the supernatant and resuspend the cell pellet in blocking buffer. Incubate for 20 min at RT (*see* **Note 23**).
2. Centrifuge the cell suspension at 450 × *g* at 4 °C for 5 min.
3. Remove the supernatant, wash the cell pellet with FACS buffer and centrifuge at 450 × *g* at 4 °C for 5 min.
4. Resuspend the cells at 1×10^7 cells/mL in FACS buffer containing the selection antibody cocktail [such as Alexa Fluor 647 anti-CD104 (1:1000), FITC anti-EpCAM (1:1000), eFluor 450 anti-CD24 (1:1000), PE anti-CD49f (1:1000), PECy7 anti-CD45 (1:1000)] and incubate in ice in the dark for 20 min (*see* **Notes 24** and **25**).
5. Add 1 mL of HBSS to the stained cells and centrifuge at 450 × *g* at 4 °C for 5 min.
6. Remove the supernatant, wash the pellet with 1 mL of HBSS and centrifuge at 450 × *g* at 4 °C for 5 min.
7. Resuspend the cells in HBSS and pass the cell suspension through a 30 μm cell strainer into a 15 mL tube to eliminate cell aggregates.
8. Adjust cell concentration to 2×10^7 cells/mL in HBSS and add 7-ADD (to the final concentration of 10 μL/mL).

Incubate for 5 min in ice in the dark and proceed with cell sorting.

9. Sort LSPCs by setting up gates for selection of single live (7-ADD^{neg}), nonhematopoietic (CD45^{neg}), EpCAM^{pos}, CD49f^{pos}, CD24^{low}, CD104^{pos} cells.
10. Collect cells in collection tubes containing 1 mL of collection medium.

3.6 Lungosphere Assay

1. To establish primary lungosphere culture, both FACS-sorted or unsorted lung epithelial cells can be used. If using FACS-sorted cells, first centrifuge them (450 × *g* at 4 °C for 5 min) to collect all the cells and resuspend the cells in fresh collection medium (at 1×10^5–5×10^6 cells/mL).
2. Seed 2.5×10^4–5×10^4 cells into polyHEMA-treated 6-well plates in 2 mL of lungosphere medium (*see* **Note 26**).
3. Incubate in a humidified atmosphere at 37 °C, 5% CO_2 while avoiding unnecessary moving or disturbing the plates (*see* **Note 27**).
4. Add fresh lungosphere medium every 3 days (*see* **Note 28**).
5. After about 10–15 days of culture count the number of primary lungospheres formed (*see* **Note 29**).

3.7 Passaging of Lungospheres

1. Collect primary lungosphere suspension from the polyHEMA-treated plate into a 50 mL tube. Wash the polyHEMA-treated plate twice with HBSS and add the HBSS wash to the 50 mL tube with the lungosphere suspension to collect all lungospheres.
2. Centrifuge at 450 × *g* at 4 °C for 5 min. Discard the supernatant and wash the lungospheres with 3 mL of HBSS.
3. Centrifuge at 450 × *g* at 4 °C for 3 min. Discard the supernatant and resuspend the lungospheres in 2 mL of HyQTase.
4. Transfer the lungosphere suspension into a 6-well plate and incubate at 37 °C for 5 min.
5. Dissociate the lungospheres into a single-celled suspension:
   (a) Using a 1000 μL pipette, pipette the suspension repeatedly (about 60 times).
   (b) Check the progress of dissociation into single cells under a microscope.
   (c) If the lungospheres are not dissociated enough (clumps of cells are remaining), incubate the suspension for 5 more minutes at 37 °C and repeat **steps 5a** and **b** until the lungospheres become single cells (*see* **Note 30**).
6. Transfer the cell suspension into a 15 mL tube. Wash the well twice with HBSS and add the HBSS wash to the 15 mL tube to collect all cells.

7. Centrifuge at 450 × *g* at 4 °C for 3 min, remove supernatant and resuspend cells in 0.5 mL of collection media. Take a small aliquot for cell counting and calculate the cell concentration.
8. Seed the cells for secondary lungosphere assay and culture them as described in Subheading 3.6, **steps 2–4**.
9. After 10–15 days count the number of secondary lungospheres.

3.8 3D Lungosphere Embedded Assay

Primary or secondary lungospheres can be embedded in 3D Matrigel and cultured in submerged culture or on ALI to model lung epithelial morphogenesis (Fig. 2).

1. Collect lungospheres from the polyHEMA-treated plate in a 50 mL tube. Wash the plates with PBS twice and add the PBS wash to the 50 mL tube with the lungosphere suspension to collect all lungospheres.
2. Centrifuge at 450 × *g* at 4 °C for 3 min. Remove supernatant, resuspend the pellet in 10 mL of DMEM/F12 and transfer the suspension into a 15 mL tube.
3. Centrifuge at 450 × *g* at 4 °C for 3 min to collect all lungospheres and cells (*see* **Note 31**).
4. Resuspend the lungospheres in basal culture medium and transfer the suspension into a 1.5 mL eppendorf tube. Centrifuge at 450 × *g* at 4 °C for 3 min.
5. Aspirate most of the medium, leaving about 20 μL and resuspend lungospheres in it. Place the tube in ice.
6. Add Matrigel to the lungosphere suspension and mix well by pipetting. Keep the tube in ice (*see* **Notes 32** and **33**).
7. Coat desired number of wells of a 24-well plate (for submerged culture) or cell culture inserts placed in 24-well plate (for ALI culture) with Matrigel:
   (a) Place a 24-well plate on ice. Add about 20 μL of Matrigel into the middle of a well and spread it with the pipette tip, creating a small patch about 8 mm in diameter. Repeat for the desired number of wells.
   (b) Alternatively, to coat the transwell insert, add about 10 μL of Matrigel to the middle of the insert membrane, creating a small patch. Repeat for the desired number of transwell inserts.
   (c) Incubate plate at 37 °C for 15 min.
8. Plate the lungospheres in Matrigel:
   (a) For submerged culture plating, place the Matrigel-coated plate on a heat block at 37 °C. Pipette 50 μL of Matrigel–lungosphere mixture per well on top of the Matrigel-coated patches in domes.

(b) For ALI culture plating, pipette 25 μL of Matrigel–lungosphere mixture per insert on top of the Matrigel-coated patches.

9. Incubate the plate at 37 °C for 30–45 min to solidify Matrigel.
10. Add pre-warmed basal culture medium (supplemented with growth factors, inhibitors or other agents as desired according to the experiment):
    (a) For submerged culture, add 1–1.5 mL of medium into each well with lungosphere–Matrigel mixture.
    (b) For ALI culture, add 600 μL of medium to the well below the transwell insert.
11. Incubate in a humidified atmosphere of cell culture incubator (37 °C, 5% CO_2).
12. Change medium every 2–3 days. Check the 3D cultures under microscope regularly.

3.9 Fixation of the 3D Embedded Cultures

When the experiment is finished, the 3D embedded cultures can be fixed and used for subsequent analysis, e.g., immunofluorescence staining.

1. Carefully aspirate the culture medium and wash the culture with PBS.
2. Fix the culture with 4% PFA for 15–30 min at RT.
3. Carefully aspirate PFA and wash the culture two to three times with PBS.
4. Fixed sample can be stored in PBS at 4 °C for weeks.

4 Notes

1. For 1 mg/mL insulin, dissolve 50 mg insulin in 50 mL of 0.1 M HCl, pH 2–3. Filter sterilize and make aliquots of 100–250 μL. Store at −20 °C.
2. For 50 mg/mL collagenase, dissolve 500 mg collagenase in 10 mL of DMEM/F12 and make 400 and 800 μL aliquots. Store at −20 °C.
3. For 50 mg/mL trypsin, dissolve 500 mg trypsin in 10 mL DMEM/F12 and make 400 and 800 μL aliquots. Store at −20 °C.
4. The basis of collagenase/trypsin solution, without the enzymes and insulin, can be prepared in advance and stored at 4 °C for 2–3 weeks. The enzymes and insulin are then freshly added to the solution on the day of tissue digestion.

5. To prepare 2 U/μL DNase I, dissolve 2000 U of DNase I in 1 mL of sterile 0.15 M NaCl. Make 40 μL aliquots and store at −20 °C.
6. To prepare Liberase stock solution (4.8 mg/mL Liberase in HBSS), dissolve 50 mg Liberase (Roche) in 10.4 mL sterile HBSS. Aliquot (40, 80 μL) and store at −20 °C.
7. Preferably, a heated (37 °C) shaker should be used. If not available, a smaller shaker can be placed into a cell culture incubator to provide heated environment.
8. Dilute the antibodies according to the manufacturer's instructions. According to the experiment, different antibodies can be used to select for specific lung epithelial cell subpopulations.
9. Thaw 10 mL vial of Matrigel in ice in refrigerator overnight. Avoid exposing Matrigel to temperatures over 10 °C because it would lead to its premature and irreversible gelation. Using chilled pipette tips, aliquot Matrigel into prechilled Eppendorf tubes on ice and store the aliquots at −20 °C.
10. PolyHEMA solution has low viscosity, care should be taken while pipetting to avoid dripping.
11. Uneven coating of the wells will result in polyHEMA-free patches that will be accessible for undesired cell attachment.
12. Seal the plates with a piece of tape to prevent accidental opening of the lids while moving the plates.
13. The polyHEMA plates can be stored for several months.
14. The mice should be sacrificed by a method approved by ethical committee for work with animals, under a valid project license. Because cervical dislocation usually leads to extensive hematomas in the neck regions, we recommend culling by anesthesia overdose.
15. To prevent contamination of the dissected tissue, the tissue should be dissected in a laminar flow hood. To avoid contamination of cell culture hoods, use a laminar hood designated specifically for dissection. If such hood is not available, the tissue can be dissected out in a clean procedure room.
16. Re-sterilize the instruments using hot bead dry sterilizer or at least wash them with 70% ethanol before reusing them in the next step.
17. Up to three lungs can be digested in one 50 mL tube in 30 mL of solution. When more than three lungs are used, we recommend dividing them into several tubes, with the amount of collagenase/trypsin solution adjusted accordingly.
18. For the following steps, pre-coat all pipettes and tubes with coating buffer to prevent lung tissue from adhering to their surfaces.

19. This step is not necessary but it helps to increase the cell yield. Alternatively, discard the supernatant from the 50 mL tube, resuspend the pellet in 8–10 mL of DMEM/F12, transfer the cell suspension to a 15 mL tube and centrifuge at 450 × *g* at 4 °C for 10 min. Continue with **step 8** (Subheading 3.3).
20. Lung stromal cells can be collected by centrifuging the suspension at 650 × *g* for 3 min. To isolate the fibroblasts, remove the supernatant and resuspend the pellet in fibroblast medium (10% FBS, 1× ITS, 100 U/mL of penicillin, and 100 μg/mL of streptomycin in DMEM). Seed the cells onto cell culture plates and incubate at 37 °C, 5% CO_2 for 30 min. During this time the fibroblasts will attach, while the epithelial cells will remain in suspension. Aspirate the medium, wash the cell culture plate with PBS to remove unattached cells. Finally, add fibroblast medium and incubate cultures at 37 °C, 5% CO_2. Change medium every 2 days and when the cell reach 85% confluence, split them onto new cell culture dishes.
21. The organoids should be dissociated into single cells within 20–25 min, after approximately three to four rounds of repetitive pipetting. Alternatively, if the repetitive pipetting is not effective enough, such as when after three rounds of the pipetting big clumps of cells are remaining, passage the sample through a 24 G needle attached to a 5 mL syringe about four to five times.
22. Use viability staining, such as Trypan blue exclusion test, to check cell viability after the cell isolation procedure and calculate number of viable cells per mL (especially needed if the cells are to be directly seeded for lungospheres, i.e., without sorting out dead cells using FACS).
23. Alternatively, anti-mouse CD16/CD32 antibody can be used to block nonspecific binding of antibodies. To do so, resuspend the cells in FACS buffer with 10 μg/mL anti-CD16/CD32 and incubate in ice for 10 min. After blocking, centrifuge the cell suspension at 450 × *g* at 4 °C for 5 min and then resuspend the cells in staining solution directly, i.e., no washing with FACS buffer is required.
24. Remember to prepare single-stained and fluorescence minus one controls as well as to save some unstained cells for setting up the FACS.
25. From this point on, avoid unnecessary exposure of the samples to the light.
26. For seeding of lower numbers of cells use smaller format of polyHEMA-treated plates. Adjust the volume of lungosphere medium and the cell seeding density accordingly.
27. Moving or disturbing the plates increases the chance of undesired spheroid aggregation or spheroid dissociation.

28. When adding the medium, take care not to disturb the cell suspension too much. Also, take this opportunity to check sphere formation using a microscope.
29. The lungosphere forming efficiency (LFE) can be calculated using this formula:

    LFE (%) = (Number of lungospheres formed)/(Number of cells seeded) × 100.
30. Usually the lungospheres can be dissociated easily by repetitive pipetting. Alternatively, 24 G needle attached to a syringe can be used.
31. To remove undesired single cells perform differential centrifugation: Resuspend the lungospheres in 10 mL of DMEM/F12, centrifuge at 450 × *g* at 4 °C for 10 s, remove supernatant. Repeat two or three more times.
32. Use prechilled pipette tips when pipetting Matrigel to prevent premature Matrigel gelation due to temperature increase. Avoid creating bubbles when pipetting or mixing Matrigel.
33. The amount of Matrigel and the number of lungospheres plated per well depend on the experimental setup. We recommend to use 50 μL of Matrigel for 50 lungospheres per one 3D culture in one well.

Acknowledgments

This work was supported by the grant no. 16-31501A form Ministry of Health of the Czech Republic, by the grant "Junior investigator 2015" from Masaryk University, Faculty of Medicine, and by the grant 16-20031Y from Czech Science Foundation (GACR).

References

1. Giangreco A, Reynolds SD, Stripp BR (2002) Terminal bronchioles harbor a unique airway stem cell population that localizes to the bronchoalveolar duct junction. Am J Pathol 161(1):173–182
2. Rawlins EL, Okubo T, Xue Y et al (2009) The role of Scgb1a1+ Clara cells in the long-term maintenance and repair of lung airway, but not alveolar, epithelium. Cell Stem Cell 4(6):525–534
3. Rock JR, Onaitis MW, Rawlins EL et al (2009) Basal cells as stem cells of the mouse trachea and human airway epithelium. Proc Natl Acad Sci U S A 106(31):12771–12775
4. Hegab AE, Ha VL, Gilbert JL et al (2011) Novel stem/progenitor cell population from murine tracheal submucosal gland ducts with multipotent regenerative potential. Stem Cells 29(8):1283–1293
5. Hegab AE, Ha VL, Darmawan DO et al (2012) Isolation and in vitro characterization of basal and submucosal gland duct stem/progenitor cells from human proximal airways. Stem Cells Transl Med 1(10):719–724
6. Zheng D, Limmon GV, Yin L et al (2012) Regeneration of alveolar type I and II cells from Scgb1a1-expressing cells following severe

pulmonary damage induced by bleomycin and influenza. PLoS One 7(10):e48451

7. Barkauskas CE, Cronce MJ, Rackley CR et al (2013) Type 2 alveolar cells are stem cells in adult lung. J Clin Invest 123(7):3025–3036
8. Tata PR, Mou H, Pardo-Saganta A et al (2013) Dedifferentiation of committed epithelial cells into stem cells in vivo. Nature 503(7475): 218–223
9. Desai TJ, Brownfield DG, Krasnow MA (2014) Alveolar progenitor and stem cells in lung development, renewal and cancer. Nature 507(7491):190–194
10. Nadkarni RR, Abed S, Draper JS (2016) Organoids as a model system for studying human lung development and disease. Biochem Biophys Res Commun 473(3):675–682
11. McQualter JL, Yuen K, Williams B et al (2010) Evidence of an epithelial stem/progenitor cell hierarchy in the adult mouse lung. Proc Natl Acad Sci U S A 107(4):1414–1419
12. Teisanu RM, Chen H, Matsumoto K et al (2011) Functional analysis of two distinct bronchiolar progenitors during lung injury and repair. Am J Respir Cell Mol Biol 44(6): 794–803
13. Chen H, Matsumoto K, Brockway BL et al (2012) Airway epithelial progenitors are region specific and show differential responses to bleomycin-induced lung injury. Stem Cells 30(9):1948–1960
14. Hegab AE, Arai D, Gao J et al (2015) Mimicking the niche of lung epithelial stem cells and characterization of several effectors of their in vitro behavior. Stem Cell Res 15(1):109–121
15. Lee JH, Bhang DH, Beede A et al (2014) Lung stem cell differentiation in mice directed by endothelial cells via a BMP4-NFATc1-thrombospondin-1 axis. Cell 156(3):440–455
16. Reynolds BA, Weiss S (1992) Generation of neurons and astrocytes from isolated cells of the adult mammalian central nervous system. Science 255(5052):1707–1710
17. Shaw FL, Harrison H, Spence K et al (2012) A detailed mammosphere assay protocol for the quantification of breast stem cell activity. J Mammary Gland Biol Neoplasia 17(2):111–117
18. McQualter JL, Bertoncello I (2015) Clonal culture of adult mouse lung epithelial stem/progenitor cells. In: Rich IN (ed) Stem cell protocols, Methods in molecular biology, vol 1235. Springer, New York, pp 231–241

Chapter 12

3D Hanging Drop Culture to Establish Prostate Cancer Organoids

Theresa Eder and Iris E. Eder

Abstract

Three-dimensional (3D) cell culture enables the growth of cells in a multidimensional and multicellular manner compared to conventional cell culture techniques. Especially in prostate cancer research there is a big need for more tissue-recapitulating models to get a better understanding of the mechanisms driving prostate cancer as well as to screen for more efficient drugs that can be used for treatment. In this chapter we describe a 3D hanging drop system that can be used to culture prostate cancer organoids as tumor epithelial monocultures and as epithelial–stromal cocultures.

Key words 3D culture, Organoids, Prostate cancer, Epithelial–stromal coculture, Hanging drops, Cell viability assay, Immunohistochemistry, Flow cytometry, Drug screening

1 Introduction

Prostate cancer (PCa) is among the most commonly diagnosed cancers in Western countries [1]. Because of the strong dependence of PCa cells on androgens, androgen deprivation therapy and/or the use of antiandrogens to inhibit the action of androgens is the gold standard systemic treatment for patients with advanced PCa. Despite good initial efficacy of this therapy, most patients invariably develop castration-resistant and further on therapy-resistant prostate cancer [2]. Hence, there is an urgent need to develop new and more efficient treatment options.

One of the major challenges in PCa research is the lack of appropriate in vitro and in vivo models. Especially in vitro, it is essential to use cell culture models that maintain the typical characteristics of PCa cells such as androgen responsiveness and sufficient AR expression, which are often lost upon extended cell culture [3]. A panel of different prostate cell lines is available so far, which represent various stages of PCa, benign prostate hyperplasia or even benign prostate (reviewed in [4]). These cell lines are commonly two-dimensionally (2D) grown on plastic culture plates

Zuzana Koledova (ed.), *3D Cell Culture: Methods and Protocols*, Methods in Molecular Biology, vol. 1612,
DOI 10.1007/978-1-4939-7021-6_12, © Springer Science+Business Media LLC 2017

and are useful tools for in vitro research due to their high replicative capacity and relatively easy handling. Despite these advantages, it must be considered that cells grown as 2D monolayers exhibit different growth characteristics and metabolic features compared to their tissue of origin [5]. Within the human prostate the epithelial cells are embedded in a microenvironment composed of extracellular matrix (ECM) and a substantial amount of stromal cells, including fibroblasts, immune cells, blood cells, and neuroendocrine cells. These stromal cells strongly influence tumor cell growth, survival, invasion and metastatic progression of PCa cells [6]. The strong impact of the tumor microenvironment emphasizes the need for a reliable prostate tumor model, which better mimics the typical growth characteristics of human PCa and which allows to study the interactions between the epithelium and the stroma in vitro.

Here we describe the establishment of PCa organoids using a 3D hanging drop culture system. This technique is used to grow cells as 3D organoids on specific 96-well plates. The starting cell numbers as well as the required medium volumes are low. Harrison et al. first used the hanging drop technique for cell or tissue culture [7]. 3D organoids can be formed with or without ECM compounds in order to study the interactions between different cell types in a more tissue-recapitulating environment. Organoid size can be controlled by seeding different cell numbers or reducing days of culture. This technique also allows conducting high throughput experiments and can be combined with any standard lab method with some modifications.

2 Materials

2.1 3D Cell Culture

1. Perfecta 3D 96-well Hanging Drop plate (3D Biomatrix).
2. 1× trypsin–EDTA (170,000 U trypsin/L, 200 mg/L EDTA).
3. Phosphate buffered saline (PBS; 1×; without Ca^{2+} and Mg^{2+}).
4. 2% agarose in distilled water (*see* **Note 1**).
5. Microwave.
6. Inverted microscope.
7. Image analysis software.
8. Prostate cell line, e.g., LNCaP.
9. Cell culture medium, e.g., LNCaP Cell culture medium: 10% fetal calf serum (FCS), 1× GlutaMAX™, 100 U/mL penicillin, 100 μg/mL streptomycin, 1 mM sodium pyruvate, 4.5 g/L D-glucose and 20 mM Hepes in DMEM.
10. Cell counting chamber, e.g., Neubauer type.
11. Centrifuge.
12. Multipette.

2.2 Cell Viability Assay

1. WST-1 Assay (Roche) (*see* **Note 2**).
2. 96-well plate.
3. 1× trypsin–EDTA.
4. Trypsin inhibitor from glycine max (soybean; 1×).
5. Spectrophotometer.
6. Centrifuge with inserts for 96-well plates.

2.3 Processing of 3D Organoids for Immunohistochemistry

1. CytoRich™ Red (Thermo Fisher Scientific).
2. PBS.
3. 15 and 50 mL centrifuge tubes.
4. 2% agarose in distilled water.
5. Microwave.
6. 4% paraformaldehyde (PFA) in PBS.
7. 70% ethanol.
8. 96% ethanol.
9. Xylol.
10. Paraffin wax.
11. Scalpel.
12. SuperFrost™ Plus slides (Thermo Fisher Scientific).
13. Rotary microtome.
14. Heating plate.

2.4 Staining for Flow Cytometry

1. 1.5 mL centrifuge tubes.
2. Trypsin–EDTA.
3. Trypsin Inhibitor (1×).
4. PBS.
5. FIX & PERM® Cell Fixation and Permeabilization Kit (An Der Grub).
6. Directly labeled antibodies, e.g., PE Mouse-Anti-Human D324 E-cadherin (BD Pharmingen™) or Anti-Human Ki-67 APC antibody (eBiosciences).
7. 0.1% FCS in PBS.
8. Centrifuge.
9. Round-bottom tubes for flow cytometry.

3 Methods

3.1 3D Cell Culture

Carry out all procedures at room temperature (RT) under a laminar flow unless otherwise specified.

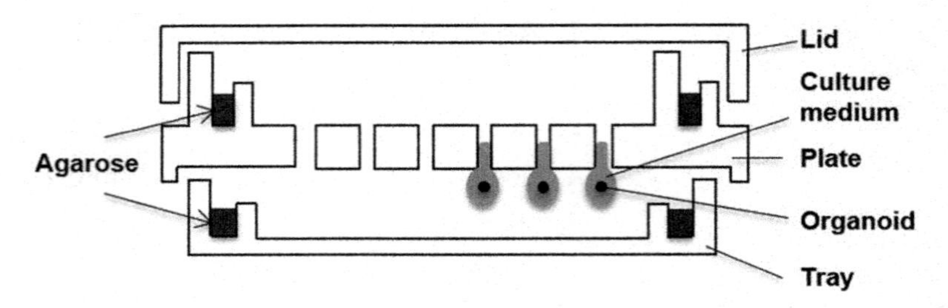

Fig. 1 Each 96-well hanging drop plate consists of a lid, a plate, and a tray. To combat evaporation of cell culture medium during long term culture, the rims of the plate and the tray are filled with 2% agarose

3.1.1 Preparation of the Hanging Drop Plate

1. To combat evaporation of cell culture medium during long term culture, the rims of the plate and the tray are filled with 2% agarose. To do this, remove the lid of the plate and place it aside, then put the plate and the tray apart from each other.
2. Heat 2% agarose in a microwave until the solution becomes clear.
3. Pipette 2 mL of lukewarm 2% agarose solution into the rim of the plate and another 1 mL into the rim of the tray by using a 1000 μL pipette tip (Fig. 1).
4. Let the agarose harden and then reassemble the plate (*see* **Note 3**).

3.1.2 Hanging Drop 3D Cell Culture

1. Prepare a single cell suspension by trypsinizing cells (either from a 2D monolayer or even from a 3D organoid culture).
2. Resuspend the cells in the appropriate culture medium and determine the cell number, e.g., via a Neubauer chamber. To establish PCa organoids you may use up to 7500 cells per drop (40 μL) depending on the growth characteristics of the cell line (*see* **Note 4**).
3. Carefully seed the cells into the 96-well plate (Fig. 1). You may use an electronic multipette (*see* **Note 5**).
4. Incubate the 96-well plate at 37 °C in a humidified atmosphere with 5% CO_2 for 4 days in order to allow the cells to form organoids. Avoid any strong movement of the plates during transport and incubation to prevent drops falling down.
5. Replenish medium at day 4 of culture and then every 48 h. To this end, remove 30 μL of the medium and add 30 μL of fresh cell culture medium (*see* **Note 5**).
6. To determine the size of the organoids, carefully place the plate under a microscope with a 4× objective. Take images and estimate organoid size using an image analysis software (*see* **Note 6**).

3.2 Cell Viability Assay

1. Put the 96-well hanging drop plate onto a standard 96-well plate and transfer the organoids into the plate by centrifugation at 290 × *g* for 5 min.
2. Add 50 μL of trypsin/EDTA per well and incubate the plate at 37 °C for 10 min.
3. Add 50 μL of trypsin inhibitor per well to stop the reaction.
4. Pipette up and down several times to get a single cell suspension.
5. Add 14 μL of WST-1 reagent per well (1:10) and incubate for 1 h at 37 °C.
6. Measure the absorbance at 450 nm (reference reading at 630 nm) using a spectrophotometer.

3.3 Processing of 3D Organoids for Immunohistochemistry

1. Carefully remove 30 μL of each drop and add 30 μL of PBS for washing.
2. Remove 25 μL of PBS and add 25 μL of CytoRich Red to fix organoids. Incubate for 45 min at RT.
3. Pool at least six fixed organoids by collecting them with 60 μL of PBS into a 15 mL centrifuge tube.
4. Centrifuge at 500 × *g* for 5 min.
5. Heat 2% agarose in a microwave until the solution becomes clear.
6. Carefully remove supernatant and add 200 μL of liquid 2% agarose.
7. Centrifuge again at 500 × *g* for 5 min to collect the organoids at the bottom of the tube (Fig. 3) (*see* **Note 7**).
8. Transfer the agarose block into a 50 mL centrifuge tube containing 10 mL of 4% PFA (*see* **Note 8**). Incubate for 30 min at RT.
9. Manually dehydrate in graded alcohols: First incubate the agarose block for 30 min in 70% ethanol, then put the block into fresh 70% ethanol and incubate at RT overnight.
10. On the following day, dehydrate the agarose block further by incubation in 96% ethanol twice for 30 min, then in 100% ethanol twice for 30 min, and in Xylol three times for 30 min (*see* **Note 9**).
11. Embed the dehydrated block in paraffin wax.
12. Cut two to 3 μm thick sections with a rotary microtome and carefully put them into a water bath.
13. Mount two to three serial sections on one SuperFrost Plus slide.
14. Dry the slides at RT and then put them on a heating plate at 70 °C for at least 20 min so that the sections firmly adhere to the slides.

15. Carry out hematoxylin–eosin staining to check organoid orientation (*see* **Note 10**).
16. Perform immunohistochemical staining with antibodies as desired.

3.4 Staining for Flow Cytometry

1. Pool eight to twelve organoids in a 1.5 mL Eppendorf tube by floating the wells with 60 μL PBS.
2. Centrifuge at 150 × *g* for 5 min.
3. Carefully remove the supernatant and wash the spheroids with 100 μL of PBS.
4. Again centrifuge at 150 × *g* for 5 min and discard supernatant.
5. Incubate the organoids with 100 μL of trypsin–EDTA for 10 min at 37 °C.
6. Add 100 μL of trypsin inhibitor and pipette the cell suspension up and down several times.
7. Centrifuge at 150 × *g* for 5 min and add 100 μL of Reagent A from Fix and Perm® Cell Fixation and Permeabilization Kit.
8. Incubate for 15 min at RT.
9. Add 500 μL of PBS for washing and centrifuge for 5 min at 300 × *g*.
10. Carefully remove the supernatant and add respective antibody diluted in 100 μL of Reagent B from Fix and Perm® Cell Fixation and Permeabilization Kit.
11. Shortly vortex (1–2 s) at low speed.
12. Incubate for 15 min at RT in darkness.
13. Wash the cells using 500 μL of PBS and centrifuge for 5 min at 300 × *g*.
14. Remove the supernatant and resuspend the cells in 100 μL of 0.1% FCS in PBS.
15. Transfer the cell suspension into a FACS tube and until measurement store the samples on ice in darkness.

4 Notes

1. Around 4 mL of 2% agarose solution is sufficient for one 96-well Hanging Drop Plate. The use of 2% agarose is recommended over distilled water or PBS to avoid the liquid to spill over into the hanging drops during movement of the plates.
2. We used the WST-1 assay within this study to determine cell viability because it is relatively cheap and the protocol is easy to handle. In particular, there are no washing steps where you

might lose relevant amounts of cells. There are of course a variety of similar methods to determine cell viability, which are also based on the conversion of tetrazolium salts to soluble formazan by cellular enzymes (e.g., MTT assay). These other assays may also be used but were not tested in this study.

3. This step can be carried out a day prior to seeding the cells. You can store the plate at 4 °C. Ensure that sterility of the plate is guaranteed.
4. This cell number has been used to establish 3D organoids of different PCa epithelial cell lines (LNCaP, DuCaP, LAPC-4, PC-3) that were maintained over 8 days in culture. Note that due to the different growth characteristics of each cell line, organoids largely differ in form and size. Therefore, seeding cell number has to be optimized in order to avoid growth limitation of the organoid over time due to limited space within the drop. To establish coculture organoids, we used a ratio of epithelial cells to fibroblasts of 1:1 (3800 cells per cell type in a final volume of 40 μL). To differentiate epithelial cells from fibroblasts, it is recommended to label at least one cell type. In our study, we used immortalized normal prostate fibroblasts and cancer-associated fibroblasts, which were stably transfected with green fluorescent protein.
5. For cell seeding as well as for adding fresh medium during medium exchange you may use an electronic multipette. Ensure that you use the setting "slow pipetting down". During medium exchange, it is crucial to remove the old medium very slowly to avoid sucking in the organoid. In general, it is recommended to pipet the medium manually when starting with this technique to get a better feeling for the whole system.
6. You can estimate the size of the organoids with ImageJ software by calculating the mean radius (R) using the formula $R = (1/2)*(ab)1/2$ [8], where a and b are the two orthogonal diameters of the organoid (Fig. 2). Consider that this formula can only be used to estimate the approximate organoid size and not for an accurate calculation of the size when the organoid is not round but for instance stellate.
7. It is important to centrifuge the samples, while the agarose is still liquid so that the organoids can be accumulated at the bottom of the centrifuge tube.
8. The organoids are usually visible within the agarose block. You may reduce the size of the agarose block by carefully removing excessive agarose with a scalpel (Fig. 3).
9. All solutions should be freshly prepared.
10. When cutting the paraffin block it is recommended to perform hematoxylin–eosin staining of every tenth slide to verify presence of organoids, which are not visible to the naked eye.

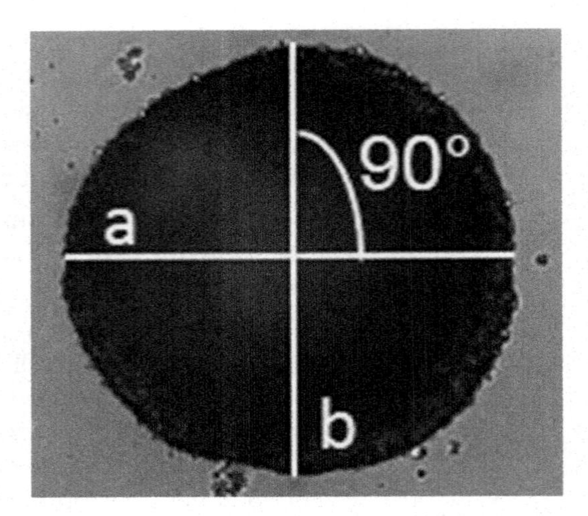

Fig. 2 The organoid size can be estimated by calculating the mean radius (R) with the formula $R = (1/2)*(ab)1/2$ where a and b are the two orthogonal diameters

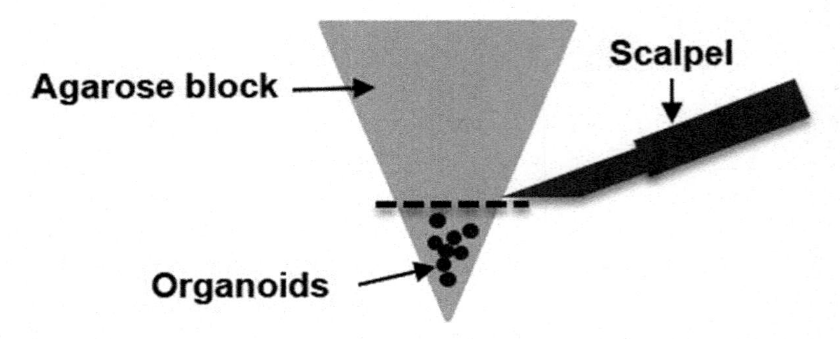

Fig. 3 Concentrate the organoids at the bottom of the centrifuge tube by centrifugation in agarose solution. Afterwards, the size of the agarose block containing the organoids can be diminished with a scalpel

Acknowledgments

This work was supported by a grant from the Medizinischer Forschungsfonds (MFF), Innsbruck, Austria, and a research grant from Astellas Pharma Ges.m.b.H. We thank Irma Sottsas and Georg Grünbacher for technical assistance and Dr. Christian Ploner and Dr. Gabriele Gamerith for helpful hints regarding 3D cell culture. We also thank Prof. Helmut Klocker for reading the manuscript.

References

1. Siegel R, Naishadham D, Jemal A (2012) Cancer statistics, 2012. CA Cancer J Clin 62(1):10–29. doi:10.3322/caac.20138
2. Cohen MB, Rokhlin OW (2009) Mechanisms of prostate cancer cell survival after inhibition of AR expression. J Cell Biochem 106(3):363–371. doi:10.1002/jcb.22022
3. Culig Z, Santer FR (2014) Androgen receptor signaling in prostate cancer. Cancer Metastasis Rev 33(2–3):413–427. doi:10.1007/s10555-013-9474-0
4. Sampson N, Neuwirt H, Puhr M et al (2013) In vitro model systems to study androgen receptor signaling in prostate cancer. Endocr Relat Cancer 20(2):R49–R64. doi:10.1530/ERC-12-0401
5. Thoma CR, Zimmermann M, Agarkova I et al (2014) 3D cell culture systems modeling tumor growth determinants in cancer target discovery. Adv Drug Deliv Rev 69–70:29–41. doi:10.1016/j.addr.2014.03.001
6. Ellem SJ, De-Juan-Pardo EM, Risbridger GP (2014) In vitro modeling of the prostate cancer microenvironment. Adv Drug Deliv Rev 79–80:214–221. doi:10.1016/j.addr.2014.04.008
7. Padmalayam I, Suto MJ (2012) Chapter twenty-four-3D cell cultures: mimicking in vivo tissues for improved predictability in drug discovery. In: MANOJ C.D. (ed.) Annual reports in Medicinal Chemistry. Wallingford, CT, USA, Academic Press
8. Zhou Y, Arai T, Horiguchi Y et al (2013) Multiparameter analyses of three-dimensionally cultured tumor spheroids based on respiratory activity and comprehensive gene expression profiles. Anal Biochem 439(2):187–193. doi:10.1016/j.ab.2013.04.020

Chapter 13

3D-Dynamic Culture Models of Multiple Myeloma

Marina Ferrarini*, Nathalie Steimberg*, Jennifer Boniotti, Angiola Berenzi, Daniela Belloni, Giovanna Mazzoleni, and Elisabetta Ferrero

Abstract

3D-dynamic culture models represent an invaluable tool for a better comprehension of tumor biology and drug response, as they accurately re-create/preserve the complex multicellular organization and the dynamic interactions of the parental microenvironment, which can affect tumor fate and drug sensitivity. Hence, development of models that recapitulate tumor within its embedding microenvironment is an imperative need. This is particularly true for multiple myeloma (MM), which survives almost exclusively in the bone marrow (BM). To meet this need, we have previously exploited and validated an innovative 3D-dynamic culture technology, based on the use of the Rotary Cell Culture System (RCCS™) bioreactor. Here, we describe, step by step, the procedures we have employed to establish two human MM ex vivo models, i.e., the culture of human BM-derived isolated cells and of MM tissues from patients.

Key words Tissue models, 3D-dynamic culture, Bioreactor, Microenvironment, Multiple myeloma

1 Introduction

Multiple myeloma (MM) is a plasma cell tumor, which accounts for 1% of all human cancers and more than 10% of all hematologic malignancies [1]. The disease develops almost exclusively within the bone marrow (BM), where MM cells home and establish dynamic interactions with their coevolving microenvironment, which delivers pro-survival signals and confers chemoresistance to neoplastic cells [2–4]. As a result, despite recent advances in treatment, and particularly the introduction of the proteasome inhibitor Bortezomib/Velcade [1], MM remains ultimately fatal.

The notion that tumor microenvironment plays a key role in tumor initiation, progression and response to therapy not only in solid tumors [5, 6] but also in hematological malignancies [7], is now widely accepted. Historically, the vast majority of our

*These authors contributed equally to the work.

Zuzana Koledova (ed.), *3D Cell Culture: Methods and Protocols*, Methods in Molecular Biology, vol. 1612,
DOI 10.1007/978-1-4939-7021-6_13,

knowledge on tumor pathogenesis and drug vulnerability has been accrued through two-dimensional (2D), often homotypic, cell culture models, which are not representative of the three-dimensional (3D) and heterotypic organization of the native tissue-specific microenvironment [8, 9]. Accordingly, 3D approaches, which better reproduce in vivo-like responses, have become a focus of intense investigation [8–13]. This is particularly true for MM, as it represents a natural paradigm of the tumor–stroma dialog (Fig. 1), which cannot be fully recapitulated by available animal models [14].

A major breakthrough in the 3D modeling of MM was achieved by Kirshner et al. [15], who proposed a new 3D culture system, enabling the clonal expansion of MM cells, as well as testing of antitumor drugs. More recently, Ghobrial and coworkers

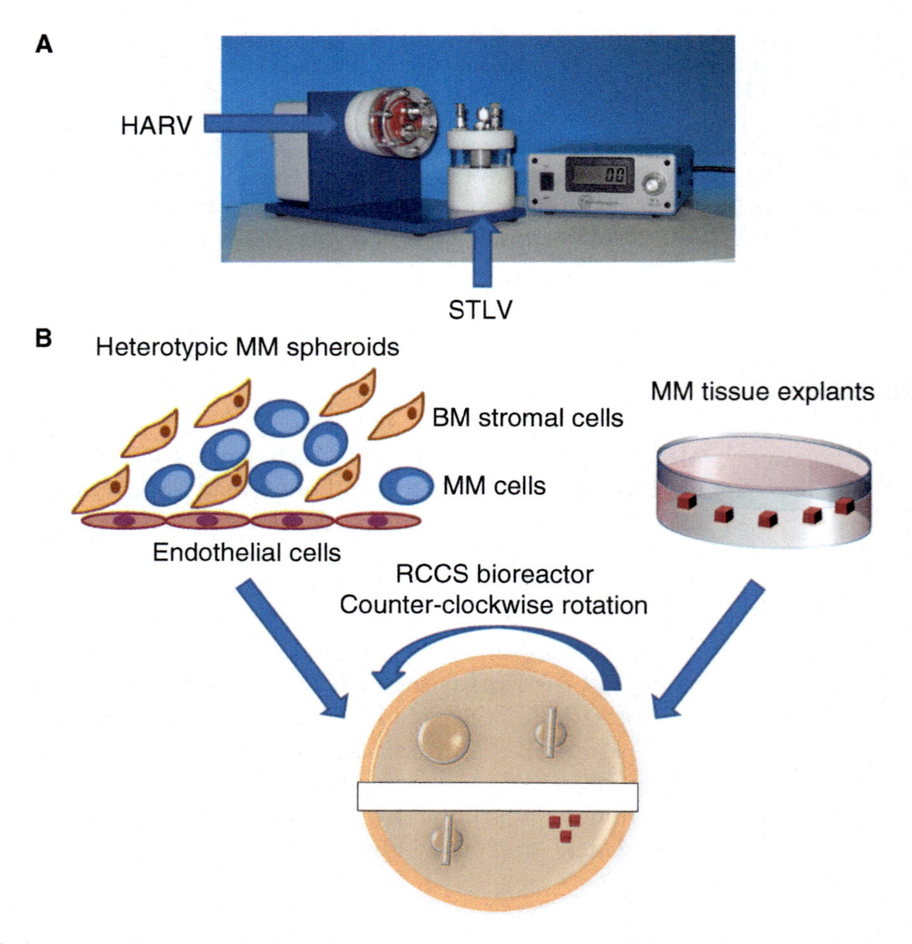

Fig. 1 3D-dynamic culture models of Multiple Myeloma. (**a**) Picture representing the RCCS™ bioreactor with High Aspect Ratio Vessel (HARV) and Slow Turning Lateral Vessel (STLV). (**b**) Two models are depicted to study human MM in a 3D context: heterotypic MM cell spheroids (*left*) and MM tissue samples (*right*) cultured in the dynamic 3D culture vessel of RCCS™ bioreactor (*below*)

developed a MM model of repopulated silk scaffolds, suitable for sustaining osteogenic differentiation and to better mimic, in such a way, the bone compartment of the in vivo MM microenvironment [16].

3D tissue models, based on the use of isolated cells, lacking original heterogeneous cell–cell and cell–matrix interactions, do not fully reproduce the complex functional architecture and the specific signaling pathways of the native tissue [17]. Culture of tissue explants would be the obvious answer. However, when kept in the traditional static condition, the basal–apical orientation of the samples and the poor mass transport (gas/nutritional support and waste removal) lead to the generation of necrotic cores, preventing its application to mid-/long-term morphofunctional analyses and drug testing.

To circumvent these limitations, we applied the innovative technology provided by the dynamic Rotary Cell Culture System (RCCS™) bioreactor to establish 3D ex-vivo models of human MM specimens. The RCCS™ bioreactor, by generating a constant dynamic fluid flow (resulting from the vessel horizontal rotation), guarantees an optimal balance between increased mass transfer and deleterious effects of shear stress, thus creating the best conditions for long-term in vitro maintenance of cell viability and function in 3D cell/tissue culture systems. This approach that we have previously validated for the culture of multicellular engineered tissue-like constructs and tissue explants [18–20] was also successfully applied to the culture of human MM tissues, which retained, for extended time periods, the morphological and functional features of the original MM tissue components, as well as their specific sensitivity to drugs [20].

In this chapter, we present two ex vivo culture models that, based on the use of the RCCS™ bioreactor (Fig. 1b), allow the study of human Multiple Myeloma biology and response to drugs (Fig. 2). The first one, representing the simplest, is based on the 3D coculture of MM cells and BM stromal cells. Even if this model presents its own limits, mainly due to the over-simplification of the BM microenvironment, it allows to mimic the specific interactions of individual cell types (autocrine and paracrine communications). This first model is based on the use of primary cells, exemplified here by the use of MM-derived cells from patients, and of the HS5 human BM stromal cell line, these two cell types representing the main actors in the tumor–stroma dialog. However, coculture can be performed with other cell types, in order to better approximate the native microenvironment and to investigate their specific, individual role [13].

The second model, which is specifically focused on the maintenance of the native MM-microenvironment, more precisely retains the characteristics of the in vivo context. The use of patient-derived tissue explants allows to better mimic hetero/homotypic cell–cell and cell–extracellular matrix (ECM) interactions, preserving also

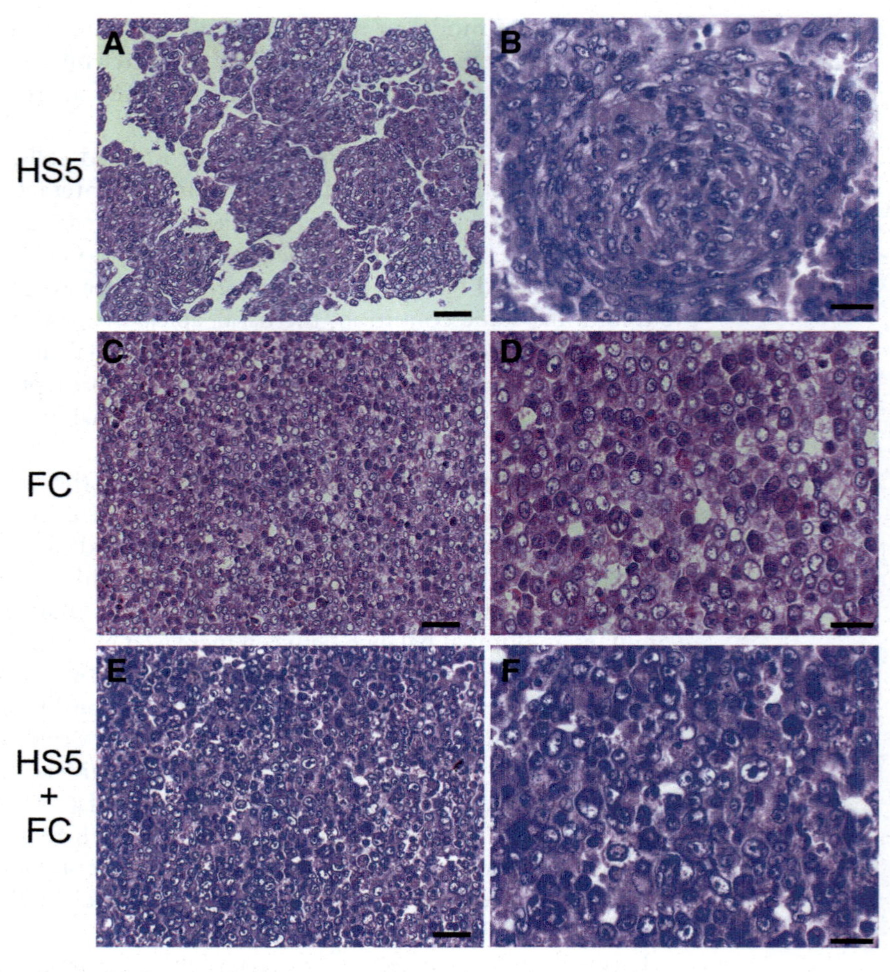

Fig. 2 3D culture of isolated cells (cytospheres). Cells, either alone (**a–d**), or in combination with other cell types (**e**, **f**), were seeded and maintained for 1 week in the dynamic 3D culture system generated by the RCCS™ bioreactor, and then processed for cryosections and stained with hematoxylin–eosin. (**a**, **b**) BM-derived HS5 cell line; (**c**, **d**) MM patient-derived FC cells. (**e**, **f**) BM-derived HS5 cell line and MM patient-derived FC cells in combination. Scale bar, 20 μm

the interindividual variability observed in patients, a fundamental parameter that needs to be taken into account for the development of patient-specific targeted therapies [20].

2 Materials

2.1 Reagents

1. Complete culture medium: 10% FBS, 100 U/mL of penicillin, 100 μg/mL of streptomycin, and 0.25 μg/mL of Fungizone® in RPMI 1640 (*see* **Note 1**). Keep sterile and store at 4 °C.
2. Hank's balanced saline solution (HBSS).

3. Trypsin (0.05%)–EDTA solution.
4. 0.4% trypan blue.
5. OCT embedding medium.
6. Liquid nitrogen.
7. Isopentane.
8. Neutral buffered 10% formaldehyde solution: In a glass bottle, mix 100 mL of formaldehyde with 900 mL of distilled water, then add 4 g of monobasic sodium phosphate and 6.5 g of dibasic sodium phosphate, and mix. Work under chemical hood. Store the solution at 4 °C (*see* **Note 2**).

2.2 Equipment

1. Sterile class II biosafety cabinet.
2. Incubator with humidified atmosphere, at 37 °C and 5% CO_2 (95% air).
3. Direct and inverted optical microscopes.
4. Refrigerated centrifuge.
5. Laboratory water bath with controlled temperature.
6. Autoclave.
7. Oven.
8. 10/20 mL sterile syringes.
9. Disposable plasticware for cell culture: 100-mm, 35-mm diameter Petri dishes, 75 cm^2 culture flasks, disposable pipettes (at least 5 mL and 10 mL), 24 well plates.
10. Hemocytometer.
11. Sterile scalpels.
12. 15 mL/50 mL sterile centrifuge tubes.
13. Pasteur pipettes.
14. Metal beaker or steel ladle.
15. Stainless steel molds.
16. Embedding rings.
17. Forceps.
18. Sterile spatulas (*see* **Note 3**).
19. Rotary Cell Culture System (RCCS™) (*see* **Note 4** and Fig. 1a) and related autoclavable 10-mL High Aspect Ratio Vessel (HARV) or 55-mL Slow Turning Lateral Vessel (STLV; *see* **Note 5**) (Synthecon Inc.).

2.3 Biological Samples

1. MM tissue biopsies from patients (*see* **Notes 6** and **7**).
2. HS5 cell line, human HPV-16 E6/E7 transformed BMSCs.
3. FC cells, a lymphoblastoid cell line, spontaneously outgrowing from isolated BM cells derived from a MM patient (*see* **Notes 6** and **7**).

3 Methods

3.1 Heterotypic MM Spheroids

The heterotypic culture of myeloma cells (spheroids) refers to the coculture of MM cells with other BM cellular components, in order to mimic the cell–cell interactions that, physiologically, take place within the MM microenvironment. In the example presented here, the choice was to use a patient-derived FC lymphoblastoid cell line (FC) and the stromal cells (HS5 cell line). Alternatively, commercially available MM cell lines can also be used.

Before starting the 3D cell coculture, it is useful to perform a cell growth study of the MM cells, in order to determine their growth properties. In such a way, the best seeding densities can be established for optimizing the cell culture conditions for the whole culture period.

All cell culture procedures must be performed under sterile conditions using sterile cell culture materials (*see* **Note 8**).

3.1.1 Cell Growth Study

1. To perform a growth study of your selected cell lines to optimize cell culture conditions, seed cells at different densities (including the mean density indicated by the cell line datasheet) in 1 mL of appropriate cell culture medium per well of a 24-well plate in triplicates for each time-point.
2. For cells growing in suspension, follow **steps 3–5**. For adherent cells, follow **steps 6–12**.
3. Every day collect a 100 μL aliquot of the cell suspension and put it in an eppendorf tube (*see* **Note 9**).
4. Add 100 μL of trypan blue (i.e., volume ratio 1:1) to assess cell viability and count viable (non-blue) cells by using a hemocytometer.
5. Calculate mean values for each cell density (from the triplicates) at each time point and plot the values against time to obtain a growth titration curve. Select the appropriate cell seeding density on the basis of the obtained results.
6. For adherent cells, e.g., mesenchymal cells, every day aspirate the culture medium from the wells and wash with HBSS (*see* **Note 9**).
7. Replace HBSS with trypsin–EDTA and incubate at 37 °C for 5 min.
8. When cells are detached, as determined by microscope observation, add complete culture medium to stop enzymatic activity and transfer the cell suspension into a centrifuge tube.
9. Centrifuge at 900 × *g* for 5 min at 4 °C.
10. Resuspend cell pellets in 1 mL of complete culture medium each.
11. Take a 100 μL aliquot of the cell suspension and proceed as described in **steps 4–5**.

3.1.2 3D Culture of Heterotypic MM Spheroids

1. Prepare cell supensions at the required densities:
   (a) For monotypic cultures of FC cells in 3D culture, seed cells at 7.0–7.5 × 10^5 cells/mL or at 1.5–1.8 × 10^5 cells/mL in complete culture medium (48-h and 1-week culture studies, respectively). For other MM cell lines, seed cells at a concentration determined upon the titration curve performed in Subheading 3.1.1.
   (b) For monotypic cultures of mesenchymal HS5 cells, seed cells at 7.0–7.5 × 10^5 cells/mL in complete culture medium.
   (c) For heterotypic cultures, seed FC cells and HS5 cells at 3 × 10^5/mL and 6 × 10^4 cells/mL, respectively, in complete culture medium, for 48-h culture assays. Reduce cell density at 1.5 × 10^5 cells/mL and at 3 × 10^4 cells/mL in complete culture medium, respectively, for the 1-week culture assays.
2. When cell suspensions are ready, introduce them into the culture vessel by using sterile syringes. Completely fill the culture vessel with culture medium (*see* **Note 10**).
3. Fix the HARV or the STLV chamber rotation speed at 7–8 rpm (*see* **Note 11**).
4. Refresh culture medium twice a week. Depending on the cell type/s used (and on its/their biological activity/proliferation rate), this procedure may be required more frequently (*see* **Note 12**).
5. At the end of the culture period, gently harvest cells/cytospheres, using sterile syringes (*see* **Note 13**).
6. Softly put the content of the syringe in a centrifuge tube, working under sterile cabinet bench.
7. Open the culture vessel with the appropriate hex key and with a sterile spatula collect the eventual remaining aggregates and, with a micropipette, the remaining culture medium (*see* **Note 14**).
8. Process samples for molecular investigations or for in situ analyses (Fig. 3). For performing histochemical/immunohistochemical analyses of monotypic and/or heterotypic cytospheres, we adapted a specific and established technique that allows to include and to process them as solid tissues (Fig. 4) [18].

3.2 3D Culture of MM Tissue Explants

The protocol was established on human-derived MM tissue explants, collected from MM patients undergoing bioptic procedures for diagnostic purposes [20].

1. After surgery: collect tissue biopsies into sterile 50 mL centrifuge tubes, containing at least 20 mL of cell culture medium, supplemented with antibiotic/antimycotic solution, and

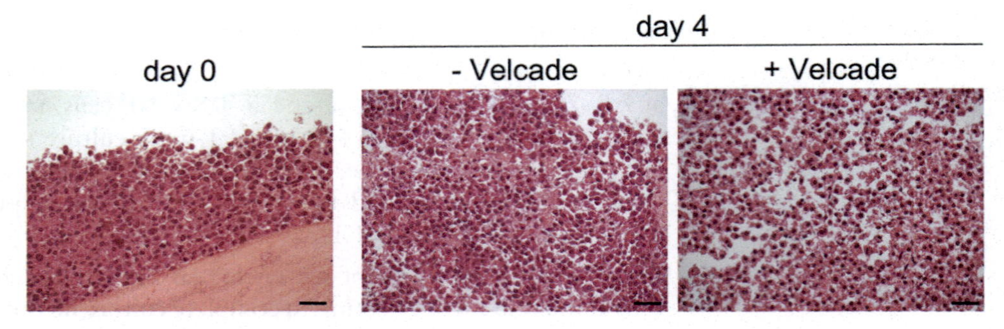

Fig. 3 Culture of MM-derived tissue explants in dynamic 3D culture conditions. MM-derived tissue explants were isolated from patients, cultured in the RCCS™ bioreactor chambers, and retrieved at day 4. Samples were then analyzed upon hematoxylin–eosin staining. Culture conditions favor high cell viability, and the preservation of the original microenvironment, including bone lamellae. The impact of drugs on MM tissues can be evaluated by comparison with parallel cultures of untreated vs. drug-treated (50 nM Velcade) samples. Scale bar, 20 μm

Fig. 4 Embedding in OCT for frozen sections. Step-by-step representation of tissue inclusion in OCT. Tissue explants are inserted into the mold and covered with OCT medium (**a**–**c**). The mold containing the samples is transferred into liquid-nitrogen chilled isopentane, and then quickly frozen (**d**, **e**). Finally, the mold is equipped with an embedding ring, filled with OCT and further frozen (**f**–**i**)

maintain the samples at 4 °C up to their arrival to the cell culture laboratory (*see* **Note 15**).

2. Place samples into a 100 mm-diameter petri dish, and wash them several times with complete culture medium, in order to eliminate erythrocytes, blood clots and cell debris.
3. Cut the sample into small fragments (volume of about 1 mm^3) by using sterile scalpels (*see* **Note 16**).
4. Open the autoclavable vessel with the hex key. Deposit MM explants on the bottom of the vessel with sterile forceps or with a spatula. Close the culture vessel with the hex key.
5. Slowly add the complete culture medium, in order to fill the whole internal volume of the vessel. It is fundamental to avoid the formation of air bubbles inside the culture chamber.
6. Fix the vessel on the rotor base and adjust the optimal rotation speed in order to avoid the sedimentation of the samples (usually, it is around the maximal rotational speed supported by the motor).
7. Refresh/change the culture medium at least twice a week.

3.3 Long-Term 3D Culture Studies of MM Tissue Explants

The RCCS™ bioreactor is very versatile and when optimal basic culture conditions are reached, it is easily applicable to a high number of investigations, including physiopathology and pharmacotoxicology studies. We validated this 3D culture model for long-term culture of human-derived MM tissue explants for predicting the impact of Velcade treatment in individual patients [20]. An example of the results obtained is illustrated in Fig. 3. MM samples can be kept in culture, maintaining their intact architecture, cellularity and cell viability; moreover, the response to Velcade treatment can be appreciated morphologically, as indicated by the diffuse cell loss and necrosis.

Taking advantage from the unique culture conditions generated by the RCCS™ bioreactor that allow a long-term tissue culture (up to 2 weeks), MM tissue explants can be harvested at different time periods, and submitted to a number of morphological, biochemical, and/or molecular analyses. For example, MM samples can be embedded in OCT embedding medium, in order to make them suitable for immunohistochemical analyses (*see* **Note 17**).

3.4 Embedding MM Explants for Frozen Sectioning

1. Deposit tissue explants in a steel mold, and eliminate as much as possible any trace of liquids with a micropipette or with a Pasteur pipette (Fig. 4a).
2. Add OCT medium up to the limit of the inferior part of the mold (Fig. 4b). Avoid bubble formation.
3. Orient and center the tissue sample in the mold (Fig. 4c).

4. Add isopentane into a metal beaker (steel ladle works also very well).
5. Dip the metal beaker into the liquid nitrogen and wait for isopentane chilling. From a milky color, isopentane becomes more transparent, and a ring of frozen opaque isopentane can be observed.
6. Dip the mold containing the samples into the isopentane (Fig. 4d, e). You can improve the orientation of the tissue sample, but it is important to quickly freeze the sample in order to avoid artifacts.
7. Place the properly labeled embedding ring on the mold, and quickly add an excess of OCT medium (fill the embedding ring almost completely). Wait 1 min for the OCT to freeze well (Fig. 4f–i).
8. Store the frozen tissue at −80 °C until use.

4 Notes

1. RPMI 1640 medium can be purchased either as a liquid or a powder product. In the latter case, dissolve the powder in purified ultrapure water, according to the manufacturer instructions. Then sterilize the culture medium by vacuum filtration on 0.22 μm polyethersulfone membrane. In order to ascertain the sterility of the final product, we recommend to incubate an aliqout of the reconstituted medium in a dry oven at 37 °C for 24 h and check for potential contamination afterwards.
2. Alternatively, ready to use neutral buffered 10% formaldehyde solution can be used. Formaldehyde is toxic and therefore must be used carefully, while using adequate collective and personal protection equipment.
3. Sterilize scalpels, spatula, forceps, and other surgical instruments by treating them in hot air oven at 160 °C for at least 180 min.
4. The Rotary Cell Culture System (RCCS™) is a bioreactor composed of: (1) a rotator base, carrying the culture vessel, screwed on a central axis directly coupled to the RCCS™ engine, and (2) a power supply, equipped with a tachometer, which allows the regulation of the rotation speed of the culture chamber. According to the rotation rate of the vessel, in its internal part it is possible to reach a sort of "microgravity state." This condition, known as "modeled microgravity," is attained when, at each moment of the culture, the rotation speed of the vessels counterbalances the gravitational force, so that the biological samples inside the vessels reach a sort of a "free-fall" condition. The absence of a liquid/air interface

inside the culture chamber, obtained by entirely filling the vessels with culture media, prevents the generation of turbulences, which can damage fragile cell/tissue samples. Moreover, the rotation speed of the culture vessel can be tuned, in order to counteract sample sedimentation, thus avoiding any deleterious contact between them and the rigid, artificial vessel walls. Under these conditions, samples are maintained in a 3D quasi-laminar, fluid-dynamic environment, where a high mass transfer (that ensures adequate nutrients/gas supply and waste removal) and low shear-stress are generated, thus warranting optimal conditions for their long-term survival, without the development of central core necrosis. Finally, culture vessels are equipped with two valves, that allow not only to refill the culture chamber with fresh culture media and to administrate drugs (or other chemicals), but also to periodically collect samples and culture media for performing kinetic studies (by harvesting medium/cells/tissue fragments on which all kind of desired investigation can be performed). Bioreactor can be kept inside a humidified incubator with internal condition of 37 °C and 5% CO_2. Depending on the experimental plan, the culture can be carried out either in normoxia or in hypoxia. Indeed, controlled distribution of gases is warranted by the presence of a membrane in the culture vessels.

5. To sterilize HARV and STLV culture chambers, autoclave them at 105 °C for 35 min.
6. MM tissue biopsies from patients are taken for diagnostic purposes, after obtaining their written informed consent, and are kept at 4 °C, in sterile culture medium, until used. MM biopsies must be processed for cell culture as soon as possible (within a few hours), in order to better preserve cell viability and functions.
7. Isolated MM cells can be obtained from BM samples upon density gradient centrifugation. BM samples are diluted in RPMI (ratio 1:3; 1:4), layered on Ficoll (density = 1.077) and centrifuged at 800 × g for 20 min, 20 °C, without brake. The resulting ring is carefully aspirated and washed twice in RPMI. Cell are then counted and seeded in six well-tissue culture plates in the presence of RPMI with 5% FBS. CD138+ plasma cells can be purified by means of different techniques (cell sorting or magnetic beads conjugated with anti-CD138 antibodies).
8. Cell culture requires a sterile environment. Use a sterile class II biosafety cabinet. Always verify the risk class of the cell lines you are manipulating and the relative level of biosafety necessary to comply with local guidelines and policies. Before beginning to work, sanitize the whole working area of the laminar

flow hood by spraying it with 70% ethanol and wiping dry with tissue; disinfect your gloves with 70% ethanol. Allow to air-dry. Keep only the necessary materials (pipettes, tips, pipette-aid, bottles with medium and other reagents, temporary waste, etc.) in the working area, in order to avoid irregular air-flow, and relative reduction of aseptic conditions. Wear adequate personal protective equipment (gloves, laboratory coats) and work in accordance with a good laboratory practice and with your institution's guidelines. Follow the standard procedures for cell culture in sterile conditions. For the sterilization of culture media, filter the solutions onto 0.22 μm syringe filters or membranes (disposable/reusable filtration systems). At the end of work, decontaminate all potentially infected materials by addition of sodium hypochlorite to both liquid and solid waste. Disinfect the working area of the cell culture cabinet with 70% ethanol and wipe dry with tissue. Close the sterile cabinet. Switch on germicidal UV irradiation for at least 30 min (depending on your UV tubes properties). Dispose waste into biohazard containers.

9. It is also useful to take into account the change of the color of the medium containing the pH indicator phenol red; its acidification (the orange–yellow coloring) testifies an increased cell metabolism and proliferation rate.
10. Whatever the model used (cytospheres or explants culture), the culture vessel must be completely filled with culture medium, by using 10 or 20 mL syringes, in order to avoid the formation of an air–liquid interface and air-bubbles, whose presence may compromise cell viability.
11. The regulation of the culture vessel rotation speed is essential for avoiding contacts among the biological samples (cytospheres/tissue explants) and between the biological samples and the vessel walls. During the first days of culture, cells require a very slow rotation speed; subsequently, since cells progressively tend to aggregate and to form cytospheres, it is necessary to up regulate the rotation speed, in order to counterbalance the increasing mass/size of the biological samples and to prevent their sedimentation.
12. At least at the beginning of the experimental procedures, in order to optimize the 3D culture conditions, media containing phenol red should be preferred, since they permit the evaluation at a glance of the metabolic activity of the cultured cells. Thus, based on the metabolic and proliferative features of the cultured samples, it may be of importance to adjust the frequency of culture medium supplementation. Moreover, during medium refresh, preserve 2–3 mL of cell culture suspension and verify, under light microscopy, the absence of excessive cell debris.

13. Biological sample harvesting must be performed carefully (by the use of 10–20 mL syringes), in order to preserve their integrity and architecture. According to the volume of the culture, use 15 or 50 mL centrifuge tubes. Even if not strictly necessary, we recommend working in sterile conditions, in order to avoid any type of contamination.
14. Autoclavable culture vessels (HARV & STLV) are provided with hex keys necessary to open/close the culture vessels. Opening the autoclavable vessels after the 3D culture allows to harvest the residual culture medium containing residual cells/ cytospheres.
15. In our hands the maximal delay between bioptic procedure and ex vivo culture of MM biospies should be inferior to 12 h.
16. With regard to the culture of tissue explants, a critical point relies on the number and volume of the fragments. In particular, the size of the fragments should be adequate, in order to counterbalance the hydrodynamic forces generated into the culture chamber, to maintain the free-fall condition, and to prevent the sedimentation of the samples. The maximal number of samples that can be kept in culture into the vessels vary according to the volume of the vessel (10 mL, 55 mL). In our experimental settings, the volume of MM samples was about 1–3 mm^3, and a maximum of 15 samples in a 10-mL culture vessel was considered optimal. Since tissue biopsies exhibit characteristics varying between individual patients (and also depending on the site of origin), the experimental conditions described above should be optimized accordingly. When different experimental conditions need to be tested at a time, parallel cultures can be performed, by adding to each vessel equal numbers of samples of comparable weight/ volume.
17. Depending on the immunohistochemical analyses planned (identification of selected structures and markers), paraffin embedding may be preferred.

Acknowledgments

This work was partially supported by the Italian Association for Cancer Research (AIRC)—Special Program Molecular Clinical Oncology AIRC 5x1000 project No. 9965 (to Prof. Federico Caligaris-Cappio).

Marina Ferrarini and Nathalie Steimberg contributed equally to this work. Giovanna Mazzoleni and Elisabetta Ferrero also contributed equally to this work.

References

1. Mimura N, Hideshima T, Anderson KC (2015) Novel therapeutic strategies for multiple myeloma. Exp Hematol 43(8):732–741. doi:10.1016/j.exphem.2015.04.010
2. Hallek M, Bergsagel PL, Anderson KC (1998) Multiple myeloma: increasing evidence for a multistep transformation process. Blood 91(1): 3–21
3. Hussein MA, Juturi JV, Lieberman I (2002) Multiple myeloma: present and future. Curr Opin Oncol 14(1):31–35
4. Hideshima T, Mitsiades C, Tonon G et al (2007) Understanding multiple myeloma pathogenesis in the bone marrow to identify new therapeutic targets. Nat Rev Cancer 7(8):585–598
5. Tlsty TD, Coussens LM (2006) Tumor stroma and regulation of cancer development. Annu Rev Pathol 1:119–150
6. Hu M, Polyak K (2008) Microenvironmental regulation of cancer development. Curr Opin Genet Dev 18(1):27–34. doi:10.1016/j.gde.2007.12.006
7. Burger JA, Ghia P, Rosenwald A et al (2009) The microenvironment in mature B-cell malignancies: a target for new treatment strategies. Blood 114(16):3367–3375. doi:10.1182/blood-2009-06-225326
8. Yamada M, Cukierman E (2007) Modeling tissue morphogenesis and cancer in 3D. Cell 130(4):601–610
9. Lee GY, Kenny PA, Lee EH et al (2007) Three-dimensional culture models of normal and malignant breast epithelial cells. Nat Methods 4(4):359–365
10. Mazzoleni G, Di Lorenzo D, Steimberg N (2009) Modelling tissues in 3D: the next future of pharmaco-toxicology and food research? Genes Nutr 4(1):13–22. doi:10.1007/s12263-008-0107-0
11. Mazzoleni G, Steimberg N (2010) 3D culture in microgravity: a realistic alternative to experimental animal use. ALTEX 27:321–324
12. Fischbach C, Chen R, Matsumoto T et al (2007) Engineering tumors with 3D scaffolds. Nat Methods 4(10):855–860
13. Steimberg N, Mazzoleni G, Ciamporcero E et al (2014) In vitro modeling of tissue-specific 3D microenvironments and possible application to pediatric cancer research. J Pediatr Oncol 2:40–76. doi:10.14205/2309-3021.2014.02.01.5
14. Ferrarini M, Mazzoleni G, Steimberg N et al (2013) Innovative models to assess multiple myeloma biology and the impact of drugs. In: Roman H (ed) Multiple myeloma—a quick reflection on the fast progress. InTech, Rijeka. doi:10.5772/54312
15. Kirshner J, Thulien KJ, Martin LD et al (2008) A unique three-dimensional model for evaluating the impact of therapy on multiple myeloma. Blood 112(7):2935–2945. doi:10.1182/blood-2008-02-142430
16. Reagan MR, Mishima Y, Glavey SV et al (2014) Investigating osteogenic differentiation in multiple myeloma using a novel 3D bone marrow niche model. Blood 124:3250–3259. doi:10.1182/blood-2008-02-142430
17. Pampaloni F, Reynaud EG, Stelzer EH (2007) The third dimension bridges the gap between cell culture and live tissue. Nat Rev Mol Cell Biol 8(10):839–845
18. Berenzi A, Steimberg N, Boniotti J et al (2015) MRT letter: 3D culture of isolated cells: a fast and efficient method for optimizing their histochemical and immunocytochemical analyses. Microsc Res Tech 78(4):249–254. doi:10.1002/jemt.22470
19. Montani C, Steimberg N, Boniotti J et al (2014) Fibroblasts maintained in 3 dimensions show a better differentiation state and higher sensitivity to estrogens. Toxicol Appl Pharmacol 280(3):421–433. doi:10.1016/j.taap.2014.08.021
20. Ferrarini M, Steimberg N, Ponzoni M et al (2013) Ex-vivo dynamic 3-D culture of human tissues in the RCCS™ bioreactor allows the study of multiple myeloma biology and response to therapy. PLoS One 8(8):e71613. doi:10.1371/journal.pone.0071613

Chapter 14

Preparation of a Three-Dimensional Full Thickness Skin Equivalent

Christian Reuter, Heike Walles, and Florian Groeber

Abstract

In vitro test systems are a promising alternative to animal models. Due to the use of human cells in a three-dimensional arrangement that allows cell–cell or cell–matrix interactions these models may be more predictive for the human situation compared to animal models or two-dimensional cell culture systems. Especially for dermatological research, skin models such as epidermal or full-thickness skin equivalents (FTSE) are used for different applications. Although epidermal models provide highly standardized conditions for risk assessment, FTSE facilitate a cellular crosstalk between the dermal and epidermal layer and thus can be used as more complex models for the investigation of processes such as wound healing, skin development, or infectious diseases. In this chapter, we describe the generation and culture of an FTSE, based on a collagen type I matrix and provide troubleshooting tips for commonly encountered technical problems.

Key words 3D culture, Collagen, Full thickness, Skin equivalent, In vitro test system, Tissue engineering, Human skin model

1 Introduction

The skin is the largest organ of the human body and serves as a first barrier that protects the internal organs from the environment and is involved in the regulation of body temperature and fluid loss [1]. For this purpose, skin is composed of three layers: the epidermis, the dermis, and the subcutis. The epidermis is the relatively thin, tough, outer layer of the skin. It is a squamous epithelium that mainly consists of human epidermal keratinocytes (hEK). These cells originate from the deepest layer of the epidermis, called the basal layer and undergo a terminal differentiation, which finally results in the assembly of the stratum corneum. Keratinocytes can trigger inflammatory responses after stimulation [2] and play an important role in the initiation, modulation, and regulation of inflammation [3]. The dermis lies between the epidermis and the

Zuzana Koledova (ed.), *3D Cell Culture: Methods and Protocols*, Methods in Molecular Biology, vol. 1612,
DOI 10.1007/978-1-4939-7021-6_14, © Springer Science+Business Media LLC 2017

subcutis and is mainly composed of human dermal fibroblasts (hDF) embedded in a collagen matrix. These cells produce collagen, elastin, and structural proteoglycans. Collagen fibers make up 70% of the dermis, giving it strength and toughness. Blood vessels, nerves, glands, and hair roots are embedded in the dermis. The subcutis is a layer of fatty tissue that helps insulate the body from heat and cold, provides a good cushioning effect, and serves as an energy storage area.

Due to the exposed position, the skin is the first site of contact of potentially harmful microbial, thermal, mechanical, and chemical influences and the target of different pharmacological and cosmetic compounds. Thus, there is great interest to evaluate the risk and efficacy of chemicals that come into contact with the skin. Especially, since the European Union (EU) banned animal experimentation for the development of cosmetic products (European Union Council Directive 76/768/EEC), substantial effort has been made to find alternative methods to animal studies. Hence, in vitro generated organ-like skin substitutes have gained a growing interest. In the past 30 years different reconstructed skin equivalents have been developed [4–6]. These skin substitutes mimic the cellular and structural properties of native skin [7] and can thus serve as an important model for consumer product testing [8]. Moreover, these in vitro skin models are applicable in dermatological research to study wound healing [9], aging [10], infectious diseases [4, 11, 12], skin cancer [13], and skin biology [14]. Highly reproducible skin equivalents can be produced in-vitro in large quantities and can provide a better predictability due to the use of human cells. Hence, experimental misinterpretations caused by species-specific differences between an animal model and the human situation can be minimized. Moreover, skin equivalents implement the concept of replacement, reduction, and refinement of animal studies (3R's principle; [15]).

A majority of the used skin models mimic the epidermal layer only and are commonly employed for hazard identification, e.g., skin irritation and skin corrosion [16, 17]. However, in these models no epidermal–dermal cross talk is possible [18]. Hence, the applicability for some research questions is limited. Alternatively FTSEs can be used, in which the dermal part is generated using different biological or synthetic matrices [19]. Since collagen is the mayor component of the dermis, most FTSE are based on collagen hydrogels. In this study, we describe the fabrication of FTSE based on a collagen type I hydrogel (Fig. 1). The 3D model consists of a stratified epithelium harboring differentiated hEK that are grown on a collagen type I matrix populated with hDF (Fig. 2).

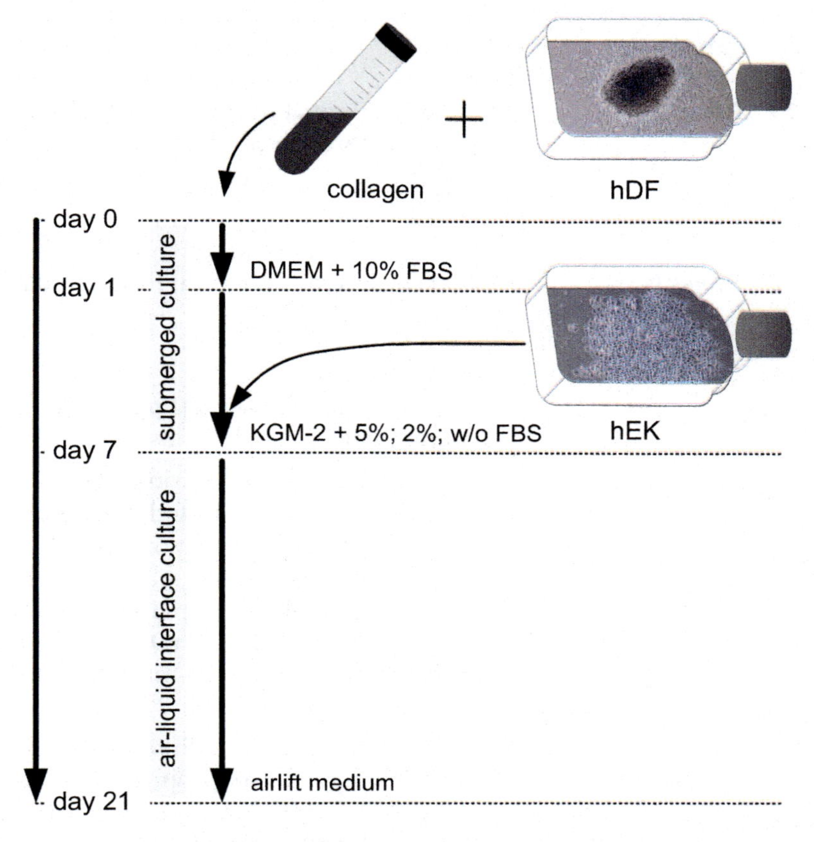

Fig. 1 Schematic drawing of the manufacturing process of a full-thickness skin equivalent. On day 1 the dermal component is generated by mixing human dermal fibroblasts (hDF) with a collagen solution. Following 1 day of culture in DMEM with 10% FBS, human epidermal keratinocytes (hEK) are seeded on top of the dermal component and the medium is switched to KGM-2 ready containing 5% FBS. During the next 6 days the FBS concentration is reduced first to 2% and then to KGM-2 ready without FBS. At day 7, air–liquid interface culture is initiated and the medium is switched to airlift medium

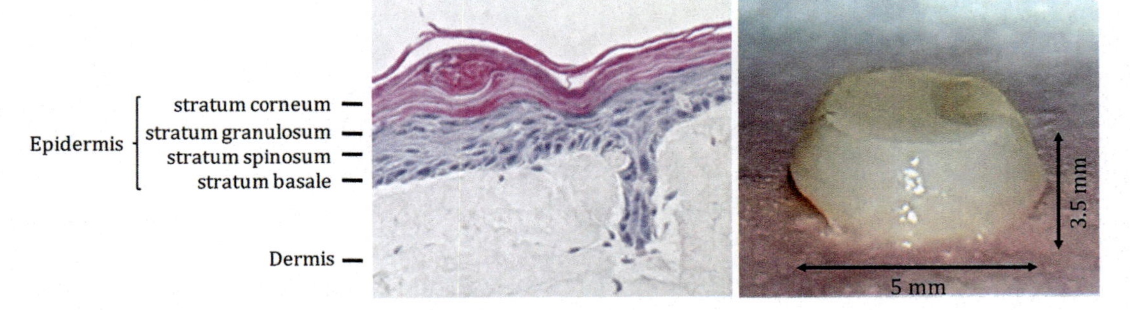

Fig. 2 HE staining (*left panel*) and macroscopic view (*right panel*) of a 3D full thickness skin equivalent cultured for 21 days. The epidermis consists of multiple layers of keratinocytes, which originate from the stratum basale and undergo a terminal differentiation, which finally results in the assembly of the stratum corneum. Due to the contraction during culture, the models shrink to approximately 5 mm and 3.5 mm in length and height, respectively

2 Materials

2.1 Cells

1. Human dermal fibroblasts (hDF) isolated from the dermis of juvenile foreskin or adult skin from different locations.
2. Human epidermal keratinocytes (hEK) isolated from the epidermis of juvenile foreskin or adult skin from different locations.

2.2 Reagents

1. Collagen, type I solution from rat tail (6 mg/mL).
2. 0.05% trypsin–EDTA.
3. Accutase.
4. Fibronectin solution: 50 μg/mL human fibronectin in ultrapure water.
5. Gel neutralization solution (GNS): For 250 mL combine 232.5 mL of DMEM High Glucose 2× concentrated, 7.5 mL of 3 M HEPES, 2.5 mL of 5 mg/mL chondroitin sulfate and 7.5 mL of fetal bovine serum (FBS) in a 250 mL laboratory glass bottle. Adjust the pH to 9 with a calibrated pH meter by adding dropwise 1 N NaOH. Filter-sterilize the final solution, prepare 10 mL aliquots in 15 mL tubes and store at 4 °C for up to 6 months.
6. Fibroblast culture medium: 10% FBS, 100 U/mL penicillin and 100 μg/mL streptomycin in High Glucose DMEM with GlutaMAX (HG-DMEM). For 500 mL, add 50 mL of FBS and 5 mL of 100× penicillin–streptomycin stock to 445 mL of HG-DMEM.
7. Keratinocyte culture medium (KGM-2 ready medium): Add one keratinocyte growth medium 2 supplement pack (Promocell) and 5 mL of 100× penicillin–streptomycin stock to the keratinocyte growth medium 2 (KGM-2, Promocell).
8. KGM-2 submersing media with 5% or 2% FBS: 5% FBS or 2% FBS in KGM-2 ready medium, respectively.
9. Airlift medium: 1.86 mM $CaCl_2$ in KGM-2 ready medium (*see* **Note 1**).

2.3 Tools and Instruments

1. Cell culture incubator (37 °C, 5% CO_2 and atmospheric O_2).
2. Centrifuge tubes, 15 mL and 50 mL.
3. Ice bucket.
4. 24-well cell culture inserts with a pore size of 3 μm, sterile (Nunc).
5. Cell culture plates, 24-well and 6-well, sterile, w/lid (Nunc).
6. Serological pipettes.
7. Swing bucket centrifuge.
8. Sterile forceps.
9. Syringes and filters.

3 Methods

3.1 Construction of Dermal Component

This section describes how to construct the dermal component of the skin model.

1. In a sterile hood, put GNS and collagen solution on ice.
2. Calculate the total amount collagen gel required for the experiment with 500 μL of gel per insert. Table 1 describes how to make one to five gels. Scale up the volumes as needed (*see* **Note 2**).
3. Put the desired volume of collagen solution in a 50 mL tube and centrifuge at 4000 × *g* for 4 min at 4 °C in a precooled centrifuge to get rid of air bubbles (*see* **Note 3**).
4. Place the desired amount of inserts into the 24-well plate.
5. Harvest the fibroblasts by trypsinization (using 0.05% trypsin–EDTA) and count the number of cells with a hemocytometer under a microscope.
6. Calculate and transfer the amount of fibroblasts needed (0.5×10^5 cells per 500 μL collagen gel) into a fresh tube and centrifuge at 270 × *g* for 5 min.
7. Aspirate the supernatant and carefully resuspend the fibroblasts in the calculated volume of cooled GNS without producing air bubbles.
8. From this point work fast to avoid premature gelation. Add the fibroblast-containing GNS to the calculated amount of collagen solution and mix carefully without producing air bubbles (*see* **Note 4**).

Table 1
Components and quantities for the construction of dermal equivalents. Resuspend the amount of fibroblasts needed in the corresponding volume of cooled GNS and mix this solution with the corresponding amount of collagen. If more gels are needed, scale up the volumes of the components

Number of gels	Amount of GNS [μL]	Amount of fibroblasts	Amount of collagen solution [μL]	Final volume [mL]
1	166.7	0.5×10^5	333.3	0.5
2	333.3	1.0×10^5	666.7	1.0
3	500.0	1.5×10^5	1000.0	1.5
4	666.7	2.0×10^5	1333.3	2.0
5	833.3	2.5×10^5	1666.7	2.5

9. Quickly fill 500 μL of the collagen–cell mixture in each insert with a 1000 μL pipette (*see* **Note 5**).
10. Incubate the gels for 1 h in the incubator (37 °C, 5% CO_2) to solidify the gels.
11. Add 2 mL of fibroblast culture medium in each well to submerse the gels and incubate for 24 h in the incubator (37 °C, 5% CO_2) (Fig. 1).

3.2 Construction of Epidermal Component

This section describes how to construct the epidermal component of the skin model.

1. After 24 h carefully aspirate the medium without touching the gels.
2. Cover each dermal equivalent with 25 μL of fibronectin solution.
3. Incubate the gels for 30 min in the incubator (37 °C, 5% CO_2).
4. In the meantime, harvest the keratinocytes using accutase and count the number of cells with a hemocytometer under a microscope.
5. Calculate and transfer the amount of keratinocytes needed (1×10^5 cells per collagen gel) into a fresh tube and centrifuge at $270 \times g$ for 5 min.
6. Aspirate the supernatant and carefully resuspend the keratinocytes in KGM-2 submersing medium with 5% FBS to the final cell concentration 1×10^6 cells/mL.
7. On each gel, seed 1×10^5 keratinocytes by adding 100 μL of cell suspension.
8. Incubate the gels for 1 h in the incubator (37 °C, 5% CO_2) to allow cells to adhere.
9. Add 2 mL of KGM-2 submersing medium with 5% FBS in each well to submerse the gels and incubate for 24 h in the incubator (37 °C, 5% CO_2) (Fig. 1).

3.3 Culturing of Skin Models

This section describes how to culture the skin models. During the first 7 days of culture, the collagen embedded fibroblasts will remodel the matrix upon production of extracellular matrix proteins and cause contraction of the gel by about 50%.

1. Culture gels for 6–7 days in KGM-2 submersing medium with descending FBS-concentration (5%, 2% and 0% FBS). Change the medium every 2–3 days with the next lower FBS concentration.
2. On day 7, the culture at the air–liquid interface is initiated. Therefore, remove the medium completely. Do not touch the gel with a pipette tip while doing so.

3. Place every insert into one well of a 6-well plate with sterile forceps.
4. Add 1.5 mL of airlift medium per well, taking care not to wet the surface of the skin models. The filling level of the medium is up to the meniscus of the gel.
5. Change medium every 2–3 days for additional 14 days (*see* **Note 6**) (Fig. 2).

4 Notes

1. To prepare the $CaCl_2$ solution, dissolve 4.4 g of $CaCl_2$ in 100 mL of ultrapure water (300 mM solution), filter-sterilize. Add 3.1 mL of 300 mM $CaCl_2$ solution to 500 mL of KGM-2 ready medium.
2. Preparing the mixture on ice prevents premature gelation. Calculate for some extra gels due to high viscosity of the gel and some unavoidable loss during pipetting.
3. Since the collagen solution is quite viscous, pipette up and down slowly. Wait for the solution to come down to the tip, and pipette it out completely. Alternatively, syringes can be used for higher accuracy. You can also use prechilled pipettes (15 min at −20 °C) for easier handling. An accurate volume of collagen solution is important during neutralization to ensure complete gelation.
4. Pipette up and down slowly and stir the collagen mixture carefully with the pipette tip until the color of the solution remains stable. When the pH changes from acidic to neutral or even slightly basic, the color changes from yellow to light pink/orange to dark pink, respectively. The desired color is light pink.
5. For easier handling of the collagen mixture cut off 0.5 cm of the pipette tip with sterile scissors.
6. During the culture at the air–liquid interface it is essential to make sure that no air bubble is trapped under the model and that the model surface is kept dry. In case medium can be seen on top of the model, remove this carefully using pieces of sterile filter paper.

Acknowledgments

The authors kindly thank the Fraunhofer Gesellschaft and the Bayrisches Staatsministerium für Wirtschaft und Medien, Energie und Technologie (Az.:VI/3-6622/453/12) for financially supporting the work.

References

1. Proksch E, Brandner JM, Jensen JM (2008) The skin: an indispensable barrier. Exp Dermatol 17(12):1063–1072
2. Steinhoff M, Brzoska T, Luger TA (2001) Keratinocytes in epidermal immune responses. Curr Opin Allergy Clin Immunol 1(5): 469–476
3. Coquette A, Berna N, Vandenbosch A et al (1999) Differential expression and release of cytokines by an in vitro reconstructed human epidermis following exposure to skin irritant and sensitizing chemicals. Toxicology In Vitro 13(6):867–877
4. MacNeil S (2007) Progress and opportunities for tissue-engineered skin. Nature 445(7130): 874–880
5. Persidis A (1999) Tissue engineering. Nat Biotechnol 17(5):508–510
6. Rheinwald JG, Green H (1977) Epidermal growth-factor and multiplication of cultured human epidermal keratinocytes. Nature 265(5593):421–424
7. Groeber F, Holeiter M, Hampel M et al (2011) Skin tissue engineering—in vivo and in vitro applications. Adv Drug Deliver Rev 63(4–5): 352–366
8. Ponec M (2002) Skin constructs for replacement of skin tissues for in vitro testing. Adv Drug Deliver Rev 54:S19–S30
9. Rossi A, Appelt-Menzel A, Kurdyn S et al (2015) Generation of a three-dimensional full thickness skin equivalent and automated wounding. J Vis Exp (96)
10. Pageon H, Zucchi H, Rousset F et al (2014) Skin aging by glycation: lessons from the reconstructed skin model. Clin Chem Lab Med 52(1):169–174
11. Green CB, Cheng G, Chandra J et al (2004) RT-PCR detection of *Candida albicans* ALS gene expression in the reconstituted human epithelium (RHE) model of oral candidiasis and in model biofilms. Microbiology 150: 267–275
12. Jannasch M, Groeber F, Brattig NW et al (2015) Development and application of three-dimensional skin equivalents for the investigation of percutaneous worm invasion. Exp Parasitol 150:22–30
13. Commandeur S, van Drongelen V, de Gruijl FR et al (2012) Epidermal growth factor receptor activation and inhibition in 3D in vitro models of normal skin and human cutaneous squamous cell carcinoma. Cancer Sci 103(12): 2120–2126
14. Gschwandtner M, Mildner M, Mlitz V et al (2012) Histamine suppresses epidermal keratinocyte differentiation and impairs the inside-out barrier function of the epidermis—a new role for mast cells in inflammatory skin diseases. J Investig Dermatol 132:S58–S58
15. Replacement, refinement & reduction of animals in research, 2014. http://www.nc3rs.org.uk. Accessed 7 July 14
16. Mallampati R, Patlolla RR, Agarwal S et al (2010) Evaluation of EpiDerm full thickness-300 (EFT-300) as an in vitro model for skin irritation: studies on aliphatic hydrocarbons. Toxicol In Vitro 24(2):669–676
17. Zhang Z, Michniak-Kohn BB (2012) Tissue engineered human skin equivalents. Pharmaceutics 4(1):26–41
18. Black AT, Hayden PJ, Casillas RP et al (2010) Expression of proliferative and inflammatory markers in a full-thickness human skin equivalent following exposure to the model sulfur mustard vesicant, 2-chloroethyl ethyl sulfide. Toxicol Appl Pharmacol 249(2):178–187
19. Brohem CA, Cardeal LB, Tiago M et al (2011) Artificial skin in perspective: concepts and applications. Pigment Cell Melanoma Res 24(1):35–50

Chapter 15

Analysis of Breast Cancer Cell Invasion Using an Organotypic Culture System

Romana E. Ranftl and Fernando Calvo

Abstract

Metastasis is the main cause of cancer patient mortality. Local tumor invasion is a key step in metastatic dissemination whereby cancer cells dislodge from primary tumors, migrate through the peritumoral stroma and reach the circulation. This is a highly dynamic process occurring in three dimensions that involves interactions between tumor, stromal cells, and the extracellular matrix. Here we describe the organotypic culture system and its utility to study breast cancer cell invasion induced by cancer-associated fibroblasts. This is a three-dimensional model that reproduces the biochemical and physiological properties of real tissue and allows for investigating the molecular and cellular mechanisms involving tumor and its microenvironment, and their contribution to cancer cell invasion. This system provides a robust, accurate, and reproducible method for measuring cancer cell invasion and represents a valuable tool to improve the mechanistic understanding of the initial steps in metastasis.

Key words Organotypic, Invasion, Metastasis, Breast cancer, Cancer-associated fibroblasts

1 Introduction

Metastatic dissemination is the major clinical complication in most types of cancer and the cause of 90% of cancer-related deaths [1]. Invasion of cancer cells into the peritumoral stroma is a key step in metastasis [2, 3]. Thus, understanding the mechanisms regulating the invasive abilities of cancer cells is imperative to inform the development of therapeutic modalities to minimize cancer dissemination and improve patient survival [4]. Local cancer cell invasion is a multifunctional and dynamic process occurring in three dimensions that is actively modulated by the peritumoral stroma [5–7]. Cancer-associated fibroblasts (CAFs) are particularly relevant as they can produce soluble factors that promote cancer cell invasion [8]. In addition, CAF-dependent matrix remodeling via focalized proteolysis activity [9] or actomyosin-dependent force generation [10, 11] can lead to the formation of tracks through the extracellular matrix (ECM) that enable

Zuzana Koledova (ed.), *3D Cell Culture: Methods and Protocols*, Methods in Molecular Biology, vol. 1612,
DOI 10.1007/978-1-4939-7021-6_15, © The Author(s) 2017

subsequent cancer cell invasion. Accordingly, three-dimensional (3D) models of cancer cell invasion that incorporate stromal components such as fibroblasts and physiologically relevant ECMs recapitulate more closely the in vivo situation. These models provide a platform for investigating the complex interactions between tumor and its microenvironment that are more likely to lead to novel insights of clinical relevance.

Here, we describe a 3D organotypic invasion assay adapted for the robust and accurate assessment of breast cancer cell invasion induced by CAFs [11–13]. This approach was originally developed as a 3D coculture model by Fusenig and colleagues [14], and further developed by other groups to study squamous cell carcinoma invasion [10, 15]. Briefly, CAFs are embedded in a dense gel composed of fibrillar collagen I and basement membrane matrix (termed Matrigel®, Cultrex®, or Engelbroth-Holm-Swarm matrix), which contains laminins, collagen IV, proteoglycans, and a broad spectrum of growth factors (*see* Fig. 1). A thin layer of gel is used to cover the breast cancer cells seeded on the surface of the CAF-containing gel to mimic the physiological condition of breast tissue. Gels are subsequently laid on a grid and maintained partially immersed in cell medium. Organotypic gels are then fixed and processed by standard histopathological procedures followed by quantitative and qualitative analysis of breast cancer cell invasion using standard image processing and analysis software.

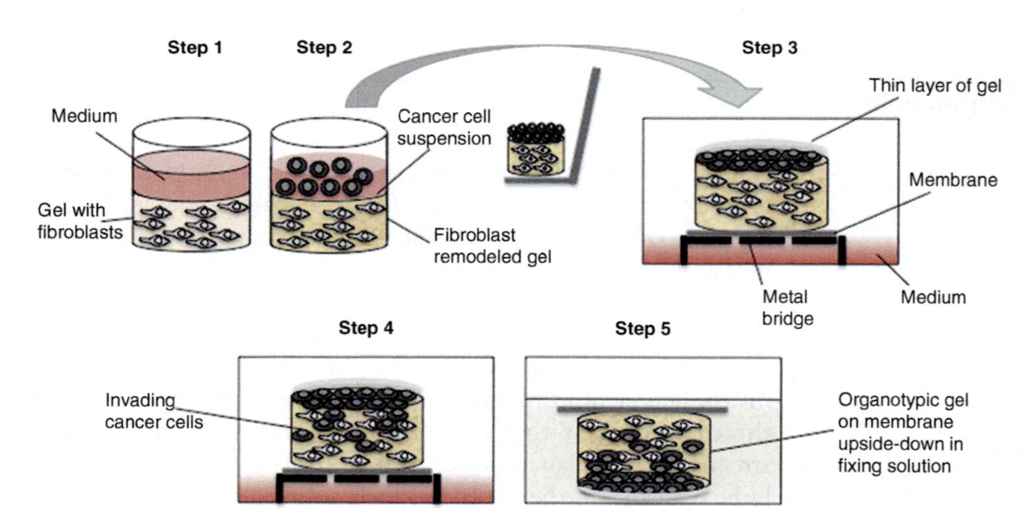

Fig. 1 Schematic representation of the workflow of an organotypic invasion assay. Step 1: Embed fibroblasts in gel and seed in 24-well dish. Step 2: Cancer cells are seeded in a single-cell suspension on top of the gel. Step 3: Once the cells have adhered, remove the medium and lift the remodeled gel onto gel-coated nylon filter on a metal bridge. Coat the cancer cells with a thin layer of gel. Step 4: Feed with complete medium up to the nylon filter. Incubate at 37 °C, 5% CO_2 for 5 days to allow for cancer cell invasion. Step 5: Terminate assay by fixing organotypic gels. Process gels for H&E staining

2 Materials

2.1 Tissue Culture

1. Cancer-associated fibroblasts (CAFs):
   (a) For the mouse model: CAFs from breast carcinomas from the FVB/n MMTV-PyMT murine model [11, 16].
   (b) For the human model: CAFs from a resection of a human breast carcinoma.
2. Normal fibroblasts (NFs):
   (a) For the mouse model: NFs from mammary glands of FVB/n wild-type siblings [11, 16].
   (b) For the human model: NFs from a resection of a reduction mammoplasty.
3. Breast cancer cells:
   (a) For the mouse model: 410.4 or 4T1 cells (ATTC®-CRL-2539™) (*see* **Note 1**).
   (b) For the human model: MDA-MB-231 (ATCC®-HTB-26) (*see* **Note 2**).
4. Fibroblast culture medium: 10% fetal bovine serum (FBS), 1× GlutaMax™ (Gibco®) and 1× insulin–transferrin–selenium (ITS, Gibco®) in Dulbecco's modified Eagle's medium (DMEM). Store at 4 °C. Warm up before use.
5. Cancer cell culture medium: 10% FBS and 1× GlutaMax™ in DMEM (*see* **Note 3**). Store at 4 °C. Warm up before use.
6. Sterile phosphate buffered saline (PBS): 3.2 mM Na_2HPO_4, 0.5 mM KH_2PO_4, 1.3 mM KCl, 135 mM NaCl, pH 7.4. Warm up before use.
7. Sterile 0.05% trypsin–0.02% EDTA. Store at 4 °C. Warm up before use.
8. If a Killing Assay is performed: selective compounds such as puromycin (puromycin dihydrochloride from *Streptomyces alboniger*, Sigma) (*see* **Note 4**).

2.2 Gel Preparation

1. 5× DMEM: 5% αDMEM powder (Gibco®), 2% $NaHCO_3$ (0.24 M $NaHCO_3$), 0.1 M Hepes pH 7.5. Store at 4 °C (*see* **Note 5**).
2. Fibroblast culture medium.
3. FBS.
4. Rat-tail collagen type I, high concentration (BD Biosciences). Store at 4 °C (*see* **Note 6**).
5. Matrigel® (BD Biosciences). Store at −80 °C in 1 mL aliquots (*see* **Note 7**).

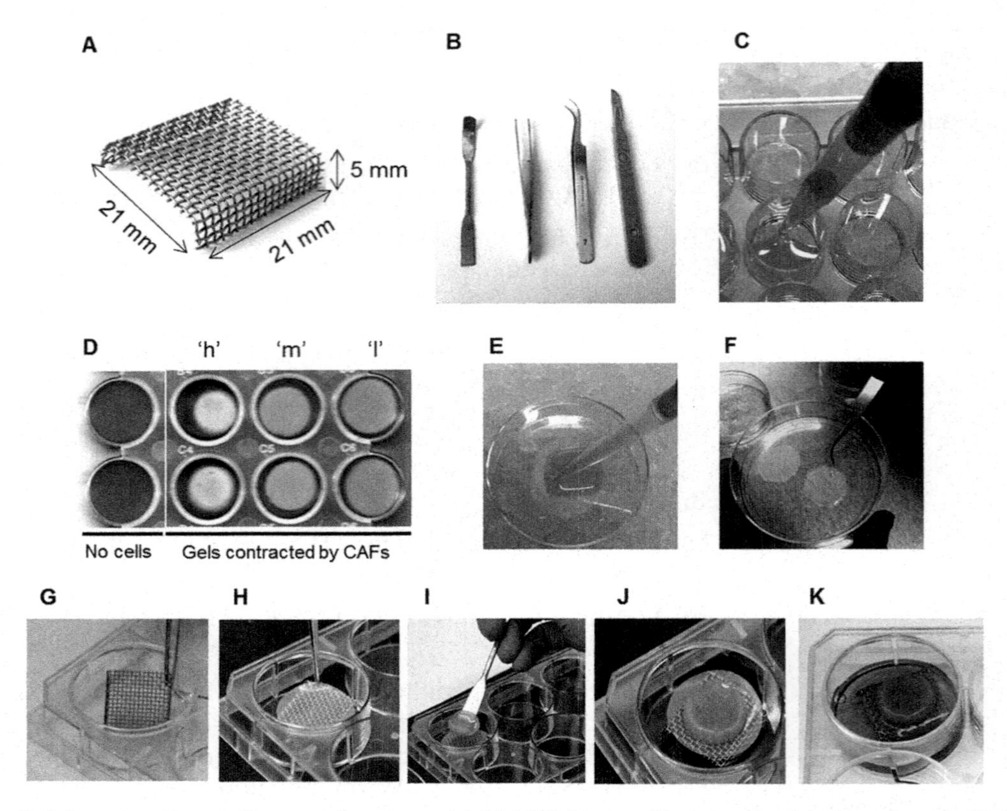

Fig. 2 Gel preparation and processing steps. (**a**) Metal bridges with approximate dimensions indicated. (**b**) Useful tools for handling the organotypic cultures, nylon filters and metal bridges. (**c**) Cell-free or fibroblast-containing gels are plated in a 24-well plate on ice. (**d**) The ability of fibroblasts to contract the collagen:Matrigel gel (a measure of their matrix remodeling capacity) can be documented prior to further processing. Images show duplicate samples of organotypic cultures free of cells and with different human CAFs that have low ("l"), medium ("m"), and high ("h") contractility, as indicated. (**e**, **f**) Nylon filters are soaked in gel (**e**) and then separated in a culture dish for setting and fixing (**f**). (**g**–**k**) Setting up the organotypic gel on the metal bridges. Sterile bridges are placed in a 6-well plate (**g**) and covered by a nylon filter (**h**). The organotypic cultures are lifted from the 24-well plate and placed over the filter:bridge using a spatula (**i**); a thin gel layer is added on the top to cover the cancer cell monolayer (**j**); and complete medium is added to the 6-well dish until soaking the nylon filter underneath the organotypic gel (**k**)

2.3 Gel Manipulation

1. Nylon NET filters 120 μm (Merck Millipore).
2. Sterile stainless steel metal bridges with a grid size of 2–3 mm. Approximate dimensions: 21 × 21 × 5 mm (Fig. 2a).
3. Sterile forceps and spatula (Fig. 2b).
4. Sterile PBS.
5. Filter fixing solution: 4% paraformaldehyde (PFA) and 0.25% glutaraldehyde in PBS.

2.4 Post Processing and Analysis

1. Gel fixing solution: 4% PFA and 1% glutaraldehyde in PBS.
2. Scalpel and forceps (Fig. 2b).

3. 70% ethanol.
4. Standard materials for paraffin embedding and hematoxylin and eosin (H&E) staining.
5. Bright field microscope with 20× (or 10×) objective and camera.
6. Image analysis software (Image J—http://imagej.nih.gov/ij—or similar).

3 Methods

This protocol is an adapted version of a previously described method for SCC12 carcinoma cells [10] (*see* Fig. 1 for a schematic representation). We have optimized the organotypic invasion assay for murine and human breast cancer cells (410.4/4T1 and MDA-MB-231) in combination with murine or human fibroblasts, respectively (*see* **Note 8**). On a general basis, cancer cells are not invasive in this setting and rely on CAF activities to invade. It is recommended to test the behavior of alternative cancer cell types in the assay by plating them on top of a fibroblast-free gel matrix. Heterogeneity is also observed in CAFs in terms of their ability to contract and remodel gel matrices. The amount of fibroblasts and cancer cells to be used, as well as the length of the protocol, need to be optimized if alternative models are used (*see* **Note 9**).

3.1 Gel Preparation

All components of the gel mix must be kept on ice. Here we describe the protocol for 1 mL gel mix containing fibroblasts or without fibroblasts. This is the volume required for one sample in one well of a 24-well plate. The total volume of gel mix needs to be scaled up according to sample size. Generally, it is recommended to prepare gel in excess to avoid pipetting errors due to the viscous nature of the components.

1. Prechill all components for the gel preparation on ice.
2. Mix 100 μL FBS, 80 μL 5× DMEM, and 120 μL of fibroblast culture medium.
3. Add 200 μL of Matrigel® and 400 μL of collagen I (*see* **Notes 10** and **11**). Keep the mixture on ice (*see* **Note 12**).
4. Trypsinize fibroblasts of interest (NFs/CAFs) from a monolayer to a single cell suspension. Count the cells, centrifuge (400 × *g* for 5 min), and resuspend the pellet at a concentration of 10^7 cells/mL in fibroblast culture medium (*see* **Notes 13** and **14**).
5. Avoiding bubble formation, add 100 μL of the cell suspension to the gel mix. If a gel without fibroblasts is required, add 100 μL of fibroblast culture medium instead of cell suspension (*see* **Note 15**).

6. Add 900 μL of the mixture in a 24 well-plate well (Fig. 2c) (*see* **Note 16**).
7. Place the plate at 37 °C and 5% CO_2 for 1 h.
8. When the gel is set, add 1 mL of appropriate medium on top and incubate overnight at 37 °C and 5% CO_2 (*see* **Note 17**).

3.2 Breast Cancer Cell Preparation

Depending on the number and matrix-remodeling ability of the fibroblasts, some of the gels may have already contracted at this point. The plate can be scanned to document this, as gel contraction can serve as an additional read-out of CAF function (Fig. 2d).

1. Trypsinize a monolayer of breast carcinoma cells (4T1, 410.4 or MBA-MD-231), count the cells and prepare a single cell suspension at 5×10^6 cells/mL in cancer cell culture medium.
2. Aspirate the medium carefully from the 24-well plate containing the gels.
3. Apply 100 μL of the cancer cell suspension on the top of each of the gels (*see* **Notes 18** and **19**).
4. Incubate at 37 °C and 5% CO_2. Leave the cells to adhere to the matrix for 6–8 h.

3.3 Coating of Nylon Filters

Given the softness of the gels, direct contact with the metal can damage the gels. Before placing the gels on the bridges, nylon filters have to be prepared and placed in-between the gels and the metal grid.

1. Prepare the adequate number of nylon filters (1 per organotypic condition) with a sterile forceps on top of each other in a culture dish.
2. Prepare 1 mL of gel without fibroblasts according to the recipe described in Subheading 3.1.
3. Coat the nylon filters using 1 mL of cell-free gel (Fig. 2e).
4. Separate the coated filters in the culture dish (Fig. 2f).
5. Incubate the coated nylon filters at 37 °C for 1 h.
6. Fix the coated nylon, using filter fixing solution for at least 2 h at room temperature (RT) or overnight at 4 °C.

3.4 Lifting the Gel and Covering the Cancer Cells

The gels are placed on the metal bridges and medium is fed from underneath to impose a direction of invasion. The breast cancer cell layer is covered with a thin layer of gel for improved simulation of in vivo conditions (*see* **Note 20**).

1. Wash the gel-coated nylon filters with sterile PBS for 10 min (×3) to remove all traces of fixing solution.
2. Add cancer cell culture medium to the filters and incubate for 30 min at 37 °C.

3. Place the sterile metal grid bridges in a 6-well plate with sterile forceps (Fig. 2g).
4. Add a coated, washed and medium-adapted nylon filter on top of each metal bridge with sterile forceps (Fig. 2h).
5. Prepare an appropriate amount of cell-free gel according to the recipe in Subheading 3.1 and keep the mixture on ice (100 μL per organotypic gel).
6. Remove the medium on the top of the gels carefully in order not to disturb the organotypic cultures.
7. Lift the gels with a sterile spatula from the 24-well and place it on a coated filter on top of a metal bridge. The cancer cell layer must be facing up (Fig. 2i, j).
8. Add medium under the bridge until it is in contact with the nylon filter. Avoid air bubbles between the interphase medium/nylon filter (Fig. 2k).
9. Add 100 μL of the cell-free gel mixture on top of the lifted gels and spread over the surface.
10. Incubate the culture for 5 days (37 °C, 5% CO_2) and change the medium daily (*see* **Note 21**).

3.5 Gel Fixing and Processing

1. After culturing for 5 days (*see* **Note 9**), remove the medium from the plate and wash with PBS or transfer the organotypic cultures to a new 6-well plate by lifting the filter and gel. Turn the gel upside-down for fixing (cancer cell layer on the bottom, Fig. 1) (*see* **Note 22**).
2. Fix the gels in 2 mL of gel fixing solution at 4 °C overnight.
3. Next day, wash the gels with PBS (2 mL) for 10 min (×3).
4. Using the forceps and the scalpel cut the gel in two halves and store one of them in 70% ethanol at 4 °C (*see* **Note 23**).
5. Embed the other half of the gel in paraffin blocks and perform standard H&E staining on sections (*see* **Notes 24** and **25**).

3.6 Data Analysis

1. Take 5–7 pictures per gel using a bright field microscope (20×/10× magnification) (*see* **Note 26** and Fig. 3a–e).
2. Measure the area of noninvasive cancer cells and the total area of cancer cells (noninvasive and invasive) using Image J or a similar software (Fig. 3f).
3. The Invasion Index is calculated by dividing the invading area (total area – invading area) by the total area (*see* **Note 27**).

3.7 Killing Assay

CAFs have been shown to form tracks in the ECM, which has been associated with increased invasiveness of cancer cells [10]. However, CAFs can also produce soluble factors that can promote cancer cell invasion [8]. The basic organotypic invasion assay described above does not allow discriminating between these two abilities.

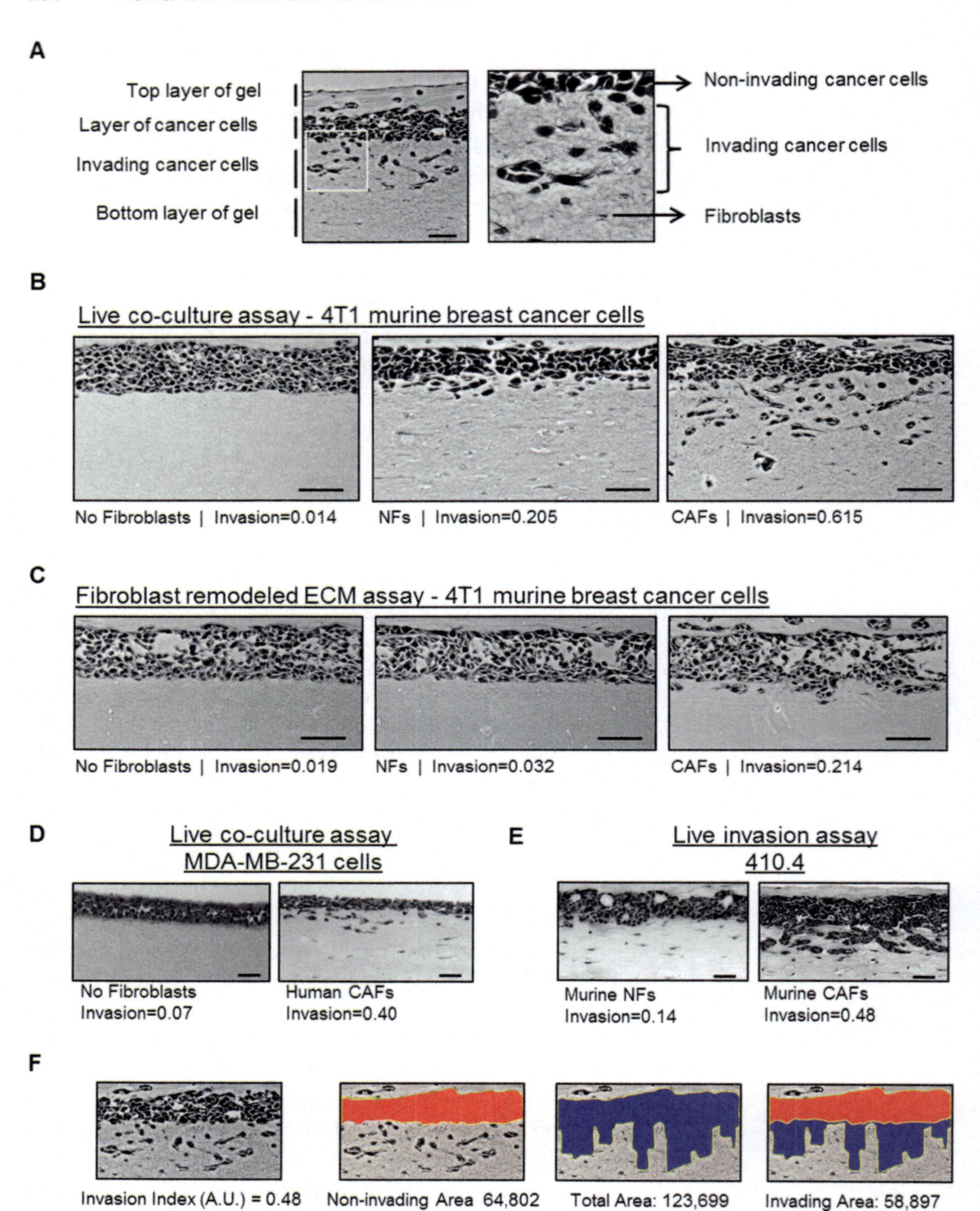

Fig. 3 Analysis of cancer cell invasion using carcinoma cells in an organotypic invasion assay. (**a**) Representative image of H&E staining of an organotypic invasion assay of 4T1 cells cultured in the presence of CAFs. The different parts of the organotypic gel are indicated. *Right panel* is a zoom up region that shows both collective and single cell invasion of 4T1 cells. Fibroblasts and ECM fibers are also observed. (**b**) Representative images of H&E staining of organotypic invasion assays of 4T1 murine breast cancer cells cultured in the absence and presence of either murine normal fibroblasts (NFs) or PyMT-CAFs. Invasion indexes of 4T1 cells for each

In order to specifically study the capacity of CAFs to form tracks in the gel that allow subsequent cancer cell invasion, as well as the potential of cancer cells to invade a gel previously remodeled by CAFs, a variation of the basic organotypic culture assay can be performed: the "killing" assay. Fibroblasts are seeded as described earlier and are allowed to remodel the gel matrix for 5 days before they are removed. When cancer cells are seeded on top of the gel, they will invade into the tracks formed by the fibroblasts (Fig. 3c). This modified version also allows for discriminating the selective effect of chemical compounds on fibroblasts/cancer cells as both compartments are never cultured together.

1. Follow the protocol as described in Subheading 3.1 (**steps 1–8**). Once the gel is set, add 1 mL of fibroblast culture medium on top.
2. Incubate at 37 °C, 5% CO_2 for 5 days. Change the medium daily (*see* **Notes 17** and **28**).
3. Fibroblast removal ("killing" step): Remove the medium from the gels and add 1 mL of fibroblast culture medium containing an appropriate selection compound (e.g., 10 μg/mL puromycin) (*see* **Note 29**).
4. Incubate at 37 °C, 5% CO_2 for 48 h.
5. Wash the gels for 1 h in fibroblast culture medium (×3). Incubate at 37 °C, 5% CO_2 overnight in fibroblast culture medium to remove all traces of selection compound (*see* **Note 30**).
6. Continue with the procedure as described in Subheadings 3.2–3.6.

Fig. 3 (continued) experimental setup are also indicated. Scale bar, 50 μm. (**c**) Representative images of H&E staining of "killing" organotypic invasion assays of 4T1 murine breast cancer cells cultured in gels previously remodeled by murine normal fibroblasts, PyMT-CAFs or mock-remodeled (no fibroblasts). Invasion indexes of 4T1 cells for each experimental set-up are also indicated. Scale bar, 50 μm. (**d**) Representative images of H&E staining of organotypic invasion assays of MDA-MB-231 human breast cancer cells cultured in the absence and presence of human breast cancer CAFs. The invasion index is indicated. Scale bar, 50 μm. (**e**) Representative images of H&E staining of organotypic invasion assays of 410.4 murine breast cancer cells with murine NFs and CAFs. Note that 410.4 cells present a "collective" mode of invasion. Scale bar, 50 μm. (**f**) Exemplars for image analysis of a representative image of H&E staining of an organotypic invasion assay of 4T1 cells cultured in the presence of CAFs (shown in panel **a**). First, the non-invaded area (*red*) and the total area (*blue*) are measured using Image J (http://imagej.nih.gov/ij/). The invading area is then calculated by subtracting the non-invading area from the total area. The invasion index is the ratio of invaded area vs. total area. Invasion index and areas are also indicated

4 Notes

1. 4T1 cells are a metastatic subpopulation of the original 410.4 murine breast cancer cell line [17].
2. For perturbation studies (i.e., RNAi), NFs/CAFs or cancer cells need to be modified before starting the protocol.
3. ATCC and other studies may suggest different media for culturing the breast cancer cell lines. However, we recommend using DMEM as the organotypic gels are based on it and it does not affect cell behavior or viability.
4. For puromycin, make aliquots of 10 mg/mL and store at −20 °C.
5. Preparation of 5× αDMEM (250 mL): dissolve 12.5 g DMEM powder (Gibco®) and 5 g $NaHCO_3$ in 50 mL of sterile MilliQ water. Add 25 mL of sterile 1 M Hepes pH 7.5 and sterile MilliQ water up to 250 mL. Mix and filter the solution using a 0.2 mm filter in sterile conditions. Aliquot in 50 mL and 10 mL aliquots and store at −20 °C. Thaw fresh aliquots for gel preparation and do not store them longer than 30 days at 4 °C.
6. Rat tail collagen type I, high concentration is supplied as a liquid in 0.02 N acetic acid with a concentration range of 8–11 mg/mL. Note the actual concentration of each particular batch when calculating the volumes to be used during gel preparation (*see* **Note 10**).
7. Matrigel® is a viscous substance at 4 °C that solidifies at RT. Store Matrigel® in sterile 1 mL aliquots at −80 °C and thaw the amount required for an experiment the night before at 4 °C. Note the actual concentration of each particular batch, as it will be required to calculate the volumes for gel preparation (*see* **Note 10**). As an alternative to Matrigel® we have successfully used Engelbroth-Holm-Swarm matrix from Sigma-Aldrich in our organotypics.
8. We have used both murine and human CAFs in this assay. In our experience, murine CAFs require higher cell numbers than human CAFs, as the latter tend to be larger in size and have higher ECM remodeling abilities (*see* **Note 13**). For comparative studies, make sure an equal amount of NFs/CAFs are used.
9. To optimize the approach, it is recommended to empirically determine the effect of two variables: (1) the amount of NFs/CAFs to be embedded on the gel matrix; and (2) the length of the protocol (i.e., how long the fibroblasts are allowed to remodel the gels and/or the cancer cells are allowed to invade).

10. The final concentrations of collagen and Matrigel® in the gel are set to 4 mg/mL and 2 mg/mL, respectively. The recipe was optimized to have almost physiological concentrations of collagen I and Matrigel®. In order to prepare reproducible gel solutions the volumes of the components have to be adapted accordingly, depending on the stock concentrations of Matrigel® and collagen I (*see* **Notes 6** and 7). For this protocol we base our volume calculations on stock concentrations of 10 mg/mL for both collagen I and Matrigel®. The volume of 5× DMEM can be adapted if larger volumes of the acidic collagen are added to retain optimal buffering.
11. Matrigel® and collagen I are very viscous and need to be pipetted with care. Avoid bubble formation (as bubbles can be detrimental to the quality of the gel) and retention of gel in the tips (as they may affect the final concentration).
12. Avoid leaving the gel mixture on ice for long periods of time as it may affect its properties.
13. For human NFs/CAFs, resuspend the pellet to a concentration of 2.5×10^6 (*see* **Note 8**).
14. To reduce the activating effect of FBS in fibroblasts, NFs/CAFs can be cultured for 5–7 days in fibroblast culture medium supplemented with 0.5% FBS (instead of 10% FBS) prior to the use in the organotypic invasion assay.
15. Cell-free gel will also be used for the nylon filter coating procedure (Subheading 3.3) and for covering the cancer cell layer on the top of the organotypic gel (Subheading 3.3).
16. Do not plate the total volume of the mix in order to avoid pipetting errors as the gel is very viscous and sticks to pipette tips. Gels can set at RT; these gels will not have the exact same properties as the ones set at 37 °C. We recommend keeping the 24-well plate on ice while adding the gel, especially when a larger number of samples are handled. This will allow the gel in all wells to set at the same time.
17. At this step, appropriate growth factors, cytokines or drugs can be added to the medium. Cell medium will buffer some of the acidity of the gels and may turn orange. For optimal results, replace with fresh medium after 2–3 h.
18. Seeding 100 μL of cell suspension on the top of highly contracted gels can be problematic. If needed, in these particular cases the empty space in the well can be covered up with fresh cell-free gel; this will allow cancer cells to be seeded on larger volumes (1 mL of 5×10^5 cell/mL suspension). However, the new gel only attaches loosely to the original gel and can sometimes rip off during the lifting process.

19. If gels have not contracted at all, cancer cells can also be applied in a volume of 1 mL per well (at 5×10^5 cell/mL).
20. Alternatively, depending on the type of carcinoma cells used in the assay, the cells can also be grown in an air–cells–liquid interface. The cells are prepared and added on top of the fibroblast-containing organotypic culture as described in Subheading 3.2. However, the cells will not be covered with a layer of gel.
21. At this step, appropriate growth factors, cytokines or drugs can be added to the medium, bearing in mind that they will affect both compartments (i.e., fibroblasts and cancer cells). The gels will become flatter during the incubation on the metal bridges.
22. Alternatively, gels can be snap-frozen for immunofluorescence analysis. Gels are placed in a plastic cuvette, covered by OCT buffer (Tissue-Tek) and immersed in liquid nitrogen. Store at −80 °C until further processing (*see* **Note 25**).
23. Keep the second half as backup until good H&E staining is obtained or use it to perform additional analysis (*see* **Notes 22**, **24** and **25**).
24. To allow for invasion analysis, gels must be embedded in paraffin in the correct orientation. Tissue sections need to be obtained from the cut side of the organotypic gel and not on the top or bottom sides. Sectioning the gel within the paraffin block can be challenging as the noninvasive cell layer can easily come off during processing. We recommend very slow trimming of the paraffin block at the first 500 μm to level the gel. Generate two to three cuts (50 μm apart) per organotypic gel per slide for analysis.
25. As an alternative to H&E staining, gels can also be processed for immunohistochemistry or immunofluorescence detection of specific cancer cell or CAF markers according to standard protocols (*see* **Note 22**). This will ascertain the cell identity of the invading cells. Alternatively, fluorescence-labeled cells can be used in the organotypic invasion assay to allow for detection without staining procedures.
26. Fibroblasts tend to concentrate on the borders of the gels, leading to artefacts on cancer cell invasion in those areas. Pictures need to be taken in the central part of the gel.
27. Other analysis can also be performed. For example, Alternative Invasive Index (invading area/non-invading area), number of invading objects, area of invading cells (without measuring gel areas), etc.

28. NFs/CAFs can be allowed to remodel the gel up to 10 days before the cancer cells are seeded on top. The length of this step has to be determined empirically if other models are used.
29. The choice of selection depends on the resistance genes expressed by the cells (e.g., selection-based stable RNAi/over-expression systems previously inserted). We recommend determining the optimal final concentration beforehand.
30. Thorough washing with PBS and medium is important; otherwise cancer cells added subsequently may be affected. If this becomes problematic, we recommend generating stable cancer cells resistant to the selective compound.

Acknowledgments

R.R. and F.C. are funded by The Institute of Cancer Research (UK). F.C. is also supported by Worldwide Cancer Research (Grant 15-0273). We thank Dr. Erik Sahai and Steven Hooper for contributing to develop the technique described here. We also thank lab members for help and advice, and for critically reading the manuscript.

References

1. Weigelt B, Peterse JL, van't Veer LJ (2005) Breast cancer metastasis: markers and models. Nat Rev Cancer 5(8):591–602
2. Gupta GP, Massague J (2006) Cancer metastasis: building a framework. Cell 127(4):679–695
3. Steeg PS (2006) Tumor metastasis: mechanistic insights and clinical challenges. Nat Med 12(8):895–904
4. Wan LL, Pantel K, Kang YB (2013) Tumor metastasis: moving new biological insights into the clinic. Nat Med 19(11):1450–1464
5. Liotta LA, Kohn EC (2001) The microenvironment of the tumour-host interface. Nature 411(6835):375–379
6. De Wever O, Mareel M (2003) Role of tissue stroma in cancer cell invasion. J Pathol 200(4):429–447
7. McAllister SS, Weinberg RA (2014) The tumour-induced systemic environment as a critical regulator of cancer progression and metastasis. Nat Cell Biol 16(8):717–727
8. Ohlund D, Elyada E, Tuveson D (2014) Fibroblast heterogeneity in the cancer wound. J Exp Med 211(8):1503–1523
9. Zigrino P, Kuhn I, Bauerle T et al (2009) Stromal expression of MMP-13 is required for melanoma invasion and metastasis. J Invest Dermatol 129(11):2686–2693
10. Gaggioli C, Hooper S, Hidalgo-Carcedo C et al (2007) Fibroblast-led collective invasion of carcinoma cells with differing roles for RhoGTPases in leading and following cells. Nat Cell Biol 9(12):1392–1400
11. Calvo F, Ege N, Grande-Garcia A et al (2013) Mechanotransduction and YAP-dependent matrix remodelling is required for the generation and maintenance of cancer-associated fibroblasts. Nat Cell Biol 15(6):637–646
12. Avgustinova A, Iravani M, Robertson D et al (2016) Tumour cell-derived Wnt7a recruits and activates fibroblasts to promote tumour aggressiveness. Nat Commun 7:10305
13. Calvo F, Ranftl R, Hooper S et al (2015) Cdc42EP3/BORG2 and septin network enables mechano-transduction and the emergence of cancer-associated fibroblasts. Cell Rep 13(12):2699–2714
14. Fusenig NE, Breitkreutz D, Dzarlieva RT et al (1983) Growth and differentiation characteristics of transformed keratinocytes from mouse and human-skin invitro and invivo. J Invest Dermatol 81(1):S168–S175

15. Nystrom ML, Thomas GL, Stone M et al (2005) Development of a quantitative method to analyse tumour cell invasion in organotypic culture. J Pathol 205(4):468–475
16. Guy CT, Cardiff RD, Muller WJ (1992) Induction of mammary tumors by expression of polyomavirus middle T oncogene: a transgenic mouse model for metastatic disease. Mol Cell Biol 12(3):954–961
17. Aslakson CJ, Miller FR (1992) Selective events in the metastatic process defined by analysis of the sequential dissemination of subpopulations of a mouse mammary tumor. Cancer Res 52(6):1399–1405

Chapter 16

3D Coculture Model of the Brain Parenchyma–Metastasis Interface of Brain Metastasis

Raquel Blazquez and Tobias Pukrop

Abstract

Central nervous system (CNS) metastasis is not only an increasing but still a very unsatisfying clinical problem, with very few treatment options nowadays and an unmet clinical need. Additionally, the patients suffer from severe neurological symptoms. Furthermore the preclinical studies are limited and thus innovative methods are needed to study the mechanism of CNS metastasis during the final steps. Especially nowadays, the most critical step during metastasis seems to be the contention of the microenvironment of host organs with the cancer cells coming from other organs. More and more data indicate that this contention often leads to apoptosis of the pre-metastatic cells and prevents successful metastasis. However, this important step is barely understood. To further improve our knowledge about this important step of metastasis we developed a new experimental tool where we coculture an organotypic brain slice with a 3D tumor cell plug embedded in Matrigel for 3–4 days.

This model especially mimics the interactions of cancer cells and glial cells at the interface of the brain parenchyma and the metastatic tissue. Therefore this coculture method with an organotypic brain slice and a tumor cell plug allows us to visualize and/or manipulate the interactions at this very important zone. Furthermore, it also allows us to use brain tissue from genetically engineered mice and/or genetically modified tumor cells to investigate genes of interest in the microenvironment or in cancer cells. Moreover this method avoids the use of a large number of animals and is especially useful in identifying the invasiveness of different tumor cell lines into the brain parenchyma as well as in studying the effect of specific treatments against brain metastasis progression during the final and most critical steps of metastasis.

Key words 3D culture, Organotypic coculture, Brain slicing, Microglia, Astrocytes, Colonization, Metastasis, Confocal microscopy

1 Introduction

The number of patients who develop CNS metastasis from a primary tumor increases every year. Despite of this fact, the therapeutical options are still very poor and insufficient. This has urged the need for development of new therapies and proper models for studying the mechanisms of metastasis and testing new drugs.

Zuzana Koledova (ed.), *3D Cell Culture: Methods and Protocols*, Methods in Molecular Biology, vol. 1612,
DOI 10.1007/978-1-4939-7021-6_16, © Springer Science+Business Media LLC 2017

CNS metastasis is a multiple step process which takes place in the primary tumor, in the blood and in the CNS itself. At the beginning of brain metastasis research the major efforts were made in order to understand the steps in the primary tumor. However, recently there has been increasing evidence that the steps in the metastatic organ are the most decisive ones [1, 2] and that the glial cells play a very important role during the final steps of metastasis, in particular astrocytes [3–7] and microglia [3, 8–10].

Unfortunately, the methods to investigate the final steps of metastasis are very limited and to date and to our knowledge, there is no suitable in vivo/in vitro model available to directly visualize glial reactions during cerebral metastasis formation.

To overcome this problem, we established a coculture system consisting of an organotypic mouse brain slice and a 3D tumor plug of epithelial cells embedded in Matrigel. In our system, the tumor plug is placed directly next to the brain slice edge in order to observe the invasion of the tumor cells into the neighboring healthy tissue. This enables us to visualize morphological changes and interactions between the stromal and carcinoma cells by confocal microscopy or even live imaging. If required, after this experiment, the brain tissue and/or the cell plug can be collected and used for further molecular analyses (e.g., qRT-PCR, IHC, or immunoblot). Our coculture method can be applied to monitor the colonization of the living brain tissue for several days avoiding laborious, costly, technically complex and ethically challenging animal experiments which result in substantial stress for the animals. Another advantage is that our model allows investigations of human carcinoma cells within immunocompetent animals. Furthermore it is a practicable alternative to in vivo approaches when testing targeted pharmacological manipulations and it reduces the number of animals needed for these tests.

2 Materials

2.1 Cell Culture and GFP Transfection

1. Cancer cell culture medium: 10% fetal calf serum (FCS) in RPMI-1640 or DMEM medium.
2. Trypsin–EDTA (1×): Add 1 mL of 10× trypsin–EDTA to 9 mL of distilled water (dH_2O) and keep at 4 °C.
3. Phosphate-buffered saline (PBS, without Ca^{2+}, Mg^{2+}).
4. Falcon tubes, 15 mL and 50 mL.
5. Mammalian GFP Expression Vector, e.g., pTurboGFP-N (FP512, Evrogen).
6. Nanofectin transfection kit (PAA Laboratories, Q051-005).
7. Geneticin® Selective Antibiotic (G418 Solution), 100–1000 μg/mL in cancer cell culture medium.

8. Cell sorter (e.g., BD FACSAria II).
9. Cells: cancer cell line of interest, e.g., MCF-7 or MDA-MB231.

2.2 Mouse Dissection

1. C57BL/6 mice (or any other mouse strain), 5–8 days old (P5–8).
2. Dubois decapitation scissors, curved end (Mopec, AA027).
3. Noyes eye scissors, straight pattern (Hermle, 532).
4. Wecker spatula (Hermle, 221).
5. Semkin standard forceps, serrated (Hermle, 6721).
6. Tissue forceps, straight pattern (Hermle, 716).
7. Sterile scalpel, #15 blade.
8. Polystyrene petri dish.

2.3 Organotypic Brain Slice Coculture System

1. Dissection medium: 4.5 mg/mL glucose, 2 mM l-glutamine, 100 U/mL penicillin, and 100 μg/mL streptomycin in MEM medium, sterile. Keep at 4 °C.
2. Cyanoacrylate glue (Renfert, 1733-0100).
3. Petri dish (92 × 17 mm) with 5% agarose, low gelling temperature (Sigma, A9414-100G): Prepare a 5% agarose solution in dH_2O. Heat it to dissolve and then add about 8 mL to a petri dish. Let it harden, cover it with Parafilm and keep at 4 °C.
4. Vibratome (e.g., Leica, VT1000S).
5. Cell culture insert, 6-well, 0.4-μm pore size (e.g., Falcon, 353090).
6. 6-well plate.
7. Incubation medium: 25% Hanks' balanced salt solution (HBSS), 25% normal horse serum (NHS) (Life Technologies, 26050-088), 4.5 mg/mL glucose, 2 mM l-glutamine, 100 U/mL penicillin and 100 μg/mL streptomycin in MEM, sterile. For one experiment prepare 100 mL of the medium. Keep at 4 °C.
8. ECM: Cultrex PathClear Basement Membrane Extract (R&D Systems, 3432-005-01), diluted in relation 2:1 in cancer cell culture medium and kept on ice. For each tumor plug, prepare 18 μL.

2.4 Immunofluorescence Staining and Confocal Microscopy

1. 4% paraformaldehyde (PFA) in PBS.
2. Wash solution (PBS-T): 0.5% Triton X-100 in PBS.
3. Blocking solution: 5% normal goat serum (NGS) (Sigma, G9023) in PBS-T.
4. Mouse monoclonal anti-glial fibrillary acidic protein (GFAP) antibody (Sigma, G3893), 1:200 dilution in PBS-T.

5. Goat anti-mouse IgG,F(ab′) 2-TRITC antibody (Santa Cruz, sc-3796), 1:100 dilution in PBS-T.
6. Isolectin GS-IB4 from *Griffonia simplicifolia*, Alexa Fluor 647 conjugate (ILB4-Alexa Fluor 647) (Life Technologies, I32450), 1:500 dilution in PBS-T.
7. DAPI solution: 1:1000 dilution of DAPI stock solution (5 mg/mL in PBS) in PBS-T.
8. Mounting medium (DakoCytomation, S-3023).
9. Microscope slides, 25 × 75 × 1 mm.
10. Cover glasses, 18 × 18 mm.
11. 12-well plate.
12. Confocal laser scanning microscope (e.g., Zeiss, LSM 510).

3 Methods

3.1 Cell Culture and GFP Transfection

1. Maintain the cell lines to be used for the plug setting in cancer cell culture medium (*see* **Note 1**).
2. Perform a stable transfection of your cell lines with a mammalian GFP expression vector using the Nanofectin transfection kit in accordance with the manufacturer's protocol.
3. Select the GFP-expressing cells with Geneticin® selective antibiotic for at least 3 weeks (*see* **Note 2**).
4. Sort the selected cells with a fluorescence-activated cell sorting method (FACS) to obtain homogeneous GFP expression.
5. After sorting the cells keep them in cell culture medium supplemented with 100 U/mL penicillin and 100 μg/mL streptomycin for at least 2 weeks to avoid contamination.

3.2 Mouse Dissection

1. Hold the mouse with one hand and sacrifice it with a decapitation scissors.
2. Transfer the head to a petri dish and disinfect it with 70% ethanol.
3. Remove the fur and the skull using chirurgical micro eye scissors (*see* **Note 3**).
4. Remove the brain rapidly from the skull under aseptic conditions using a spatula.
5. Remove the frontal pole and the cerebellum from the whole brain section with a scalpel.
6. Transfer the brain to ice-cold dissection medium (*see* **Note 4**).

3.3 Organotypic Brain Slice Coculture

1. Using a scalpel, cut the 5% agarose in petri dish into blocks (approximately 5 × 5 mm).

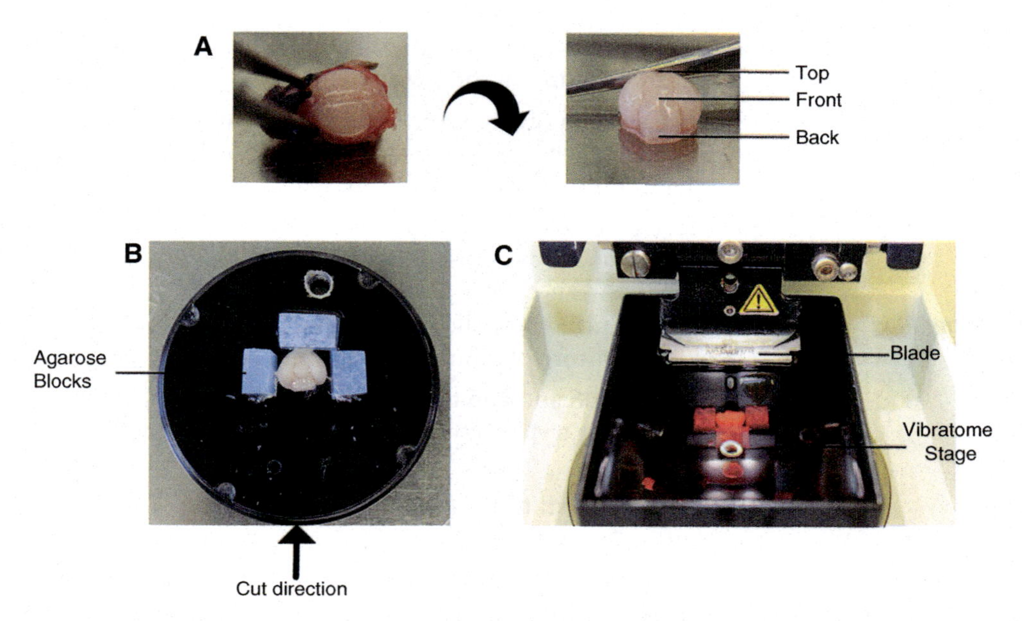

Fig. 1 Schema showing the correct setting of the mouse brain in the vibratome. After removing the brain from the skull, turn it over (**a**) and fixate it in the vibratome stage using agarose blocks (**b**). Lastly set the stage in the vibratome tray in the cutting direction (**c**)

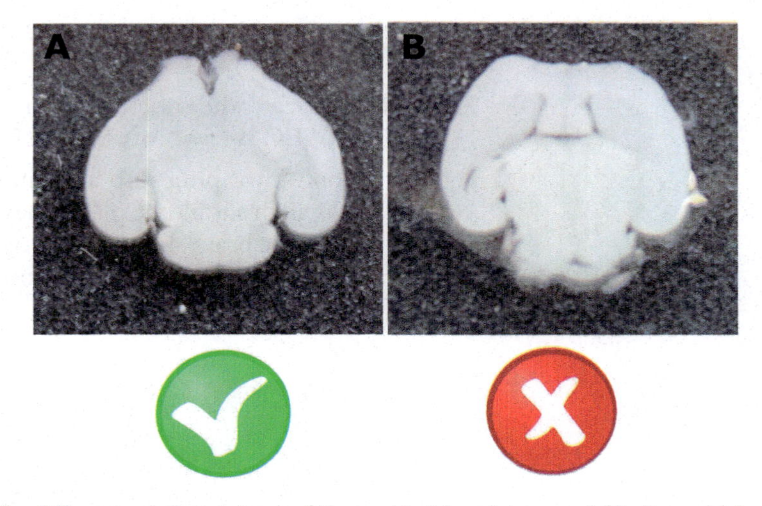

Fig. 2 Representative pictures of the usable (**a**) and damaged (**b**) slices obtained during the slicing process. Collect only the intact slices as any damaged ones would be destroyed during the following culture steps

2. Transfer the brain to the vibratome stage and remove any medium excess using a filter paper. Fix and stabilize the brain with cyanoacrylate glue and 5% agarose blocks according to the schema in Fig. 1.
3. Slice the brain sections horizontally to a 350 μm thickness by using a vibratome. Discard the first 2–3 slices until a usable and complete slice is obtained (Fig. 2). Collect the usable ones in a petri dish with ice-cold dissection medium (*see* **Note 5**).

4. Prepare a 6-well plate with cell culture inserts and add 1 mL of culture medium in the lower well.
5. Transfer each organotypic brain slice to an insert. To avoid the brain slice getting dried add about 50–100 μL of culture medium on the insert (*see* **Note 6**).
6. Incubate the organotypic brain slices overnight in an incubator with humidified atmosphere of 5% CO_2 at 37 °C.

3.4 Tumor Plug Setting

1. To trypsinize and collect GFP-transfected tumor cells, discard old cell culture medium and wash the cells with PBS.
2. Discard PBS and add trypsin–EDTA (1 mL for 100 mm dish). Ensure that trypsin covers the entire surface on which cells are adherent.
3. Allow 2–3 min for trypsin to work (*see* **Note 7**).
4. Add 9 mL of cancer cell culture medium to stop the reaction and ensure cell disaggregation by repetitive pipetting.
5. Collect the cells in a 15 mL Falcon tube and use 10 μL of the cell suspension to determine the cell number using a Neubauer chamber or a similar method.
6. Take the volume containing 10^5 GFP-transfected tumor cells (*see* **Note 8**) and centrifuge it at 200 × g for 5 min at room temperature (RT).
7. Carefully discard the supernatant and resuspend the cell pellet in 18 μL of ECM (*see* **Notes 8** and **9**).
8. Pipette the tumor plug into a sterile metallic spacer (3.8 mm diameter) which should be placed next to the cortical region of the organotypic brain slice (*see* Fig. 3) and let it set for approximately 1 h in an incubator with humidified atmosphere of 5% CO_2 at 37 °C.

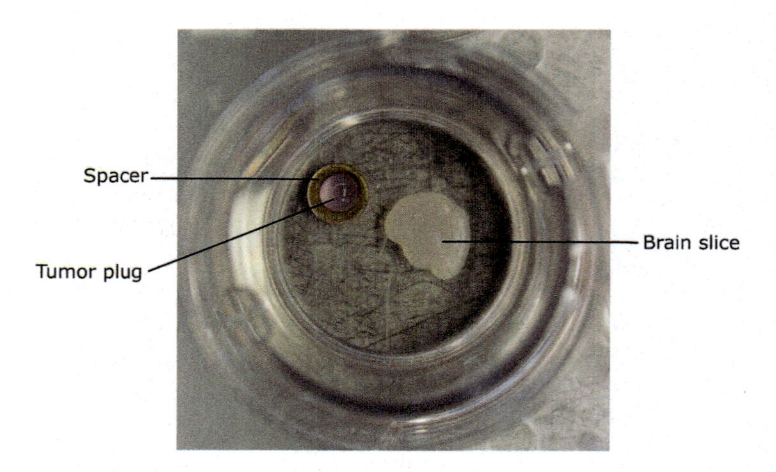

Fig. 3 Representative picture of the plug setting

Fig. 4 Representative picture of the organotypic brain slice and the tumor plug

9. Carefully remove the spacer and place the tumor plug in contact with the brain slice (*see* **Note 10**) as shown in Fig. 4. Change the medium. Allow the 3D tumor spheroid to coculture with the organotypic slice for 48–96 h.
10. Change the culture medium every 48 h.

3.5 Immunofluorescence Staining and Confocal Microscopy

1. Discard the incubation medium under and on the insert and fix the organotypic brain slice coculture with 4% PFA overnight at 4 °C. Add 1 mL of PFA under the insert and 200–250 μL on the insert without disrupting the brain slice.
2. Wash the slice coculture with PBS-T three times for 5 min.
3. Cut the membrane including the coculture with a scalpel and transfer it to a 12-well plate.
4. Block the samples with 200–250 μL of blocking solution at RT for 1 h (*see* **Note 11**).
5. Stain the astrocytes by incubating the brain slice coculture with 200–250 μL of mouse anti-GFAP antibody solution in PBS-T for 36 h at 4 °C (*see* **Note 12**).
6. Wash the samples with PBS-T three times for 5 min.
7. Add 200–250 μL of anti-mouse-TRITC antibody solution in PBS-T and incubate 1 h at RT in darkness (*see* **Note 13**).
8. Wash the samples with PBS-T three times for 5 min.

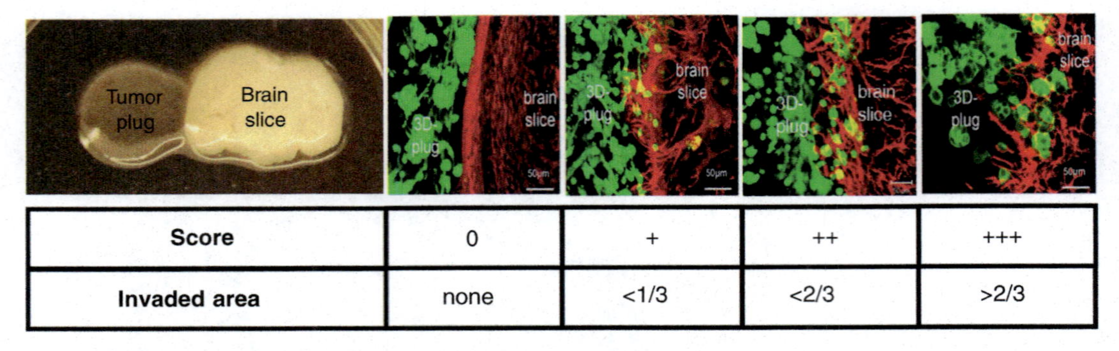

Score	0	+	++	+++
Invaded area	none	<1/3	<2/3	>2/3

Fig. 5 Representative pictures and scoring system for the evaluation of the tumor cell invasion: "+++" >2/3 of the brain slice is invaded by cancer cells, "++" <2/3, "+" <1/3, and "0" no invasion except some single cancer cells. At least ten plugs per condition must be analyzed by confocal microscopy in a blind manner. Reproduced from Siam et al. [11] with permission under the terms of the Creative Commons Attribution License

9. Stain the microglial cells with 200–250 μL of ILB4-Alexa Fluor 647 in PBS-T for 1 h at RT in darkness (*see* **Note 14**).
10. Counterstain the brain slice coculture with DAPI (1:1000 in PBS-T) for 5 min at RT in darkness.
11. Mount and coverslip the brain slice coculture with mounting medium (*see* **Notes 15** and **16**).
12. Image the stained samples using a confocal microscope.
13. Evaluate the grade of tumor invasion according to the area of the brain slice invaded by the tumor cells. For the evaluation of the invasion, the scoring system shown in Fig. 5 can be used.

4 Notes

1. Cells should be regularly tested for mycoplasma.
2. Our cell lines are sensitive to G418 in a range between 100 and 1000 μg/mL. This range can be used as a reference but a previous cytotoxicity test (MTT assay) is always recommended in order to set the sensitivity range for each cell line.
3. It is very important not to damage the brain while removing it from the skull. To this purpose cut the thin skull in the middle starting from the back part of the brain (neck). Using jeweler forceps try to remove both parts of the skull setting them aside.
4. To avoid the brain getting dry and/or being damaged due to oxidation keep it in ice-cold medium during the whole process.

5. Depending on the mouse species and the age, four to six whole brain slices can be obtained from a single mouse brain.
6. Do not to pipette the medium directly onto the slice to avoid it being destroyed.
7. Trypsinization time varies depending on the cell type. If detachment is low, place flask in the incubator and inspect it microscopically every 5 min. To facilitate cell detachment, "bang" the flask on a hard surface or tap on the side of the flask with your hand. Be careful not to break the flask.
8. We usually prepare the amount necessary for ten plugs (10^6 tumor cells in 180 μL of ECM). It is also recommended to prepare about 20% more volume than needed as you may lose some microliters by ECM pipetting.
9. The ECM gel must always be kept in ice to avoid it becoming hardened.
10. The tumor plug must be placed in contact with the brain slice but both entities should not overlap.
11. If a shorter staining protocol is wanted, for example by only staining the microglia, it can be done without previous blocking.
12. Putting some Parafilm around the plate in order to avoid the evaporation of the staining solution is recommended.
13. In case the cells are RFP-transfected, the astrocytes can also be stained in green using goat anti-mouse-FITC antibody.
14. If staining only the microglial cells, another color can be chosen, for example ILB4-Alexa Fluor 568 (red).
15. Handle the slice gently while transferring it to the glass slide. Add a drop of mounting media to the coverslip and place it very carefully on the slice. Do not press the coverslip!
16. Keep the slices at 4 °C in darkness until evaluation on the confocal microscope. Under these conditions the slices can be stored for 6–12 months.

Acknowledgments

The authors thank Han-Ning Chuang, Raphaela Lohaus, Faramarz Dehghani, and Uwe-Karsten Hanisch for their excellent technical advice and support. This work is funded by the German Research Council (DFG) in Project 2 of Forschergruppe 942 (FOR942 BI 703/3-1) and by the German Federal Ministry of Education and Science (BMBF) project MetastaSys in the platform Medical Systems (0316173).

References

1. Chambers AF, Groom AC, Macdonald IC (2002) Dissemination and growth of cancer cells in metastatic sites. Nat Rev Cancer 2:563–572
2. Kienast Y, Von BL, Fuhrmann M et al (2009) Real-time imaging reveals the single steps of brain metastasis formation. Nat Med 16:116–122
3. Chuang HN, Van Rossum D, Sieger D et al (2013) Carcinoma cells misuse the host tissue damage response to invade the brain. Glia 61:1331–1346
4. Kim SJ, Kim JS, Park ES et al (2011) Astrocytes upregulate survival genes in tumor cells and induce protection from chemotherapy. Neoplasia 13:286–298
5. Lin Q, Balasubramanian K, Fan D et al (2010) Reactive astrocytes protect melanoma cells from chemotherapy by sequestering intracellular calcium through gap junction communication channels. Neoplasia 12:748–754
6. Valiente M, Obenauf AC, Jin X et al (2014) Serpins promote cancer cell survival and vascular co-option in brain metastasis. Cell 156:1002–1016
7. Zhang L, Zhang S, Yao J et al (2015) Microenvironment-induced PTEN loss by exosomal microRNA primes brain metastasis outgrowth. Nature 527:100–104
8. Pukrop T, Dehghani F, Chuang HN et al (2010) Microglia promote colonization of brain tissue by breast cancer cells in a Wnt-dependent way. Glia 58:1477–1489
9. Rietkotter E, Bleckmann A, Bayerlova M et al (2015) Anti-CSF-1 treatment is effective to prevent carcinoma invasion induced by monocyte-derived cells but scarcely by microglia. Oncotarget 6:15482–15493
10. Rietkotter E, Menck K, Bleckmann A et al (2013) Zoledronic acid inhibits macrophage/microglia-assisted breast cancer cell invasion. Oncotarget 4:1449–1460
11. Siam L, Bleckmann A, Chaung HN et al (2015) The metastatic infiltration at the metastasis/brain parenchyma-interface is very heterogeneous and has a significant impact on survival in a prospective study. Oncotarget 6(30):29254–29267

Part III

Micropatterning

Chapter 17

3D Neural Culture in Dual Hydrogel Systems

J. Lowry Curley and Michael J. Moore

Abstract

3D in vitro culture systems may yield physiological outcomes that more closely approximate in vivo behavior. A number of fabrication techniques and hydrogel scaffold materials are available to researchers, but often their implementation is complex and seemingly prohibitive. Herein, we describe a simplistic and adaptable dual hydrogel photolithography method utilized to engineer advanced in vitro systems for studies of neuronal development and characterization.

Key words 3D culture, Hydrogel, Tissue engineering, In vitro technique, Neural model, Electrophysiology, Neuronal characterization, Biomimetic scaffolding

1 Introduction

The increasing interplay between tissue engineering and biology has unlocked a wealth of new techniques geared towards in vitro modeling. Scientists are beginning to understand some limitations of conventional 2D cell culture techniques, while simultaneously the benefits of 3D culture environments have become more apparent over the last decade [1, 2]. Traditional in vitro techniques will still be critical moving forward, but an increased recognition of the need for advanced biomimetic microenvironments is driving advances in 3D scaffold systems [3]. In the context of disease models and therapeutic development, focus has been shifting from merely improving the throughput of culture systems to also advancing the content of data available for analysis [4]. A number of techniques exist for 3D cell culturing, but scaffold and matrix based approaches arguably represent the closest representation to the in vivo environment [5]. Particularly in neural applications, the ability to engineer distinct tissue architecture into cell-based models will be critical to allow for functionally relevant outputs [6].

Herein, we describe a 3D cell culture system optimized for use with neuronal cells in which a dual hydrogel system is utilized in order to mechanically direct axonal outgrowth. This is achieved by

Zuzana Koledova (ed.), *3D Cell Culture: Methods and Protocols*, Methods in Molecular Biology, vol. 1612,
DOI 10.1007/978-1-4939-7021-6_17,

spatially patterned fabrication of a stiff, cell-restrictive hydrogel, which forms a micro-mold into which a compliant, cell-permissive gel can be added [7]. This system serves to constrain growth into a highly dense structure resembling a neural fiber tract. Choice of gels, particularly for the permissive scaffolding, depends highly on the intended application, and notes are included to assist in guiding readers on gel choices. One example is given for a self-assembling permissive gel and for a photo-cross-linked permissive gel, and typically either one or the other would be used for a certain application. Instructions are given for three potential measurements: cell viability, axon growth, and electrophysiology. Though outside of the scope of this chapter, it is important to point out that we have used the same system to incorporate both immobilized and soluble guidance cues, to further direct axonal outgrowth [8, 9].

2 Materials

Prepare all solutions using sterile materials and deionized water. If necessary, use a 0.2 μm filter to sterilize solutions after combining components. Store all reagents at 4 °C unless otherwise noted.

2.1 Polymer Solutions

1. Photo-cross-linkable polyethylene glycol (PEG) solution: 10% (w/v) PEG diacrylate (MW 1000; Sigma Aldrich) and 0.5% (w/v) Irgacure 2959 (Ciba Specialty Chemicals) in Neurobasal medium. This hydrogel is used as a cell restrictive component to define the geometry of the cell permissive gel (*see* **Note 1**).
2. 0.30% Puramatrix solution (self-assembling permissive hydrogel, for inclusion of dissociated cells): 0.30% Puramatrix, 1 μg/mL laminin, and 1 μg/mL collagen in sterile deionized water (*see* **Note 2**).
3. Glycidyl methacrylate hyaluronic acid (GMHA) solution (photo-cross-linking permissive hydrogel, an alternative to self-assembling hydrogels): Dissolve 32% methacrylated GMHA (Sigma Aldrich) overnight to 4% in phosphate buffered saline (PBS) without Ca^{2+} and Mg^{2+} supplemented with 1% Irgacure 2959 and 0.03% (v/v) N-vinyl-pyrrolidone (*see* **Note 3**).

2.2 Cell Culture Components

1. Permeable cell culture support inserts, 0.4 μm polystyrene.
2. Hydrophobic solution: Rain-X (Sopus Products).
3. Adhesion medium: 10% (v/v) fetal bovine serum, 100 U/mL penicillin and 100 μg/mL streptomycin, 0.5 mM L-glutamine, and 20 μg/mL NGF in Neurobasal medium.
4. Growth medium: 1× B-27, 100 U/mL penicillin, 100 μg/mL streptomycin, 0.5 mM L-glutamine, and 20 μg/mL NGF in Neurobasal medium.

5. Wash solution: 100 U/mL penicillin and 100 μg/mL streptomycin in phosphate buffered saline (PBS) with Ca^{2+} and Mg^{2+}.
6. Sucrose solution: 20% sucrose in sterile deionized water.
7. Cells: Embryonic day 15 (E15) primary embryonic dorsal root ganglion (DRG) cells are primarily used, though any explant or cell line can be utilized.

2.3 Electrophysiology Components

1. Artificial cerebrospinal fluid (ACSF): 124 mM NaCl, 5 mM KCl, 26 mM $NaHCO_3$, 1.23 mM NaH_2PO_4, 4 mM $MgSO_4$, 2 mM $CaCl_2$, and 10 mM glucose in water.
2. Borosilicate glass capillary tubes with thin walls, outer diameter (OD) of 1.5 mm and inner diameter (ID) of 1.16 mm (Thomas Scientific).
3. Concentric bipolar electrode, platinum-iridium (FHC, Inc.).

2.4 Cell Staining Components

1. Viability testing solution: Live/Dead® Viability/Cytotoxicity Kit with 2 μM calcein AM and 4 μM EthD-1 working solution (Invitrogen).
2. CellTracker™ CMTPX Dye (Thermo Scientific) (*see* **Note 4**).
3. Fixative: 4% paraformaldehyde in PBS.
4. Blocking solution: 0.1% (v/v) Triton X-100 and 2% (w/v) bovine serum albumin in PBS.
5. Staining wash solution: 0.1% (v/v) Triton X-100 in PBS.
6. Primary antibody: Mouse monoclonal Anti-beta III Tubulin antibody [2G10] (Abcam, ab78078).
7. Secondary antibody: Goat Anti-Mouse IgG H&L (Alexa Fluor® 488) preadsorbed (Abcam, ab150117) (*see* **Note 5**).
8. Inverted fluorescence or confocal microscope.
9. Image processing and analysis software, such as ImageJ.

2.5 Digital Projection Photolithography

Projection photolithography may be achieved simply by attaching a digital spatial light modulator (e.g., Mosaic by Andor or Polygon by Mightex) to an upright microscope. A UV light source in conjunction with a low power objective lens will achieve micropatterning of hydrogel solutions.

Alternatively, a custom projection lithography apparatus may be constructed according to Fig. 1, which utilizes a digital micromirror device OEM kit (Digital Light Innovations or Wintech Digital) and UV-grade optics for projecting a black-and-white image onto the surface of a permeable cell culture support insert containing a hydrogel solution. A dichroic beam splitter may be used to enable alternative projection of both visible and UV light.

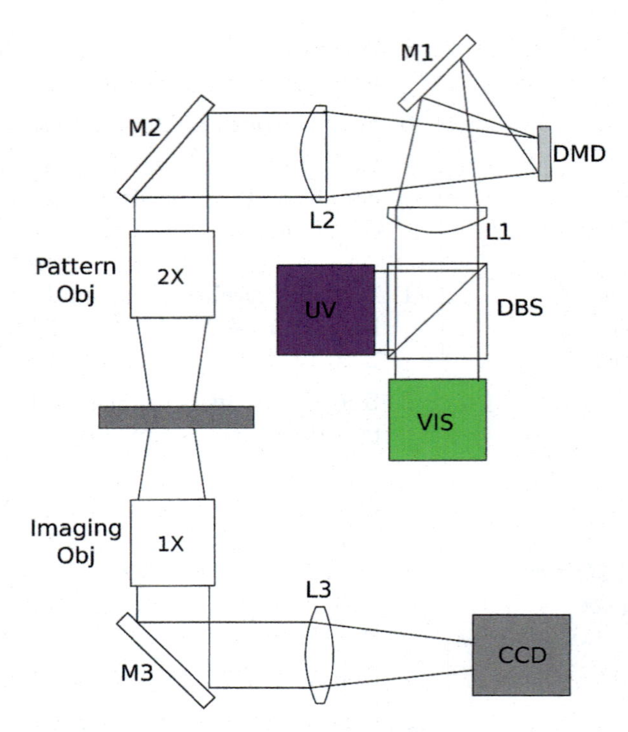

Fig. 1 Diagram of digital projection photolithography apparatus. A 4F optical system was constructed to translate the image of the DMD at ½ size to the specimen stage, where the projected image could be visualized with a CCD. *DBS* dichroic beam splitter (409 nm), *L1 and L2* = 2-in. UV-grade plano-convex lenses (FL = 100 mm), *M1* 2-in. rotating UV-reflective mirror, *M2* 2-in. UV fold mirror, *M3* 1-in. visible-reflective fold mirror, and *L3* = 1-in. biconvex lens (FL = 1 mm). Diagram not to scale

3 Methods

3.1 Dual Hydrogel Fabrication

3.1.1 Restrictive Gel Fabrication by Digital Projection Photolithography

1. Load an appropriate mask on the digital micromirror device setup (Fig. 2).
2. Coat Rain-X on a microscope slide as directed by manufacturer.
3. Apply Rain-X to cell culture insert walls with a cotton swab.
4. Place cell culture insert onto microscope slide beneath digital micromirror device.
5. Add 500 μL of PEG solution to the insert (*see* **Note 6**).
6. Expose PEG solution to UV light for between 40 and 60 s, depending on light intensity at the projection focal plane.
7. Rinse three times with wash solution.
8. Using a carefully rolled Kimwipe, remove excess unpolymerized solution from the well and inside the voids in the polymerized PEG.

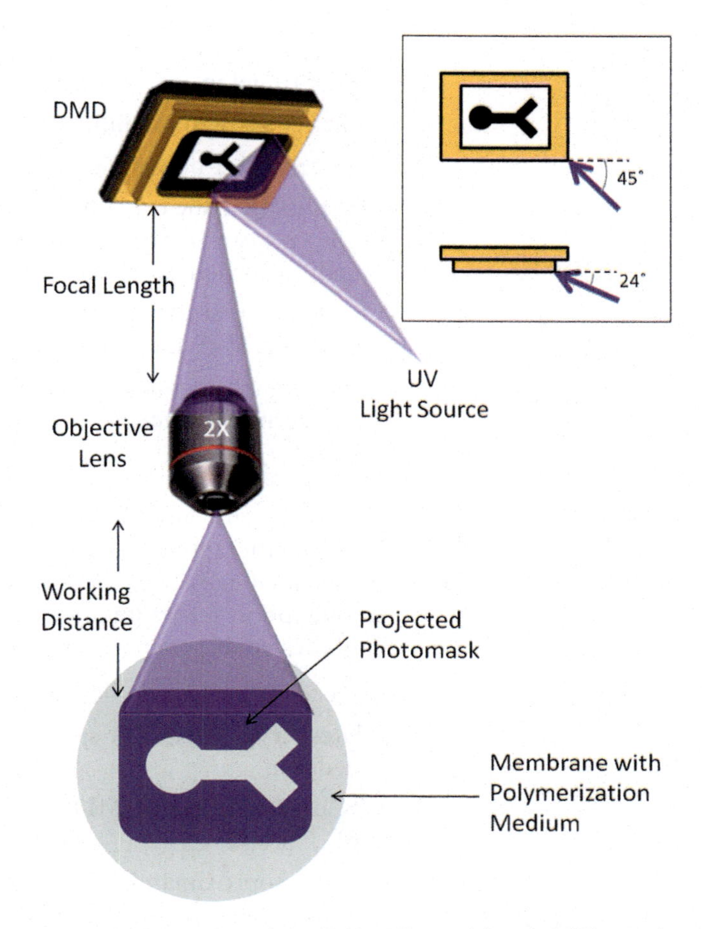

Fig. 2 Schematic illustration of the light path used for photolithography. *Inset*: The UV light illuminates the DMD at an angle of 45°, and 24° below the plane of the mirror array. Reproduced from Curley et al. [13] with permission from JoVE

3.1.2 Permissive Gel Fabrication by Self-Assembly Cross-Linking

Neuronal cells and/or supportive cells will be added two ways, during permissive gel polymerization and after. Combinations of the two techniques enable even more complex environments.

1. To add cells during polymerization:
   (a) After trypsinization, spin supportive cells down and be careful to remove all medium (*see* **Note 7**).
   (b) Mix 0.30% Puramatrix solution and 20% sucrose solution at the ratio 1:1, resulting in 0.15% Puramatrix/10% sucrose solution.
   (c) Resuspend cells in the 0.15% Puramatrix/10% sucrose solution at approximately 1×10^6 cells/mL.
   (d) Inject approximately 5 μL of the Puramatrix/cell/sucrose mixture inside voids in PEG using gel loading tips under a microscope.

   (e) Add 1.5 mL of growth medium underneath the insert, and allow to gel inside incubator for 1 h.
   (f) Change growth medium after 1 h to equilibrate pH.
2. To add cells after polymerization:
   (a) If utilizing tissue explants, dissect desired tissue. If utilizing cells, trypsinize them and resuspend in the appropriate medium at desired concentration.
   (b) Physically place the cells/tissue explants within pre-cross-linked 0.15% Puramatrix from **step 1**.
3. Change growth medium every 2–3 days until axonal growth fills the permissive gel or as required depending on desired application.

3.1.3 Permissive Gel Fabrication by Photo-cross-linking

This represents an alternative to self-assembling hydrogels. Depending on the application, one type or the other should be chosen. Neuronal cells and/or supportive cells will be added two ways, during photo-cross-linking or after. Combinations of the two techniques enable even more biomimetic environments.

1. To add cells during polymerization,
   (a) After trypsinization, spin cells down and remove all medium.
   (b) Resuspend cells in GMHA at approximately 1×10^6 cells/mL.
   (c) Inject approximately 5 μL of the GMHA/cell mixture inside voids in PEG using gel loading tips under a microscope.
   (d) Expose the GMHA solution to either visible light or UV light for between 40 and 60 s, depending on light intensity, to cross-link. The minimal exposure time is preferred, as excessive light exposure can potentially damage cells.
   (e) Add 1.5 mL of growth medium underneath the insert, and allow to equilibrate inside incubator for 1 h.
   (f) Change growth medium after 1 h to equilibrate pH.
2. To add cells after polymerization,
   (a) If utilizing tissue explants, dissect desired tissue. If utilizing cells, trypsinize them and resuspend in the appropriate medium at desired concentration.
   (b) Physically place the cells/tissue within pre-cross-linked GMHA gel from **step 1**.
3. Change growth medium after 24 h and every 2–3 days after until axonal growth fills the permissive gel or as required depending on desired application.

3.2 Cell Characterization

3.2.1 Cell Viability

After an appropriate amount of time, typically at least 24 h and depending on application, cell viability can be assessed or the cells can be tracked by staining.

1. Add 2 mL of the Live/Dead Viability testing solution underneath the constructs.
2. Allow approximately 30 min for the solution to diffuse into the gel and label the cells.
3. Use confocal or fluorescent microscopy to visualize subpopulations of cells and cell processes. Choice of microscopy technique depends on hydrogel thickness used.
4. Combine image sections from confocal microscopy using the Z-project function in ImageJ.
5. Count the number of live or dead labeled cells manually, with Image J plugins, or using custom Matlab scripts (Fig. 3).

3.2.2 Cell Tracking

1. Dilute the stock CellTracker™ CMTPX Dye in serum free medium to a working concentration of ~5 μM and warm the solution to 37 °C.
2. Add approximately 2 mL underneath the cell culture insert for 30 min to label cells.
3. Alternatively, cells can be labeled prior to addition into the 3D environment to track individual cell types.
4. Use epifluorescence or confocal microscopy to visualize cell migration, proliferation, and axonal outgrowth.

3.2.3 Antibody Staining

1. Fix constructs in fixative for 1 h to accommodate for diffusion.
2. To reduce nonspecific binding of the antibody, block the construct with the blocking solution for 2 h at 4 °C.
3. Add the primary antibody diluted in the blocking solution and incubate overnight at 4 °C.
4. Wash three times for 30 min each at 4 °C with staining wash solution.
5. Add the secondary antibody diluted in the blocking solution and incubate for 1 h at 4 °C.
6. Wash again three times for 30 min each at 4 °C with staining wash solution.
7. Visualize with confocal or epifluorescence microscopy (Fig. 4).

3.2.4 Axon Growth Measurement

Because neuronal growth is typically very dense in 3D environments, traditional axon counting is difficult. Instead using a volumetric pixel analysis is beneficial for quantification of extension and growth.

1. Project confocal z-stacks into a 2D images using Image-J for a pseudo-volumetric analysis of pixel density.

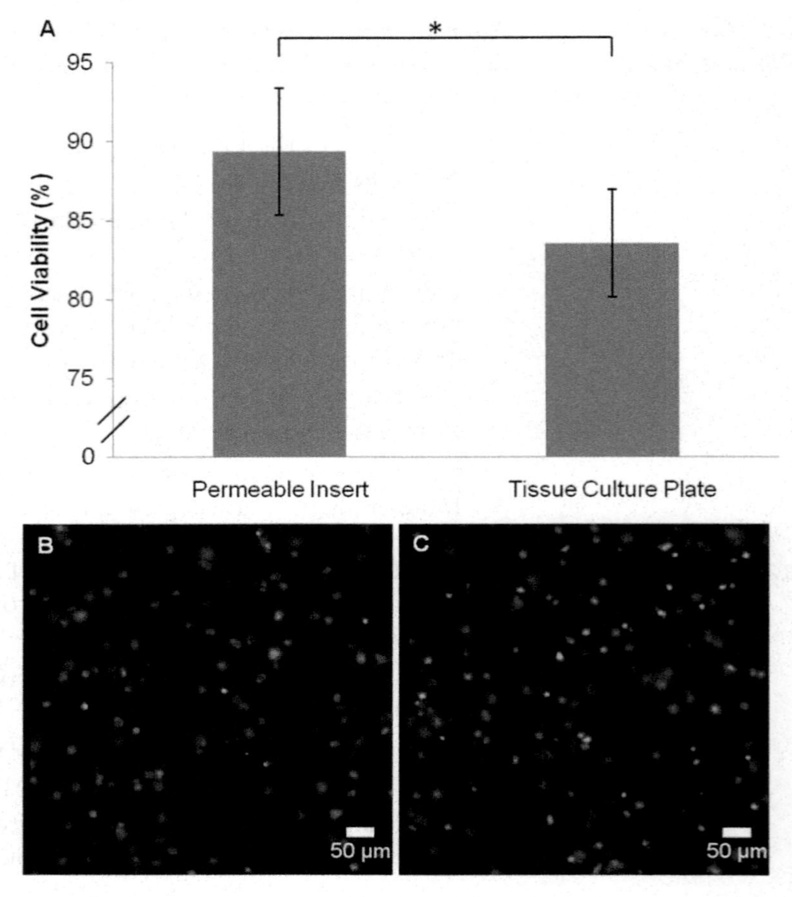

Fig. 3 Cell viability in Puramatrix constructs on tissue culture plates versus permeable supports. (**a**) Graph representing the average percentage of live cells for both conditions. Representative images of Live/Dead assay on (**b**) 24-well permeable insert and (**c**) 24-well tissue culture plate with live cells represented in *green* and dead cells in *red*. Median images are shown as representatives of each experimental group. Reproduced from Curley and Moore [7]. © 2011 Wiley Periodicals, Inc. Used by permission

2. Threshold the image.
3. Count the area or density of fluorescence (*see* **Note 8**).

3.2.5 Electrophysiological Testing

Electrophysiological techniques could be an entire chapter on their own, but described below are the basics of integrating the 3D environments with traditional extracellular and intracellular recording.

1. To accommodate perfusion chambers dimensions, individual hydrogel constructs must be cut out of the cell culture insert membranes.
2. Perfuse with ACSF at 37 °C during electrophysiological testing.

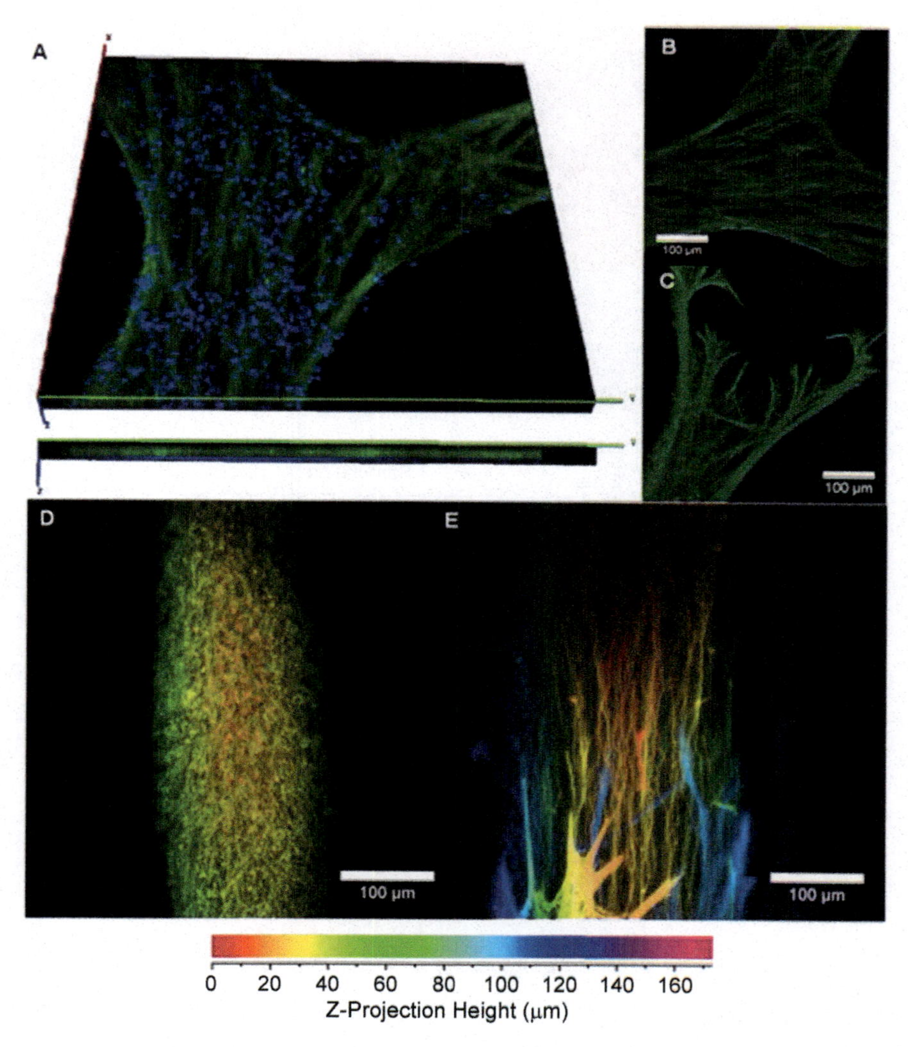

Fig. 4 Confocal micrographs of β III tubulin (*green*) and DAPI (*blue*, A only) stained constructs. (**a**) 3D representation of growth near bifurcation point, showing both an orthographic view, and a side view to demonstrate thickness. Image slices were interpolated to account for distance between slices. (**b**) Merged Z-stack projection of neurite growth in dual hydrogel construct. (**c**) Merged Z-stack projection of neurite growth in PEG construct without Puramatrix. (**d**) Depth-coded Z-stack projection of neurite growth in PEG construct without Puramatrix. (**e**) Depth-coded Z-stack projection of neurite growth in dual hydrogel construct. Reproduced from Curley and Moore [7]. © 2011 Wiley Periodicals, Inc. Used by permission

3. Measure population level field recordings by placing a field electrode near the neuronal cell bodies and applying the stimulating electrode at the distal axons (Fig. 5).
4. For field recording electrodes, use borosilicate glass capillary tubes pulled to obtain resistances between 3 and 8 MΩ.
5. Backfill the glass capillary tube 3/4 full with ACSF, and use a micromanipulator to place the recording electrode inside the

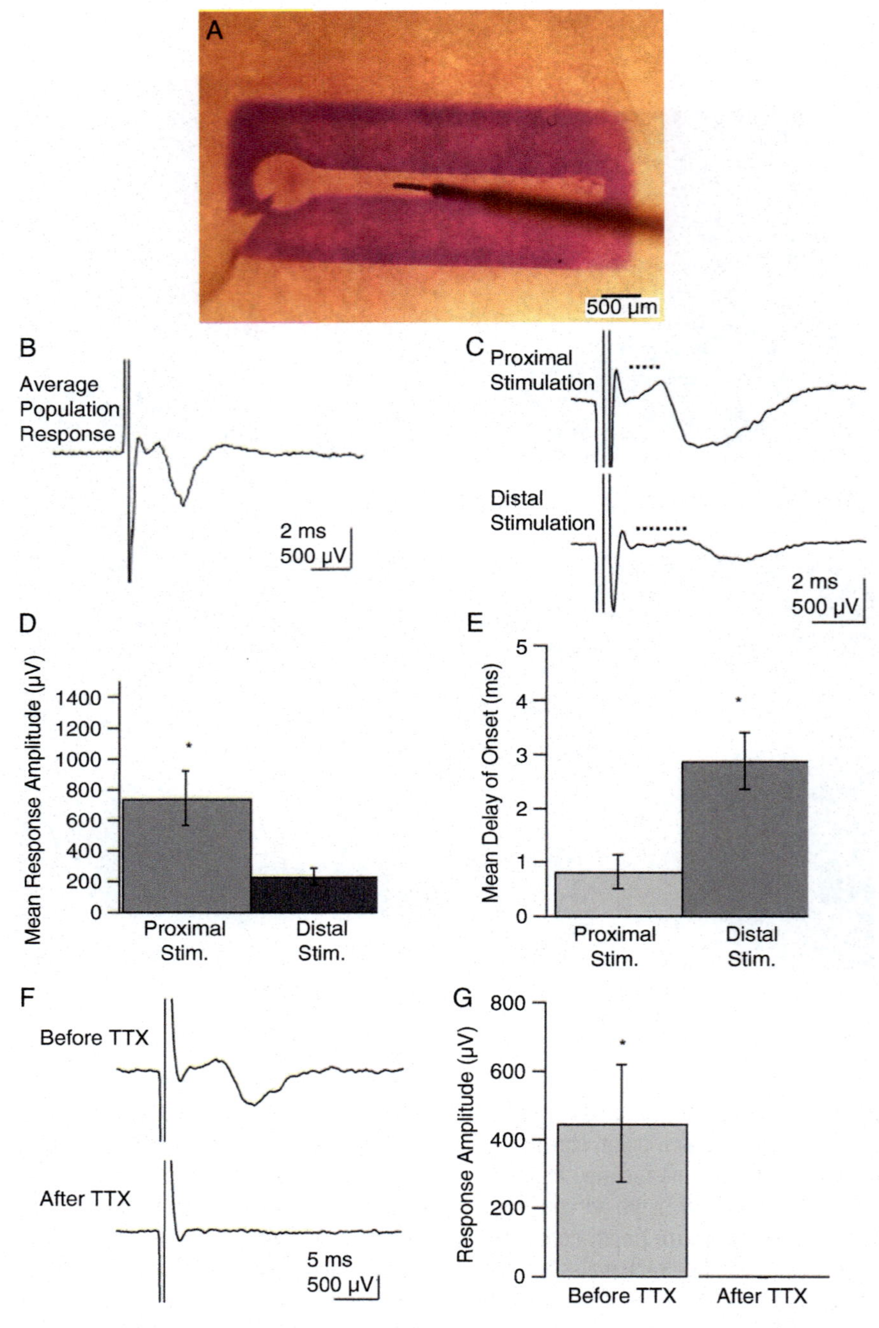

Fig. 5 Dual hydrogel construct with placement of recording (*left*) and stimulating (*right*) electrodes placed within ganglion and neural tract in channel, respectively, for field recording (**a**). Example trace of population response demonstrating successful field potential recordings in 3D neural constructs and waveform properties characteristic of a compound action potential (**b**, CAP). Field potential evoked in ganglion of 3D neural cultures from proximal (1.5 mm) and distal (2.25 mm) locations, $n = 4$. Marked by *dotted lines*, average traces highlight the increase in delay to onset when stimulating distally (**c**). Distal stimulation caused a significant decrease ($p = 0.03$) in average response amplitude (**d**), as well as a significant increase ($p = 0.02$) in latency (**e**). Stimulation distances were measured from the start of the straight channel to the point of stimulation. Delay of onset was measured as the time between the return of the stimulus artifact to baseline to the positive peak of the response. Blockade of Na+ channel activity using 0.5 μM tetrodotoxin (TTX) resulted in abolishment of population response, as demonstrated by average traces (**f**). Response amplitudes were significantly different ($p = 0.029$) after TTX wash-in (**g**). Amplitudes were measured peak-to-peak. Reproduced from Huval et al. [6] with permission from The Royal Society of Chemistry

gel, resting just on top of the middle of the DRG or aggregate of neuronal cell bodies.

6. Use a concentric bipolar electrode to stimulate the axons distally. Place the electrode at the desired distance down the fabricated channels using a micromanipulator, usually ½ to ¾ of the length of the channel.
7. Use a stimulus generator to deliver approximately 10 V through the stimulating electrode. You will see a corresponding response in your recording electrode (*see* **Note 9**).
8. Alternatively, single cell recordings are obtained in living cultures using a traditional patch clamp setup. Apply positive pressure to the recording electrode as it is introduced to the hydrogel environment.
9. Position the recording electrode adjacent to the neuron of interest using a micromanipulator.
10. After carefully placing the tip of the electrode on the cell surface, use negative pressure to break the cell membrane to gain intracellular access to the neuron, ensuring a GOhm seal.

4 Notes

1. Alternative gels can be used, so long as the composition and/or stiffness of the hydrogel resists cell adhesion and migration. We used a photopolymerizable gel in conjunction with digital projection lithography for hydrogel patterning. Alternative methods such as static mask, stamping or mold-based lithography can be used [10].
2. Again, alternative gels can be used. These gels must permit cell migration, proliferation and growth, according to specific applications. Matrigel and collagen are alternatives that use chemical and temperature dependent mechanisms for cross-linking [11].
3. The gel described here cross-links under both visible and UV light exposure. If cell encapsulation is desired, visible light exposure has been shown to be less damaging to neuronal cells (unpublished data). This is just one example of a photo-cross-linkable gel, while dextran represents another commonly used gel.
4. A variety of fluorophores are available, choose according to need and appropriate imaging wavelength. Dilute CellTracker™ dye in DMSO to a concentration of 10 mM to obtain a stock solution.
5. Depending on your application, primary antibodies to stain dendrites, or particular cell receptors can also be used in

conjunction with the more general beta III tubulin stain. Secondary fluorescence can be adjusted as necessary as well, but ensure no species cross-reactivity occurs if multiple primary and secondary antibodies are used.

6. Depending on the thickness of 3D environment desired, this volume can be adjusted.
7. Puramatrix self-assembles upon introduction to physiological salt concentrations. Avoid any interaction between uncross-linked Puramatrix and medium or PBS.
8. Use image analysis software or automated Matlab scripts to count number of pixels expressing the fluorescent antibody of choice (*see* Curley et al. [12] for a detailed example).
9. If you do not see the response in your recording electrode, try varying the placement to ensure that both the axons are receiving the stimulus and that your recording electrode is in close proximity to the neuronal bodies themselves. The voltage used for stimulation can also be adjusted, but be sure not to go too high, as the cells can become damaged.

Acknowledgments

This work was supported in part by Tulane University, NIH R21-NS065374, an NSF CAREER Award to M.J.M. (CBET-1055990), and the DoD (W82XWH-12-1-0246).

References

1. Abbott A (2003) Cell culture: biology's new dimension. Nature 424(6951):870–872
2. Cushing MC, Anseth KS (2007) Hydrogel cell cultures. Science 316(5828):1133–1134
3. DeForest CA, Anseth KS (2012) Advances in bioactive hydrogels to probe and direct cell fate. Annu Rev Chem Biomol Eng 3:421–444
4. Astashkina A, Grainger DW (2014) Critical analysis of 3-D organoid in vitro cell culture models for high-throughput drug candidate toxicity assessments. Adv Drug Deliv Rev 69:1–18
5. Breslin S, O'Driscoll L (2013) Three-dimensional cell culture: the missing link in drug discovery. Drug Discov Today 18(5): 240–249
6. Huval RM, Miller OH, Curley JL et al (2015) Microengineered peripheral nerve-on-a-chip for preclinical physiological testing. Lab Chip 15(10):2221–2232
7. Curley JL, Moore MJ (2011) Facile micropatterning of dual hydrogel systems for 3D models of neurite outgrowth. J Biomed Mater Res A 99(4):532–543
8. Horn-Ranney EL, Curley JL, Catig GC et al (2013) Structural and molecular micropatterning of dual hydrogel constructs for neural growth models using photochemical strategies. Biomed Microdevices 15(1):49–61
9. Horn-Ranney EL, Khoshakhlagh P, Kaiga JW et al (2014) Light-reactive dextran gels with immobilized guidance cues for directed neurite growth in 3D models. Biomater Sci 2(10): 1450–1459
10. Tsang VL, Bhatia SN (2004) Three-dimensional tissue fabrication. Adv Drug Deliv Rev 56(11):1635–1647
11. Lee KY, Mooney DJ (2001) Hydrogels for tissue engineering. Chem Rev 101(7): 1869–1880

12. Curley JL, Catig GC, Horn-Ranney EL et al (2014) Sensory axon guidance with semaphorin 6A and nerve growth factor in a biomimetic choice point model. Biofabrication 6(3): 035026
13. Curley JL, Jennings SR, Moore MJ (2011) Fabrication of micropatterned hydrogels for neural culture systems using dynamic mask projection photolithography. J Vis Exp 48:e2636

Chapter 18

3D Cell Culture in Micropatterned Hydrogels Prepared by Photomask, Microneedle, or Soft Lithography Techniques

Seyedsina Moeinzadeh and Esmaiel Jabbari

Abstract

Despite the advantages of three-dimensional (3D) hydrogels for cell culture over traditional 2D plates, their clinical application is limited by inability to recapitulate the micro-architecture of complex tissues. Micropatterning can be employed to modify the homogenous micro-architecture of hydrogels. Three techniques for cell encapsulation in 3D micropatterned gels are described. The photomask and micromold techniques are used for cell encapsulation in relatively shallow patterns like disks or short rectangles but due to the presence of PDMS mold, the resolution of micromold technique is potentially higher than the photomask. The microneedle technique is often used for cell encapsulation in relatively deep microchannels within any geometry.

Key words Cell culture, Hydrogel, Micropatterning, Photomask, Micromold, Soft lithography

1 Introduction

Two-dimensional (2D) cell culture on adherent plates is conventionally used for studying cells physiology and pathophysiology in vitro. A number of studies have reported a significant difference between the morphology, shape, and expression of cells cultured on 2D plates versus those in 3D microenvironment of the native tissue [1, 2]. For example, chondrogenic differentiation of embryonic stem cells increased by cultivation of the cells in 3D embryoid bodies compared to 2D culture [3]. An exciting alternative to 2D cell culture is cell encapsulation in natural or synthetic hydrogels [4]. Hydrogels are 3D hydrophilic macromolecular or polymeric networks that retain a significant fraction of water in their structure in physiological solution [5]. Nutrient molecules, proteins and waste products readily diffuse through the hydrogels mesh [5]. Further, resorbable hydrogels provide temporary mechanical support to the cells for in vivo applications and degrade concurrent with the production of extracellular matrix (ECM) by the cells [6].

Zuzana Koledova (ed.), *3D Cell Culture: Methods and Protocols*, Methods in Molecular Biology, vol. 1612,
DOI 10.1007/978-1-4939-7021-6_18, © Springer Science+Business Media LLC 2017

However, hydrogel networks with a homogenous microstructure do not recapitulate the architecture of complex tissues with spatial heterogeneity on the order of 100 μm [7]. Micropatterning has been used to geometrically confine normal or cancer cells to specific zones or to mimic the architecture of complex tissues [8, 9].

Geometrical confinement affects self-renewal, differentiation, and gene expression of normal and cancer cells [10]. For example, tumor spheroid size and expression of cancer stem cells significantly depends on their geometrical confinement [11]. Micropatterning of hydrogels is used extensively in tissue engineering for geometrical confinement of the encapsulated cells. As an example, the extent of bone formation depends strongly on pattern and distribution of mesenchymal stem cells (MSCs) and endothelial progenitor cells within a hydrogel matrix [12]. In intramembranous ossification during development, bone is formed by invasion of microcapillaries into a soft MSC-laden matrix [12]. Endothelial cells (ECs) lining the inner surface of microcapillaries secrete bone morphogenetic protein-2 (BMP-2) that diffuse through the capillary wall and surrounding tissue to differentiate MSCs to osteoblasts that form a mineralized matrix [13]. The differentiating MSCs on the other hand secrete vascular endothelial growth factors (VEGF) that promote the formation of microvessels [13]. Therefore, micropatterned hydrogels with ECs in the patterns and MSCs in the matrix mimic the process of bone formation during development which is a promising strategy for treatment of skeletal injuries [14]. In this chapter, three techniques for micropatterning cell-laden hydrogels are described. In the first part, the photomask technique is used for encapsulation of MDA-MB-231 breast cancer cells in a micropatterned hydrogel. In the second part, the microneedle technique is used to form a cell-laden multicellular patterned construct with MSCs and endothelial colony-forming cells (MSCs/ECFCs) in fast-degrading low-stiffness microchannels and MSCs in the slow-degrading high-stiffness matrix. In the third part, the soft lithography micromold technique is used to encapsulate ECFCs in microchannels of rectangular geometry [15, 16].

2 Materials

2.1 Synthesis of Star Acrylate-Functionalized Lactide-Chain-Extended Polyethylene Glycol (SPELA) and Polyethylene Glycol Diacrylate (PEGDA) Macromers

1. Star polyethylene glycol (SPEG, MW = 5 kDa, Sigma-Aldrich).
2. Lactide monomer (LA, Ortec).
3. Tin (II) 2-ethylhexanoate (TOC, Sigma-Aldrich).
4. Polyethylene glycol (PEG, MW = 4.6 kDa).
5. Dichloromethane (DCM).
6. Methanol.
7. Ethyl ether.

8. Hexane.
9. Acryloyl chloride.
10. Triethylamine.
11. Dimethyl sulfoxide (DMSO).
12. Deionized water.
13. Three-necked reaction flask with an overhead stirrer.
14. Oil bath.
15. Reaction flask.
16. Dialysis tubes with 3.5 kDa MW cutoff (Spectrum Laboratories).

2.2 Synthesis of GelMA Macromer

1. Gelatin (porcine skin gelatin, type A, 300 bloom).
2. Phosphate buffer saline (PBS).
3. Methacrylic anhydride (MA).
4. Dialysis tubes with 3.5 kDa MW cutoff (Spectrum Laboratories).
5. Deionized water.

2.3 Cell Culture

1. MDA-MB-231, human breast cancer cells.
2. MDA-MB-231 culture medium: 10% FBS, 100 units/mL penicillin and 100 μg/mL streptomycin in RMPI-1640.
3. Human mesenchymal stem cells (MSCs).
4. Basal medium: 10% FBS, 100 units/mL penicillin and 100 μg/mL streptomycin in high-glucose Dulbecco's modified Eagle's medium.
5. Human endothelial colony-forming cells (ECFCs, Boston Children Hospital).
6. 1% gelatin-coated flasks (*see* **Note 1**).
7. ECFC medium: 20% FBS in full EGM-2 medium (BulletKit, Lonza).
8. Sterile phosphate buffered saline (PBS).
9. 0.25% trypsin/0.53 mM EDTA.
10. Cell culture incubator (5% CO_2, 37 °C).
11. Inverted microscope.
12. Centrifuge tubes, 15 mL and 50 mL.
13. Centrifuge.
14. Computer-aided design (CAD) software.
15. Transparent sheets (CAD/Art Services).
16. Irgacure-2959 photo-initiator (CIBA).
17. 0.2 μm filter.
18. 3 M adhesive tape or similar biomedical grade adhesive tape.

19. Glass slides (75 × 25 × 1 mm).
20. UV lamp, long wavelength (365 nm) (e.g., BLAK-RAY 100-W mercury, Model B100-AP; UVP).
21. 12 sterile 400-μm diameter needles (G-20).
22. Teflon cylinder, 5 mm diameter, 3 mm height.
23. Osteogenic medium: 100 nM dexamethasone, 50 μg/mL ascorbic acid, 10 mM β-sodium glycerophosphate in basal medium.
24. Vasculogenic medium: complete EBM-2 medium (Lonza, containing VEGF, human FGF2, IGF-1, human EGF, ascorbic acid and hydrocortisone).
25. Polydimethylsiloxane (PDMS) elastomer and curing agent.
26. 3-(trimethoxysilyl)propyl methacrylate (TMSMA, Sigma).
27. 70% ethanol.
28. Distilled water.
29. Vacuum chamber.
30. Oven.
31. Plasma chamber.
32. SU-8 photoresist (SU-8 2100, MicroChem).
33. Propylene glycol methyl ether acetate (PGMEA, Sigma).
34. Isopropanol.
35. Acetone.
36. Methanol.
37. Isopropyl alcohol (IPA).
38. Spin coater.
39. Silicon wafers (4-in. diameter, 500 μm thickness).
40. Hot plate.

3 Methods

3.1 Synthesis of SPELA Macromer

1. Azeotropically distil SPEG from toluene to remove residual moisture.
2. Dry lactide monomer under vacuum at 40 °C for 12 h.
3. Add 20 g of dry SPEG and 5.8 g of dry lactide monomer in a three-necked reaction flask equipped with an overhead stirrer. Heat with an oil bath to 120 °C under steady flow of dry nitrogen to melt the reactants.
4. Add 1 mL of TOC to the reaction mixture, raise the reaction temperature to 135 °C and allow the reaction to continue for 8 h.

5. Allow the reaction mixture to cool down for 4–5 h under the steady flow of nitrogen.
6. Dissolve the reaction mixture in DCM.
7. Precipitate the reaction mixture first in ice cold methanol, followed by precipitation in ethyl ether and hexane to fractionate and remove the unreacted monomer and initiator.
8. Vacuum dry the star lactide-chain-extended poly(ethylene glycol) (SPEL) product to remove residual solvent. Store the product at −20 °C.
9. To functionalize the reaction product, dissolve 10 g of SPEL in a reaction flask with 50 mL of dry DCM and immerse the flask in an ice bath.
10. Dissolve each 0.8 mL of acryloyl chloride and 1.4 mL of triethylamine in 20 mL of dry DCM.
11. Add the two solutions simultaneously drop-wise to the reaction mixture with stirring.
12. Allow the reaction to proceed for 12 h under nitrogen flow at 0 °C.
13. After completion of the reaction, remove solvent by rotary evaporation and dissolve the residue in anhydrous ethyl acetate to precipitate the by-product triethylamine hydrochloride salt.
14. Remove the ethyl acetate by vacuum distillation.
15. Dissolve the residue in dry DCM and precipitate it twice in ice-cold ethyl ether.
16. Dissolve the precipitate in DMSO and dialyze against distilled deionized water for 24 h to remove the by-products (*see* **Note 2**).
17. Freeze-dry the purified SPELA macromer and store at −20 °C.

3.2 Synthesis of PEGDA Macromer

1. Azeotropically distil PEG from toluene to remove residual moisture.
2. Dissolve 20 g of dried PEG in 100 mL of dry DCM in a reaction flask and immerse the flask in an ice bath to cool the polymer solution and limit temperature rise during the acrylation reaction.
3. Dissolve 1 mL of acryloyl chloride and 1.7 mL of triethylamine each in 20 mL of dry DCM.
4. Add the two solutions simultaneously drop-wise to the reaction flask with stirring.
5. Allow the reaction to proceed for 12 h under nitrogen flow at 0 °C.
6. Follow **steps 13–17** from Subheading 3.1.

3.3 Synthesis of GelMA Macromer

1. Dissolve 1 g of gelatin in 10 mL of phosphate buffer saline (PBS) at 50 °C with rigorous stirring.
2. Add 180 μL of MA to the gelatin solution and allow the methacrylation reaction to continue for 1 h at 50 °C under stirring.
3. After completion of the reaction, dilute the mixture with 40 mL of warm PBS (50 °C) and dialyze it against deionized water at 40 °C for 3 days using dialysis tubes with 3.5 kDa MW cutoff (*see* **Note 2**).
4. Freeze-dry the dialyzed solution to obtain GelMA as white foam and store at −20 °C.

3.4 Cell Culture

1. Culture the MDA-MB-231 cells in MDA-MB-231 culture medium, MSCs in basal medium and ECFCs on 1% gelatin-coated flasks in ECFC medium.
2. When cells reach 70% confluency, remove the medium by aspiration and wash the cell monolayer with PBS.
3. Add trypsin/EDTA to the cells (1 mL of trypsin–EDTA solution per T75 flask).
4. Incubate the cells at 37 °C for 2 min to detach the cells. Check cell detachment using a microscope.
5. Add 1 mL of culture medium to the flask and transfer the cell suspension to a 15 mL centrifuge tube and centrifuge at 220 × *g* for 5 min.
6. After centrifugation, remove supernatant and resuspend the cell pellet in 50 μL of culture medium (1×10^8 cells/mL) and maintain at 37 °C for use in cell encapsulation in patterned hydrogels.

3.5 Engineering Micropatterned Hydrogels Using Photomasks

1. Design micropatterns using a CAD software, e.g., sets of circular micropatterns with 50, 100, 150, and 200 μm diameter as shown in Fig. 1a.
2. Print the designed micropatterns on transparent sheets (CAD/Art Services) for hydrogel patterning with photomask.
3. Dissolve 10 mg Irgacure-2959 photo-initiator in 1 mL of PBS with vortexing and heating to 50 °C to aid dissolution.
4. Dissolve 100 mg of PEGDA macromer in 900 μL of the photo-initiator solution in an Eppendorf tube to form the hydrogel precursor solution (*see* **Notes 3** and **4**).
5. Sterilize the hydrogel precursor solution by filtration with a 0.2 μm filter.
6. Cover the edges of a glass slide with a layer of biomedical grade adhesive tape (Fig. 2) (*see* **Note 5**).
7. Transfer 50 μL of the sterilized hydrogel precursor solution on the glass slide (*see* **Note 5**).

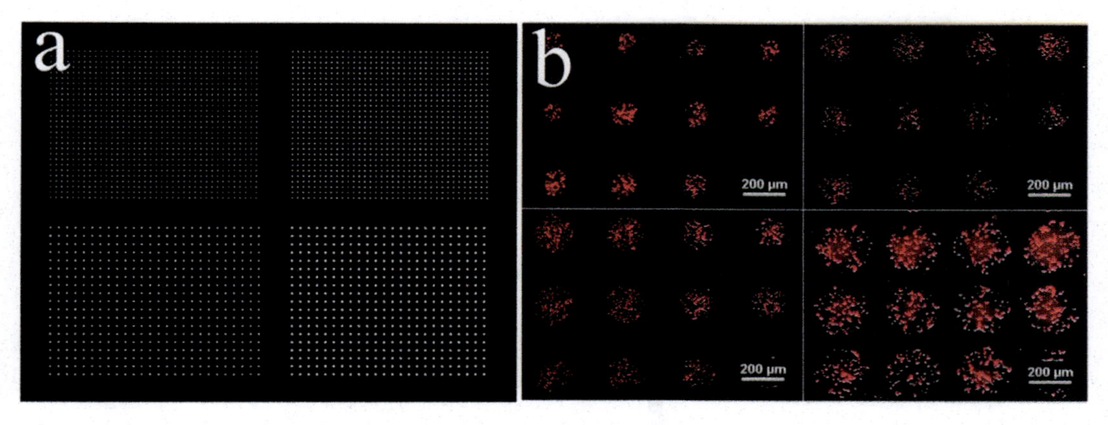

Fig. 1 (**a**) Circular micropatterns with 50, 100, 150, and 200 μm diameter printed on a transparent sheet. (**b**) MDA-MB-231 cancer cells encapsulated in micropatterned PEGDA hydrogel with pattern diameters of 50, 100, 150, and 200 μm

8. Irradiate the hydrogel precursor solution with a long wavelength UV lamp for 8 min (*see* **Note 6**) to form the base layer that will separate the cell-laden hydrogel layer from the glass slide (Fig. 2).
9. Cover the glass edges with a second layer of adhesive tape.
10. Add 5 μL of the MDA-MB-231 cell suspension to 500 μL of PEGDA hydrogel precursor solution to get cell-suspended hydrogel precursor solution with 1×10^6 cells/mL cell density (*see* **Note 7**).
11. Transfer 40 μL of the cell-suspended hydrogel precursor solution on top of the base gel layer (Fig. 2).
12. Place the UV mask over the cell-suspended hydrogel precursor solution and press against the adhesive tape with a glass slide.
13. Irradiate the cell-suspended hydrogel precursor solution with the UV lamp for 5 min to form the pattern of the photomask on the hydrogel.
14. Remove the mask, wash the cell-encapsulated gel with PBS and UV irradiate for an additional 3 min to complete gelation (*see* **Note 6**).
15. Wash the patterned cell-laden gel with PBS (*see* **Note 8**).
16. To fill the void space around the patterns, inject 30 μL of the acellular hydrogel precursor solution on the micropatterned hydrogel, UV-irradiate the solution for 5 min (*see* **Note 6**) and wash the gel with PBS.
17. Using sterile forceps peel the micropatterned hydrogel from the glass slide, place it into a cell culture dish and culture the hydrogel–cell construct in the MDA-MB-231 medium at 37 °C and 5% CO_2.

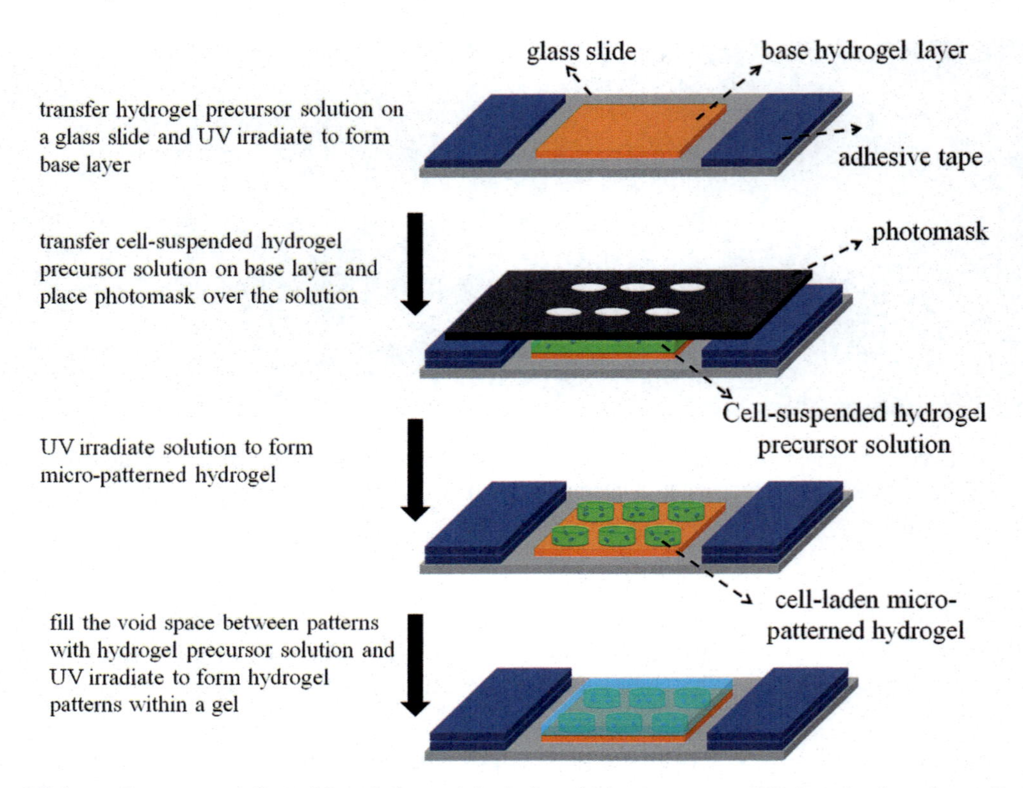

Fig. 2 Schematic representation of the photomask technique to form cancer cell-laden circular micropatterns in a hydrogel matrix

3.6 Patterning Deep Microchannels in a Hydrogel Matrix

1. Insert 12 needles with a needle-to-needle distance of 500 μm through the end-caps of a Teflon cylinder (Fig. 3) (*see* **Note 4**).
2. Dissolve 20 mg Irgacure-2959 photo-initiator (CIBA) in 2 mL of PBS with vortexing and heating to 50 °C to aid dissolution.
3. Dissolve 200 mg of SPELA macromer in 800 μL of the photo-initiator solution in an Eppendorf tube to form the hydrogel precursor solution (*see* **Notes 3** and **4**).
4. Sterilize the SPELA hydrogel precursor solution by filtration with a 0.2 μm filter.
5. Add 10 μL of the human MSC cell suspension to 500 μL of the sterilized SPELA hydrogel precursor solution to get MSC-suspended hydrogel precursor solution with 2×10^6 cells/mL cell density (*see* **Note 7**).
6. Transfer 50 μL of the MSC-suspended SPELA gel precursor solution inside the Teflon cylinder.
7. Irradiate the cell-suspended hydrogel precursor solution inside the Teflon cylinder with the UV lamp for 8 min (*see* **Note 6**) to form the MSC-laden SPELA matrix (Fig. 3).

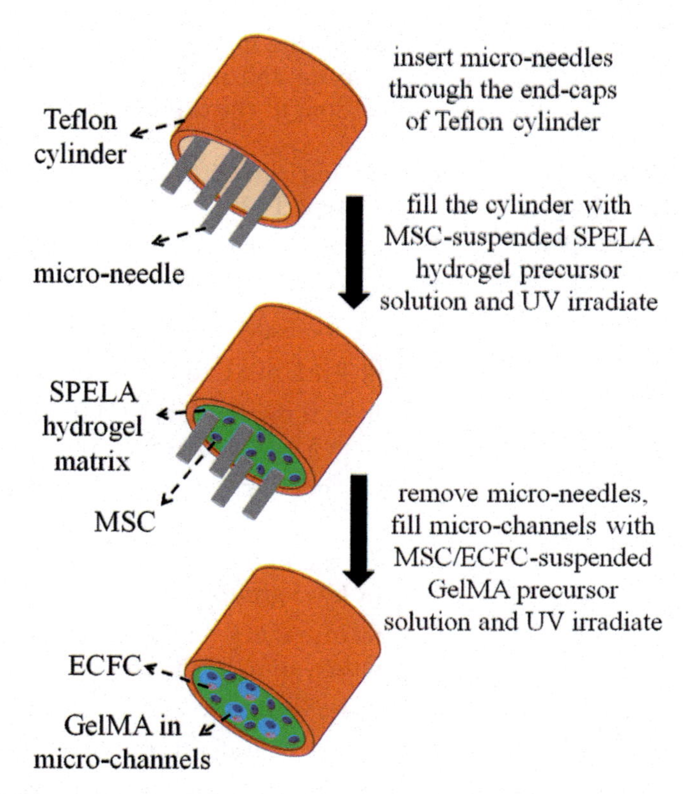

Fig. 3 Schematic representation of the microneedle technique to form a micropatterned hydrogel composed of MSC-laden slow-degrading stiff matrix (SPELA) and MSC/ECFC laden fast-degrading soft microchannels (GelMA)

8. Remove the needles from the cell-laden hydrogel and wash the hydrogel with microchannels with PBS.
9. Dissolve 50 mg of GelMA macromer in 950 μL of the photoinitiator solution with vortexing and heating to 50 °C to aid dissolution (*see* **Notes 3** and **4**).
10. Sterilize the GelMA hydrogel precursor solution by filtration with a 0.2 μm filter.
11. Add 5 μL of MSC cell suspension and 5 μL of ECFC cell suspension to 500 μL of the sterilized GelMA hydrogel precursor solution to generate MSC/ECFC-suspended hydrogel precursor solution with 2×10^6 cells/mL cell density (*see* **Note 7**).
12. Fill the microchannels in the cell-laden hydrogel within the Teflon cylinder with MSC/ECFC-suspended GelMA precursor solution.
13. Irradiate the MSC/ECFC suspended GelMA precursor solution in the microchannels with UV lamp for 8 min (*see* **Note 6**) to form MSC/ECFC-laden GelMA microchannels and wash the patterned cellular constructs with PBS (Fig. 3).

14. Transfer the micropatterned cell-laden hydrogel to a cell culture plate and incubate in osteogenic/vasculogenic medium according to the experiment at 37 °C and 5% CO_2.

3.7 Patterning Cell-Laden Hydrogels by Soft Lithography

1. Print micropatterns on a transparent sheet (*see* Subheading 3.5, **steps 1–2**).
2. Clean the wafer with acetone followed by methanol and IPA.
3. Bake the wafer at 200 °C for 5 min.
4. Place the wafer on the chuck center of a spin coater.
5. Add 4 mL of SU-8 to center of the wafer.
6. Spin coat the SU-8 at 500 rpm for 5 s (100 rpm/s ramp) and then at 510 × *g* for 45 s (300 rpm/s ramp). Then ramp down.
7. Place the wafer on a hot plate at 65 °C for 5 min followed by 95 °C for 20 min and then cool down to 65 °C.
8. Place the photo-mask on top of the wafer substrate. Irradiate with UV light for 90 s.
9. Place the wafer on the hot plate at 65 °C and ramp the temperature up to 95 °C and hold the wafer at 95 °C for 11 min.
10. Dip the wafer into a PGMEA bath placed on a shaker for 90 s.
11. Remove the wafer from PGMEA bath and rinse that with IPA.
12. Place the wafer on a hot plate at 150 °C for 20 min and then cool down.
13. Mix PDMS elastomer and curing agent in a weight ratio of 10:1 to form the PDMS rubber precursor mixture and degas the viscous liquid in a vacuum chamber to remove bubbles.
14. Transfer the PDMS precursor mixture onto the micropatterned silicon wafer.
15. Transfer the silicon wafer to an oven at 60 °C for 4 h to cure the PDMS precursor mixture to form a micropatterned PDMS mold.
16. Carefully peel the PDMS mold from the silicon wafer and wash the mold in 70% ethanol.
17. Dry the mold at 65 °C for 10 min.
18. Transfer the PDMS mold to a plasma chamber for 3 min and 60 W power to clean and make the mold surface hydrophilic.
19. Clean a glass slide in a plasma chamber cleaner for 5 min at 60 W.
20. Coat the glass slide with TMSMA.
21. Incubate the TMSMA coated glass slide at 100 °C for 30 min followed by 110 °C for 10 min to make a methacrylated glass slide to prevent adhesion of the hydrogel layer to the PDMS mold.

22. After surface coating, rinse the glass slide with distilled water and dry.
23. Dissolve 10 mg of Irgacure-2959 photo-initiator in 1 mL of PBS with vortexing and heating to 50 °C to aid dissolution.
24. Dissolve 50 mg of GelMA macromer in 950 μL of the photo-initiator solution with vortexing and heating to 50 °C to form the GelMA hydrogel precursor solution (*see* **Notes 3** and **4**).
25. Sterilize the GelMA hydrogel precursor solution by filtration with a 0.2 μm filter.
26. Add 10 μL of ECFC suspension to 500 μL of the sterilized GelMA hydrogel precursor solution to generate ECFC-suspended hydrogel precursor solution with 2×10^6 cells/mL cell density (*see* **Note** 7).
27. Transfer 20 μL of cell-suspended GelMA precursor solution onto the PDMS mold (Fig. 4).

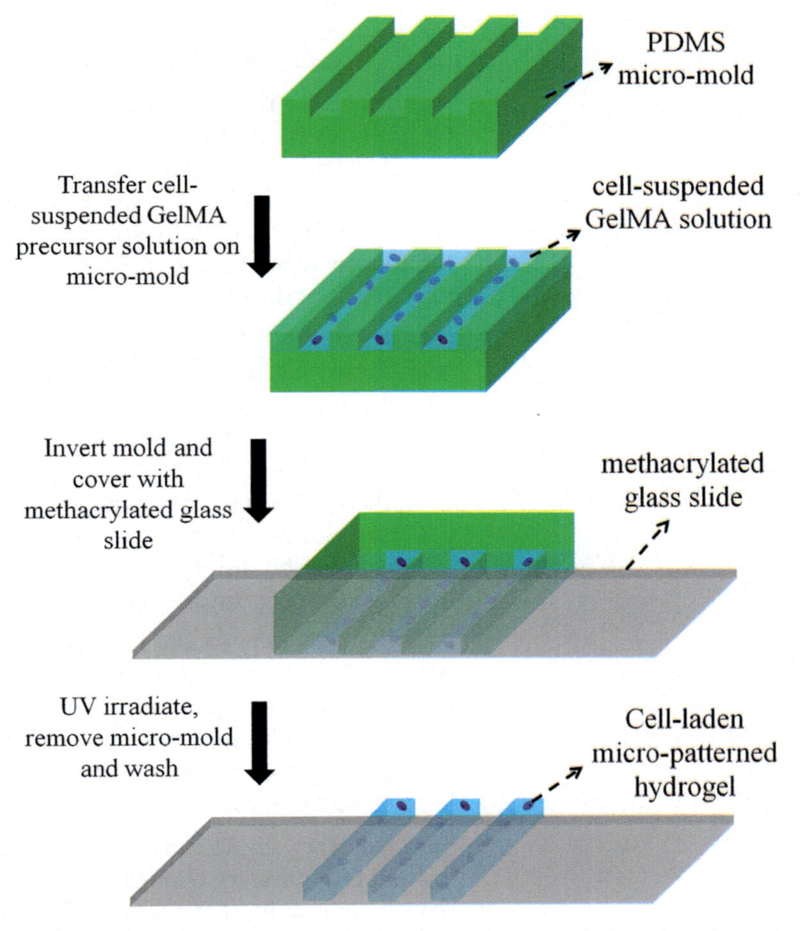

Fig. 4 Schematic representation of the soft lithography technique for micropatterning a cell-laden hydrogel

28. Invert the mold and cover it with the methacrylated glass slide (Fig. 4).
29. Irradiate the hydrogel solution with a UV lamp for 8 min (*see* **Note 6**) to crosslink the cell-laden precursor solution.
30. Peel off the PDMS mold from the hydrogel, remove the hydrogel from the coated glass, and wash the cell-laden micropatterned hydrogel with PBS.
31. Transfer the micropatterned cell-laden hydrogel to a cell culture plate and incubate in vasculogenic medium at 37 °C under 5% CO_2.

4 Notes

1. In order to coat the flasks with gelatin, pipette 1% gelatin in PBS into the flask, spread it to cover the entire cell culture surface and incubate for 24 h at room temperature. Then remove the gelatin solution from the flask by aspiration. Use fresh gelatin-coated flasks to culture ECFCs.
2. For dialysis of the synthesized macromers, hydrate the membrane in distilled water before loading the sample. Make sure that two ends of the dialysis membrane are clamped properly and there is no leak. It is recommended to use a volume of distilled water that is 200 times larger than the volume inside the dialysis tube. For example, dialyze 10 mL of polymer solution in 2 L of distilled water. Change the distilled water at least four times to ensure purification of the polymers.
3. Make sure that macromers (PEGDA, SPELA or GelMA) are completely dissolved in the photoinitiator solution and the solution is clear before filtration. If there are polymer aggregates larger than 0.2 μm (filter pore size) in the precursor solution, the filtration will be difficult, some of the polymer cannot pass through the filter, and the concentration of polymer after filtration will be different than the desired concentration. Use heating and vortexing to aid dissolution of the macromers.
4. Add the polymers to the photoinitiator solution immediately before making the photo-patterned gels. If you intend to make multiple patterned gels or a multilayered gel, cover the tube containing hydrogel precursor solution with aluminum foil during the experiment to prevent premature gelation of the light sensitive macromers.
5. Sterilize every single tool that is needed for the micropatterning experiments to eliminate contamination. Autoclave the spatula, micropipette tips, and microneedles. Sterilize the glass slides, adhesive tapes, and photo-masks with 70% ethanol solu-

tion. Filter the polymer solution as well as the cell media using 0.2 μm filters. Do all the steps under the laminar hood.

6. The gelation time of PEGDA, SPELA, and GelMA polymer solutions depend on the degree of acrylation of macromers which vary from batch to batch. Therefore, the optimal gelation time for photo-patterning may be slightly different from those mentioned in the text. Therefore, it is recommended that for each batch the gelation time is tested before conducting the photo-patterning experiments.
7. Make the patterned gels immediately after suspending cells in the polymer solution. Keeping the cells in the polymer solution for a long time negatively affects cells viability.
8. After photopatterning, make sure that excess polymer solution is washed with PBS and the gel surface is dried before filling the void space or adding another layer of gel.

Acknowledgments

This work was supported by research grants to E. Jabbari from the National Science Foundation under Award Number CBET-1403545 and the National Institute of Arthritis and Musculoskeletal and Skin Diseases of the National Institutes of Health under Award Number AR063745. The content is solely the responsibility of the authors and does not necessarily represent the official views of the National Institutes of Health.

References

1. Hess MW, Pfaller K, Ebner HL et al (2010) 3D versus 2D cell culture: implications for electron microscopy. Methods Cell Biol 96:617–637
2. Ravi M, Paramesh V, Kaviya SR et al (2015) 3D cell culture systems: advantages and applications. J Cell Physiol 230(1):16–26. doi:10.1002/jcp.24683
3. Tanaka H, Murphy CL, Murphy C et al (2004) Chondrogenic differentiation of murine embryonic stem cells: effects of culture conditions and dexamethasone. J Cell Biochem 93(3):454–462. doi:10.1002/jcb.20171
4. Tibbitt MW, Anseth KS (2009) Hydrogels as extracellular matrix mimics for 3D cell culture. Biotechnol Bioeng 103(4):655–663. doi:10.1002/bit.22361
5. Peppas NA, Hilt JZ, Khademhosseini A et al (2006) Hydrogels in biology and medicine: from molecular principles to bionanotechnology. Adv Mater 18(11):1345–1360. doi:10.1002/adma.200501612
6. Drury JL, Mooney DJ (2003) Hydrogels for tissue engineering: scaffold design variables and applications. Biomaterials 24(24): 4337–4351. doi:10.1016/S0142-9612(03)00340-5
7. Gore JC, Xu JZ, Colvin DC et al (2010) Characterization of tissue structure at varying length scales using temporal diffusion spectroscopy. NMR Biomed 23(7):745–756. doi:10.1002/nbm.1531
8. Applegate MB, Coburn J, Partlow BP et al (2015) Laser-based three-dimensional multiscale micropatterning of biocompatible hydrogels for customized tissue engineering scaffolds. Proc Natl Acad Sci U S A 112(39):12052–12057. doi:10.1073/pnas.1509405112
9. Khetan S, Burdick JA (2011) Patterning hydrogels in three dimensions towards controlling cellular interactions. Soft Matter 7(3): 830–838. doi:10.1039/c0sm00852d

10. Chen Z, Xu WR, Qian H et al (2009) Oct4, a novel marker for human gastric cancer. J Surg Oncol 99(7):414–419. doi:10.1002/jso.21270
11. Jabbari E, Sarvestani SK, Daneshian L et al (2015) Optimum 3D matrix stiffness for maintenance of cancer stem cells is dependent on tissue origin of cancer cells. PLoS One 10(7):e0132377. doi:10.1371/journal.pone.0132377. ARTN
12. Kanczler JM, Oreffo ROC (2008) Osteogenesis and angiogenesis: the potential for engineering bone. Eur Cell Mater 15:100–114
13. Kempen DHR, Creemers LB, Alblas J et al (2010) Growth factor interactions in bone regeneration. Tissue Eng Part B Rev 16(6):551–566. doi:10.1089/ten.teb.2010.0176
14. Barati D, Shariati SRP, Moeinzadeh S et al (2016) Spatiotemporal release of BMP-2 and VEGF enhances osteogenic and vasculogenic differentiation of human mesenchymal stem cells and endothelial colony-forming cells co-encapsulated in a patterned hydrogel. J Control Release 223:126–136. doi:10.1016/j.jconrel.2015.12.031
15. Khademhosseini A, Eng G, Yeh J et al (2006) Micromolding of photocrosslinkable hyaluronic acid for cell encapsulation and entrapment. J Biomed Mater Res A 79A(3):522–532. doi:10.1002/jbm.a.30821
16. Hansen AS, Hao N, O'Shea EK (2015) High-throughput microfluidics to control and measure signaling dynamics in single yeast cells. Nat Protoc 10(8):1181–1197. doi:10.1038/nprot.2015.079

Chapter 19

3D Stem Cell Niche Engineering via Two-Photon Laser Polymerization

Michele M. Nava, Tommaso Zandrini, Giulio Cerullo, Roberto Osellame, and Manuela T. Raimondi

Abstract

A strategy to modulate the behavior of stem cells in culture is to mimic structural aspects of the native cell–extracellular matrix (ECM) interaction. An important example of such artificial microenvironments for stem cell culture is the so-called "synthetic niche." Synthetic niches can be defined as polymeric culture systems mimicking at least one aspect of the interactions between stem cells and the extracellular surroundings, including biochemical factors (e.g., the delivery of soluble factors) and/or biophysical factors (e.g., the microarchitecture of the ECM). Most of the currently available approaches for scaffold fabrication, based on self-assembly methods, do not allow for a submicrometer control of the geometrical structure of the substrate, which might play a crucial role in stem cell fate determination. A novel technology that overcomes these limitations is laser two-photon polymerization (2PP). Femtosecond laser 2PP is a mask-less direct laser writing technique that allows manufacturing three dimensional arbitrary microarchitectures using photosensitive materials. Here, we report on the development of an innovative culture substrate, called the "nichoid," microfabricated in a hybrid organic–inorganic photoresist called SZ2080, to study mesenchymal stem cell mechanobiology.

Key words Two-photon polymerization, Engineered niche, Nichoid, Mechanobiology

1 Introduction

The cellular environment is complex and plays an important role in cellular processes [1]. Besides two-dimensional (2D) culture substrates developed to investigate stem cell fate by confining the surface available for the cell adhesion (e.g., microislands, micropatterned/nanopatterned surfaces) [2], there is an increasing interest in the development of three-dimensional (3D) scaffolds to mimic the native architecture of the extracellular environment [3]. An important example of such artificial microenvironments for stem cell culture is the so-called "synthetic niche" [4, 5]. Synthetic niches can be defined as polymeric culture systems mimicking at least one aspect of the interactions between stem cells and the

Zuzana Koledova (ed.), *3D Cell Culture: Methods and Protocols*, Methods in Molecular Biology, vol. 1612,
DOI 10.1007/978-1-4939-7021-6_19,

extracellular surroundings, including biochemical factors (e.g., the delivery of soluble factors) and/or biophysical factors (e.g., the stiffness or architecture of the extracellular matrix).

A novel technology employed for niche manufacturing is laser two-photon polymerization (2PP) [6]. This is a mask-less direct laser writing technique that allows manufacturing arbitrary micro-architectures with a spatial resolution down to 100 nm, thus better than the light diffraction limit. In 2PP, photopolymerization occurs by nonlinear two-photon absorption induced by femtosecond laser pulses in transparent materials. Besides active research into the synthesis of new biocompatible and biodegradable materials [7–9], most of the groups have used hybrid inorganic–organic resins because they provide an excellent compromise between ease of use and mechanical robustness of the fabricated structures [10]. Moreover, the biocompatibility of these materials was extensively demonstrated [11, 12]. The use of laser in the biomedical field enables on scaffold manufacturing for hard tissue engineering and for the investigation on several aspects of cell behavior. This may include cell morphology, viability, proliferation, and differentiation [13–26]. Here, we report on the development of an innovative culture substrate (i.e., the nichoid substrate), microfabricated by 2PP in a hybrid organic–inorganic photoresist (SZ2080) for stem cell culture.

2 Materials

2.1 Photoresist Drop Casting and Baking

1. Glass coverslip, 12 mm diameter, thickness 1.5 (*see* **Note 1**).
2. Hot plate.
3. Acetone.
4. Forceps.
5. Objective lens tissue.
6. Photoresist: SZ2080 with 1% Irgacure 369 (i.e., zirconium/silicon hybrid sol–gel with photoinitiator) [27].

2.2 Two-Photon Polymerization of the 2PP-Nichoid Substrate

1. Infrared femtosecond laser system focused with high NA oil immersion objective, e.g., a cavity-dumped Yb:KYW laser with 1030 nm wavelength, 300 fs pulses and 1 MHz repetition rate, with a 100 × 1.4 NA oil immersion focusing objective [28] (*see* **Note 2**).
2. Objective immersion oil.
3. Computer-controlled mechanical shutter.
4. Computer-controlled, three-axis motion stage with high speed (more than 10 mm/s), high accuracy (better than 100 nm) and long travel (more than 50 mm), e.g., ANT130 mechanical stages (Aerotech).

5. Software: automated control program to synchronously move the three-axis stage and operate the mechanical shutter according to a user defined pattern, e.g., A3200 CNC Operator Interface (Aerotech) (*see* **Note 3**).
6. Gimbal mount.
7. Machine vision system for the laser setup composed by beam-splitter, CCD camera and red light source.
8. Power meter.
9. Objective/lens paper.
10. Methanol.
11. Coverslip holder (*see* **Note 4**).
12. Tape.
13. Forceps.

2.3 Nichoid Development, Observation, and Processing for SEM

1. Pipette controller.
2. 20 mL sterile pipettes.
3. Forceps.
4. 100 mL beaker.
5. pH meter.
6. UV filter.
7. Standard optical microscope.
8. Sputter coater with gold target (*see* **Note 5**).
9. Scanning electron microscope (SEM).
10. Developing solution: 50% (v/v) 3-pentanone, 50% (v/v) isopropyl alcohol.
11. SEM fixing solution: 1.5% (v/v) glutaraldehyde and 0.1 M sodium cacodylate (pH 7.1–7.2) in water.
12. SEM buffer solution: 0.1 M sodium cacodylate (pH = 7.1–7.2) in PBS.
13. Graded dehydration solution series, consisting of 20%, 30%, 40%, 50%, 60%, 70%, 90%, and 100% (v/v) ethanol in deionized water.

2.4 Mesenchymal Stem Cell (MSC) Isolation and Nichoid Culture

1. Bone marrow aspirate from tibias and femurs (n = 4) of a Sprague-Dawley (CD) rat. Keep in ice.
2. Centrifuge tubes, 2 mL and 50 mL.
3. 10 mL syringe with a 21 Gauge (G) needle.
4. Sterile pipettes, 2 mL and 10 mL.
5. 100 µm cell strainer.
6. 75 cm^2 flasks.

7. Centrifuge.
8. Cell counting device, e.g., Neubauer cytometer.
9. Cryovial.
10. Freezing container filled with isopropyl alcohol, e.g., MrFosty.
11. Serum-free (SF) culture medium: 100 U/mL penicillin and 0.1 mg/mL streptomycin in α-Minimum essential medium (α-MEM).
12. MSC complete culture medium: 20% fetal bovine serum (FBS), 100 U/mL penicillin, 0.1 mg/mL streptomycin, and 2 mM L-Glutamine (*see* **Note 6**) in α-MEM.
13. Phosphate buffered saline (PBS).
14. Trypsin–EDTA (0.25%).
15. Freezing medium: 30% FBS and 5% dimethyl sulfoxide in α-MEM.
16. Ultra-low attachment 24-well plate.
17. UV lamp.
18. Sterile deionized water.
19. 70% ethanol.
20. Forceps.

2.5 Fluorescence Staining and Image Acquisition

1. 2 mL tube.
2. Fixing solution: 2% paraformaldehyde (PFA) in PBS.
3. 1 μg/mL 4′,6′-diamidino-2-phenylindoledihydrochloride (DAPI) in PBS (*see* **Note 7**).
4. Aqueous fluorescence mounting medium.
5. Microscope slide.
6. Laser confocal microscope.

3 Methods

3.1 Photoresist Drop Casting and Baking

Carry out all procedures in a fume hood and out of the reach of UV light sources (yellow light can be used) unless otherwise specified.

1. Take a glass coverslip with forceps.
2. Clean the glass coverslip with the objective lens tissue wetted with acetone.
3. Using forceps, place the glass coverslip on a hot plate.
4. Drop 100 μL of SZ2080 photoresist on the glass coverslip (*see* **Note 8**).

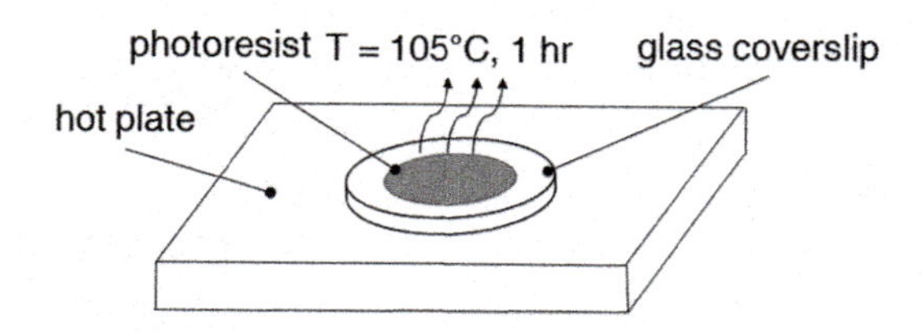

Fig. 1 Drop casting and baking of the SZ2080 photoresist

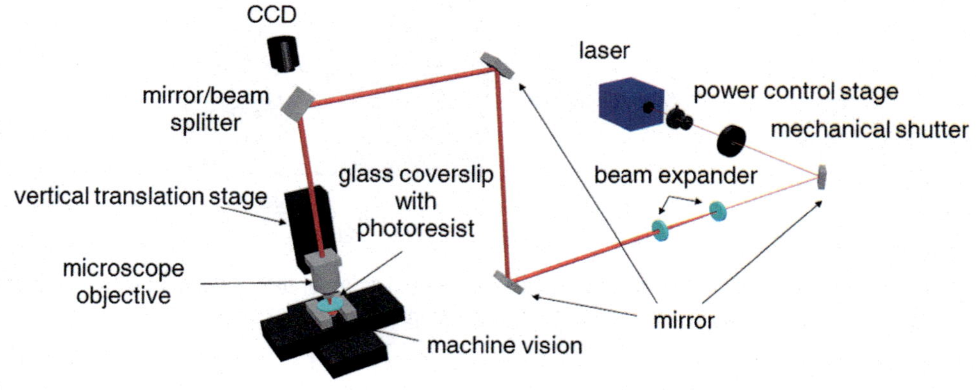

Fig. 2 Sketch of the experimental fabrication set-up

5. Bake the sample for 1 h at 105 °C (Fig. 1) to increase the viscosity of the photoresist by letting the solvent evaporate (*see* **Note 9**).
6. Using forceps remove the coverslip from the hot plate by and let it cool down for 10 min.

3.2 Two-Photon Polymerization of the 2PP-Nichoid Substrate

Carry out all procedures out of the reach of UV light sources (yellow light can be used) and wearing personal protective equipment. A sketch of the experimental setup is shown in Fig. 2.

1. Align the beam pathway (Fig. 2).
2. Place the glass coverslip on the metal holder (Fig. 3) (*see* **Note 4**).
3. Using adhesive tape, anchor the coverslip holder on the gimbal mount, mounted on the computer-controlled, three-axis motion stage (*see* **Note 10**).
4. Focus the laser at the air–photoresist interface (*see* **Note 11**).
5. Align the sample in the x–y plane by means of the gimbal to avoid tilting.
6. Move the objective far from the sample, and then deposit an oil drop on the glass coverslip, directly under the objective.
7. Move the objective back to the previous position. The focus position has been moved due to the oil insertion; now the laser beam is focused inside the glass coverslip.

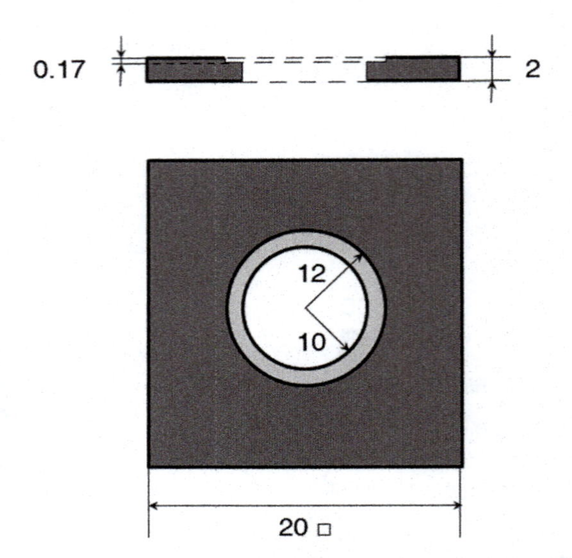

Fig. 3 The custom-made aluminum holder for the 12 mm diameter, thickness 1.5 glass coverslip. Dimensions are in mm

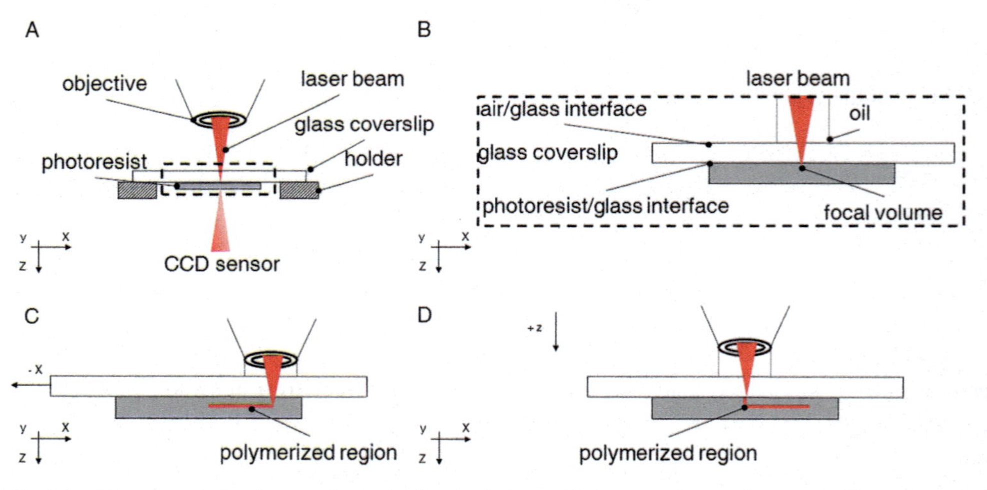

Fig. 4 Sketch of the two-photon laser polymerization technique. (**a**) Beam pathway focused with a 1.4 numerical aperture oil immersion objective on to the sample (SZ2080 photoresist in *gray*). (**b**) Zoomed view showing the laser the focal volume in the bulk of the photosensitive material. (**c**, **d**) Translation of the computer-controlled motion stage along *–x* and *+z*, respectively and the resulting polymerized region

8. Check the laser power by means of a power meter and set the scan speed (*see* **Note 12**).
9. Trace single lines by moving the objective vertically towards the coverslip with the shutter closed, then translating the stage horizontally with the shutter open, to find the laser focus at the glass–photoresist interface (Fig. 4a, b). Polymerized lines can be seen in the CCD when the glass–photoresist interface is reached.

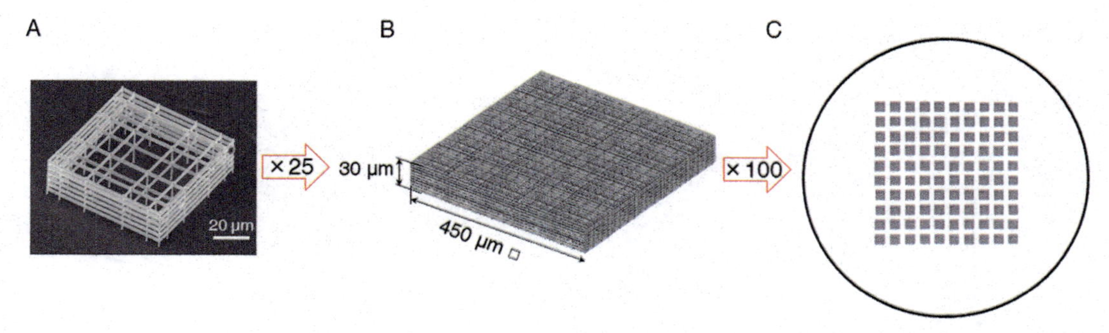

Fig. 5 The nichoid culture system fabricated by 2PP. (**a**) SEM picture of the nichoid repetitive unit forming the matrix of nichoids. (**b**) Computer aided design of the matrix of nichoids. (**c**) Top view of the glass coverslip covered by 10 × 10 matrices of nichoids

10. Run the software (Fig. 4c, d) (*see* **Notes 3** and **13**) to manufacture the 3D arbitrary geometry (Fig. 5a–c) (*see* **Note 14**).
11. At the end of the manufacturing process, remove the tape from the sample holder.
12. Using forceps, remove the sample from the holder.
13. Gently remove the oil both from the objective and from the sample by holding with forceps the objective lens tissue wetted with methanol.
14. Observe the sample using an inverted optical microscope (i.e., transmitted light) equipped with a red filter to avoid accidental polymerization of the unpolymerized area of the negative photoresist.

3.3 Nichoid Development, Observation, and Processing for SEM

Carry out all procedures under chemical hood and without UV light sources (yellow lights can be used) unless otherwise specified.

1. Pour 20 mL of the developing solution into a beaker and immerse the manufactured sample in it for 15 min at room temperature (RT) to dissolve the unpolymerized material (*see* **Note 15**).
2. Remove the sample from the solution and let it air-dry for 15 min at RT. Once the unpolymerized photoresist has been removed, the sample can be observed without the need for a UV filter.
3. Observe the manufactured sample using an optical microscope.
4. For SEM analysis of the manufactured nichoid samples,
   (a) Glue the sample onto SEM stubs.
   (b) Gold coat the sample in a sputter coater (*see* **Note 5**).
   (c) Observe the manufactured sample by means of a SEM (Fig. 6).

Fig. 6 Fabrication results. SEM pictures of the manufactured nichoid culture substrate

3.4 Mesenchymal Stem Cell (MSC) Isolation

Carry out all procedures in a biological safety hood for cell culture.

1. Centrifuge the freshly obtained bone marrow aspirate in a 50 mL tube at 276 × *g* for 8 min at 4 °C.
2. Remove the supernatant and resuspend the pellet in 3 mL of cold (4 °C) SF culture medium.
3. Using a 2 mL pipette resuspend and dissociate the bone marrow.
4. Using a 10 mL syringe equipped with a 21G needle dissociate further the small bone marrow fragments by aspirating and ejecting twice the cellular suspension.
5. Filter the cell suspension through a 100-µm cell strainer.
6. Using SF culture medium wash out any residual tissue from the syringe, the 50 mL tube and the 2 mL pipette used in **steps 1–4**, in the specified order (*see* **Note 16**), and filter the suspension through the 100 µm cell strainer.
7. Centrifuge at 276 × *g* for 10 min at 22 °C.
8. Remove the supernatant and resuspend the cells in 20 mL of MSC complete medium.
9. Plate the cells in two 75-cm^2 flasks.
10. After 24 h, change the medium to remove the non-adherent cell fraction.
11. Change medium every 3 days until confluence (*see* **Note 17**).
12. Cells at this passage (P0) can be expanded (1:3–1:5) to obtain cells at passage 1 (P1). MCSs (P0 or P1) can be directly seeded into nichoid or cryopreserved.
13. To cryopreserve cells, detach cells by trypsin–EDTA and neutralize with MSC complete medium.
14. Centrifuge the cell suspension at 192 × *g* for 5 min.

15. Remove the supernatant and resuspend the cells in MSC complete medium.
16. Count the cells, e.g., by means of a Neubauer cytometer.
17. Centrifuge the cells at 192 × *g* for 5 min.
18. Remove the supernatant and resuspend the cells in MSC freezing mix at the desired concentration (*see* **Note 18**).
19. Aliquot the cells in MSC freezing mix into cryovials and put the cryovial in a freezing container at −80 °C.
20. Next day, move the cryovials to liquid nitrogen for long-term storage.

3.5 Nichoid Preparation and Mesenchymal Stem Cell (MSC) Culture

Carry out all procedures in a biological safety hood for cell culture.

1. Using forceps, place the nichoid substrates in an ultra-low attachment 24-well plate.
2. Wash thoroughly the substrates with deionized water: three times for 5 min and then overnight (*see* **Note 19**).
3. Thoroughly disinfect the nichoid substrates by washing with 70% ethanol: three times for 5 min and then overnight (*see* **Note 19**).
4. Wash thoroughly the nichoid substrates with sterile deionized water (three times for 5 min and overnight) (*see* **Note 19**).
5. Let the nichoid substrates air-dry for 30 min (*see* **Note 20**).
6. UV-sterilize the nichoid substrates for 30 min (*see* **Note 21**).
7. Prepare MSC suspension for seeding:
   (a) Detach MSCs by trypsin–EDTA (0.25%) and neutralize with complete medium.
   (b) Centrifuge the cell suspension at 192 × *g* for 5 min.
   (c) Remove the supernatant and resuspend the cells in MSC complete medium.
   (d) Count the cells and adjust cell concentration as needed.
8. Seed MSCs directly on the nichod substrates at a density of 20,000 cells/cm^2.
9. Incubate the cell-nichoid construct for a week at 37 °C, 5% CO_2, while changing medium every 2 days.

3.6 Fluorescence Staining and Confocal Microscopy

1. To process the cell-nichoid construct for confocal microscopy, remove the culture medium.
2. Wash twice with PBS.
3. Fix the samples with the fixing solution for 10 min under chemical hood.
4. Wash twice for 5 min with PBS.
5. Stain the cell nuclei using DAPI solution for 15 min.

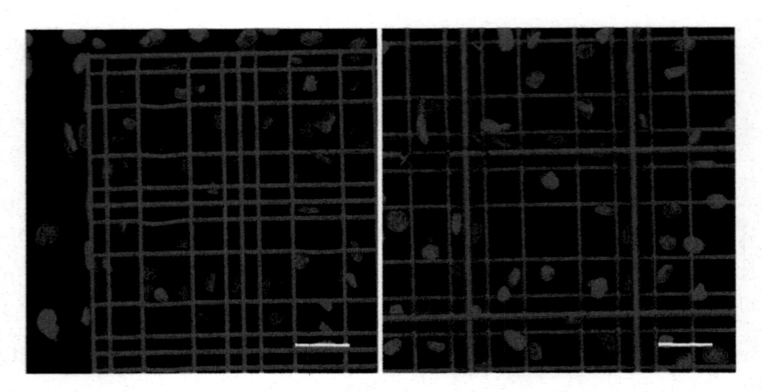

Fig. 7 Confocal microscopy results. *Z*-projection of the nichoid cellularized substrates. Cell nuclei are stained by DAPI (*blue*). The scale bar is 30 μm

6. Rinse several times with PBS.
7. Mount the glass coverslips onto a microscope slide with mounting medium.
8. Acquire Z-stack images using a confocal microscope (Fig. 7).

3.7 SEM Analysis of Cell-Populated Nichoid Samples

1. Fix the samples with the SEM fixing solution for 2 h at RT (*see* **Note 22**).
2. Rinse with the SEM buffer solution twice for 5 min (*see* **Note 23**).
3. Dip the samples twice for 5 min in the dehydration graded series solution (*see* **Note 24**).
4. Glue the sample onto SEM stubs and coat it in a vacuum ion coater.
5. Observe the manufactured sample by SEM.

4 Notes

1. As an alternative, a chambered cover glass, 1.5 mm thickness with cover or 35 mm glass bottom dishes, 1.5 mm thickness with cover can be employed as substrates.
2. Other laser systems can obtain similar results, with adequate parameters tuning [29].
3. The software interface we employed provides a multi-axis readout with position, velocity shown in user units, immediate command, program scan during execution, and a multi-axis jog screen. Typical cycle start, stop, feedhold, and system stop buttons are included. It is able to control the movements of the three-axis stages, as well as the opening and closing of a mechanical shutter. The software runs G-code scripts, programmed by the user according to the desired geometry of the

3D structure. Some fabrication software programs can also automatically generate the irradiation pattern from a given CAD file, e.g., DeScribe+NanoWrite (Nanoscribe) or 3DPoli (Femtika).

4. We manufactured a custom-made holder in aluminum for the 12-mm diameter, 1.5 thickness glass coverslips. Size and dimensions are depicted in Fig. 3.
5. To investigate the quality of the nichoid sample in the absence of cells, fixation followed by dehydration with a graded series of ethanol is not required. However, an electrically conductive coating can be applied to electrically insulating samples to improve the image resolution. The thickness of the gold layer prior to SEM investigation ranges from 15 to 20 nm.
6. When using α-Minimum essential medium, Gibco®, code 22571-020 with GlutaMAX™, L-glutamine does not to need to be added. It is already supplied with the basal medium.
7. Keep DAPI solution protected from light.
8. SZ2080 photoresist is a viscous liquid. Avoid bubble formation by carefully pipetting several times prior to deposition of the photoresist on the glass coverslip. Air bubbles negatively affect the manufacturing procedure.
9. The increase in viscosity leads to solidification of the photoresist. This allows the material to be placed as a droplet on any transparent substrate (e.g., a glass coverslip) instead of placing the material in between two glass plates separated by a distance holder (i.e., "sandwich" configuration of the sample).
10. The sample must be mounted on the translational stage with the photoresist at the opposite side of the objective (Fig. 4a, b).
11. To focus the laser beam on the air–glass interface, the experimental system is equipped with a beam splitter which deviates the retro reflected beam on a CCD camera connected to a computer. In such a way, the beam spot can be displayed on the computer monitor (Fig. 4a, b). By translating vertically (i.e., z-axis) the objective towards the anchored sample, an intense beam spot can be visualized on the monitor at the air–glass interface. By further translating in the vertical direction the objective towards the anchored sample, a second beam spot can be displayed on the monitor at the photoresist–glass interface (Fig. 4c, d). This should be avoided, since it can lead to undesired photoresist polymerization.
12. Optimum fabrication conditions were 1.5 mm/s writing speed, 12 mW average power (before the objective).
13. By translating the three-axis motion stages relative to the objective, polymerization occurs in focal volume in the bulk of the photoresist (Fig. 4c, d).
14. Here we designed and fabricated a new substrate with 10 × 10 matrices of nichoids per sample, covering 30% of the available

culture surface. Each matrix was composed of 5 × 5 elementary nichoids. Each nichoid was 30 μm high and 90 × 90 μm in transverse dimensions and consisted of a lattice of interconnected lines, with a graded spacing between 10 and 30 μm transversely and a uniform spacing of 15 μm vertically (Fig. 5a). This was found in a previous study to be the optimum geometry for MSC homing and proliferation [24]. The overall size of each matrix was 30 μm high and 450 × 450 μm in transverse dimensions. The spacing between matrices was set to 80 μm (Fig. 5b, c). Each elementary nichoid, as well as each matrix was surrounded by four outer confinement walls formed by horizontal lines spaced by 5 μm, resulting in gaps of 1 μm (Fig. 5a).

15. The development process may range from 15 to 30 min according to the volume of the material deposited on the glass coverslip surface and to the microfabricated architecture.
16. Following this order is highly recommended.
17. Primary bone marrow MSCs reach confluency in 8–10 days.
18. For example, 1.5 mL for 1 million of cells.
19. Rinse the nichoid substrate gently to avoid the detachment of the nichoid blocks.
20. The air-drying time ranges from 30 min to 1 h. Prior to UV-sterilization, check the nichoid substrate using an inverted light microscope.
21. Alternatively, sterilization via hydrogen peroxide plasma may be feasible due to the low temperature (i.e., up to 50 °C) attained in the process. However, at the best of our knowledge, we are not aware of possible degradation effects on the material after hydrogen peroxide plasma sterilization.
22. Carry out this step procedures under chemical hood.
23. If necessary, samples can be stored in buffer solution at 4 °C for 1 week.
24. If necessary, samples can be stored in 70% ethanol overnight or storage at 4 °C.

Acknowledgments

This project has received funding from the European Research Council (ERC) under the European Union's Horizon 2020 research and innovation program (grant agreement No. 646990—NICHOID). These results reflect only the authors' view and the Agency is not responsible for any use that may be made of the information contained.

References

1. Kress S, Neumann A, Weyand B et al (2012) Stem cell differentiation depending on different surfaces. Adv Biochem Eng Biotechnol 126:263–283
2. Nikkhah M, Edalat F, Manoucheri S et al (2012) Engineering microscale topographies to control the cell-substrate interface. Biomaterials 33(21):5230–5246
3. Kraehenbuehl T, Langer R, Ferreira L (2011) Three-dimensional biomaterials for the study of human pluripotent stem cells. Nat Methods 8(9):731–736
4. Peerani R, Zandstra P (2010) Enabling stem cell therapies through synthetic stem cell niche engineering. J Clin Invest 120(1):60–70
5. Joddar B, Ito Y (2013) Artificial niche substrates for embryonic and induced pluripotent stem cell cultures. J Biotechnol 106(2):218–228
6. Maruo S, Fourkas J (2008) Recent progress in multiphoton microfabrication. Opt Lett 2(1–2):100–111
7. Claeyssens F, Hasan EA, Gaidukeviciute A et al (2009) Three-dimensional biodegradable structures fabricated by two-photon polymerization. Langmuir 25(5):3219–3223
8. Turunen S, Käpylä E, Terzaki K et al (2011) Pico- and femtosecond laser-induced crosslinking of protein microstructures: evaluation of processability and bioactivity. Biofabrication 3(4):045002
9. Ovsianikov A, Malinauskas M, Schlie S et al (2011) Three-dimensional laser micro- and nano-structuring of acrylated poly(ethylene glycol) materials and evaluation of their cytoxicity for tissue engineering applications. Acta Biomater 7(3):967–974
10. Ovsianikov A, Mironov V, Stampfl J et al (2012) Engineering 3D cell-culture matrices: ultiphoton processing technologies for biological and tissue engineering applications. Expert Rev Med Devices 9:613–633
11. Raimondi MT, Eaton SM, Nava MM et al (2012) Two-photon laser polymerization: from fundamentals to biomedical application in tissue engineering and regenerative medicine. J Appl Biomater Function Mater 10(1):56–66
12. Danilevicius P, RekŽtyte S, Balciunas E et al (2013) Laser 3D micro-nanofabrication of polymers for tissue engineering applications. Opt Laser Technol 45:518–524
13. Correa DS, Tayalia P, Cosendey G et al (2009) Two-photon polymerization for fabricating structures containing the biopolymer chitosan. J Nanosci Nanotechnol 9(10):5845–5849
14. Malinauskas M, Danilevicius P, Baltriukiene D et al (2010) 3D artificial polymeric scaffolds for stem cell growth fabricated by femtosecond laser. Lithuan J Phys 50(1):75–82
15. Klein F, Richter B, Striebel T et al (2011) Two-component polymer scaffolds for controlled three-dimensional cell culture. Adv Mater 23(11):1341–1345
16. Koroleva A, Deiwick A, Nguyen A et al (2015) Osteogenic differentiation of human mesenchymal stem cells in 3-D Zr-Si organic-inorganic scaffolds produced by two-photon polymerization technique. PLoS One 10(2):e0118164
17. Marino A, Filippeschi C, Genchi GG et al (2014) The Osteoprint: a bioinspired two-photon polymerized 3-D structure for the enhancement of bone-like cell differentiation. Acta Biomater 10(10):4304–4313
18. Marino A, Filippeschi C, Mattoli V et al (2015) Biomimicry at the nanoscale: current research and perspectives of two-photon polymerization. Nanoscale 7(7):2841–2850
19. Ovsianikov A, Schlie S, Ngezahayo A et al (2007) Two-photon polymerization technique for microfabrication of CAD-designed 3D scaffolds from commercially available photosensitive materials. J Tissue Eng Regen Med 1(6):443–449
20. Psycharakis S, Tosca A, Melissinaki V et al (2011) Tailor-made three-dimensional hybrid scaffolds for cell cultures. Biomed Mater 6(4):045008
21. Tayalia P, Mendonca CR, Baldacchini T et al (2008) 3D cell-migration studies using two-photon engineered polymer scaffolds. Adv Mater 20(23):4494–4498
22. Terzaki K, Kissamitaki M, Skarmoutsou A et al (2013) Pre-osteoblastic cell response on three-dimensional, organic–inorganic hybrid material scaffolds for bone tissue engineering. J Biomed Mater Res Part A 101(8):2283–2294
23. Kapyla E, Aydogan DB, Virjula S et al (2012) Direct laser writing and geometrical analysis of scaffolds with designed pore architecture for three-dimensional cell culturing. J Micromech Microeng 22(11):115016
24. Raimondi MT, Eaton SM, Laganà M et al (2013) 3D structural niches engineered via two-photon laser polymerization promote stem cell homing. Acta Biomater 9(1):4579–4584

25. Raimondi MT, Nava MM, Eaton SM et al (2014) Optimization of femtosecond laser polymerized structural niches to control mesenchymal stromal cell fate in culture. Micromachines 5(2):341–358
26. Nava MM, Raimondi MT, Credi C et al (2015) Interactions between structural and chemical biomimetism in synthetic stem cell niches. Biomed Mater 10(1):015012
27. Ovsianikov A, Viertl J, Chichkov B et al (2008) Ultra-low shrinkage hybrid photosensitive material for two-photon polymerization microfabrication. ACS Nano 2(11): 2257–2262
28. Killi A, Steinmann A, Dörring J et al (2005) High-peak-power pulses from a cavity-dumped Yb:KY(WO4)2 oscillator. Opt Lett 30(14): 1891–1893
29. Malinauskas M, Farsari M, Piskarskasa A et al (2013) Ultrafast laser nanostructuring of photopolymers: a decade of advances. Phys Rep 533(1):1–31

Part IV

Microfluidic Approaches for 3D Cell Culture

Chapter 20

Microfluidic-Based Generation of 3D Collagen Spheres to Investigate Multicellular Spheroid Invasion

Fabien Bertillot, Youmna Attieh, Morgan Delarue, Basile G. Gurchenkov, Stephanie Descroix, Danijela Matic Vignjevic, and Davide Ferraro

Abstract

During tumor progression, cancer cells acquire the ability to escape the primary tumor and invade adjacent tissues. They migrate through the stroma to reach blood or lymphatics vessels that will allow them to disseminate throughout the body and form metastasis at distant organs. To assay invasion capacity of cells in vitro, multicellular spheroids of cancer cells, mimicking primary tumor, are commonly embedded in collagen I extracellular matrix, which mimics the stroma. However, due to their higher density, spheroids tend to sink at the bottom of the collagen droplets, resulting in the spreading of the cells on two dimensions. We developed an innovative method based on droplet microfluidics to embed and control the position of multicellular spheroids inside spherical droplets of collagen. In this method cancer cells are exposed to a uniform three-dimensional (3D) collagen environment resulting in 3D cell invasion.

Key words Multicellular spheroids, 3D model, Cancer cells invasion, Droplet microfluidics, Extracellular matrix, Collagen

1 Introduction

The metastatic cascade is a multistep process that requires cancer cells to overcome many obstacles: cells need to breach the basement membrane they rest on, invade the surrounding stroma, find their way to the circulation that will allow them to travel throughout the body before they arrest in capillaries, extravasate, and colonize secondary organs [1]. Each step is crucial for cancer cells to move on to the next one. However, the study of cancer cell invasion and cell dynamics is a complex task, mostly because of the lack of proper model systems.

In 2003, Nature published an editorial article entitled "Goodbye, flat Biology?" that highlighted the necessity of switching from two-dimensional (2D) to 3D cell cultures, as it became evident that 3D models recapitulate the complexity of in vivo cell's behavior more faithfully. This article also predicted that it

Zuzana Koledova (ed.), *3D Cell Culture: Methods and Protocols*, Methods in Molecular Biology, vol. 1612,
DOI 10.1007/978-1-4939-7021-6_20, © Springer Science+Business Media LLC 2017

was "only a matter of time before 3-D techniques become standardized and cost-benefit ratios become irresistible in many areas of biology."

It is also becoming accepted that genetic and epigenetic modifications of cancer cells are not sufficient to drive metastasis formation. During their metastatic journey, cancer cells constantly interact with their microenvironment and modify it. In turn, microenvironment plays an active role in throughout the metastatic tumor progression, stimulating tumor growth, survival, and invasion capacity. However, "normal" microenvironment can also have the ability to revert cancer cells to a "normal" phenotype [2].

The tumor microenvironment consists of different cell types, such as endothelial cells, pericytes, immune cells, and fibroblasts; and the extracellular matrix (ECM), a 3D mesh of proteins that constitutes a scaffold for those cells [1]. Tumor-associated ECMs are mostly composed of collagen I fibers. Other matrix proteins such as fibronectin, laminin, or tenascin C are present in different amounts [3]. The ECM is generated and deposited by stromal cells, primarily fibroblasts.

Self-derived matrices generated by stimulating fibroblasts by hyaluronic acid to deposit their own matrix are commonly used as model of the tumoral ECM [4]. Fibroblasts are then removed and cancer cells are plated on those decellularized matrices. The major limitation of this model is that those matrices are rather thin, thus cells are usually not fully embedded [5].

Alternative model is based on artificial hydrogels. Synthetic scaffolds such as poly(ethylene glycol) (PEG) hydrogels allow embedding of the cells while polymerizing the matrix. Some hydrogels can reproduce the fibrillary structure of biological matrices and, most importantly, offer complete control over its mechanical properties such as stiffness and pore size [6]. However, the drawback of the artificial gels is that they lack signals necessary for cell survival, proliferation, and migration.

In order to overcome this issue, it is possible to use biological component of in vivo matrices, such as collagen I that can be polymerized in vitro. If the right polymerization conditions are used, these matrices resemble the matrices found in vivo [7, 8]. Even though their composition is simplified, they allow generation of truly 3D scaffolds.

In order to model tumor invasion, cell biologists either use single cancer cells or cell aggregates (spheroids), aiming to mimic small tumors [8, 9]. Cell aggregates can be easily generated using nonadhesive substrates, such as agarose or by more sophisticated methods, for example in alginate capsules made using microfluidic strategies [10]. Spheroids could be then embedded in collagen I gels, mimicking a small tumor invading the ECM [9–11]. However, because spheroids are objects denser than a water phase solution, when deposited in a non-polymerized collagen, they have the tendency

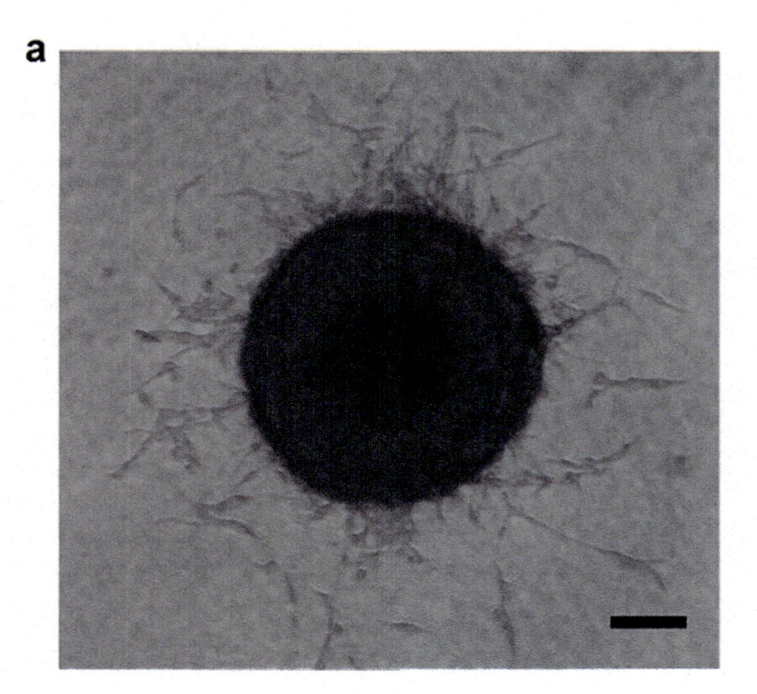

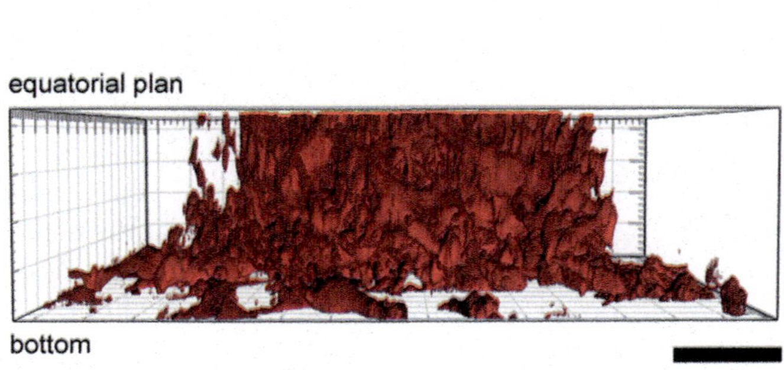

Fig. 1 2D spreading of a CT26 spheroid embedded in collagen droplet. Spheroid sank onto the glass surface of the dish and cells migrate in 2D. Scale bar, 50 μm. (**a**) Phase image of cancer cell invasion. The spheroid appears to invade in 3D but all the cells that have escaped are in the same focus plan, possibly, migrating on the glass surface. (**b**) Side view of a reconstructed 3D stack of a spheroid (LifeAct-Cherry, *red*) embedded in collagen (*not shown*). All cells that have escaped the spheroid migrated on the glass surface. The spheroid was imaged using an inverted two-photon microscope with a 40× oil objective

to sink onto the plastic dish while the collagen is polymerizing. Once on the hard and flat surfaces, cells tend to invade only in 2D (Fig. 1). To prevent spheroids' sinking, it is possible to polymerize collagen faster, for example by raising the collagen polymerization temperature. Nevertheless, when polymerized at higher temperatures, resulting gels are made of thin collagen fibers with small pores that do not resemble the structure of the ECM in vivo [7, 8]. Alternatively, collagen can be still polymerized at room temperature, and sinking of the spheroids could be prevented by flipping the plastic dish continuously upside/down until collagen fibers are formed and position of the spheroid locked in the center of the collagen matrix. Although robust, this method requires an extensive amount of work and time. In order to overcome these limitations, we have developed an innovative strategy based on droplet microfluidics to embed cancer cells spheroids in the center of the collagen spherical droplets. In particular, droplet microfluidics mainly represents the possibility of generating stable and regular emulsion of two or more fluidic phases. Additionally, due to its large spreading during the last 10 years, droplet microfluidics is considered one of the most promising candidates for the generation of innovative and useful tools for new generation of biological experiments [12]. The capability of encapsulating cells in droplets has been already used for spheroids production [10, 13]; however, there are no examples concerning their encapsulation in collagen matrix especially with the final aim to control their position within the collagen matrix. Therefore, we have established a microfluidic platform that generates trains of non-polymerized collagen droplets containing spheroids (one spheroid per collagen droplet) flowing in a spiral capillary (Fig. 2). During collagen polymerization, which is performed at room temperature, spheroids are continuously moving inside the collagen droplet miming the flipping gesture of the plastic dish described above. Additionally, we worked with a confined droplet (droplet diameter is larger than the capillary size) [14] allowing the control of the final shape of the collagen droplet. Therefore, adjusting the droplet volume, at the end of the spiral, spheres of polymerized collagen containing spheroids in its center can be easily collected in conventional culture medium.

The validation of our approach has been performed with two different cell lines (CT26 and NIH 3T3), monitoring the spheroids position in the collagen droplet, their proliferation and invasion in the 3D matrix (Fig. 3).

2 Materials

2.1 Cell Culture

1. CT26-LifeAct-GFP cells: CT26 cells (ATCC CRL-2638) stably transfected with LifeAct-GFP (courtesy of S. Geraldo, Institut Curie).
2. NIH 3T3-GFP cells: NIH 3T3 cells (Cell Biolabs) stably transfected to express cytoplasmic green fluorescent protein (GFP).
3. Cell culture medium: 10% fetal bovine serum, 100 units/mL penicillin, and 100 μg/mL streptomycin in Dulbecco's Modified Eagle's Medium (DMEM) with GlutaMAX.
4. 0.05% Trypsin/EDTA.

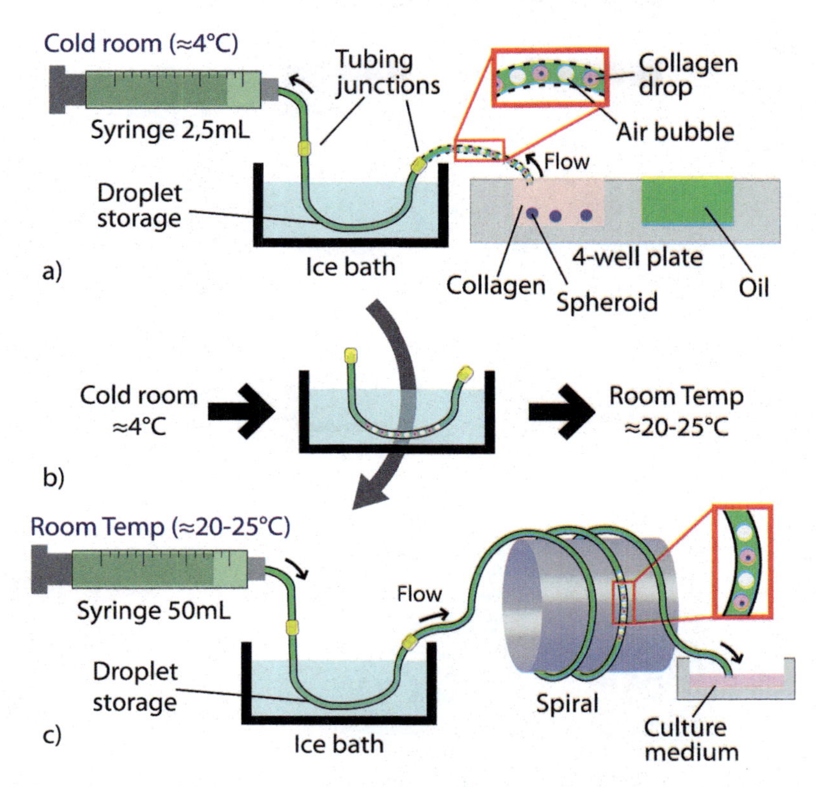

Fig. 2 Scheme of the microfluidic experimental setup for collagen droplet generation and polymerization. The *continuous black line* represents the larger PTFE capillary, *dashed black line* is the smaller capillary, and the *yellow connections* are silicon capillaries used as junctions. (**a**) Droplets are generated by alternating sucking of the oil solution (*green*), collagen containing spheroids (*light red* and *violet*, respectively) and the air, with the aspiration step performed by the syringe pump. The train of droplets is stored in the portion of capillary between *yellow connections* and it is immersed in the ice bath. (**b**) This part is detached from the rest of the setup, transferred to the room temperature and connected between the second syringe and the spiral (**c**). The train of collagen/spheroids droplets is pushed into the spiral. Collagen is polymerized during the flow and the collagen droplets containing spheroids positioned centrally are collected in culture medium

5. Cell culture dishes or flasks (T25).
6. Inverted light microscope.

2.2 Multicellular Cancer Cell Spheroids

1. Ultrapure agarose (Life Technologies).
2. 96-well flat-bottom plates.
3. 0.4% Trypan blue.

2.3 Collagen Mix

1. Rat tail collagen type I (Corning, 354236).
2. 10× phosphate buffered saline (PBS): Combine 2 g KCl, 2.4 g KH_2PO_4, 80 g NaCl, and 11.45 g Na_2HPO_4 and add H_2O to 1 L. Autoclave.

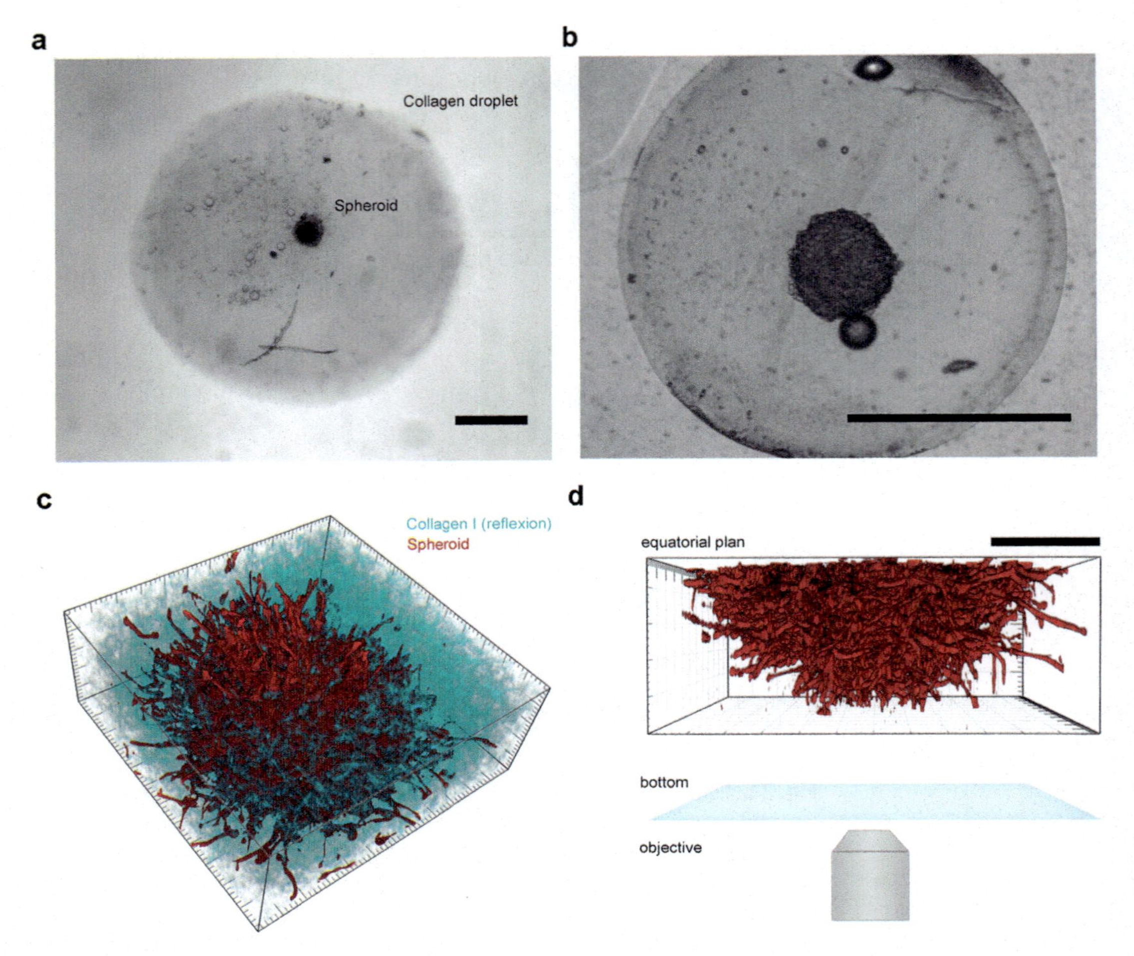

Fig. 3 CT26 and NIH 3T3 spheroids embedded in collagen droplets, before and after invasion. Scale bar, 500 mm. (**a**) 2 mm droplet with a spheroid of NIH 3T3 cells. (**b**) 800 μm droplet with a spheroid of CT26. (**c**) 3D reconstruction of a spheroid (LifeAct-Cherry CT26 cells, *red*) invading collagen (reflection, *cyan*) droplet in 3D. The spheroid was imaged using an inverted two-photon microscope with a 40× oil objective. Only one half of the spheroid has been imaged. (**d**) Side view of the invading spheroid. Spheroid is not in contact with the glass bottom of the dish and all cells invade in 3D

3. 1 N NaOH, sterile.
4. DMEM.

2.4 Microfluidic Setup

1. Syringe Pump neMESYS (Cetoni GmbHS).
2. Syringes of 2.5 mL and 50 mL of maximum capacity.
3. Capillaries (not necessarily sterilized, *see* **Note 1**):
   (a) About 6 m of PTFE capillary inner/outer diameter of 2 mm/4 mm.
   (b) About 50 cm of PTFE capillary inner/outer diameter of 1.5 mm/2 mm.

(c) About 10 cm of silicon capillary inner/outer diameter of 3 mm/4 mm.

4. Oil solution: 2% (w/w) 1H,1H,2H,2H–perfluorodecan-1-ol (Fluorochem) in FC-40 oil (3 M).
5. 4-well plate.
6. Plastic rectangular bar (about 4 × 2 × 30 cm).
7. 24-well cell culture plate.

3 Method

3.1 Cell Culture

1. Maintain the NIH 3T3-GFP and CT26-LifeAct-mCherry cells in cell culture medium at 37 °C in 5% CO_2 humidified air.
2. Passage cells at a ratio of 1:10 every 3 days to maintain their exponential growth.
3. To prepare single-cell suspension for cells for spheroid preparation, trypsinize a subconfluent culture of cells for 5 min at 37 °C (in cell culture incubator).
4. Stop trypsinization by adding fresh cell culture medium. Pipette the cell suspension repeatedly to create a single-cell suspension.
5. Pellet the cells by centrifugation at 500 × *g* for 3 min.
6. Resuspend the cells in fresh cell culture medium, count the cells, and adjust cell number per mL as needed.

3.2 Multicellular Spheroid Preparation

1. Prepare a fresh 1% agarose solution in deionized H_2O and boil it.
2. Immediately after boiling, add 50 μL of the hot agarose solution to each well of a 96-well plate in a laminar hood.
3. Let the agarose in plate cool down at room temperature for 30 min and by eyes check the gelation of agarose.
4. Prepare a single-cell suspension of 5000 cells/mL (*see* Subheading 3.1 and **Note 2**).
5. Add 200 μL of cell suspension to each well of a 96-well plate; one spheroid per well is generated.
6. Incubate the plate for 24 h (NIH 3T3) or 72 h (CT26) at 37 °C in 5% CO_2 humidified air (*see* **Note 3**).

3.3 Collagen Mix Preparation

Collagen polymerization mix is prepared including multicellular spheroids. The final concentration of the collagen mix is 2 mg/mL. From the initial stock collagen concentration ci and for a final volume of 1 mL of solution, calculate the volume of stock collagen necessary.

1. Transfer spheroids from the 96-well to a centrifuge tube.
2. Let spheroids sediment to the bottom of the centrifuge tube. Check by eyes.
3. Aspirate the medium without aspirating spheroids and resuspend in 315 μL of DMEM (*see* **Note 4**). Incubate in ice bath for 5 min.
4. In a 2 mL eppendorf tube, add 100 μL (1/10 of the final volume) of 10× PBS stocked at 4 °C.
5. Add 12 μL of 1 N NaOH stocked at 4 °C.
6. Add 315 μL of the suspension of spheroids.
7. Add 573 μL of collagen I. Gently mix the solution, avoiding formation of air bubbles.
8. Check the pH of the final solution with a pH paper indicator to ensure a pH of 7.5. Incubate the solution into an ice bath until its use.

3.4 Collagen Droplets Generation

1. Set the microfluidic setup for the droplet generation as represented in Fig. 2a in a cold room to avoid premature collagen polymerization. The setup is composed by the 2.5 mL syringe connected to the large PTFE capillary (represented by a continuous black line in Fig. 2) that will be used as droplets storage, which is connected to the small PTFE capillary (dashed black line) that needs for the droplets generation. The connection between the two capillaries is assured by the silicon capillary (yellow parts in Fig. 2) (*see* **Note 5**). Additionally, the storage part is immersed in an ice bath to prevent the collagen polymerization during the transfer to the second part of the experiment outside the cold room. After the setup installation, the droplet generation is achieved by pipetting as follows.
2. Prefill the whole fluidic system with the oil solution (green solution in Fig. 2).
3. Fill one of the wells of the 4-well plate with the oil solution and another one with the mix of non-polymerized collagen with the spheroids. Droplet generation is observed by a conventional inverted microscope.
4. Set a negative flow rate of 1.5 μL/s and a volume of 5 μL.
5. Pipette repetitive sequences of "oil solution - collagen - oil solution - air bubbles-…" as shown in Fig. 2a to generate droplet of collagen containing spheroids separated with air bubbles. During the pipetting of the collagen droplet, take care of placing the capillary close to a selected spheroid (*see* **Note 6**). The air is used as a spacer to avoid any contact between the droplets during flow (*see* **Note 7**) [14].
6. After the generation of a desired number of droplets (typically we make 30–40), store the droplets in the portion of the tube kept in the ice bath by sucking.

7. Detach this portion of tube from the rest of the setup in correspondence of the junctions (yellow parts in Fig. 2) and transfer it to room temperature for the second part of the protocol, as shown in Fig. 2b.

3.5 Collagen Droplets Polymerization

The second part of the experimental setup (Fig. 2c) is composed by a second syringe pump (the same pump could be used) mounted with the 50 mL syringe filled with the oil solution. About 6 m of the large PTFE capillary is rolled up on a plastic rectangular bar forming a spiral of about 4 cm of high, 2 cm of width and 30 cm of length. This spiral assures that the droplets are continuously moved up and down in order to center the spheroids during the collagen polymerization.

1. After the completion of the protocol described in Subheading 3.4, connect the capillary part immersed in the ice bath, in between the syringe and the spiral as shown in Fig. 2c.
2. Set the syringe at 18 μL/s; this flow allows keeping the droplets in the spiral for about 17 min. Once the collagen is completely polymerized, the position of spheroids is fixed.
3. At the end of the process collect droplets in a 24-well plate filled with cell culture medium (*see* **Note 8**).
4. Wash three times with the medium to remove the oil (*see* **Note 9**).
5. Incubate samples at 37 °C in 5% CO_2 humidified air.
6. Observe cell invasion after 24–72 h by light microscopy.

4 Notes

1. The capillaries used for droplet generation and transportation do not need to be sterilized because, due to the presence of the oil solution, collagen droplets are never in contact with the capillary surface. As a matter of fact, there is always a small layer of oil between the water phase droplet and the capillary. For the same reason, the same capillaries can be reused without any particular precaution.
2. The initial spheroid size will mostly depend on the number of cells N and the size of your cell line. To estimate roughly the diameter D of spheroids use $D = N^{1/3}d$ with d the mean diameter of your cell type. For example, for NIH 3T3 cell line we calculated roughly that 1000 cells/well are needed to form a spheroid of 200 μm.
3. Aggregation time of the cells highly depends on the cell type. For the NIH 3T3 cells aggregation occurs within 10 h while for the CT26 it occurs within 3 days.
4. Sticking of spheroids to the pipette tip can occur occasionally. Coating the tip with 0.1% bovine serum albumin in PBS avoids sticking.

5. All the capillary connections must be done by putting in close contact two extremities of the tubes. The silicon capillary is used as a joint between the two parts.
6. While aspirating the collagen make sure not to transfer the capillary too early, even if the syringe software indicates that the sucking step is completed. A delay of 1 s must be adopted before transferring the capillary between the collagen and oil wells. This is due to the presence of the air that acts as a dampener in the fluidic system. This effect increases with the increasing number of the air bubbles in the capillary.
7. Spacing the droplets with air is fundamental because during the second part of the protocol, the collagen is polymerizing during the flow and therefore the viscosity of the droplet phase changes. As a consequence, the droplets flow at different speeds in the capillary and touch each other.
8. We recommend limiting the number of collected droplets in the same well to prevent sticking and fusion of the droplets.
9. Complete filtration of oil is not possible. Small leftovers are still present on the surface.

Acknowledgments

This work was supported by ERC grant STARLIN, 311263 and Invadors project (by French National Research Agency—ANR) (D.M.V); Digidiag project (by French National Research Agency—ANR) and ARC foundation for young researcher fellowship (D.F.), ITMO cancer PhD fellowship (F.B). We thank Sara Geraldo and Marine Verhulsel for advices on collagen preparation and cell lines.

References

1. Hanahan D, Weinberg RA (2011) Hallmarks of cancer: the next generation. Cell 144:646–674
2. Bissell MJ, Hines WC (2011) Why don't we get more cancer? A proposed role of the microenvironment in restraining cancer progression. Nat Med 17:320–329
3. Naba A, Clauser KR, Whittaker CA et al (2014) Extracellular matrix signatures of human primary metastatic colon cancers and their metastases to liver. BMC Cancer 14:518
4. Grinnell F, Fukamizu H, Pawelek P et al (1989) Collagen processing, crosslinking, and fibril bundle assembly in matrix produced by fibroblasts in long-term cultures supplemented with ascorbic acid. Exp Cell Res 181:483–491
5. Abbott A (2003) Cell culture: biology's new dimension. Nature 424:870–872
6. Trappmann B, Chen CS (2013) How cells sense extracellular matrix stiffness: a material's perspective. Curr Opin Biotechnol 24:948–953
7. Wolf K, Te Lindert M, Krause M et al (2013) Physical limits of cell migration: control by ECM space and nuclear deformation and tuning by proteolysis and traction force. J Cell Biol 201:1069–1084
8. Geraldo S, Simon A, Elkhatib N et al (2012) Do cancer cells have distinct adhesions in 3D collagen matrices and in vivo? Eur J Cell Biol 91:930–937
9. Geraldo S, Simon A, Vignjevic DM (2013) Revealing the cytoskeletal organization of invasive cancer cells in 3D. J Vis Exp 80:e50763
10. Alessandri K, Sarangi BR, Gurchenkov VV et al (2013) Cellular capsules as a tool for multicel-

lular spheroid production and for investigating the mechanics of tumor progression in vitro. Proc Natl Acad Sci U S A 110:14843–14848
11. Kopanska KS, Bussonnier M, Geraldo S et al (2015) Quantification of collagen contraction in three-dimensional cell culture. Methods Cell Biol 125:353–372
12. Kintses B, van Vliet LD, Devenish SRA et al (2010) Microfluidic droplets: new integrated workflows for biological experiments. Curr Opin Chem Biol 14:548–555
13. Yu L, Chen MCW, Cheung KC (2010) Droplet-based microfluidic system for multicellular tumor spheroid formation and anticancer drug testing. Lab Chip 10:2424–2432
14. Nightingale AM, Phillips TW, Bannock JH et al (2014) Controlled multistep synthesis in a three-phase droplet reactor. Nat Commun 5:3777

Chapter 21

High-Throughput Cancer Cell Sphere Formation for 3D Cell Culture

Yu-Chih Chen and Euisik Yoon

Abstract

Three-dimensional (3D) cell culture is critical in studying cancer pathology and drug response. Though 3D cancer sphere culture can be performed in low-adherent dishes or well plates, the unregulated cell aggregation may skew the results. On contrary, microfluidic 3D culture can allow precise control of cell microenvironments, and provide higher throughput by orders of magnitude. In this chapter, we will look into engineering innovations in a microfluidic platform for high-throughput cancer cell sphere formation and review the implementation methods in detail.

Key words Microfluidics, Cancer sphere, High-throughput, PDMS, 3D culture

1 Introduction

In a seminal paper, Dr. Mina J. Bissell first highlighted the importance of three-dimensional (3D) culture, especially for understanding the roles of extracellular matrix (ECM) in tissue physiology and cancer pathology [1]. In cancer studies, to bridge the different microenvironments and drug responses between in-vitro 2D cell culture and in-vivo outcomes, more and more experiments have been performed in 3D culture assimilating in-vivo conditions [2, 3]. 3D models can mimic cancer microenvironments and also recapitulate drug distribution and hypoxia inside the tumor [4–6]. Due to the fact that differences in the number of cells aggregated into spheres (small or large) may induce different drug and nutrition distribution [7], the unregulated cell aggregation can significantly skew the results when 3D culture can be performed in low-adherent petri dish/well-plate [8]. In the past decade, the development of microfluidic devices provided a new tool for cell culture in a revolutionary way, handling a small volume of samples down to pL with high accuracy and high spatial resolution [9]. In this manner, the microfluidic system does not only enable the precise

Zuzana Koledova (ed.), *3D Cell Culture: Methods and Protocols*, Methods in Molecular Biology, vol. 1612, DOI 10.1007/978-1-4939-7021-6_21, © Springer Science+Business Media LLC 2017

control of cell microenvironments, but also increase the throughput and reduce the waste of reagents by orders of magnitude. In this chapter, we introduce a microfluidic high-throughput 3D culture platform that has the capabilities of: (1) reliable formation of cancer cell spheres, (2) a high sphere-formation rate, (3) a uniform sphere size, (4) capability to culture spheres for more than 2 weeks, and (5) capability to retrieve spheres from the device for further analyses [10].

The presented high-throughput sphere formation platform is composed of an array of 1024 (32 columns by 32 rows) non-adherent microwells connected by single inlet and outlet, respectively (Fig. 1) [10]. To facilitate cell aggregation, each microwell is coated with polyHEMA, which is a biocompatible hydrogel used as a non-adherent coating material for suspension cell culture [11, 12]. The microwells are designed to be 300 μm in depth, so that the introduced cells can settle into each microwell and then aggregate into a sphere inside the well (Fig. 2).

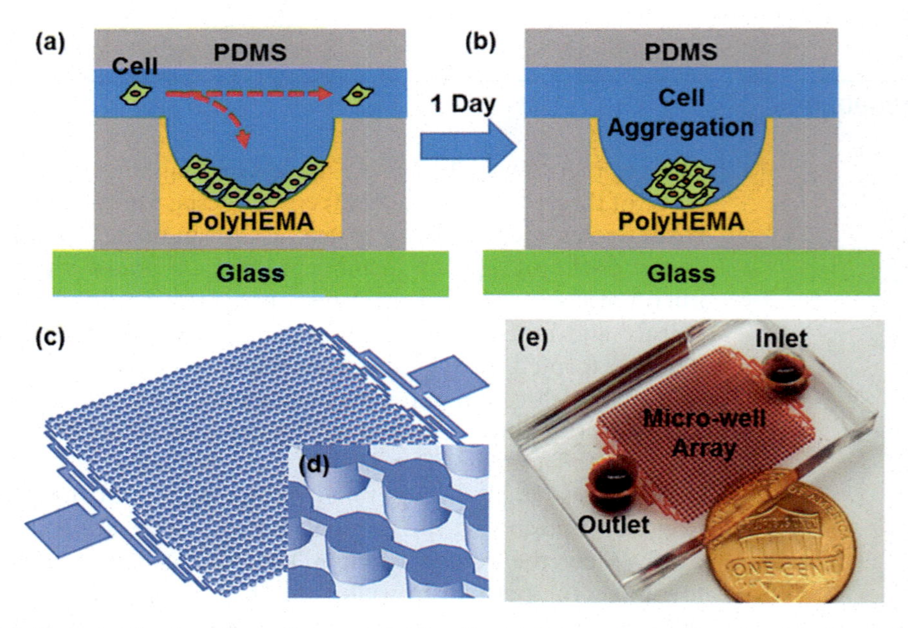

Fig. 1 High-throughput sphere formation microfluidic chip. (**a**) Schematic drawing shows cells settling into a microwell during loading. (**b**) Schematic drawing shows cell aggregation and sphere formation 1 day after cell loading. (**c**) 3D schematic view of the microfluidic chip. (**d**) Enlarged 3D schematic of micro-wells. (**e**) The fabricated device that contains 1024 micro-wells within a core area of 2 cm by 2 cm. The cells are loaded into the inlet (*top right*) and then flow through the microwell array to the outlet (*bottom left*). Reproduced from Chen et al. [10] with permission from Macmillan Publishers Limited (Springer Nature)

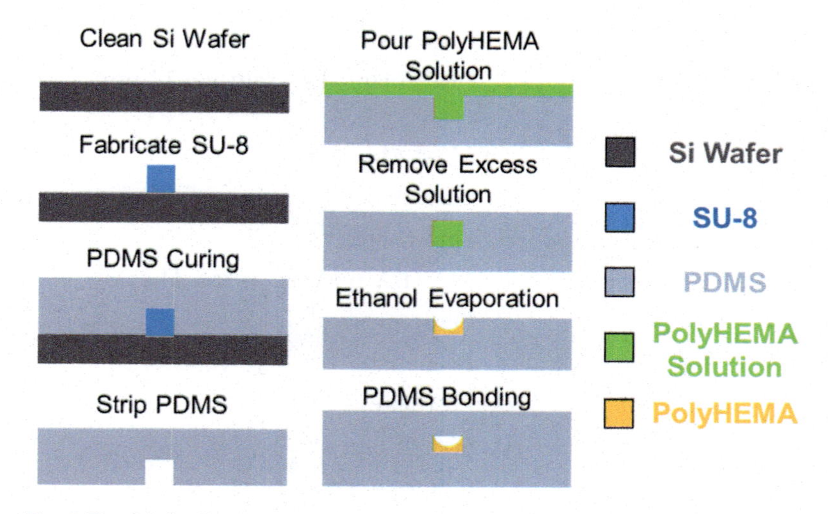

Fig. 2 The fabrication process of the microfluidic 3D cell culture chip

2 Materials

2.1 Reagents

1. Chromium Etch 1020.
2. MF-319.
3. PRS2000.
4. Nanostrip.
5. Deionized (DI) water.
6. Silicon wafer.
7. Hexamethyldisilazane (HMDS).
8. SU-8 2100.
9. SU-8 developer.
10. Isopropyl alcohol (IPA).
11. Trichloro(1H,1H,2H,2H-perfluorooctyl)silane.
12. SYLGARD® 184 Silicone Elastomer (PDMS) Base and Curing Agent.
13. PolyHEMA solution (60 mg/mL in 95% ethanol).
14. 1% (w/v) Pluronic F-108 solution in DI water.
15. Sterile PBS.
16. Cell culture medium: 10% fetal bovine serum, 100 U/mL penicillin and 100 μg/mL streptomycin in RPMI.
17. 0.05% Trypsin/EDTA.
18. LIVE/DEAD staining kit.

2.2 Tools and Equipment

1. AutoCAD.
2. Mask plate, 127 mm × 127 mm, coated with chrome on one side.
3. Mask Maker.
4. Glen Asher.
5. Spin coater.
6. MA/BA-6 Mask/Bond Aligner.
7. Desiccator.
8. Biobench.
9. Hotplate.
10. Scotch tape.
11. Compressed air blow gun.
12. Razor blade.
13. Biopsy punch, 6 mm in diameter.

3 Methods

3.1 Fabrication of the Microfluidic Sphere-Formation Platform

3.1.1 Mask Making

1. Create a layout of the microfluidic chip using AutoCAD.
2. Pattern a mask plate, coated with chrome on one side of the surface, using Mask Maker.
3. Develop the mask plate in MF-319 for 1 min.
4. Etch the chrome on mask plate in Chromium Etch 1020 for 2 min.
5. Soak the mask plate in PRS2000 or Nanostrip to remove the residual photoresist.
6. Wash the mask plate in DI water for 5 min and dry it with compressed air blow gun.

3.1.2 SU-8 Master (100-μm-Thick Channel Layer)

1. Remove the organics on a silicon wafer by applying Glen Asher plasma treatment to the wafer surface with 100 sccm O_2 and 12 sccm Ar at 200 mT with 300 W for 120 s (*see* **Note 1**).
2. Dehydrate for 5 min on a hotplate at 150 °C.
3. Pour HMDS on the wafer to enhance the adhesion of SU-8.
4. Put the wafer on a spin coater and spin at 4000 rpm for 10 s to dry HMDS.
5. Pour 10.0 g of SU-8 2100 resist on the wafer (*see* **Note 2**).
6. Spread the photoresist at 500 rpm for 10 s. This step may be repeated multiple times until the resist fully covers the whole wafer surface.
7. Spin at 3000 rpm to achieve a thick SU-8 layer of 100 μm.

8. Perform soft bake for 5 min at 65 °C, 30 min at 95 °C, and 5 min at 65 °C, consecutively (*see* **Note 3**).
9. Expose the wafer for 50 s using MA/BA-6 Mask/Bond Aligner through an optical filter to eliminate UV radiation under 350 nm in wavelength. The aligner UV supply should be calibrated to 20 mW/cm^2.
10. Perform post-exposure bake for 5 min at 65 °C, 8 min at 95 °C, and 5 min at 65 °C, consecutively (*see* **Note 3**).
11. Develop SU-8 in a SU-8 developer bath for 8 min and rinse with SU-8 developer for 15 s after taking out from the bath.
12. Rinse with IPA and dry the wafer with compressed nitrogen.
13. Perform hard bake for 30 min at 150 °C.

3.1.3 SU-8 Master (300-μm-Thick Microwell Layer)

1. Pour 10.0 g of SU-8 2100 resist on the wafer.
2. Spread the resist at 500 rpm for 10 s. This step may be repeated multiple times until the resist fully covers the whole wafer surface.
3. Spin at 900 rpm to achieve a thick SU-8 layer of 300 μm.
4. Perform the soft bake for 5 min at 65 °C, 60 min at 95 °C, and 5 min at 65 °C, consecutively.
5. Align the mask to the pattern of the first layer of SU-8 which is 100 μm in thickness. Then, expose the wafer for 60 s using MA/BA-6 Mask/Bond Aligner through an optical filter to eliminate the UV radiation under 350 nm in wavelength. The aligner UV supply should be calibrated to 20 mW/cm^2.
6. Perform the post-exposure bake for 5 min at 65 °C, 15 min at 95 °C, and 5 min at 65 °C, consecutively.
7. Develop SU-8 in a SU-8 developer bath for 12 min and rinse with SU-8 developer for 15 s after taking out from the bath.
8. Rinse with IPA and dry the wafer with compressed nitrogen.
9. Perform hard bake for 30 min at 150 °C.

3.1.4 Release Agent Coating

1. Place the wafer in a desiccator.
2. Load 1 mL of release agent, Trichloro(1H,1H,2H,2H-perfluorooctyl)silane, into a disposable boat inside the desiccator (*see* **Note 4**).
3. Evaporate the silane-based release agent under low pressure (0.2–0.4 atm) in a desiccator for 5 min.
4. Seal the desiccator after reducing the pressure down to 0.2–0.4 atm for 2 h to coat the release agent on the wafer (*see* **Note 4**).

3.1.5 PDMS Soft Lithography

1. Mix 40.0 g SYLGARD® 184 Silicone Elastomer (PDMS) Base and 4.0 g Curing Agent in a disposable weighing boat thoroughly with a stirrer (*see* **Note 5**).
2. Then, place the mixed silicone elastomer in a low-pressure (~0.2–0.4 atm) chamber for 1 h to remove bubbles generated during the mixing.
3. Place the SU-8 master wafer in a 4 in. circular foil pan and pour the PDMS into the foil pan (*see* **Note 6**).
4. Place the foil pan on the hotplate at 100 °C overnight to cure the elastomer.
5. After PDMS is cured, cut the cured PDMS along the edge of the wafer using a razor blade and then peel off the PDMS layer from the SU-8 master.

3.1.6 PolyHEMA Stamping Process

1. Apply plasma treatment to make the patterned PDMS layer hydrophilic with 100 sccm O_2 and 20 sccm Ar for 60 s at 250 mT with power of 80 W.
2. Pipet 1 mL of polyHEMA solution on the patterned PDMS layer.
3. Place a piece of dummy blank PDMS layer on the top of the patterned PDMS layer and then apply pressure (~13 k Pa) to squeeze out excess polyHEMA solution [13].
4. Put both PDMS layers on a hot plate at 110 °C for 2 h. As ethanol evaporates, polyHEMA is deposited in the indented microwells on the patterned PDMS.
5. To remove the residual polyHEMA, apply plasma treatment with 100 sccm O_2 and 20 sccm Ar for 30 s at 250 mT with power of 800 W.

3.1.7 Device Assembly

1. Cut the patterned PDMS layer into devices using a razor blade and punch inlets and outlets by using a biopsy punch of 6 mm in diameter.
2. Clean (remove particles from) the PDMS surface using scotch tape (*see* **Note 7**).
3. Apply plasma treatment to activate the surface of the patterned PDMS layer as well as a thin (~2 mm thick) blank PDMS with 100 sccm O_2 and 20 sccm Ar for 60 s at 250 mT with power of 80 W.
4. Bond the patterned PDMS layer onto the blank PDMS.
5. Enhance the bonding strength by placing the bonded PDMS device on a hotplate at 80 °C for 15 min.

3.2 Priming of the Fabricated Microfluidic Device for Cell Culture

1. Place the device in a desiccator under low pressure (~0.2–0.4 atm) for 15 min.
2. Place the device into biobench and turn on UV light for 3 min for sterilization.

3. Pipet 100 μL of 1% (w/v) Pluronic F-108 solution into inlet (*see* **Note 8**).
4. Apply negative pressure to the outlet using a pipette bulb (~1000 Pa) for 10 s (*see* **Note 9**).
5. After 10 min, observe the device under a microscope to check if there are any air bubbles left in microchannels/chambers. Air bubbles should be completely eliminated.
6. If there are any air bubbles, repeat **steps 4** and **5** until air bubbles are completely removed.
7. Load Pluronic F-108 solution into the device through the inlet for coating the sidewall for 1 h.
8. Pipet 100 μL of PBS to the inlet and let it flow through to wash the microchannels and microwells inside the device for 1 h.

3.3 Cell Loading and Cell Culture in the Microfluidic Device

1. After the devices are prepared, harvest cells from culture plates for loading into the device:
   (a) Remove culture medium from a cell culture dish.
   (b) Pipet 5 mL of PBS to wash the culture dish, then aspirate the PBS.
   (c) Pipet 2 mL of trypsin into the dish and incubate at 37 °C for 5 min.
   (d) Shake the dish several times to promote cell detachment, add 5 mL of cell culture medium to the dish to neutralize the trypsin, and pipet several times.
   (e) Transfer the cell suspension into a 15 mL centrifuge tube and centrifuge at 100 × *g* for 5 min.
2. Remove the supernatant and resuspend the cells to a concentration of 5×10^6 cells/mL in cell culture medium.
3. Pipet 100 μL of this cell suspension in the inlet and apply negative pressure in the outlet using a pipette bulb (~1000 Pa) for 10 s.
4. Remove the cell suspension medium from both inlet and outlet.
5. Repeat **steps 3** and **4** twice.
6. Wash both inlet and outlet with cell culture medium three times.
7. Pipet 100 μL of fresh cell culture medium in the inlet.
8. Place the device in a cell culture dish together with a small dish filled with water (*see* **Note 10**).
9. Put the device back to an incubator.
10. Start drug testing or other treatments after aggregated spheres are formed. Depending on the cell types, cells can aggregate to sphere in 1–2 days (Fig. 3).

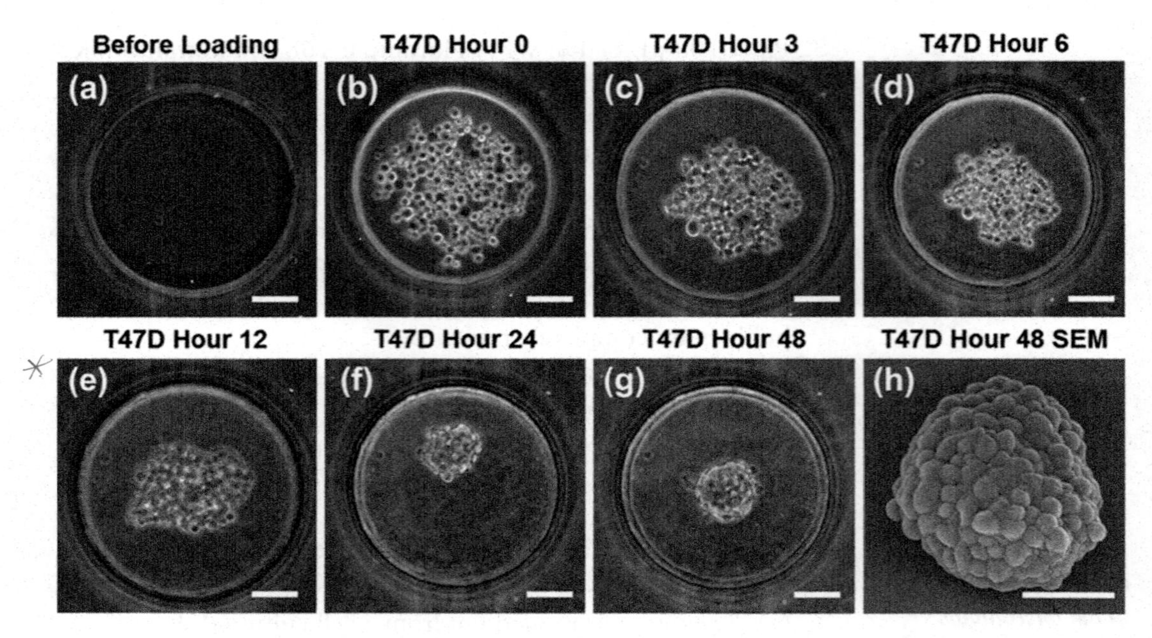

Fig. 3 Sphere formation on chip. (**a**–**g**) Sphere formation of T47D breast cancer cells in a large microwell: (**a**) before cell loading, (**b**) right after cell loading, (**c**) 3 h, (**d**) 6 h, (**e**) 12 h, (**f**) 24 h, and (**g**) 48 h after cell loading. Scale bar, 100 μm. (**h**) SEM image of a sphere retrieved from the chip after 48 h culture. Scale bar, 50 μm. Reproduced from Chen et al. [10] with permission from Macmillan Publishers Limited (Springer Nature)

11. Culture the aggregated spheres in the device up to 2 weeks. Exchange medium daily to guarantee enough nutrition supply for sphere culture (*see* **Note 11**). To this end, aspirate the solution both in the inlet and in the outlet. Then, add 100 μL of fresh cell culture medium in the inlet, but not in the outlet.

3.4 Cell Viability Staining and Imaging

1. Wash the device by flowing PBS for 10 min.
2. Add 100 μL of the LIVE/DEAD staining reagent into the inlet and incubate at 37 °C for 30 min.
3. Wash with PBS for 10 min.
4. Use a microscope with automated *xyz* stage, acquire images of the whole device (both phase-contrast and fluorescent images).
5. Assess efficacy of treatment by quantifying the intensity of green (live) and red (dead) cells. The results can also be measured by counting live and dead cells using co-focal microscopy.

4 Notes

1. Adhesion of SU-8 on a silicon wafer should be carefully characterized because the thermal expansion coefficients of SU-8 and silicon are quite different. The thermal stress induced

from thermal cycles can easily make the SU-8 layers peel off from the silicon wafer. To avoid this problem, thorough cleaning of the wafer prior to bonding is required to enhance the adhesion.

2. Though viscous SU-8 2100 (viscosity: 45,000 cSt) is suitable for making a thick SU-8 layer, air bubbles can be easily trapped inside the SU-8 layer during pouring. This may result in defects in device fabrication. To avoid trapping the bubbles, SU-8 should be poured quickly at once rather than slowly. If air bubbles are trapped inside the SU-8 layer, it is recommended to wash away the SU-8 using acetone and restart the process.
3. The adhesion of SU-8 and silicon is critical. To minimize the thermal stress, temperature should be ramped up and down slowly both during soft and post-exposure hard bakes. It is recommended to control slow ramping (up and down) by less than 10 °C per min.
4. The release agent helps in releasing PDMS from SU-8 master. Since the release agent is highly toxic, this step must be performed in a ventilated hood with great caution.
5. Uniform mixing of PDMS and curing agent is critical to guarantee uniform curing of PDMS on the whole wafer. PDMS should be mixed carefully for 1–2 min. The homogeneity of viscosity in the mixed PDMS can be a good indicator of the homogeneity of mixing.
6. As PDMS is viscous, it should be poured quickly to avoid trapping bubbles. If bubbles are trapped, the foil pan (containing the wafer and PDMS) should be placed in a low-pressure desiccator again for 15 min. The bubbles can then be removed.
7. Depending on the cleanness of fabrication environment, some dust can fall on the PDMS layer. The dust can affect the bonding between two pieces of PDMS layers. In a relatively clean environment (clean room of class 1000), the amount of dust is relatively low; therefore, using scotch tape to remove the dust should be enough. If the environment is not controlled by laminar flow, the PDMS layer can be washed by cleanroom-level IPA or acetone and dried by compressed nitrogen before bonding.
8. Pluronic F-108 is used to prevent nonspecific adhesion of cells to the PDMS sidewalls while loading.
9. The introduction of medium into the microfluidic channels is critical. The entire channels should be filled. If air bubbles are present, flow can be interrupted. In the freshly bonded device, PDMS is hydrophilic because of the last step of plasma treatment in fabrication. Therefore, it is easier to fill the channels with cell-suspension medium. However, for the aged devices, in which

PDMS becomes more hydrophobic, it becomes easier to introduce air bubbles inside the channel. It is recommended to give a mild plasma treatment to the device before introducing the medium or the priming of the device should be performed carefully.

10. Although incubator is a humid environment, evaporation of medium is inevitable. Since the volume of the inlet is small (100 μL), evaporation can lead to changes in osmolarity. Mild changes in osmolarity can affect the viability of cultured cells. To avoid evaporation, the device should be placed in a cell culture dish together with a small dish filled with water to guarantee saturated humidity. This can significantly reduce the evaporation of medium and changes in osmolarity.
11. Load the medium to make a different height in the inlet and outlet, so that the culture medium can spontaneously flow from inlet to outlet overnight until the next medium exchange.

Acknowledgments

This work was supported in part by the Department of Defense (W81XWH-12-1-0325) and in part by the National Institute of Health (1R21CA17585701). The Lurie Nanofabrication Facility of the University of Michigan (Ann Arbor, MI) is greatly appreciated for device fabrication.

References

1. Bissell MJ, Hall HG, Parry G (1982) How does extracellular matrix direct gene expression? J Theor Biol 99:31–68
2. Baker EL, Bonnecaze RT, Zaman MH (2009) Extracellular matrix stiffness and architecture govern intracellular rheology in cancer. Biophys J 97:1013–1021
3. Lee GY, Kenny AP, Lee AE et al (2007) Three-dimensional culture models of normal and malignant breast epithelial cells. Nat Methods 4:359–365
4. Pickl M, Ries CH (2009) Comparison of 3D and 2D tumor models reveals enhanced HER2 activation in 3D associated with an increased response to trastuzumab. Oncogene 28: 461–468
5. Francesco P, Emmanuel GR, Ernst HKS (2007) The third dimension bridges the gap between cell culture and live tissue. Nat Rev Mol Cell Biol 8:839–845
6. Perche F, Torchilin VP (2012) Cancer cell spheroids as a model to evaluate chemotherapy protocols. Cancer Biol Ther 13(12):1205–1213
7. Lin RZ, Chang HY (2008) Recent advances in three-dimensional multicellular spheroid culture for biomedical research. Biotechnol J 3(9–10):1172–1184
8. Tung YC, Hsiao AY, Allen SG et al (2011) High-throughput 3D spheroid culture and drug testing using a 384 hanging drop array. Analyst 136:473–478
9. Neuži P, Giselbrecht S, Länge K et al (2012) Revisiting lab-on-a-chip technology for drug discovery. Nat Rev Drug Discov 11:620–632
10. Chen YC, Lou X, Zhang Z et al (2015) High-throughput cancer cell sphere formation for characterizing the efficacy of photo dynamic therapy in 3D cell cultures. Sci Rep 5:12175
11. Re F, Zanetti A, Sironi M et al (1994) Inhibition of anchorage-dependent cell spreading triggers apoptosis in cultured human endothelial cells. J Cell Biol 127:537–546
12. Chen YC, Ingram P, Lou X, et al. (2012) Single cell suspension culture using PolyHEMA coating for anoikis assay and sphere formation.

Paper presented at the 16th international conference on miniaturized systems for chemistry and life sciences (MicroTAS '12), Okinawa, 28 October–1 November 2012.

13. Ingram PN, Chen YC and Yoon E (2013) PolyHEMA soft lithography for selective cell seeding, migration blocking, and high-throughput suspension cell culture. Paper presented at the international conference on miniaturized systems for chemistry and life sciences (MicroTAS '13), Freiburg, 27–31 October 2013.

Chapter 22

High-Throughput 3D Tumor Culture in a Recyclable Microfluidic Platform

Wenming Liu and Jinyi Wang

Abstract

Three-dimensional (3D) tumor culture miniaturized platforms are of importance to biomimetic model construction and pathophysiological studies. Controllable and high-throughput production of 3D tumors is desirable to make cell-based manipulation dynamic and efficient at micro-scale. Moreover, the 3D culture platform being reusable is convenient to research scholars. In this chapter, we describe a dynamically controlled 3D tumor manipulation and culture method using pneumatic microstructure-based microfluidics, which has potential applications in the fields of tissue engineering, tumor biology, and clinical medicine in a high-throughput way.

Key words Microfluidic platform, 3D tumor, High-throughput, Pneumatic manipulation, Recyclable system

1 Introduction

Three-dimensional (3D) tumors cultures have proved to be an excellent in vitro 3D model, mimicking the in vivo-like growth of human solid tumors [1]. Over the past several decades, 3D tumor cultures have gained attention in various fields of tumor-relevant research, like tumor biology, tumor engineering, drug screening, and discovery [2]. Currently, plenty of in vitro 3D tumor culture methods are available. The typical methodological members include anti-adhesive surfaces, hanging drops, 3D hydrogel scaffolds, and rotating bioreactors [3–5], and are frequently applied in pathophysiological studies and preclinical drug trials. However, the process of using these techniques is frequently slow, cumbersome, and laborious. Meanwhile, the technological improvements over scalable control and high-throughput production of 3D tumors are still necessary.

Microfluidic technology features the controllable manipulation of fluids, and has shown considerable promise for improving biological research at microscale [6]. Various properties of microfluidic

Zuzana Koledova (ed.), *3D Cell Culture: Methods and Protocols*, Methods in Molecular Biology, vol. 1612,
DOI 10.1007/978-1-4939-7021-6_22,

technologies, such as rapid sample processing and the precise control of both fluids and mammalian cells, have made them attractive candidates to replace conventional experimental approaches [7]. Moreover, the application of pneumatically activated valves with the ability to restrict fluidic channels promotes the advancement of microfluidic large-scale integration for high-throughput and spatio-temporal sample manipulation [8]. In 3D tumor culture, there are two main microfluidic approaches by either mechanical trapping or micro-droplet generation tools [9]. The mechanical trapping methods are usually based on the fixed barriers such as micro-wells and micro-pillars set in the microfluidic chambers to specifically localize the tumor cells [10, 11]. However, these manipulations are passive and the methods lack the ability to perform active operations, which are important for advanced sample retrieval methods and for developing reusable systems. The microfluidic droplet methods can produce large quantities of size-controlled microdroplets for tumor cell embedding [12, 13], which increases quantitative accuracy and analytical convenience. Nevertheless, 3D tumor culture using these cell-loaded micro-droplets methods is always performed in off-chip containers like petri dishes, similar to traditional culture methods. Moreover, the recyclable micro-system with the capacity for robust 3D culture and high-throughput manipulation of tumors remains largely out of reach.

Here, we present a high-throughput 3D tumor culture approach using pneumatic microfluidics. The polydimethylsiloxane-fabricated microfluidic devices with the arrayed pneumatic microstructures (PμSs) allow dynamic arrangement of tumor cells, and mass formation of 3D tumors. The processes of tumor cell localization and 3D tumor recovery can be actively controlled by simple on- and off-switches of PμSs in the recyclable microfluidic platform, being useful to perform quantity-steady cell trapping and size-controlled 3D tumor production [14, 15].

2 Materials

2.1 Photolithography

1. Silicon wafers, 3 in. in diameter, 250 μm thick, orientation <111> (Shanghai Xiangjing Electronic Technology Ltd).
2. AZ 50XT photoresist and developer (AZ Electronic Materials).
3. SU-8 2025 photoresist and developer (Microchem).
4. Spin coater (Laurell Technologies Corporation, WS-400B-6NPP).
5. Mask aligner (No. 45 Research Institute of China Electronics Technology Group Corporation, BG-401A).
6. Heating plates, 200 × 300 mm, capable of heating to 200 °C.

7. Plastic Petri dish, 100 mm in diameter.
8. Wafer forceps.

2.2 Microfluidic Devices

1. PDMS prepolymer (RTV615 A and B, Momentive Performance Materials) (*see* **Note 1**).
2. Electronic balance.
3. Centrifugal Mixer (THINKY, AR-100).
4. Vacuum pump linked with a vacuum chamber dryer.
5. Electric drying oven.
6. Glass slide, 25 × 76 × 1 mm.
7. Scotch tape.
8. Puncher.
9. Stereo microscope.
10. Syringe pump (Longer pump, LSP01-1A).
11. Disposable syringe (1 mL).
12. Tygon tubing. The tubing should match the puncher size (*see* **Note 2**).
13. Autoclaved ultra-purified water.
14. 10 mg/mL Pluronic F127 in ultra-pure water.
15. Phosphate buffered saline (PBS), pH 7.4, sterile.

2.3 Cell Preparation

1. Human glioma (U251) cells. Other types of tumor cells can also be used to produce 3D tumors in the microfluidic devices.
2. Cell culture medium: 10% fetal bovine serum, 100 units/mL penicillin, and 100 μg/mL streptomycin in Dulbecco's modified Eagle's medium.
3. 0.25% Trypsin-EDTA.
4. Sterile PBS.
5. Micropipette.
6. Hemocytometer.
7. Centrifuge.
8. Cell culture Petri dish, 100 mm in diameter.
9. Cell culture incubator (37 °C, 5% CO_2).

3 Methods

3.1 Fabrication of Device Molds by Photolithography

1. Design the microfluidic device using AutoCAD software and print three photomasks from the company (*see* **Note 3**).
2. Place two silicon wafers on a heating plate with wafer forceps. Heat at 200 °C for 5 min.

3. For the fluidic mold, spin-coat a 100 μm thick layer of SU-8 photoresist onto a silicon wafer.
4. Bake the SU-8 coat on a heating plate at 95 °C for 6 min, and then expose it to UV light through the first photomask (*see* **Note 3**) using a BG-401A mask aligner.
5. Bake the exposed coat on the heating plate at 95 °C for 6 min, and develop it in SU-8 developer.
6. For the control mold, spin-coat a 20 μm thick layer of SU-8 photoresist onto another silicon wafer.
7. Bake the coated SU-8 photoresist on the heating plate at 95 °C for 4 min, and expose it to UV light through the second photomask (*see* **Note 3**) using the BG-401A mask aligner.
8. Bake the exposed SU-8 photoresist on the heating plate at 95 °C for 4 min, and develop it in SU-8 developer.
9. Spin-coat a 40 μm thick AZ 50XT photoresist onto the same silicon wafer.
10. Bake the AZ coat on the heating plate at 95 °C for 5 min, and expose it to UV light through the third photomask (*see* **Note 3**).
11. Develop the exposed AZ 50XT photoresist in AZ developer.
12. Place two micropatterned silicon wafers on the heating plate. Heat at 95 °C for 10 min.

3.2 Fabrication of PDMS Microfluidic Devices

1. Prepare PDMS prepolymer (RTV 615 A and B, *see* **Note 1**) at a ratio of 5:1 (w/w) for the fluidic layer, 30:1 (w/w) for the control layer, and 10:1 (w/w) for the supporting layer. The device layers are presented in Fig. 1.

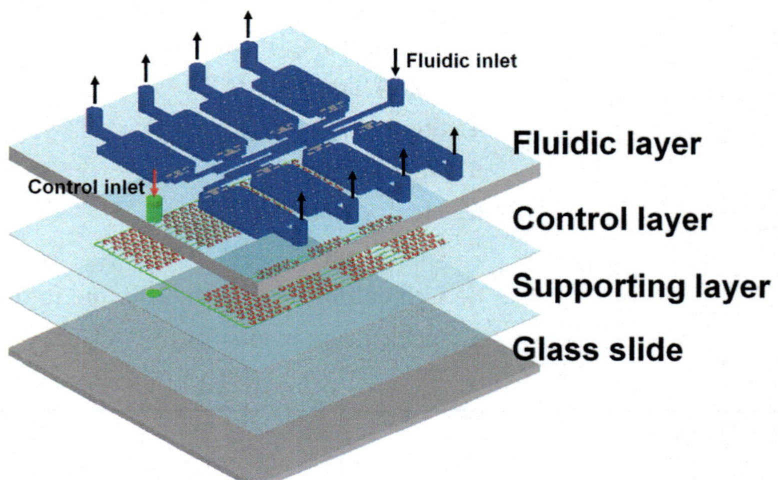

Fig. 1 Composition of the microfluidic device. The device is composed of four layers, sequentially from top to bottom: the fluidic layer, control layer, supporting layer, and glass slide. The fluidic layer contains one inlet, eight outlets, and eight individual chambers organized symmetrically. In the control layer, a channel network incorporates the arrayed PμSs at their centers or terminals

2. Mix RTV 615 A and B using a centrifugal mixer (*see* **Note 4**).
3. For the fluidic layer, pour the PDMS mixture (30 g, mixing ratio 5:1) onto the fluidic mold placed in the Petri dish surrounded by tin foil.
4. Place the Petri dish in a vacuum chamber to degas PDMS mixture for 30 min to remove excess air bubbles from the mixture.
5. Place the fluidic mold with PDMS into an oven at 90 °C, and allow to cure for 30 min (*see* **Note 5**).
6. For the control layer, spin-coat the PDMS mixture (3.1 g, mixing ratio 30:1) onto the control mold at 61 × *g* for 65 s. Place the control mold with PDMS into the oven at 90 °C, and allow to cure for 50 min.
7. For the supporting layer, spin-coat the PDMS mixture (3.3 g, mixing ratio 10:1) onto a glass slide at 43 × *g* for 60 s. Bake the coated glass slide at 90 °C for 10 min.
8. Peel off the fluidic layer from the fluidic mold, and punch holes in the fluidic layer for cell and nutrient supply access, chamber purging, and waste exclusion (Fig. 1).
9. Trim the fluidic layer using a surgical blade, clean the fluidic layer using a scotch tape, and align the fluidic layer onto the control layer using a stereo microscope. Place the aligned two layers into the oven at 90 °C, and bake for 12 h.
10. After baking, peel off the assembled layers from the control mold and punch a hole for accessing the channels and PμSs in the control layer in the microfluidic device.
11. Place the assembled layers on the supporting layer coated on the glass slide, and bake at 90 °C for 3 days before use.

3.3 Pretreatment of Microfluidic Devices

1. Irradiate the microfluidic device with UV for 3 h (*see* **Note 6**).
2. Use autoclaved tygon tubing to connect the syringe with the fluidic inlet, and the pneumatic controller (*see* **Note 7**) with the control inlet of the device, respectively.
3. Rinse the microfluidic device with autoclaved water (*see* **Note 8**). To this end, load water using the syringe.
4. Introduce the filtered anti-adhesive reagent Pluronic F127 (*see* **Note 9**). This treatment is to modify the surface of microfluidic channels and chambers.
5. Rinse the channels and chambers of the microfluidic device with PBS to remove any excess of anti-adhesive reagent.

3.4 Cell Culture

1. Culture U251 cells in cell culture medium at 37 °C in a humidified atmosphere with 5% CO_2 and passage the cells at a ratio of 1:3 every 3 days to maintain their exponential growth.

2. To harvest cells for 3D tumor culture in microfluidic device, wash the cells twice with PBS and then trypsinize with 0.25% trypsin-EDTA at 37 °C.
3. Stop trypsinization with the addition of fresh cell culture medium and pipette the cell suspension repeatedly to create a single-cell suspension.
4. Centrifuge the cell suspension at 200 × *g* for 3 min.
5. Discard supernatant and resuspend the cells in fresh cell culture medium.
6. Count the cells using a hemocytometer and adjust cell density to 5.0 × 10^6 cells/mL.

3.5 Repeatable 3D Tumor Culture in Microfluidic Devices

1. Load tumor cells at a density of 5.0 × 10^6 cells/mL from the fluidic inlet into the device at a flow rate of 10 μL/min using a normal syringe pump (*see* **Note 10**). When the tumor cells flow into the microfluidic chambers homogeneously (Fig. 2a), the preparation of cell introduction for precise cell localization is ready.
2. Activate the PμSs. Gas pressure (20 psi, a compressed nitrogen source) makes the PμSs transform from 2D to 3D spatial structures and the tumor cells flowing into the activated PμSs are trapped mechanically (Figs. 2b and 3a). Allow the trapping process to proceed for 10–20 min.
3. After cell localization, culture the trapped cells for 5–14 days using fresh cell culture medium at a slow perfusion rate of

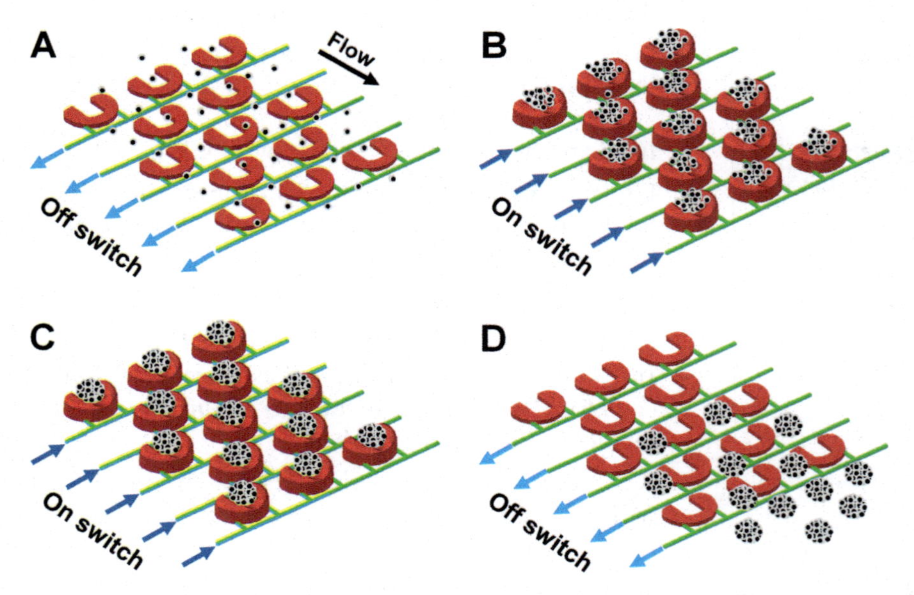

Fig. 2 Microfluidic manipulations of 3D tumor production in the device with arrayed PμSs. (**a**) Cell loading. (**b**) Cell trapping. (**c**) 3D tumor culture and formation. (**d**) 3D tumor recovery. The PμSs can be controlled dynamically based on the use of the pneumatic on/off switch

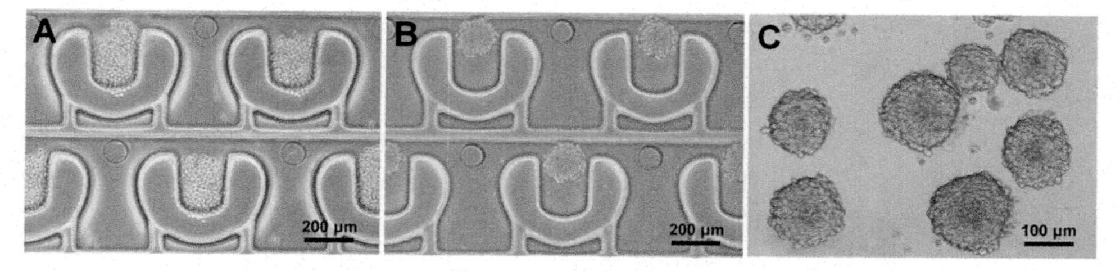

Fig. 3 Optical images of 3D tumor production at different manipulations. (**a**) The trapped tumor cells in the activated PμSs. (**b**) 3D tumor formation in the activated PμSs. (**c**) 3D tumors recovered from the microfluidic device

5 μL/min (*see* **Note 11**). Tumor cells will self-assemble into 3D tumors (Figs. 2c and 3b) that can be used for biological analysis or experimental treatment.

4. After tumor culture and experiment (e.g., drug delivery), recover the 3D tumors by cancelling the activation of PμSs (switch off) to release the tumors from the microfluidic chambers into the outlets of the device (Figs. 2d and 3c). These 3D tumors can be selectively used for off-chip analysis of different biological items, such as ultrastructure by electron microscopy or apoptosis by flow cytometry.
5. Rinse the chambers of the microfluidic devices with PBS and water.
6. The microfluidic device can be used repeatedly (*see* **Note 12**).

4 Notes

1. PDMS is an inert, nontoxic, nonflammable, optically transparent, and elastomeric material. It is a mineral-organic polymer (a structure containing carbon and silicon) of the siloxane family. For the fabrication of microfluidic devices, PDMS elastomer (RTV615 A) mixed with a crosslinker (RTV615 B) is poured onto a microstructured mold and heated to obtain an elastomeric replica of the mold (PDMS cross-linked).
2. The inner diameter of the puncher should be a little smaller than the outer diameter of the tygon tubing, which is important to prevent the leakage of cell suspension and other reagents from either the inlet or the outlet of the microfluidic device.
3. High-resolution transparency photomasks containing the design of microfluidic components in the device are fabricated by Shenzhen MicroCAD Photo-mask LTD. The first, second, and third photomasks correspond to the fluidic layer, the channel network part of the control layer, and the PμSs part of the control layer in the device, respectively (Fig. 1).

Next, the first mask is used for the fabrication of the fluidic mold, and the other two masks are used for the fabrication of the control mold.

4. Thorough mixing of PDMS elastomer and crosslinker is critical for acquiring homogeneous PDMS mixture before curing.
5. Higher temperature accelerates curing of the PDMS mixture. However, to prevent operation at high temperature, we and others commonly bake the PDMS mixture at 80–90 °C.
6. Because PDMS material is UV-transparent, the microfluidic devices can be conveniently sterilized by UV irradiation.
7. An external solenoid valve or hand-push valve can be used as the pneumatic controller to activate the PμSs. The controller is connected with both a N_2 cylinder and the control inlet of the device using the tygon tubing.
8. Ultra-purified water is prepared by an Ultra Pure Water Systems (Sartorius) purifying deionized water to attain an electrical resistivity of 18 MΩ cm.
9. Pluronic F127 is highly effective in preventing protein adsorption and cell attachment. The modified chamber surface can effectively inhibit the access of tumor cells.
10. The density of tumor cells ranging from 1×10^6 cells/mL to 1×10^7 cells/mL is suitable for cell loading. However, the low cell density can result in a longer cell loading process.
11. To reduce hydrodynamic shear stress, the medium flow rate used for tumor culture is very slow. Also, excessive tumor cells remaining at the nonspecific trapping region can be removed by this medium flow due to the anti-adhesive pretreatment of chamber surface with Pluronic F127 solution.
12. Because the PμSs can be switched on and off repeatedly, the microfluidic device can be reused for 3D tumor culture. The microfluidic system can provide remarkable long-term (over 1 month) and cyclic stability.

Acknowledgments

This work was supported by the National Natural Science Foundation of China (No. 31470971, No. 21375106, No. 21175107, No. 31100726), the program of China Scholarships Council (No. 201208610047), the Fundamental Research Funds for the Central Universities (No. 2452015439), and Northwest A&F University.

References

1. Lee GY, Kenny PA, Lee EH et al (2007) Three-dimensional culture models of normal and malignant breast epithelial cells. Nat Methods 4:359–365
2. Mehta G, Hsiao AY, Ingram M et al (2012) Opportunities and challenges for use of tumor spheroids as models to test drug delivery and efficacy. J Control Release 164:192–204
3. Fennema E, Rivron N, Rouwkema J et al (2013) Spheroid culture as a tool for creating 3D complex tissues. Trends Biotechnol 31: 108–115
4. Thurber GM, Wittrup KD (2008) Quantitative spatiotemporal analysis of antibody fragment diffusion and endocytic consumption in tumor spheroids. Cancer Res 68:3334–3341
5. Laguinge LM, Lin S, Samara RN et al (2004) Nitrosative stress in rotated three-dimensional colorectal carcinoma cell cultures induces microtubule depolymerization and apoptosis. Cancer Res 64:2643–2648
6. Whitesides GM (2006) The origins and the future of microfluidics. Nature 442:368–373
7. Liu W, Li L, Wang X et al (2010) An integrated microfluidic system for studying cell-microenvironmental interactions versatilely and dynamically. Lab Chip 10:1717–1724
8. Gómez-Sjöberg R, Leyrat AA, Pirone DM et al (2007) Versatile, fully automated, microfluidic cell culture system. Anal Chem 79:8557–8563
9. Sung KE, Beebe DJ (2014) Microfluidic 3D models of cancer. Adv Drug Deliv Rev 79–80:68–78
10. Kwapiszewska K, Michalczuk A, Rybka M et al (2014) A microfluidic-based platform for tumour spheroid culture, monitoring and drug screening. Lab Chip 14:2096–2104
11. Yu L, Chen MC, Cheung KC (2010) Droplet-based microfluidic system for multicellular tumor spheroid formation and anticancer drug testing. Lab Chip 10:2424–2432
12. Kim C, Chung S, Kim YE et al (2011) Generation of core-shell microcapsules with three-dimensional focusing device for efficient formation of cell spheroid. Lab Chip 11:246–252
13. Alessandri K, Sarangi BR, Gurchenkov VV et al (2013) Cellular capsules as a tool for multicellular spheroid production and for investigating the mechanics of tumor progression in vitro. Proc Natl Acad Sci U S A 110:14843–14848
14. Liu W, Li L, Wang J-C et al (2012) Dynamic trapping and high-throughput patterning of cells using pneumatic microstructures in an integrated microfluidic device. Lab Chip 12:1702–1709
15. Liu W, Xu J, Li T et al (2015) Monitoring tumor response to anticancer drugs using stable three-dimensional culture in a recyclable microfluidic platform. Anal Chem 87: 9752–9760

Chapter 23

High-Throughput Microfluidic Platform for 3D Cultures of Mesenchymal Stem Cells

Paola Occhetta*, Roberta Visone*, and Marco Rasponi

Abstract

The design of innovative tools for generating physiologically relevant three-dimensional (3D) in vitro models has been recently recognized as a fundamental step to study cell responses and long-term tissue functionalities thanks to its ability to recapitulate the complexity and the dimensional scale of the cellular microenvironment, while directly integrating high-throughput and automatic screening capabilities.

This chapter addresses the development of a poly(dimethylsiloxane)-based microfluidic platform to (1) generate and culture 3D cellular microaggregates under continuous flow perfusion while (2) conditioning them with different combinations/concentrations of soluble factors (i.e., growth factors, morphogens or drug molecules), in a high-throughput fashion. The proposed microfluidic system thus represents a promising tool for establishing innovative high-throughput models for drug screening, investigation of tissues morphogenesis, and optimization of tissue engineering protocols.

Key words Microfluidics, High-throughput screening, Mesenchymal stem cells, Perfused micromasses, Soft-lithography

1 Introduction

In vivo, cells reside in a complex multifactorial microenvironment, which provides them with tissue-specific biochemical and physical cues, finally regulating their behavior and functions [1]. Biochemical signals, which are mainly a combination of cytokines, hormones, growth factors and diffusive biomolecules, contribute to activate complex signaling pathways, which modulate intrinsic regulatory mechanisms and, in turn, determine cell fates [2]. This soluble molecules-mediated signaling is fundamental for both autocrine (affecting the secreting cell) and paracrine (acting on neighboring cells) intercellular communication. In addition to biochemical mediators, also the extracellular matrix (ECM) together with

*These authors equally contributed to the work.

Zuzana Koledova (ed.), *3D Cell Culture: Methods and Protocols*, Methods in Molecular Biology, vol. 1612,
DOI 10.1007/978-1-4939-7021-6_23, © Springer Science+Business Media LLC 2017

neighboring cells, forming the solid-state microenvironment, plays a crucial role in regulating cell behavior through the activation of solid cues-mediated signaling pathways [3]. The orchestrated interplay of these matrix-bound molecules, soluble factors, cell–cell interaction, and cell–ECM connections, contributes to establish the complexity of the 3D cellular microenvironment, characterized by spatiotemporally dynamic gradients, eventually affecting cellular specification [4]. Considering this complexity and the intricate pathway interconnectivity characterizing the native microenvironment, the design of in vitro models able to recapitulate the dynamic combinatorial role of biochemical and physical factors on cell behavior becomes fundamental to study cell responses and long-term tissue functionalities.

In the last decades, two-dimensional (2D) standard culture methods have been recognized to be poorly representative of such in vivo complexity. They indeed expose cells to non-physiological cues, in terms of mechanical stiffness of the substrate, cell spatial organization (long distances among cells, cell–cell interactions rarely controlled) and culture condition (static culture, high ratio between medium and cell volume and uncontrolled chemical gradients) [5]. Nowadays, different strategies have been developed to address some of these issues and generate more physiologically relevant in vitro models for cell culture. At the macroscale, these systems mainly rely on the use of innovative three-dimensional (3D) scaffold biomaterials, able to offer a wide range of 3D architectures, combined with dynamic culture within bioreactors, which provide a better control over environmental conditions [6]. However, these 3D macroscale models are still hampered by dimensional scales that are orders of magnitude bigger than the native microenvironment, thus lacking in accurately controlling biological phenomena at a cellular level. Moreover, the necessity of modeling the concurrent effect of multiple factor impedes their exploitation for high-throughput studies.

Microscale engineering technologies, conversely, provide unprecedented opportunities for modeling cellular microenvironment going beyond traditional 2D and 3D in vitro systems. The integration of microfabricated substrates with microfluidic principles, indeed, enables to scale the culture conditions down to the cell size level (i.e., micrometer scale), offering the chance to monitor and recapitulate unique features of native tissue microenvironments [7]. Microfluidic platforms can be indeed designed for culturing cells under continuous perfusion of medium and the highly predictable laminar fluid dynamics inside microchannels allows an unprecedented tuning over local environmental conditions (medium factor and dissolved gas concentration, temperature, pH, shear stress). Moreover, this ability to predict fluid dynamics allows for precisely guiding the spatial positioning of soluble factors and particles (i.e., cells, nanoparticles and microparticles) within chambers and channels [8–10]. The manipulation of fluid and particle flow within microchannels has been extensively exploited in cell culture applications for generating linear or complex patterns of soluble and solid factors [11–16], offering the chance to recapitulate in a high-throughput fashion spatiotemporal gradients characterizing the in vivo microenvironment. Regarding the fabrication of these microfluidic platforms for cell culture, soft lithography is recognized to be one of the most robust strategies. It exploits poly(dimethylsiloxane) (PDMS), a biocompatible and optically transparent elastomeric rubber, cheap and easy to prototype, and thus suitable for cell culture studies [17]. Thanks to the optical transparency of PDMS, immunofluorescence analysis has been by far the most commonly adopted readout for microfluidic cell cultures due to the difficulties in the extraction of biological matter. However, alternative fabrication techniques have recently allowed to apply traditional techniques, such as histology [18] or electrical recording [19, 20]. Moreover, in the last few years, microfluidic principles have started to be extended to 3D culture systems, either combining innovative hydrogel compositions [21, 22] with microstructuration techniques [23–26], or taking advantages of the high area-to-volume ratio achievable at the microscale to favor intrinsic tendency of cells to self-organize in 3D [27–29]. Given these premises, microfluidic technologies constitute an innovative approach for precise in vitro tailoring of different features of the 3D cell microenvironment within a single model, contemporarily allowing automatic and parallelized experimentations and enhancing the throughput of analyses.

In this chapter, we describe the development of a PDMS microfluidic platform as innovative tool for investigating and modeling the effect of different soluble cues of cellular microenvironment on cell fate in a 3D configuration. In particular, the platform is designed to generate and culture 3D cellular microaggregates under continuous flow perfusion, while conditioning them with different sequences, combinations and/or concentrations of soluble factors in a high-throughput fashion. The system consists of two functional units (Fig. 1): (1) a 3D culture area and (2) a serial dilution generator (SDG). The culture area was specifically designed to favor the condensation of tens of mesenchymal stem cells (MSCs) within fluidically connected microchambers located in spatially defined configurations, accomplishing the formation of microaggregates with uniform size and shape [30]. The device layout is compatible with the implementation of differently shaped SDG networks [15] (i.e., logarithmic to base 10 and linear), thus allowing the investigation of effects of soluble factors over wide concentration ranges (logarithmic patterns) or fine-tuning of narrower concentration windows (linear patterns). The presented device has been exploited to perform studies on limb bud development and to investigate processes involved in MSC chondrogenic differentiation [29].

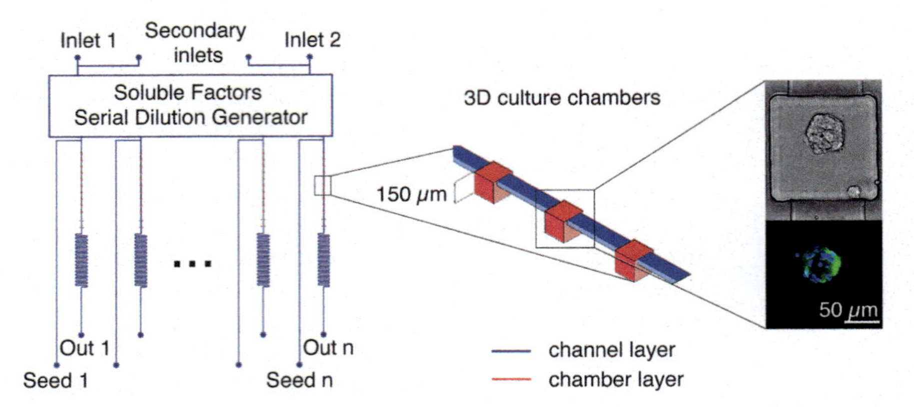

Fig. 1 Layout of a microfluidic platform for 3D cellular micromasses generation and culturing under spatially defined patterns of soluble factors. The chip layout consists of a serial dilution generator and a 3D culture area. Dilutions of chemicals are generated from two main inlets (Inlet 1 and 2) and delivered to downstream culture units, each consisting of cubic culture chambers sequentially connected by a fluidic channel (Out_1, Out_n). Furthermore, each unit is provided with a secondary channel ($Seed_1$, $Seed_n$). Two additional inlets (secondary inlets) facilitate the medium change operations, defining a by-pass for the device

However, the platform offers high versatility and its design can be easily extended to different cell models, where initial cell condensation and the onset of a 3D structure are phenomena expected to mediate a physiological response to exogenous signals. The proposed microfluidic system thus represents a promising tool for establishing innovative high-throughput models for drug screening, investigation over tissues morphogenesis and optimization of tissue engineering protocols.

2 Materials

2.1 Equipment for Device Fabrication

1. CAD software (AutoCAD, Autodesk).
2. Mask aligner (Karl Suss, MA56), located in a cleanroom facility.
3. Quartz glass compatible with the mask holder of the mask aligner.
4. Two flat hotplates, located in a cleanroom facility.
5. Fume-hood, located in a cleanroom facility.
6. Spincoater, located in a cleanroom facility.
7. Optical microscope, located in a cleanroom facility.
8. Nitrogen gas, from nitrogen gas line located in a cleanroom facility.
9. Stereomicroscope.
10. Plasma cleaner (HarrickPlasma Inc.) (*see* **Note 1**).

11. Oven.
12. Vacuum chamber.
13. Vacuum pump.

2.2 Equipment for MSC Culture and Readout

1. Humidified 5% CO_2 incubator.
2. Programmable syringe pump Harvard NE-100-240 (*see* **Note 2**).
3. Optical microscope.
4. Confocal/epifluorescence microscope.

2.3 Materials for Device Fabrication

1. 4-in. polished silicon wafers.
2. Negative photoresists (SU-8 2050 and SU-8 2100, Microchem).
3. SU-8 developer (Microchem).
4. PDMS kit (Sylgard 184, Dow Corning): prepolymer (component A) and curing agent (component B).
5. Round coverslips diameter 50 mm, thickness n.1.
6. 100% ethanol.
7. Isopropyl alcohol (IPA).
8. Acetone.
9. Sharpened biopsy puncher, diameter 500 μm.
10. Scalpel.
11. Stainless steel spatula.
12. Tape (Magic Tape, 3M).
13. Disposable cell culture non-treated petri dish, 100 mm and 120 mm diameter.
14. Tweezers for silicon wafer handling.
15. Stainless steel forceps.
16. Stainless steel couplers, 23 Gauge (G) and 22 G, 12 mm or more long.
17. Stainless steel plugs, 23 G (*see* **Note 3**), sterilized by autoclaving.
18. Blunt needles, 23 and 22 G.
19. Polyethylene tubing, inner diameter 0.6 mm, outer diameter 0.97 mm, fitting 22 G.
20. Tygon tubing, inner diameter 0.51 mm, outer diameter 1.52 mm, fitting 23 G.
21. 1 mL syringes.

2.4 MSC Culture and Readout

1. Plating medium: 10% fetal bovine serum (FBS), 4.5 mg/mL D-glucose, 0.1 mM nonessential amino acids, 1 mM sodium pyruvate, 100 mM HEPES buffer, 100 U/mL penicillin,

100 μg/mL streptomycin, and 0.29 mg/mL L-glutamine, and 5 ng/mL fibroblast growth factor 2 (FGF2) in alpha-MEM (*see* **Note 4**).

2. Serum-free culture medium: 1 mM sodium pyruvate, 100 mM HEPES buffer, 100 U/mL penicillin, 100 μg/mL streptomycin, 0.29 mg/mL L-glutamine, 1× ITS (10 μg/mL insulin, 5.5 μg/mL transferrin, 5 ng/mL selenium), 5 μg/mL human serum albumin, and 4.7 μg/mL linoleic acid in DMEM (*see* **Note 5**).
3. Phosphate buffered saline (PBS).
4. Fungizone solution: 0.25 μg/mL Fungizone in sterile PBS (*see* **Note 6**).
5. 4% paraformaldehyde (PFA) solution prepared according to standard protocols [31].
6. Diluting buffer (DB): 1.5% goat serum and 0.5% Triton X-100 in PBS (*see* **Note 7**).
7. Primary antibodies, e.g., Anti-human Collagen type II, Anti-human Collagen type X, Anti-human Vinculin, Actin, and Anti-human Ki67.
8. Secondary antibodies, e.g., Goat anti-mouse 488, Goat anti-rabbit 546.
9. 10% glycerol (*see* **Note 8**).

3 Methods

3.1 Device Fabrication

This section describes preparation of the device, as schematically shown in Fig. 2, in several steps, including fabrication of the master mold by a sequential process of channel layer and chamber layer fabrication (Fig. 2a, b), fabrication of the bottom (Fig. 2c i) and top PDMS layers (Fig. 2c ii), and device assembly (Fig. 2c iii–v).

3.1.1 Silicon Master Mold Fabrication

1. Using a CAD software, design the chip layout (Fig. 1) composed of two distinct layers, namely a channel layer and a chamber layer (*see* **Note 9**).
2. For each layer, print out a photomask at high-resolution (greater than 20,000 dpi) on a transparency sheet (deep black for unexposed regions) (Fig. 2a) (*see* **Note 10**).
3. Perform the following **steps 4–16** under clean room operations standards.
4. To prepare the channel layer, clean a new wafer by rinsing it with acetone and IPA, and subsequently dry it with nitrogen gas and heating for 20 min on a hotplate, preheated to 180 °C.
5. Spin-coat SU-8 2050 to a thickness of 70 μm onto the wafer (*see* **Note 11**).

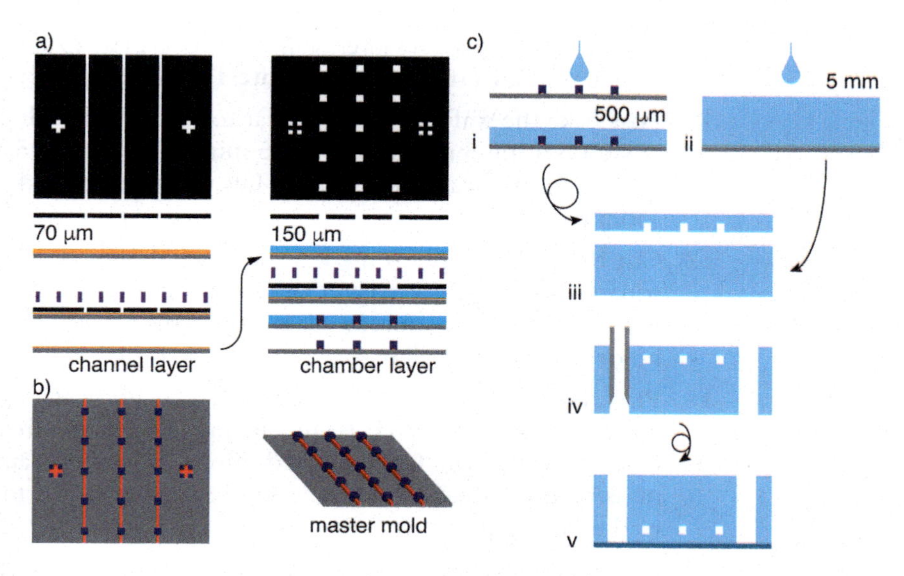

Fig. 2 The fabrication process of the microfluidic bioreactor is schematically represented here only through a triplet of culture units. (**a**) Photolithographic process to fabricate the silicon master mold. Two negative photomasks are produced, defining channels (*top left*) and chambers (*top right*) respectively. At first, negative photoresist is spin-coated on a clean wafer to a height of 70 μm. After exposure, a second round of thicker resist is spin-coated to a height of 150 μm. Upon alignment of the corresponding photomasks to the channel layer on the wafer (typical alignment marks are showed here), exposure provides channels with chambers at the desired locations. (**b**) Top and lateral view of the master molds, highlighting different feature heights with different colors: channels at 70 μm in orange, and chambers at 220 μm in *blue*. (**c**) The microfluidic bioreactor consists of two PDMS layers. (*i*) The layer containing micrometer features is casted by pouring a thin layer of PDMS on the silicon master mold. (*ii*) A thicker PDMS slab is casted on a clean silicon wafer. (*iii*) The two layers are permanently bonded together upon plasma activation, so that microchannels result embedded. (*iv*) A small biopsy puncher is used to make through holes at all inlet/outlet ports locations. (*v*) A final plasma treatment allows to finally bond the microfluidic bioreactor to a glass surface

6. Soft-bake the wafer through a three-step (two-hotplate) procedure: Move the coated wafer with tweezers on a 65 °C preheated hotplate. After 2 min, move rapidly the wafer on a 95 °C preheated hotplate. Bake it for 7 min and move again on a 65 °C preheated hotplate. After 2 min let it cool down to room temperature.
7. Cut the channel photomask around the designed wafer to a square shape, slightly greater than wafer diameter, and tape it upside-down to a quartz glass compatible with the mask holder of the mask aligner (*see* **Note 12**).
8. Place the wafer in the mask aligner, move it against the mask in soft-contact mode and expose to a dose of 200 mJ/cm^2 in the i-line region (λ = 365 nm) (*see* **Note 13**).
9. Post-bake the wafer in a fashion similar to the soft-bake described above (**step 6**): 2 min at 65 °C, 7 min at 95 °C, and 2 min at 65 °C.

10. To prepare the chamber layer, spin-coat SU-8 2100 to a thickness of 150 μm onto the wafer (*see* **Note 14**).
11. Post-bake the wafer in a way similar to the soft-bake described in **step 6** using 5 min at 65 °C, 25 min at 95 °C, and 5 min at 65 °C (on preheated hotplates). Let the wafer cool down to room temperature.
12. Cut the chamber photomask around the designed wafer to a square shape, slightly greater than wafer diameter, and tape it upside-down to a quartz glass compatible with the mask holder of the mask aligner.
13. Place the wafer in the mask aligner. Align the wafer and the chamber photomask correctly, taking advantage from the alignment markers. Once aligned, move the wafer against the mask in soft-contact mode and expose to a dose of 375 mJ/cm^2 in the i-line region (*see* **Note 15**).
14. Using the method described in **step 6**, bake the wafer for 5 min at 65 °C, then for 11 min at 95 °C and 5 min at 65 °C.
15. Develop the wafer in a SU-8 developer bath for about 18 min.
16. Rinse briefly the wafer with IPA then dry it with a gentle stream of nitrogen gas.
17. Place the wafer (master mold) in a petri dish and tie it with tape for further use.

3.1.2 Bottom PDMS Layer Fabrication

1. Pour PDMS components into a plastic cup, in a 10:1 ratio (w/w), prepolymer to curing agent (*see* **Note 16**).
2. Mix vigorously the PDMS components using a steel spatula, until a uniform bubbly suspension is achieved.
3. Spin the PDMS on the silicon master mold with the spin coater up to a thickness of 500 μm (Fig. 2c i).
4. Place in a vacuum chamber until all bubbles are removed.
5. Cure in an oven at 80 °C for 2 h.

3.1.3 Top PDMS Layer Fabrication

1. Pour PDMS components in a plastic cup, in a 10:1 ratio (w/w), prepolymer to curing agent.
2. Mix vigorously the PDMS components using a steel spatula, until a uniform bubbly suspension is achieved.
3. Degas the mixture in a vacuum chamber for about 30 min.
4. Pour the PDMS mixture into the petri dish to achieve a thickness of about 5 mm (*see* **Note 16**) (Fig. 2c ii).
5. Place in a vacuum chamber until all bubbles are removed.
6. Cure in an oven at 80 °C for 2 h.

3.1.4 Device Assembly

1. Cut out the device contours directly on the silicon master mold using a scalpel, and peel off the resulting stamp (bottom layer) (*see* **Note 17**).
2. Place the stamp (features facing up) together with the petri dish containing the top layer inside the plasma cleaner.
3. Turn on the vacuum pump and wait until the pressure within the vacuum chamber drops below 100 mTorr (*see* **Note 18**).
4. Turn on the RF power to ignite the plasma and keep flowing air for 1 min to activate the surface (*see* **Note 19**).
5. After the plasma treatment is over, take out the device and place it into a biosafety cabinet.
6. Bring in contact the PDMS stamp (bottom layer) and the PDMS top layer, taking care of removing all possible air trapped (*see* **Note 20**).
7. Place the assembled device in an oven at 80 °C for at least 20 min to make the plasma bonding permanent.
8. Cut the assembled device contour following the features, and peal it off.
9. Punch input and output access ports with the biopsy puncher (*see* **Note 21**).
10. Place the stamp (bottom layer on top) and a glass coverslip inside the plasma cleaner, and repeat **steps 3–7**.
11. Use magic tape to protect the PDMS device and store it in a petri dish for future use.

3.2 Device Preparation for Cell Seeding

1. Remove the tape from the PDMS device and sterilize it by autoclaving (121 °C, 20 min) (Fig. 3a).
2. Place the sterile device inside a sterile petri dish.
3. Incubate it in the oven overnight at 80 °C (*see* **Note 22**).
4. Take the petri dish containing the sterile device out of the oven and store in sterile conditions (*see* **Note 23**).
5. Fill the petri dish containing the device with Fungizone solution.
6. Put the filled petri dish in a vacuum chamber (*see* **Note 24**).
7. Turn the vacuum pump on and degas for at least 15 min (*see* **Note 25**).
8. Turn the vacuum pump off and allow the air to exit the device's channels for at least 15 min (Fig. 3a) (*see* **Note 25**).
9. Take the petri dish out of the vacuum chamber and check under an optical microscope that the device is completely filled with the solution. In case it is needed, repeat **steps 7–9**.

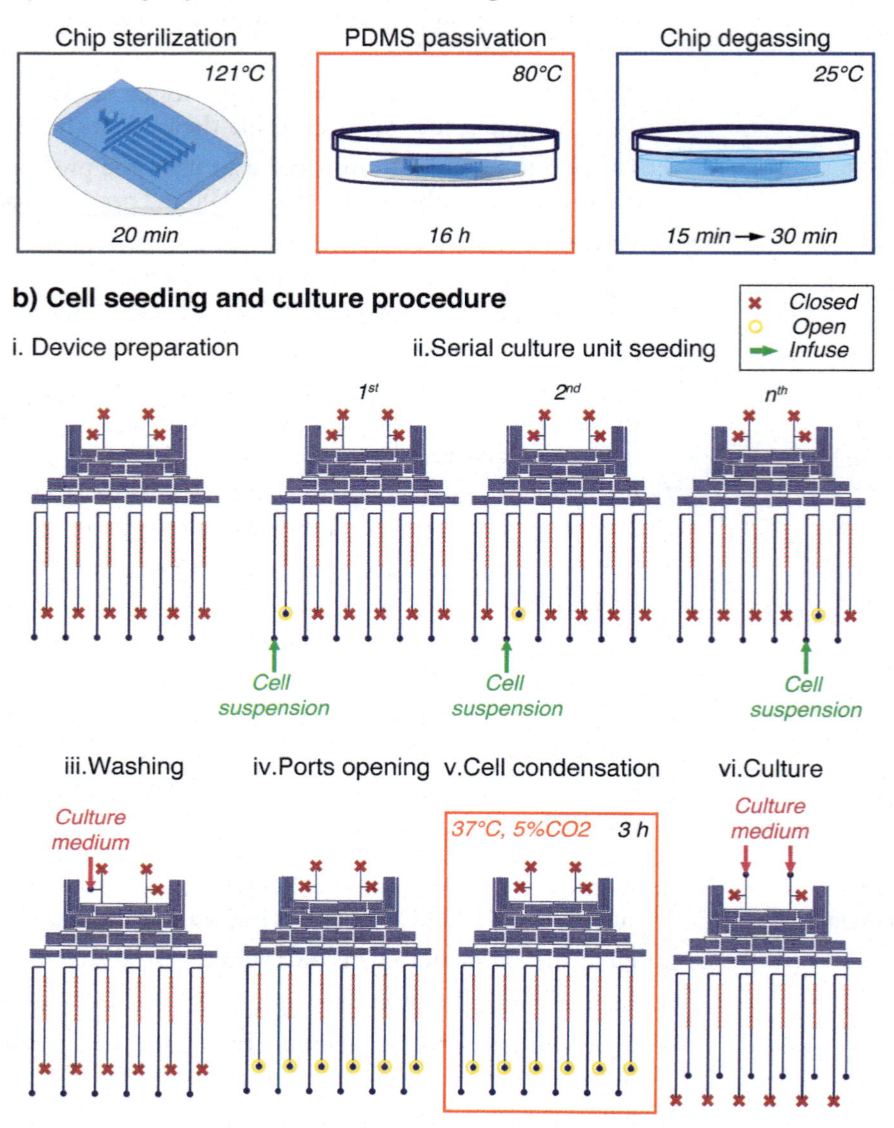

Fig. 3 Schematic representation of the main sequential steps needed to prepare and exploit the micro-bioreactor for cell culture. (**a**) Before coming in contact with cells, the chip is sterilized in an autoclave with a high pressure saturated steam at 121 °C for 20 min. Then, PDMS surfaces are made hydrophobic and non-adherent to cells by placing the chip into the oven overnight at 80 °C. Lastly, all the channels and chambers are filled with sterile Fungizone solution by applying vacuum to the submerged chip for around 15–30 min. (**b**) Six steps are then required for cell seeding. (*i*) First, all four inlet ports and six culture outlet ports are closed by inserting sterile stainless plugs (*red cross*). (*ii*) The serial seeding procedure is performed by sequentially removing the plug from the culture outlet port (*yellow circle*), injecting the cell suspension from the corresponding seeding inlet port (*green arrow*) and closing again the culture outlet port (*red cross*). (*iii*) After seeding all the n culture units, a washing step is performed to avoid cell accumulation inside the channels, by perfusing culture medium (*pink arrow*) from one of the primary inlet port while all the outlet ports are kept closed. To allow cell aggregations (*iv*) all the outlet ports are opened (*yellow circles*) again and (*v*) the chip is placed inside a humidifier incubator (37 °C, 5% CO_2) for 3 h. (*vi*) To start the dynamic culture, culture medium is perfused from two of the primary inlets (inlet 1 and 2) (*pink arrows*), while maintaining closed all the seeding inlet ports

10. Close all four inlet ports (two primary and two secondary) and six culture outlet ports (Fig. 3b i) by inserting sterile stainless plugs using sterile tweezers (*see* **Note 26**).
11. The device is now ready for cell seeding.

3.3 MSC Seeding in the Microfluidic Platform

1. Prepare the Tygon tubing for the cell seeding. Cut 5 cm long Tygon tube; assemble a 23 G stainless coupler on one side and a 23 G blunt needle on the other side of the tube. Wash the assembled tubing by perfusing 100% ethanol and subsequently sterile PBS. Store the seeding tubing within a sterile petri dish until used.
2. Prepare MSCs for seeding according to supplier's instructions or following standard procedure [32] (*see* **Note 27**).
3. In an Eppendorf tube, prepare cell suspension at the concentration of 7.5×10^6 cells/mL in serum-free culture medium as necessary (*see* **Notes 28** and **29**).
4. Connect the seeding tube to a 1 mL syringe, previously filled with serum-free culture medium. Fill the tube with medium by applying a pressure to the syringe and, once the tube is free from bubbles, aspirate some air to form a visible gas–fluid interface. Place the blunt needle terminal of the seeding tube directly into the Eppendorf tube containing the cell suspension, and withdraw the syringe piston to fill the tube with 5 μL of the cell suspension (*see* **Note 30**).
5. Remove the plug from the first culture outlet port (Fig. 3b ii).
6. Connect the cell-laden tube to the first seeding inlet port and inject the cell suspension into the device (*see* **Notes 31** and **32**). Afterwards, close the corresponding culture outlet port by inserting a sterile stainless plug (Fig. 3b ii).
7. Disconnect the tube from the specific seeding inlet port of the device.
8. Sequentially repeat the operation for all culture units (*see* **Note 33**).
9. Manually perfuse the device with serum-free culture medium through one of the primary inlet port (while maintaining the other closed) to wash cells away from the fluidic channels (Fig. 3b iii) (*see* **Note 34**).
10. Remove all plugs to open the culture outlet ports (Fig. 3b iv).
11. Place the petri dish containing the device in a humidified incubator (37 °C, 5% CO_2) for 3 h to allow cell aggregation under static condition (Fig. 3b v) (*see* **Notes 35** and **36**).
12. Check with an optical microscope that 3D cellular micromasses have generated in all chambers.
13. The device is now ready for subsequent culture and conditioning steps (Fig. 3b vi).

3.4 MSC 3D Micromasses Culture

1. Prepare polyethylene tubing (culture tubing) for cell culture (*see* **Note 37**). Cut two pieces of polyethylene tubing, each 60 cm long (*see* **Note 38**). Assemble a 22 G stainless coupler on one side and a 22 G blunt needle on the other side of each tube. Wash the assembled tubing by perfusing 100% ethanol and subsequently sterile PBS. Store the culture tubing within a sterile petri dish until use.
2. Connect the culture tubing to syringes previously filled with serum-free culture medium. Fill both tubes with medium, pay attention not to incorporate air bubbles into the circuits.
3. Mount the syringes onto the syringe pump.
4. Turn the syringe pump on and set the perfusion rate at 3 μL/min.
5. Take the seeded device, together with the petri dish, out of the incubator.
6. Use sterile tweezers to open the four inlet ports by removing the stainless steel plugs.
7. Connect both culture tubes to the primary inlet ports.
8. Perfuse for 5 min at a flow rate of 3 μL/min. During this phase, secondary inlet ports serve as bypass to avoid bubble to enter the device main circuit (gradient generator).
9. Use sterile tweezers to close the secondary inlets by inserting sterile plugs.
10. Reduce the flow rate down to 3 μL/h (*see* **Note 39**) and wait 10 min to allow the flow partitioning to reach steady-state conditions.
11. Use sterile tweezers to close the seeding inlets by inserting sterile plugs: in this way the flow will be uniformly distributed among the culture units.
12. Aspirate the PBS contained in the petri dish (*see* **Note 40**).
13. Place the device inside a humidified incubator (37 °C, 5% CO_2), and leave it there throughout the culture period (*see* **Note 41**).

3.5 Analysis and Readout

3.5.1 Phase Contrast Microscopy

1. Stop medium perfusion by pausing or switching off the syringe pump.
2. Take the device out of the incubator and put it in the housing of an optical microscope to monitor the micromasses' morphology in the culture units of interest (Fig. 4a) (*see* **Notes 42** and **43**).
3. Put the device back into the incubator and restart the pump.

3.5.2 Micromasses Fixation

1. Prepare two pairs of Tygon tubes to wash and fix micromasses, respectively. Cut four pieces of 15-cm long Tygon tubes; assemble a 23 G stainless coupler on one side and a 23G blunt needle on the other side of each tube (*see* **Note 44**).

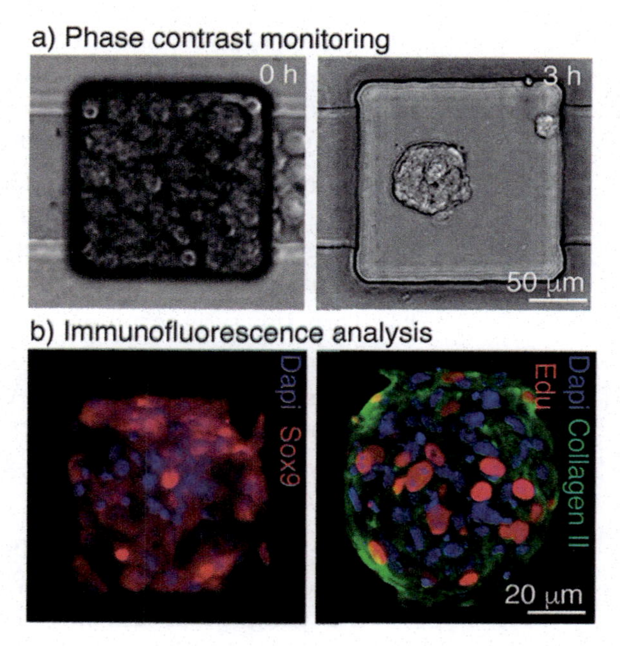

Fig. 4 Examples of possible readouts within the microfluidic platform. (**a**) Phase-contrast monitoring of MSC micromasses generation upon cell seeding within chambers. (**b**) Immunofluorescence staining of MSC micromasses upon 7 days in culture under chondrogenic medium perfusion: the expression of Sox9 and Collagen II and cell proliferation (Edu positivity) were assessed

2. Assemble two syringes filled with sterile PBS onto the syringe pump and connect them with the Tygon tubing while being careful not to introduce any air bubbles into the circuit.
3. Turn the syringe pump on and set the perfusion rate at 3 μL/min (resulting in 1 μL/min for each line).
4. Take the device out of the incubator.
5. Fill the petri dish with sterile Fungizone solution. Use enough liquid to fully cover the device, thus avoiding possible air influx and bubble formation inside the channels during the subsequent tubing connection procedures.
6. Disconnect the culture tubing from primary inlet ports using sterile tweezers and then open the two secondary inlet ports by removing the stainless steel plugs.
7. Connect the primary inlet ports of the device to the PBS syringes with sterile tweezers.
8. Perfuse PBS at a flow rate of 3 μL/min for 1 min through the open secondary inlets, serving as bypass to avoid bubbles to enter the device main circuit.
9. Use sterile tweezers to close the secondary inlets with sterile plugs.

10. Maintaining the flow rate at 3 μL/min, wash the culture units with PBS for 10 min.
11. Disconnect the PBS tubing from the primary inlet ports, stop the syringe pump and open the secondary inlet ports.
12. Assemble two syringes filled with 4% PFA onto the syringe pump and connect them with the second pair of Tygon tubes, while being careful not to introduce any air bubbles into the circuit.
13. Turn the syringe pump on and set the perfusion rate at 3 μL/min.
14. Connect the device to the PFA syringes by inserting the tubing to the primary inlet ports.
15. After 1 min of perfusion at 3 μL/min, close the secondary inlets with plugs.
16. Perfuse the PFA in the culture units at 3 μL/min for 30 min.
17. Disconnect the PFA tubing from the primary inlet ports, stop the syringe pump and open the secondary inlets ports.
18. Repeat the PBS perfusion **steps 7–11** to wash out the PFA (*see* **Note 45**).

3.5.3 Immunofluorescence Staining Within the Microfluidic Device

1. Prepare new Tygon tubing to stain the micromasses. Cut 15-cm long Tygon tubing; assemble a 23 G stainless coupler on one side and a 23 G blunt needle on the other side of the tube.
2. Repeat **steps 2–11** from Subheading 3.5.2 using the DB solution instead of PBS and perfusion time of 45 min.
3. Prepare the primary antibody mixture in DB (*see* **Note 46**).
4. Repeat **steps 2–11** from Subheading 3.5.2 using the primary antibody solution instead of PBS and a perfusion time of 60 min.
5. Perfuse DB at 3 μL/min for 15 min to wash out the primary antibody solution.
6. Prepare the secondary antibody mixture in DB supplemented with DAPI (*see* **Note 47**).
7. Repeat **steps 2–11** from Subheading 3.5.2 using the secondary antibody solution instead of PBS and a perfusion time of 45 min.
8. Wash the culture units by perfusing PBS at 3 μL/min for 15 min.
9. Perfuse 10% glycerol at 3 μL/min for 5 min (*see* **Notes 48** and **49**).
10. Move the stained device into the housing of a confocal microscope to acquire images of the micromasses directly in the culture unit of interest (Fig. 4b).

3.6 Exploitations: High-Throughput Screening of Soluble Factor Concentration Effects on 3D MSC Micromasses

1. Choose the microfluidic device layout that best fits your experimental design (*see* **Note 50**). For example, the logarithmic configuration can be proficiently used to screen soluble factors over a wide concentration range, while the linear layout is recommended for finer tunings within narrower concentration windows.
2. Fabricate the corresponding microfluidic device and seed it with MSCs as described in Subheadings 3.1–3.3.
3. Upon the generation of 3D MSC micromasses, the biochemical stimulation with a concentration pattern of a specific soluble factor (SF) can be started immediately or anytime during the cell culture within the bioreactor.
4. To generate a concentration pattern of the SF, prepare the two different culture media: basal serum-free culture medium (i.e., buffer solution) and serum-free culture medium supplemented with the SF to be tested at the maximum concentration of interest (i.e., SF-containing solution).
5. Fill one syringe with the buffer solution and one syringe with the SF-containing solution. Use two syringes with the same volume for the linear SDG, and two syringes with a volume ratio of 3:1 (buffer solution: SF-containing solution) for the logarithmic SDG, in order to have a uniform partition of the flow among the culture units.
6. Assemble the filled syringes onto the syringe pump and connect them with the previously prepared sterile culture tubing. Be careful not to introduce air bubbles into the circuit.
7. Turn the syringe pump on and set the perfusion at 3 μL/min.
8. Take the previously seeded device out of the incubator.
9. Open the four inlet ports by removing the stainless plugs with sterile tweezers.
10. Connect the device using the buffer-filled tubing to the 0% primary inlet port, and using the SF-filled tubing to the 100% primary inlet port. Use the open secondary inlets as a bypass to prevent bubbles from entering the device.
11. After 5 min of perfusion, close secondary inlets with sterile plugs.
12. Reduce the flow rate down to 3 μL/h (*see* **Note 51**) and wait 10 min to allow the flow partitioning to reach steady-state conditions (*see* **Note 52**).
13. Close the seeding inlets with sterile plugs: in this way the flow will be uniformly distributed among the culture units.
14. Aspirate the excess of PBS contained in the petri dish.
15. Place the device inside a humidified incubator (37 °C, 5% CO_2) throughout the SF screening period.
16. At the end of the experiment perform the required readout according to instructions from Subheadings 3.5.2 and 3.5.3.

4 Notes

1. Either an air or oxygen plasma cleaner can be used. Oxygen plasma is effective in less time than air plasma treatment.
2. Any syringe pump is suitable, provided that it can impose infusion rates ranging from 0.5 μL/h to 200 μL/min.
3. Instead of stainless steel plugs, you can also use fitting tubing closed with clamps.
4. The composition of the plating medium has to be adjusted depending on the specific cell type.
5. The composition of the cell culture medium depends on the specific application, but it is mandatory to use a serum free-based medium to prevent the PDMS device walls from protein coating that would promote 2D cell adhesion.
6. Concentration can vary between 0.25 and 2.5 μg/mL without affecting cell behavior and viability.
7. Different serum can be used, depending on the species in which the implied secondary antibodies were raised and different permeabilization solution can be used, e.g., Tween 20. The concentration of serum and permeabilization solution may require an optimization step depending on the specific staining to be performed.
8. Others water-based mounting solutions can be used. Glycerol is indeed required here to reduce optical reflection and avoid evaporation during confocal acquisition. This mounting solution has to be injected within the device, thus it cannot be too viscous.
9. Alignment marks need to be designed properly to allow subsequent alignment steps.
10. Outsourcing services for mask productions are available at low cost, e.g., Micro Lithography Services Ltd.
11. Use a two-ramp spin-coating protocol. Dispense about 4–5 mL of SU-8 on the wafer. Unify the wafer using a first slow step, consisting of 10 s at 500 rpm with an acceleration ramp of 100 rpm/s. Afterwards coat the SU-8 at the desired thickness with a spin rate of 2600 rpm for 30 s with an acceleration ramp of 300 rpm/s.
12. To achieve high resolution, the emulsion ink should be placed downwards, i.e., towards the coated wafer.
13. The recommended wavelength to expose SU-8 2000 photoresist is 365 nm, which is obtained using an Hg lamp with a spectral line at 365 nm, named “i-line”.
14. Use a two-ramp spin-coating protocol. Dispense about 4–5 mL of SU-8 on the wafer. Unify the wafer using a first slow step,

consisting of 10 s at 500 rpm with an acceleration ramp of 100 rpm/s. Afterwards coat the SU-8 at the desired thickness with a spin rate of 2900 rpm for 30 s with an acceleration ramp of 300 rpm/s.

15. To decrease the risk of cracks in the SU-8, it is advisable to provide the required energy dose in multiple exposure steps rather than a single one.
16. The PDMS reticulation starts as soon as the two components (prepolymer and curing agent) are mixed and it takes from 15 min to 24 h to solidify completely, depending on the temperature, from 100 °C to 25 °C, respectively. Consider to prepare fresh PDMS mixture each time. To spin the PDMS on the master mold (4-in. wafer), 5 mL of mixture is sufficient, while to achieve a thickness of about 5 mm, pour 56 mL of PDMS into the petri dish (diameter 120 mm).
17. The scalpel tip needs to gently touch the silicon master mold, without exerting an excessive force, as the wafer is very fragile.
18. For the Harrick Plasma system, the time required to reach a pressure level compatible with plasma activation ranges from 1 to 5 min, depending on the vacuum pump extraction rate.
19. In case of oxygen plasma this value can be decreased to 15–20 s.
20. The air removal can be assisted by pressing a clean spatula on top of the stamp.
21. Different dimensions of inlets ports and consequently needles, couplers and plugs can be used. Suitable puncher diameters range from 500 μm (for smaller ports that would favor cell clotting) to 1 mm (for bigger ports that would create undesired dead volumes).
22. This step serves to make the PDMS hydrophobic and non-adherent to cells. The temperature range tested to be effective without melting the petri dish and affecting the sterility of the device goes from 50 to 120 °C.
23. You can store the sterile device up to 6 months.
24. Be careful to preserve the sterility, do not overfill the dish but make sure the PBS fully covers the device.
25. The time depends on the level of vacuum into the chamber and on the dimension/complexity of the features.
26. Adding secondary inlet ports, connected to the main one through a bifurcation, can minimize the risk of bubble injection into the device. The secondary inlet will be maintained clamped during the culture, while serving as outlet for possible bubbles during medium change phases.
27. Any other cell type for which the aggregation into 3D micro-masses is known to mediate a specific physiological or pathological effect can be used.

28. The concentration of the cell suspension depends on the target number of cells required in a chamber, which depends on cell type and dimension. For an estimation of the relationship between cell number and aggregate size, refer to the literature [29, 32].
29. A total of 5 μL *number of units is enough to seed a device.
30. The charged volume can be varied depending on cells availability. The minimum volume of cell suspension sufficient to fully seed a culture unit is 5 μL.
31. Maintain the loaded tube vertical to avoid cell sedimentation on the tube walls.
32. This operation can successfully be performed manually.
33. The seeding of a culture unit at a time ensures a better uniformity.
34. You can seed different cell types in each culture unit without having cross-contamination. In this case, change the seeding tubing for each unit and perform the washing step after seeding each unit.
35. At this point the device is still submerged in PBS.
36. The time required for aggregation may vary depending on cell type. During this incubation maintain the device submerged in PBS to avoid bubble formation due to the absence of perfusion.
37. Polyethylene was chosen because it has been reported to be more biocompatible; moreover, conversely to Tygon, it does not induce debris formation during long-term cultures [33].
38. The length of the tubes has to be adjusted depending on the specific incubator: the tubes need to be long enough to connect the syringe pump, placed outside of the incubator, to the device placed inside.
39. The optimal flow rate needs to be optimized depending on the specific application [34, 35].
40. At this point the fluidic circuit is closed and the petri dish is no longer necessary. However, it represents a valid holder for the device.
41. The device has been tested up to 14 days in culture showing high viability of 3D MSC micromasses.
42. 3D MSC micromasses should not stay out of incubator for more than 1 h.
43. Live cell imaging and time lapse analysis can be performed under a microscope equipped with environmental chamber.
44. The length of the tubing can be adjusted according to the distance between the pump and the device during the staining procedure.

45. At this stage you can either proceed with the staining or store the sample submerged in PBS supplemented with 5% of sodium azide at 4 °C up to 6 months.
46. The primary antibodies should be diluted in DB according to manufacturer's instructions and they can be mixed together to stain different proteins simultaneously only if they have been raised in different species or if they are of different isotypes. The maximum number of proteins that can be stained simultaneously depends on the number of fluorescence channels that the microscope have (standard fluorescence microscope are usually equipped with 3–4 fluorescence channels.
47. The secondary antibodies have to be chosen to match the source species of the primary antibodies used and they must not be raised in the same species to avoid cross-reactivity. Moreover, any other compatible counterstain than DAPI can be used.
48. At this stage you can either proceed with the imaging on confocal microscope or store the sample protected from light submerged in PBS at 4 °C up to 1 month. Consider to stain the sample as close as possible to the confocal acquisition time.
49. The 10% glycerol solution can be also injected within the device manually from the primary inlets.
50. We designed linear or logarithmic (base 10) layout, but other fluidic network configurations can be designed and integrated within the same device configuration [13].
51. The optimal flow rate depends on application. Consider the range of working for the generation of the soluble factor gradient—usually up to few tens of μL/h. [13]—and the minimum flow rate for guarantying cell feed up [34].
52. The logarithmic gradient is more sensitive to small perturbations.

Acknowledgments

This study was partially supported by Fondazione Cariplo, grant no. 2012-0891.

References

1. Abo A, Clevers H (2012) Modulating WNT receptor turnover for tissue repair. Nat Biotechnol 30(9):835–836
2. Keung AJ, Kumar S, Schaffer DV (2010) Presentation counts: microenvironmental regulation of stem cells by biophysical and material cues. Annu Rev Cell Dev Biol 26:533–556
3. Sasai Y, Eiraku M, Suga H (2012) In vitro organogenesis in three dimensions: self-organising stem cells. Development 139(22): 4111–4121

4. Sant S, Hancock MJ, Donnelly JP et al (2010) Biomimetic gradient hydrogels for tissue engineering. Can J Chem Eng 88(6):899–911
5. Griffith LG, Swartz MA (2006) Capturing complex 3D tissue physiology in vitro. Nat Rev Mol Cell Biol 7(3):211–224
6. Vunjak-Novakovic G (2013) Biomimetic platforms for tissue engineering. Isr J Chem 53(9–10):767–776
7. Huh D, Torisawa YS, Hamilton GA et al (2012) Microengineered physiological biomimicry: organs-on-chips. Lab Chip 12(12):2156–2164
8. Occhetta P, Glass N, Otte E et al (2016) Stoichiometric control of live cell mixing to enable fluidically-encoded co-culture models in perfused microbioreactor arrays. Integr Biol (Camb) 8(2):194–204
9. Pennella F, Rossi M, Ripandelli S et al (2012) Numerical and experimental characterization of a novel modular passive micromixer. Biomed Microdevices 14(5):849–862
10. Occhetta P, Malloggi C, Gazaneo A et al (2015) High-throughput microfluidic platform for adherent single cells non-viral gene delivery. RSC Adv 5(7):5087–5095
11. Titmarsh D, Cooper-White J (2009) Microbioreactor array for full-factorial analysis of provision of multiple soluble factors in cellular microenvironments. Biotechnol Bioeng 104(6):1240–1244
12. Li Jeon N, Baskaran H, Dertinger SK et al (2002) Neutrophil chemotaxis in linear and complex gradients of interleukin-8 formed in a microfabricated device. Nat Biotechnol 20(8):826–830
13. Lee K, Kim C, Ahn B et al (2009) Generalized serial dilution module for monotonic and arbitrary microfluidic gradient generators. Lab Chip 9(5):709–717
14. Sahai R, Martino C, Castrataro P et al (2011) Microfluidic chip with temporal and spatial concentration generation capabilities for biological applications. Microelectron Eng 88(8):1689–1692
15. Kim C, Lee K, Kim JH et al (2008) A serial dilution microfluidic device using a ladder network generating logarithmic or linear concentrations. Lab Chip 8(3):473–479
16. Piraino F, Camci-Unal G, Hancock MJ et al (2012) Multi-gradient hydrogels produced layer by layer with capillary flow and crosslinking in open microchannels. Lab Chip 12(3):659–661
17. Xia Y, Whitesides GM (1998) Soft lithography. Annu Rev Mater Sci 28(1):153–184
18. Rasponi M, Piraino F, Sadr N et al (2010) Reliable magnetic reversible assembly of complex microfluidic devices: fabrication, characterization, and biological validation. Microfluid Nanofluid 10(5):1097–1107
19. Biffi E, Menegon A, Piraino F et al (2012) Validation of long-term primary neuronal cultures and network activity through the integration of reversibly bonded microbioreactors and MEA substrates. Biotechnol Bioeng 109(1):166–175
20. Biffi E, Piraino F, Pedrocchi A et al (2012) A microfluidic platform for controlled biochemical stimulation of twin neuronal networks. Biomicrofluidics 6(2):24106–2410610
21. Khademhosseini A, Langer R, Borenstein J et al (2006) Microscale technologies for tissue engineering and biology. Proc Natl Acad Sci U S A 103(8):2480–2487
22. Occhetta P, Visone R, Russo L et al (2015) VA-086 methacrylate gelatine photopolymerizable hydrogels: a parametric study for highly biocompatible 3D cell embedding. J Biomed Mater Res A 103(6):2109–2117
23. Lopa S, Piraino F, Kemp RJ et al (2015) Fabrication of multi-well chips for spheroid cultures and implantable constructs through rapid prototyping techniques. Biotechnol Bioeng 112(7):1457–1471
24. Marsano A, Conficconi C, Lemme M et al (2016) Beating heart on a chip: a novel microfluidic platform to generate functional 3D cardiac microtissues. Lab Chip 16(3):599–610
25. Occhetta P, Sadr N, Piraino F et al (2013) Fabrication of 3D cell-laden hydrogel microstructures through photo-mold patterning. Biofabrication 5(3):035002
26. Zervantonakis IK, Hughes-Alford SK, Charest JL et al (2012) Three-dimensional microfluidic model for tumor cell intravasation and endothelial barrier function. Proc Natl Acad Sci U S A 109(34):13515–13520
27. Kim JY, Fluri DA, Marchan R et al (2015) 3D spherical microtissues and microfluidic technology for multi-tissue experiments and analysis. J Biotechnol 205:24–35
28. Mathur A, Loskill P, Shao K et al (2015) Human iPSC-based cardiac microphysiological system for drug screening applications. Sci Rep 5:8883
29. Occhetta P, Centola M, Tonnarelli B et al (2015) High-throughput microfluidic platform for 3D cultures of mesenchymal stem cells, towards engineering developmental processes. Sci Rep 5:10288
30. DeLise AM, Stringa E, Woodward WA et al (2000) Embryonic limb mesenchyme micromass culture as an in vitro model for chondrogenesis and cartilage maturation. Methods Mol Biol 137:359–375

31. Harlow E, Lane D (2006) Fixing attached cells in paraformaldehyde. CSH Protoc 2006(3): 4294–4296. doi:10.1101/pdb.prot4294
32. Martin I, Muraglia A, Campanile G et al (1997) Fibroblast growth factor-2 supports ex vivo expansion and maintenance of osteogenic precursors from human bone marrow. Endocrinology 138(10):4456–4462
33. Jiang X, Jeffries RE, Acosta MA et al (2015) Biocompatibility of Tygon(R) tubing in microfluidic cell culture. Biomed Microdevices 17(1):20
34. Titmarsh D, Hidalgo A, Turner J et al (2011) Optimization of flowrate for expansion of human embryonic stem cells in perfusion microbioreactors. Biotechnol Bioeng 108(12): 2894–2904
35. Young EW, Beebe DJ (2010) Fundamentals of microfluidic cell culture in controlled microenvironments. Chem Soc Rev 39(3):1036–1048

Chapter 24

3D Anastomosed Microvascular Network Model with Living Capillary Networks and Endothelial Cell-Lined Microfluidic Channels

Xiaolin Wang*, Duc T.T. Phan*, Steven C. George, Christopher C.W. Hughes, and Abraham P. Lee

Abstract

This protocol describes detailed practical procedures for generating 3D intact and perfusable microvascular network that connects to microfluidic channels without appreciable leakage. This advanced 3D microvascular network model incorporates different stages of vascular development including vasculogenesis, endothelial cell (EC) lining, sprouting angiogenesis, and anastomosis in sequential order. The capillary network is first induced via vasculogenesis in a middle tissue chamber and then EC linings along the microfluidic channel on either side serve as artery and vein. The anastomosis is then induced by sprouting angiogenesis to facilitate tight interconnection between the artery/vein and the capillary network. This versatile device design and its robust construction methodology establish a physiological microcirculation transport model of interconnected perfused vessels from artery to vascularized tissue to vein.

Key words 3D microvascular network, Microfluidic chip, Vasculogenesis, EC lining, Sprouting angiogenesis, Anastomosis, Non-physiological leakage, Organ-on-a-chip

1 Introduction

The cardiovascular system plays a vital role in maintaining homeostasis of the human body, which allows blood to circulate throughout body for gas exchange and mass transportation to maintain organ viability. In order to construct a more physiological organ-on-a-chip, it is critical to be able to construct a functional 3D microvasculature system with a closed network of arteries, veins, and capillaries [1, 2].

On that basis, different research groups around the world have applied microfluidics to develop 3D in vitro microvascular models

*These authors contributed equally to this work.

Zuzana Koledova (ed.), *3D Cell Culture: Methods and Protocols*, Methods in Molecular Biology, vol. 1612,
DOI 10.1007/978-1-4939-7021-6_24, © Springer Science+Business Media LLC 2017

that physiologically mimic the vascular biology [3]. For cell and tissue engineering applications, microfluidic technologies were first used to confine cells in patterned, restricted regions to create microenvironments with well-controlled cellular interactions. More recently, microfluidic researchers have demonstrated control of the complex microenvironment at physiological levels by regulating chemical factors (molecular gradients etc.) or mechanical factors (interstitial flow etc.). Currently, there are two strategies to develop a 3D microvascular model. One strategy is to create microfabricated vessel scaffolds with specific geometries and dimensions, and line their inner surface with endothelial cells (ECs). The advantage of this method is that the vessel diameter can be precisely controlled, and the tightness of EC junctions can be flexibly adjusted by imposing different shear stress parameters on these lined ECs [4–6]. The other strategy is to seed cells in 3D extracellular matrices (ECMs) to allow spontaneous formation and remodeling of vascular networks through vasculogenesis and angiogenesis, which can closely mimic vascular development in vivo [7–12]. However, the microvascular network pattern embedded in 3D ECMs cannot be controlled and easily perfused [13].

In this protocol, we introduce an advanced 3D microvascular network model with intact perfused physiological blood vessels. By combining the two strategies mentioned above, we have engineered a system with tight interconnections between artery/vein and the capillary microvasculature network without non-physiological leakage. The major procedures involved in this advanced 3D microvascular network model are: (1) The capillary network formation is induced via vasculogenesis in a middle tissue chamber, and ECs are lined along the microfluidic channel on both sides to serve as the artery and vein. (2) Anastomosis is induced bidirectionally through sprouting angiogenesis to guarantee a good connection between the lined ECs along the microfluidic channels and the capillaries inside the tissue chamber [14]. Flow of fluorescent microparticles confirms perfusability of the lumenized microvascular network. In addition, minimal leakage of 70 kDa FITC-dextran confirms physiologic integrity of the interconnections between the artery/vein and the capillary network, which is critically important for drug screening applications [15]. This model can provide a promising avenue for future integration of multiple organs-on-a-chip through the vascular interface between arteries and veins [16, 17].

Building on our previous publications [18–22], this protocol seeks to provide detailed practical procedures to develop a complete and contiguous 3D perfused microvascular network without non-physiological leakage. The enabling technologies detailed here include microfluidic device design and microfabrication, cell-seeded hydrogel preparation and loading, maintenance of cell culture in the microfluidic chip, flow control for different stages of vascular development, and immunostaining of the vascular network.

2 Materials

2.1 Reagents

1. 5% hydrofluoric acid (HF).
2. Photoresist SU-8 2050 (Microchem).
3. Photoresist developer (Microchem).
4. Isopropyl alcohol (IPA).
5. Poly(dimethylsiloxane) (PDMS, Sylgard 184 Silicone Elastomer Kit, Dow Corning).
6. Trichloro(1H,1H,2H,2H-perfluorooctyl)silane.
7. 10 mg/mL fibrinogen: Weight out 28.8 mg of fibrinogen (>75% Clottable Bovine Fibrinogen, Sigma-Aldrich) (*see* **Note 1**) and dissolve in 2 mL of warmed DPBS by flicking the tube (*see* **Note 2**). Put the tube back to water bath for 10–15 min. After fully dissolved, filter fibrinogen solution using 0.22 μm syringe filter. This solution should be made fresh for each experiment.
8. Dulbecco's phosphate-buffered saline (DPBS) with Ca^{2+} and Mg^{2+}.
9. 50 U/mL thrombin (from bovine plasma, Sigma-Aldrich) in phosphate-buffered saline (PBS).
10. Full supplement endothelial cell growth medium (EGM-2, Lonza).
11. EGM-2 without VEGF and bFGF. To prepare EGM-2 without VEGF/bFGF, remove the VEGF and bFGF growth factor bullets from the kit, and then add the remaining growth factor bullets. Save the VEGF and bFGF growth factor bullets at −20 °C for future use.
12. 10% fetal bovine serum (FBS) in Dulbecco's Modified Eagle's Medium (DMEM).
13. Human endothelial colony forming cell-derived ECs (ECFC-ECs), passages 4–7.
14. Normal human lung fibroblasts (NHLFs, Lonza), passages 5–8.
15. Hanks' balanced salt solution (HBSS) without Ca^{2+} and Mg^{2+}.
16. 0.05% trypsin–EDTA solution.
17. 1 mg/mL laminin solution: Thaw laminin stock solution (mouse natural laminin, Thermo Fisher) slowly at 4 °C and aliquot into 100 μL tubes (*see* **Note 3**). Keep 1–2 aliquots at 4 °C for experiment and short-term storage.
18. 4% paraformaldehyde (PFA).
19. Sterile PBS.
20. 0.05% Tween 20 in PBS.

21. DAPI solution: 0.5–1 μg/mL DAPI in sterile PBS.
22. Blocking solution: 5–10% bovine serum albumin (BSA) or goat serum in PBS, sterile filtered.
23. 25 μg/mL 70 kDa-FITC dextran (Sigma-Aldrich) in sterile PBS.
24. Deionized water.
25. Primary and secondary antibodies, as desired.
26. 15 μm fluorescent microspheres (Thermo Fisher), supplied in 10^6 beads/mL concentration.

2.2 Equipment and Tools

1. Silicon wafers, 3 in. in diameter.
2. Spin coater.
3. Vacuum desiccator.
4. Petri dish.
5. Tweezers.
6. Oven.
7. Hotplate.
8. UV lamp with long pass filter (eliminate UV radiation below 350 nm).
9. Oxygen/air plasma cleaner.
10. Glass slides (75 × 50 mm) and glass coverslips.
11. Hole punchers, tip inner diameter of 1 mm.
12. Disposable scalpels.
13. 1.5 mL cryo vials.
14. Autoclave.
15. Incubator with 37 °C, 5% CO_2, 20% O_2.
16. Sterile syringe, 1 mL.
17. 21 gauge blunt end needle.
18. Sterile syringe filter, 0.22 μm.
19. Water bath.
20. T75 flask.
21. Hemacytometer.
22. Benchtop centrifuge.
23. 0.5 mL and 1.5 mL snap cap tube.
24. Double-sided tape.
25. Transparent tape.
26. BSL-2 laminar flow hood.
27. Micropipettors.
28. Inverted epifluorescence microscope.
29. Inverted microscope.
30. High speed cooled CCD camera.

3 Methods

3.1 Fabrication of SU-8 Silicon Mold and Its Silanization

1. After simulation of designed chip (Fig. 1), print out the photomask with high resolution.
2. Put a 3-in. silicon wafer into 5% HF for cleaning its surface, followed by a deionized water rinse, and dehydrate the wafer by placing it into the oven maintained at 120 °C for 10 min.
3. Place the silicon wafer on the spinner chuck of spin coater and apply vacuum (*see* **Note 4**).
4. Pour proper amount of SU-8 2050 on the center of the wafer (*see* **Note 5**).
5. Start the spin coater, and spin the wafer at 500 rpm for 20 s and then increase the speed to 130 × *g* for 30 s to achieve the desired thickness of 100 μm.
6. Place the photoresist-coated wafer into the oven or on the hotplate at 65 °C for 5 min.
7. Transfer the wafer to a hotplate/oven at 95 °C for 20 min (*see* **Note 6**).
8. Remove the baked wafer from the hotplate/oven and cool down to room temperature (RT) gradually.
9. Cover the photoresist-coated wafer with the photomask and bring them in tight contact (*see* **Note 7**).
10. Expose the wafer under UV light at 230 mJ/cm^2.
11. Move the exposed wafer to oven at 65 °C for 5 min, and then bake the wafer at 95 °C for 10 min.
12. Place the wafer into photoresist developer for proper time (*see* **Note 8**).
13. Spray the wafer with developer to remove the undeveloped photoresist, especially at the small features, and then wash the wafer with IPA.
14. Dry the wafer with pressurized filter nitrogen or air.
15. Place the SU-8 silicon mold into the desiccator and put a small drop of silane on coverslip close to the wafer, and evacuate the chamber to induce the evaporation of silane (*see* **Note 9**).
16. Remove the silicon mold from desiccator after overnight silanization.

3.2 Fabrication and Assembly of PDMS Microfluidic Device

1. Pour silicone elastomer base and curing agent at a weight ratio of 10:1 (base: curing agent) into an empty disposable plastic cup.
2. Mix the base and curing agent vigorously using a disposable plastic knife until the PDMS mixture appears opaque.
3. Attach the silicon mold to the bottom surface of petri dish with double-sided tape.

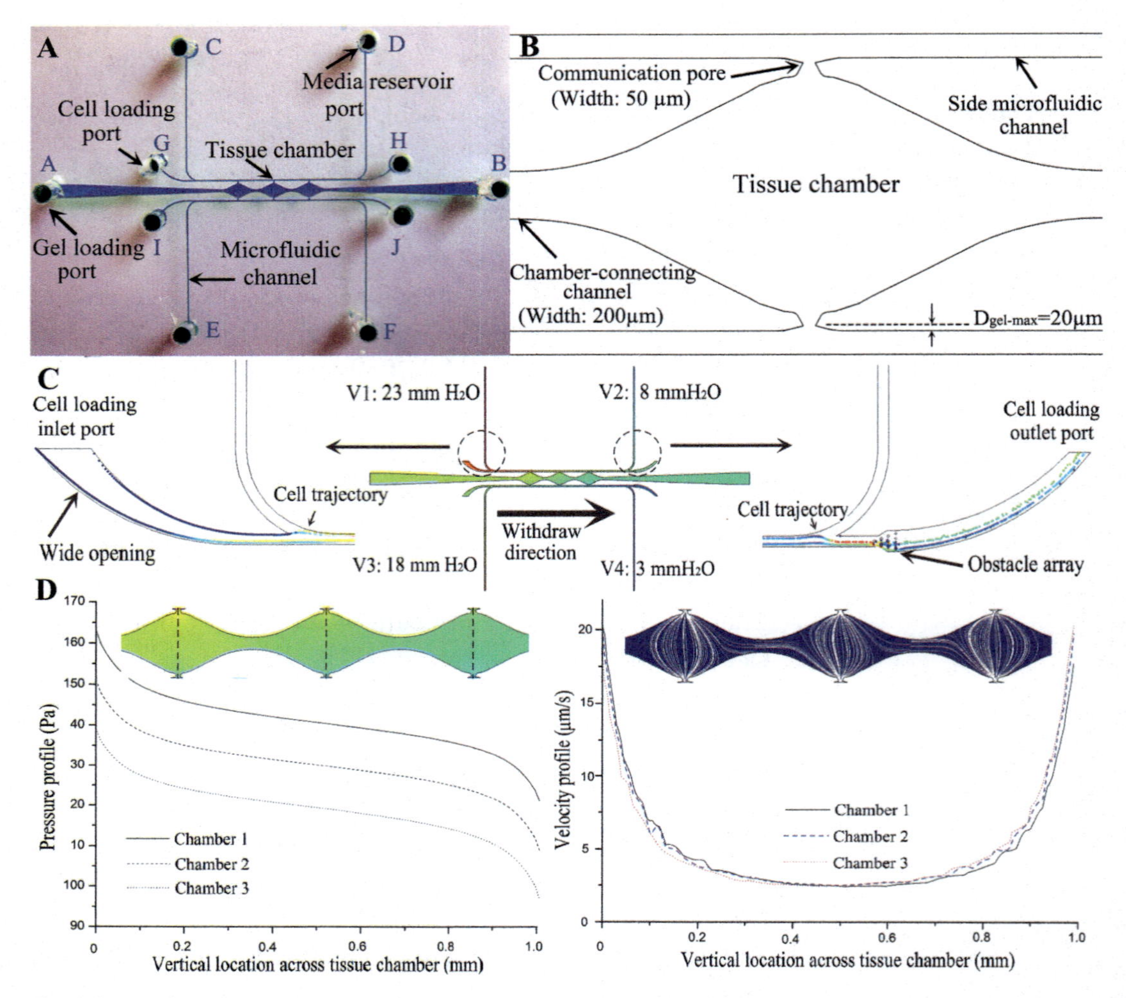

Fig. 1 Chip design and simulation. (**a**) Top-view of channel structure and ports for gel loading, cell loading and media reservoir. The entire device structure consists of three central millimeter-sized diamond tissue chambers (1 × 2 mm) and two side square cross-sectional microfluidic channels (100 × 100 μm) that connect to the tissue chambers through a series of communication pores. (**b**) Schematic and dimensions of one tissue chamber with optimized communication pore design to prevent gel bursting (or leaking) into side microfluidic channels from pressure built-up during the gel loading process. (**c**) Simulation result on pressure distribution inside the entire device, and the particle tracing from cell loading inlet port with wide opening to cell loading outlet port with obstacle array through withdrawal mode. (**d**) Simulation results on pressure drop and velocity profile of interstitial flow across three tissue chambers. Due to the fully symmetrical configuration of communication pores along both side microfluidic channels, the pressure drop across each tissue chamber from the top communication pore to the bottom one is almost the same ($\Delta P_{chamber} \sim 5$ mmH_2O), which induces a uniform interstitial flow profile inside each tissue chamber to stimulate vasculogenesis. Reproduced from Wang et al. [21] with permission from Royal Society of Chemistry (RSC)

4. Use compressed nitrogen to gently blow off dust particles from the silicon mold.
5. Pour approximately 20–25 g of the PDMS mixture onto the silicon mold inside the petri dish, and approximately 5 g of the PDMS mixture into an empty petri dish to fabricate a thin PDMS sheet.
6. Place both petri dishes into a vacuum desiccator to degas until the PDMS mixture becomes transparent.
7. Transfer the petri dishes covered with lid to the oven at 60 °C for 4–5 h to fully cure PDMS.
8. Remove the petri dishes from the oven and cool down to RT.
9. Carefully remove the PDMS microfluidic chip from the silicon mold and the PDMS sheet from the petri dish by cutting around the perimeter of the device using a scalpel and trim to proper size.
10. Punch holes at inlets and outlets of the PDMS microfluidic chip using a hole puncher.
11. Remove debris inside the punched holes using a compressed nitrogen gun.
12. Clean both sides of the PDMS microfluidic chip and PDMS sheet with transparent tape.
13. Treat the cleaned PDMS microfluidic chip and PDMS sheet inside a plasma cleaner for 3 min at 250–300 mTorr.
14. Quickly remove these two plasma-treated pieces from the plasma cleaner and bond them together.
15. Place the bonded two-layer device into the oven at 120 °C for 10 min.
16. Treat the PDMS sheet of bonded device and a glass slide (or a glass coverslip) inside the plasma cleaner for 1 min at 250–300 mTorr.
17. Quickly bond the PDMS sheet and the glass slide to make a three-layer microfluidic device (Fig. 2a).
18. Place the bonded three-layer device into the oven at 120 °C for 10 min.
19. Place pipette tips into all inlets and outlets of bonded microfluidic device to prevent PDMS seeping into these punched holes.
20. Cut off the bottom of plastic cryo vials using a scalpel.
21. Immerse the edge of bottomless plastic vials into PDMS mixture and attach them to all medium inlets and outlets of the microfluidic device.

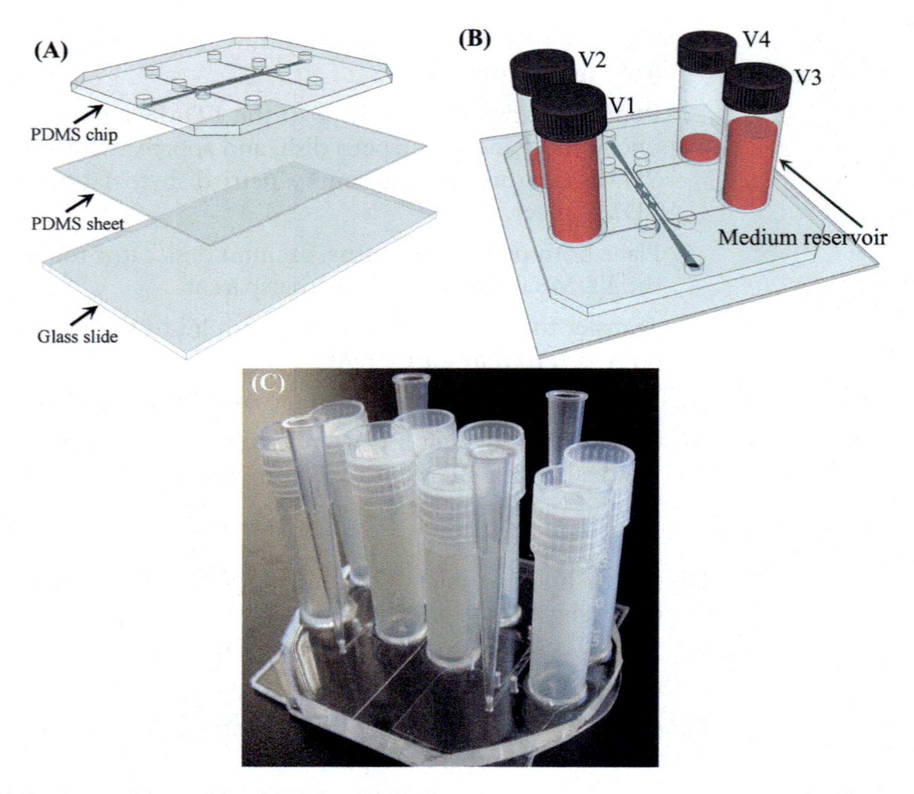

Fig. 2 Fabrication and assembly of PDMS microfluidic device. (**a**) Three layer microfluidic device design. *Top layer* is the PDMS microfluidic chip, the *middle layer* is the thin PDMS sheet, and the *bottom layer* is the glass slide. (**b**) Schematic of decoupling design with four media reservoirs containing different medium volume that could separately control flow to the tissue chamber for vasculogenesis and along the microfluidic channel for EC lining. (**c**) Prototype of assembled microfluidic device with plastic vials. Reproduced from Wang et al. [21] with permission from RSC

22. Place the assembled microfluidic device into the oven at 60 °C overnight to fully cure PDMS (Fig. 2b).
23. Sterilize the microfluidic device by autoclaving at 121 °C for 30 min before experiment (Fig. 2c).

3.3 Cell Harvesting and Cell-Seeded Hydrogel Preparation

1. Warm EGM-2, DPBS, and trypsin in water bath at 37 °C for 20–30 min before experiment.
2. Wash ECFC-ECs and NHLFs with 5–10 mL of HBSS 1–2 times.
3. Aspirate wash solution and add 2 mL of trypsin for each T75 flask to harvest cells.
4. Agitate and place the flasks in incubator for 20–30 s.
5. Remove the flasks from incubator and check under microscope for cell detachment.

6. Stop trypsin with 4 mL of EGM-2 per flask, transfer the cell suspensions into separate conical tubes and rinse the flasks with another 4 mL of EGM-2. Combine the rinsed media into conical tubes for a final volume of 10 mL.
7. Count both ECFC-ECs and NHLFs using a hemocytometer to determine the amount of cells harvested.
8. Centrifuge cells at 300 × *g* for 3 min.
9. Aspirate the supernatants and reconstitute cells in EGM-2 at a density of 10^6 cells/mL.
10. Prepare cell matrix solution: The desired cell concentration for ECFC-ECs and NHLFs is both 5×10^6 cells/mL of fibrinogen solution, for a combined concentration of 10^7 cells/mL. For example, after harvesting each cell type, resuspend the cell pellet in medium at 10^6 cells/mL, take out 500 μL of ECFC-ECs and 500 μL of NHLFs and combine in a single tube.
11. Centrifuge cells at 300 × *g* for 3 min and resuspend the combined pellet in 100 μL of sterile filtered fibrinogen solution (*see* **Note 10**).

3.4 Cell-Seeded Hydrogel and Medium Loading into Device

1. Aliquot 2 μL of thrombin into a 0.5 mL snap cap tube for each device planned to load (*see* **Note 11**).
2. Put a filter pipette tip of 200 μL at one end of gel loading pore to prevent spilled out hydrogel.
3. Gently mix and take out 15 μL of cell-seeded fibrinogen suspension without any air bubble (*see* **Note 12**).
4. Quickly mix with the thrombin droplet 2–3 times in the 0.5 mL snap cap tube without generating air bubbles (*see* **Note 13**).
5. Quickly inject the mixed suspension into the loading pore and gently push the gel through to the other side with steady pipette pressure (*see* **Note 14**).
6. Slowly place the microfluidic device down and twist the pipette tip out of the micropipettor to eject (*see* **Note 15**).
7. Leave the device inside the hood for 1–2 min before bringing out to check under the microscope. During this time, load the next device.
8. After checking under microscope, transfer the device into incubator at 37 °C and incubate for 10–15 min to let the gel fully polymerize (Fig. 3). While waiting for gel polymerization, thaw laminin solution from 4 °C to RT 30–45 min before loading medium.
9. After incubation, pipette 3–4 μL of laminin solution into the cell lining inlets to coat the inner surface of both microfluidic channels adjacent to the tissue chamber (*see* **Note 16**).

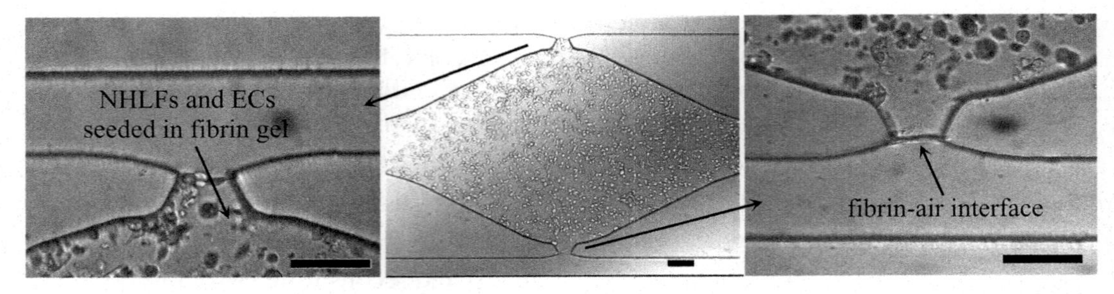

Fig. 3 Cell-seeded fibrinogen gel loading. The optimized communication pore design can precisely control the gel at the vertexes of communication pores with a flat gel-air interface without bursting into the microfluidic channels. This enabled the formation of a smooth EC monolayer on the gel interface at the communication pore, a critical condition for EC lining and sprouting angiogenesis. Scale bars, 100 μm. Reproduced from Wang et al. [21] with permission from RSC

10. Block all cell lining inlets and outlets with PDMS plugs (*see* **Note 17**).
11. Take 200 μL of full supplement EGM-2 and load through one medium inlet inside the plastic vial (*see* **Note 18**).
12. Wait until medium droplet spills out to the medium outlet inside the other plastic vial.
13. Repeat **steps 11** and **12** to load medium into the other side channel.
14. Slowly twist pipette tips out of the medium inlets and let medium form droplets.
15. Add appropriate amount of EGM-2 into each plastic vial to achieve desire pressure drop (V1: 23 mm H_2O, V2: 8 mm H_2O, V3: 18 mm H_2O, V4: 3 mm H_2O) and mark the medium levels with a marker (*see* **Note 19**).
16. Place caps onto all plastic vials and unscrew halfway to allow gas exchange.
17. Place device into a secondary container with sterile water underneath to maintain humidity and transfer to a 37 °C, 20% O_2, 5% CO_2 incubator (*see* **Note 20**).
18. After 1 h, check back to see if there is any air bubble in both side channels. If there is, remove it by pipetting medium through the medium inlet to push air bubble out (*see* **Note 21**).

3.5 Medium Changing and Vasculogenesis

1. Keep the device inside incubator with full supplement EGM-2 for 2 days before changing medium.
2. On day 2 post-embedding, warm EGM-2 without VEGF/bFGF for 20–30 min in the water bath.
3. Aspirate old medium, leaving a small amount of liquid inside each plastic vial to avoid air bubble.

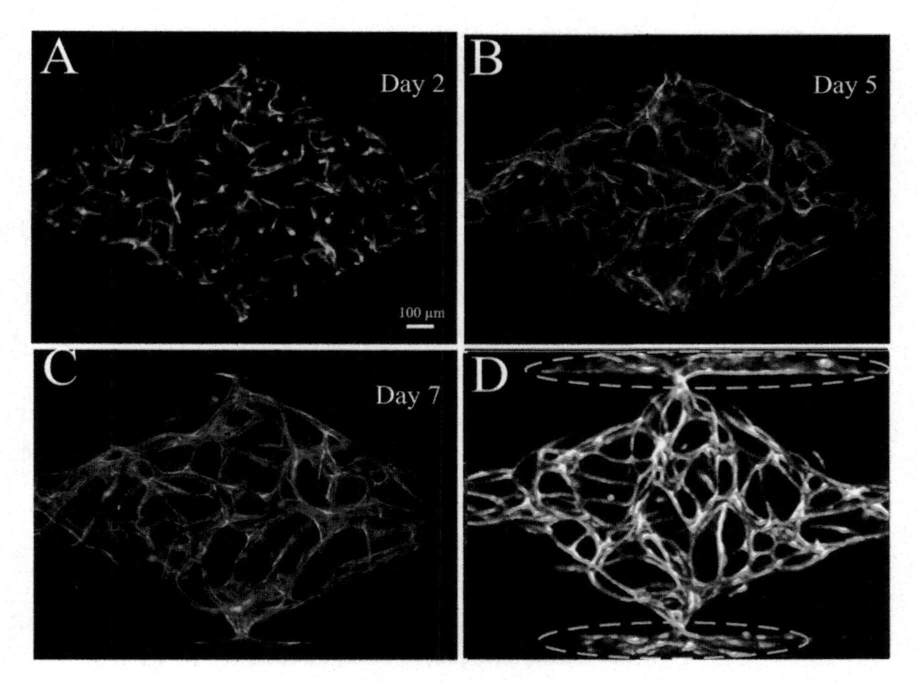

Fig. 4 Progression of vasculogenesis over time. (**a**) After establishing interstitial flow across the tissue chamber, ECFC-ECs formed vascular fragments as early as day 2. (**b**) A capillary network continued to develop through day 5. (**c**) By day 7, the capillary network was lumenized and interconnected. (**d**) The laminin coating inside the microfluidic channels promoted migration of ECs out from the tissue chamber through the communication pore and onto the surface of the channels, thereby facilitating interconnection of the capillary network with the outer channels. Reproduced from Wang et al. [21] with permission from RSC

4. Add fresh medium (EGM-2 without VEGF/bFGF) to each vial with the same height as that of **step 15** in Subheading 3.4.
5. Change medium for the device in similar manner every other day.
6. ECFC-ECs will form vascular fragments as early as day 2, and continue to develop into a capillary network through day 5. By day 7, the capillary network is lumenized and interconnected (Fig. 4a–c).
7. Laminin coating inside the microfluidic channels promotes ECs from the tissue chamber migrating through the communication pore outward and partially lining the inner surface of the channels (Fig. 4d).

3.6 EC Lining, Sprouting Angiogenesis, and Anastomosis

1. Harvest ECFC-ECs and resuspend cells in EGM-2 for a concentration of 2×10^7 cells/mL.
2. Remove medium from the plastic vials, leaving a small amount to avoid air bubble.

3. Block medium inlet and outlet of one side microfluidic channel with plugs.
4. Slowly remove plugs from cell lining inlet and outlet and add 1–2 μL of medium to cover the inlet and outlet and prevent air bubbles.
5. Mix and take out 5 μL of cell suspension using a micropipettor and insert the pipette tip into the cell lining inlet.
6. Insert a bent syringe needle connected with an empty 1 mL syringe into the cell lining outlet (*see* **Note 22**).
7. Place the device under the microscope.
8. Slowly pull syringe to withdraw cells from cell lining inlet toward cell lining outlet. Cells will flow through and accumulate inside the side microfluidic channel by the obstacle array near the cell lining outlet (*see* **Note 23**).
9. Monitor under the microscope and stop withdrawing when cells have filled up a whole microfluidic channel.
10. Alternatively, cells can be withdrawn through the cell lining outlet using a micropipettor instead of the bent syringe needle. Set micropipettor for 5 μL volume, insert an empty pipette tip to the cell lining outlet, and withdraw cells by releasing the micropipettor button.
11. Bring device into the incubator and incubate for 2 h to allow ECs to adhere to all inner side microfluidic channel walls.
12. Reverse flow gently from the cell lining outlet toward cell loading inlet to remove non-adhering ECs by gently pipetting 1–2 μL of medium through the cell lining outlet blocked with a filtered, empty pipette tip.
13. Remove pipette tips at the cell lining inlet and outlet and block with plugs.
14. Reestablish the same medium flow as that of **step 15** in Subheading 3.4.
15. Adhered ECs elongate to form a monolayer in response to shear stress generated by the pressure drop along the side microfluidic channel (Fig. 5).
16. After the EC lining process, a confluent EC monolayer will form on the gel interface at the communication pore (Fig. 6a).
17. Reverse the interstitial flow direction across tissue chamber by changing the hydrostatic pressure distribution (V1: 18 mm H_2O, V2: 3 mm H_2O, V3: 23 mm H_2O, V4: 8 mm H_2O) to induce the basal-to-apical transendothelial flow across tissue chamber.
18. Switch medium supplied to the microfluidic channel with higher pressure to full supplement EGM-2 to induce a positive VEGF gradient that promotes sprouting angiogenesis.

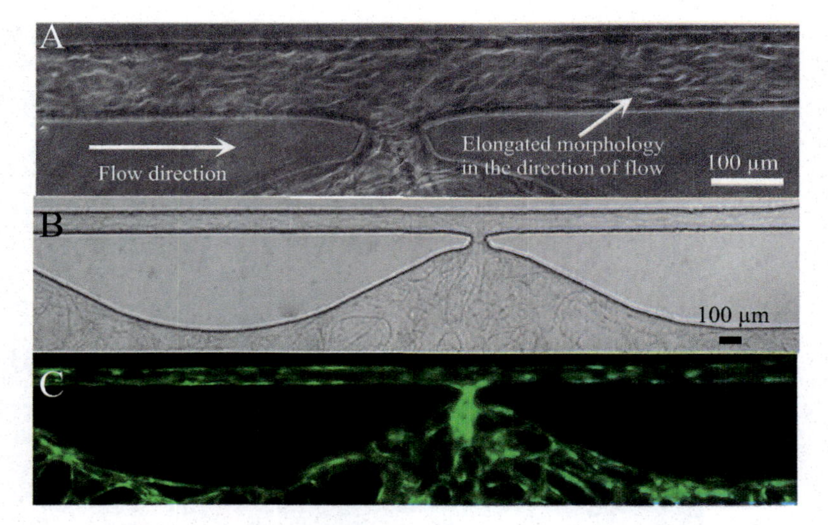

Fig. 5 EC lining along microfluidic channel. (**a**) ECs with elongated morphology in the direction of fluidic flow. (**b**) Bright field image of EC lining along entire microfluidic channel after vasculogenesis, which represent the artery (high pressure) and vein (low pressure) of our 3D microvascular model. (**c**) Corresponding fluorescent image. Reproduced from Wang et al. [21] with permission from RSC

19. Lined ECs at the cell–matrix interface invade into the ECM and form microvascular sprouts as early as 24 h post-lining (Fig. 6b).
20. Highly branched and abundant sprouts extend into ECM (Fig. 6c).
21. Anastomosis with tight interconnection is a bidirectional process. From inside the tissue chamber, ECs migrate outward to the laminin-coated microfluidic channel to connect with the EC-lined microfluidic channel during vasculogenesis. From the outer microfluidic channel, the lined ECs invade into the ECM and form sprouts to connect with the capillaries inside the tissue chamber during sprouting angiogenesis (Fig. 6d).
22. Repeat the EC lining process for the other side microfluidic channel 2 days after lining the first channel. An intact and perfusable microvascular network of artery/vein and capillaries can be developed in this microfluidic device via a multistep process through 12 days (Fig. 7).

3.7 Microparticle Perfusion

1. Remove medium from all plastic vials, leaving a small amount at the bottom to avoid air bubble.
2. Add microparticle solution to the plastic vial with the highest hydrostatic pressure (23 mm H_2O) and sterile PBS to the other plastic vials.
3. Monitor microparticle moving inside vessel lumen under an inverted (epifluorescent) microscope (Fig. 8a).

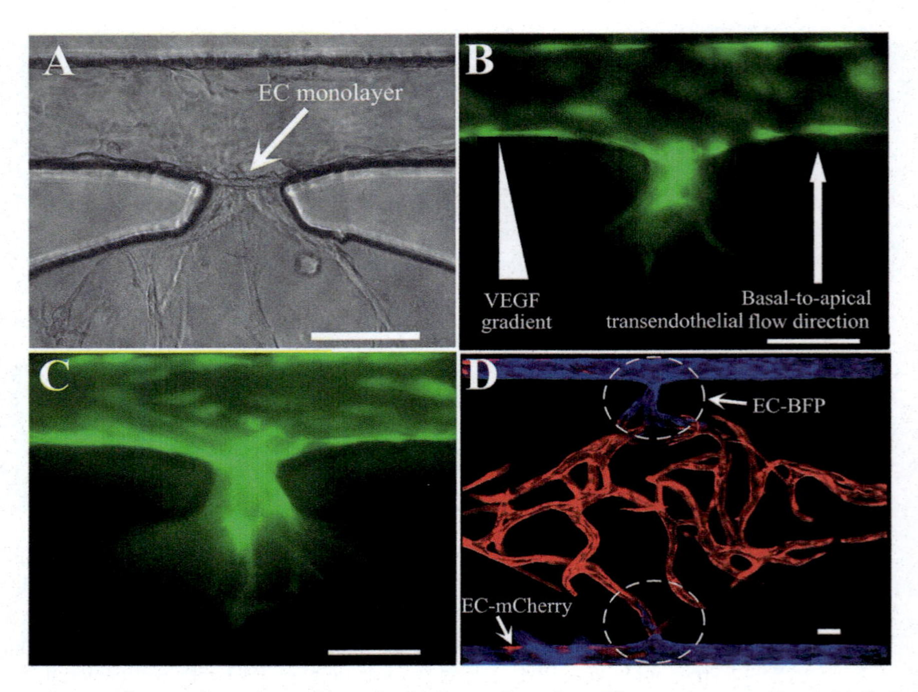

Fig. 6 Sprouting angiogenesis and anastomosis. (**a**) Formation of an EC monolayer on the gel interface at the communication pore after EC lining. (**b**) Invasion of microvascular sprouts from the lined EC monolayer into the gel after 24 h. (**c**) More sprouts and deep invasion after 48 h post-lining, which provided a precondition for anastomosis with the capillary network inside the tissue chamber. (**d**) Tight interconnection between EC lining along the outer channel and the capillary network inside tissue chamber by anastomosis with bidirectional migration. From inside the tissue chamber, mCherry-expressing (red) ECs (ECs-mCherry) migrated outward to the laminin-coated microfluidic channel to connect with the lined BFP-labeled ECs (ECs-BFP) in the channels. From the outer microfluidic channels, the lined ECs-BFP invaded into the 3D gel and formed sprouts to connect with the capillaries in the tissue chamber. Scale bars, 100 μm. Reproduced from Wang et al. [21] with permission from RSC

4. After microparticle perfusion, replace old PBS containing microparticles with sterile PBS and allow flow through for 20 min to flush away microparticles.
5. Replace sterile PBS with fresh medium and continue culture in incubator.

3.8 70 kDa-FITC Dextran Perfusion

1. Remove medium from all plastic vials, leaving a small amount at the bottom to avoid air bubble.
2. Add dextran solution to the plastic vial with the highest hydrostatic pressure (23 mm H_2O) and sterile PBS to the other plastic vials.
3. Monitor dextran flow under an inverted epifluorescence microscope (*see* **Note 24**).
4. After dextran perfusion, replace old PBS containing dextran with sterile PBS and allow flow through for 20 min (Fig. 8b).
5. Replace sterile PBS with fresh medium and continue culture in incubator.

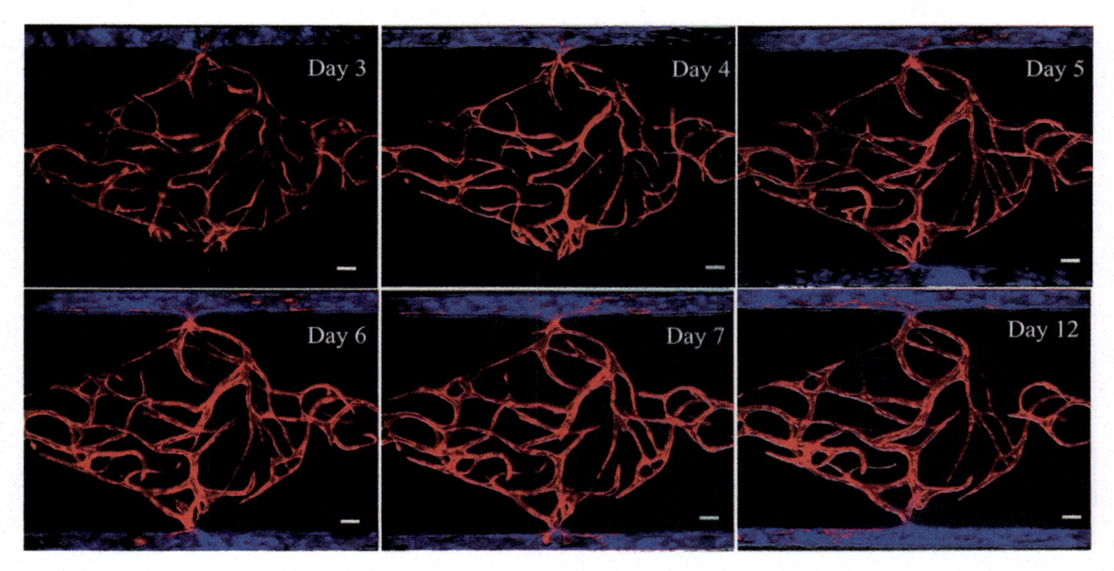

Fig. 7 Formation of intact and perfusable microvascular network with artery/vein and capillary network by multistep processes throughout 12 days. Vasculogenesis of ECs-mCherry started to occur 3 days after introducing cell-seeded fibrin gels into the tissue chamber. ECs-BFP were added to an outer channel at this point. After 24 h post-lining, these lined ECs-BFP began to elongate and form connections with ECs-mCherry through the communication pore. On day 5, the other channel was lined with ECs-BFP and the direction of interstitial flow, which was also the transendothelial flow direction, was reversed to promote remodeling of the capillaries inside the tissue chamber and to stimulate angiogenic sprouting of lined ECs. In addition, ECs-mCherry also migrated out from the tissue chamber and interlaced with ECs-BFP lining the microfluidic channel. ECs-BFP along both microfluidic channels continued to proliferate, elongate, and form connections with the microvascular network inside the tissue chamber thereafter. By day 12, a fully lumenized capillary network inside the tissue chamber had anastomosed with both EC-lined microfluidic channels to form the intact microvascular network. Scale bars, 100 μm. Reproduced from Wang et al. [21] with permission from RSC

3.9 Immunostaining of Vascular Network

1. Replace medium in all plastic vials with freshly thawed 4% PFA (*see* **Note 25**).
2. Fix device with 4% PFA for 15–20 min at RT inside a chemical fume hood.
3. Replace 4% PFA with sterile PBS and wash the device at RT for 1–2 h.
4. Replace washed PBS with sterile PBS (*see* **Note 26**).
5. For intracellular staining that requires cell membrane permeabilization, replace PBS with 0.05% Tween 20 in PBS and permeabilize for 1 h at RT.
6. After permeabilization, wash the device with sterile PBS.
7. Replace PBS with blocking solution and leave at 4 °C overnight.
8. Prepare primary antibody staining solution, 100–150 μL per device. Typically, start with antibody dilution 1:200 in blocking solution and optimize if needed.

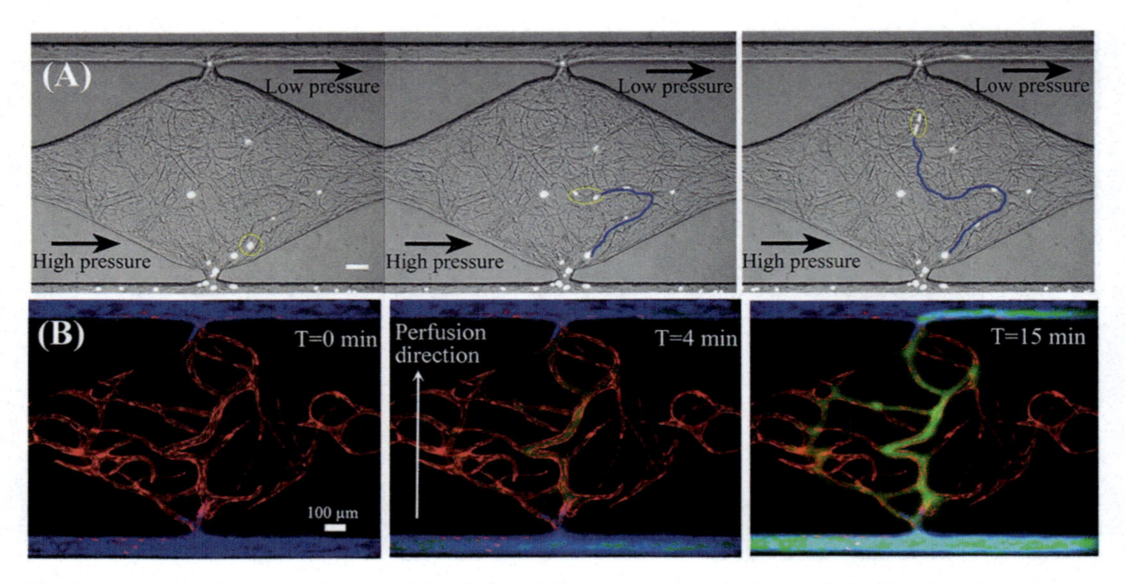

Fig. 8 Particle and dextran perfusion. (**a**) Perfusion of 15 μm fluorescent microparticles into the high pressure outer channel and these then flowed through the microvascular network from one communication pore to the other and out into the low pressure outer channel. (**b**) Fluorescence images showing 70 kDa FITC-dextran perfusion in the absence of non-physiological leakage. After loading dextran from left to right in an EC-lined microfluidic channel for 4 min (the lower channel in the figure), it quickly flowed into the capillary network through the communication pore. Over time, the fluorescence intensity in the lumen of microvessels increased due to the intensive dextran influx. Finally, dextran passed through the capillaries and flowed into the low pressure side microfluidic channel and exited within 15 min, validating the connectivity and perfusability of the microvascular network. More importantly, no dextran was observed outside the vessel during the 15 min perfusion, including the communication pore region where lined ECs connect to the microvascular network. This demonstrated the strong barrier property of the microvascular network, and the tightness of the anastomosis. Reproduced from Wang et al. [21] with permission from RSC

9. Remove blocking solution from all plastic vials, leaving a small amount at the bottom to avoid air bubbles.
10. Carefully remove the plastic vials without disrupting the blocking solution droplets.
11. Insert pipette tips with antibody staining solution to the medium inlets. Place empty pipette tips to the medium outlets (*see* **Note 27**).
12. Gently withdraw antibody dilution from the outlets using a micropipettor. Monitor for a small amount of liquid moving upward through the empty pipette tips.
13. Aspirate excess blocking solution at the inlets and outlets and leave the device at 4 °C overnight.
14. Reverse the flow direction by switching antibody dilution from the inlet to the outlet and leave at 4 °C for 4–6 h.
15. After primary antibody staining, wash the device with sterile PBS for 4–6 h (Fig. 9a).

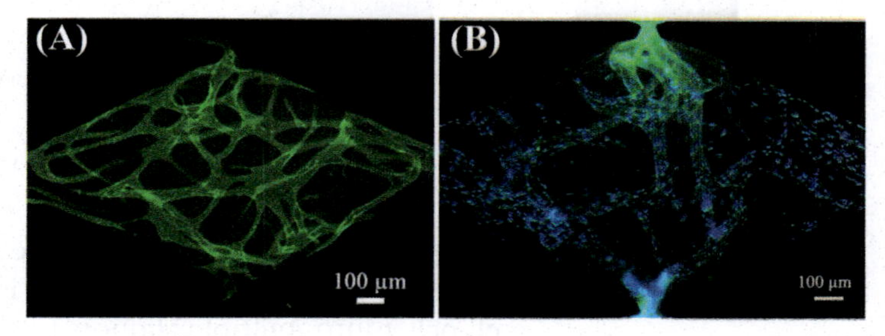

Fig. 9 Immunostaining image of a capillary network inside a tissue chamber. (**a**) CD31 immunostaining. CD31 (PECAM-1) is a cellular adhesion and signaling receptor that is highly expressed at endothelial cell–cell junctions, and has been shown to play an important role in maintenance of vascular barrier integrity. This CD31 immunostaining highlighted the integrity of the vascular network formed in our device. (**b**) DAPI nuclear staining. Reproduced from Wang et al. [21] with permission from RSC

16. Prepare secondary antibody staining solution, 100–150 μL per device. Typically for secondary antibody, start with 1:300 or 1:400 dilution in blocking solution.
17. Repeat the antibody staining process from **steps 11–13**.
18. After secondary antibody staining, wash device multiple times with sterile PBS to remove background.
19. For nuclei staining, replace washing PBS with DAPI solution and leave at 4 °C overnight.
20. Wash device multiple times with sterile PBS to remove excess DAPI (Fig. 9b).

4 Notes

1. Make extra fibrinogen to account for sterile filter after dissolved in DPBS.
2. Do not vortex the tube to dissolve the fibrinogen powder.
3. Do not thaw laminin stock solution directly to RT. For long-term storage, keep laminin at −80 °C.
4. Place silicon wafer at the center of spinner chuck to ensure a uniform thickness of the photoresist layer.
5. Slowly perform this step to prevent air bubble formation in the poured photoresist layer.
6. Keep the hotplate/oven on a flat surface to prevent the unwanted changes of photoresist film.
7. Ensure the ink printed side of photomask is in contact with the photoresist layer on the wafer.

8. Placing the container with the developer and wafer into an ultrasonic instrument will highly accelerate the development speed.
9. Perform the silanization procedure inside a ventilated chemical fume hood.
10. Prepare no more than 200 μL of cell–fibrinogen mixture per aliquot. If more cell-matrix volume is needed, divide combined cell pellet into multiple aliquots to assure desired cell density.
11. Ensure thrombin droplet is at the bottom of the tube.
12. Change pipette tips between mixing the solution and loading the solution into the device.
13. Perform this step quickly to avoid damaging the gel or creating uneven polymerization, as the cell matrix solution will begin polymerizing quickly upon mixing with thrombin.
14. Press the push-button smoothly and evenly.
15. Do not eject pipette tip by pushing the eject button on the micropipettor. Otherwise, gel will burst out.
16. Pre-coating side microfluidic channels with laminin before adding culture medium is crucial to ensure EC adherence to PDMS for EC lining, and stimulate EC migration outward from the tissue chamber to anastomose with the EC-lined microfluidic channels.
17. PDMS plug is fabricated by dipping pipette tip into PDMS mixture and allowing to fully cure at 60 °C overnight.
18. Perform this step slowly to avoid damaging the gel inside tissue chamber.
19. Calculate the volume based on the diameter of vial.
20. To minimize the medium evaporation from the plastic vials and maintain humidity, prepare a large pipette tip box with sterile water and a smaller pipette tip rack (or a small plastic petri dish) to place the device on top and separate from the water. Sterile water should be changed every other day.
21. Removing air bubbles formed inside the microfluidic channel as soon as detected is crucial after loading device. Air bubbles can block medium flow and damage gel inside the tissue chamber.
22. Bend a 21-gauge blunt end needle and attach to the syringe. Do not use slant end needle.
23. Withdraw slowly to avoid introducing air bubble into the microfluidic channel.
24. Do not perfuse dextran for more than 90 min under fluorescent microscope.
25. Thaw fresh 4% PFA for optimal fixing. Do not freeze and thaw 4% PFA multiple times.

26. Device with sterile PBS can be stored at 4 °C for 1–2 weeks after fixing. Seal storage container with Parafilm to prevent evaporation.
27. Prepare 100–150 μL of antibody dilution per device. Set up a temporary hydrostatic pressure to allow antibodies flowing through the device. Typically, 70–80 μL of antibody dilution is for the highest pressure inlet, and 30–40 μL is required for the second highest pressure inlet.

Acknowledgments

This work was supported by grants from the National Institutes of Health: UH3 TR00048 and PQD5 CA180122. C.C.W.H. receives support from the Chao Family Comprehensive Cancer Center (CFCCC) through an NCI Center Grant award P30A062203. X.W. receives support from National Natural Science Foundation of China (No. 31600781). We would also like to thank the permission of The Royal Society of Chemistry (RSC) for reproduction of materials from Lab on a Chip journal.

References

1. Lee H, Chung M, Jeon NL (2014) Microvasculature: an essential component for organ-on-chip systems. MRS Bull 39(1):51–59
2. Schimek K, Busek M, Brincker S et al (2013) Integrating biological vasculature into a multi-organ-chip microsystem. Lab Chip 13(18): 3588–3598
3. Hasan A, Paul A, Vrana NE et al (2014) Microfluidic techniques for development of 3D vascularized tissue. Biomaterials 35(26): 7308–7325
4. Esch MB, Post DJ, Shuler ML et al (2011) Characterization of in vitro endothelial linings grown within microfluidic channels. Tissue Eng A 17(23–24):2965–2971
5. Bischel LL, Young EWK, Mader BR et al (2013) Tubeless microfluidic angiogenesis assay with three-dimensional endothelial-lined microvessels. Biomaterials 34(5):1471–1477
6. Booth R, Noh S, Kim H (2014) A multiple-channel, multiple-assay platform for characterization of full-range shear stress effects on vascular endothelial cells. Lab Chip 14(11): 1880–1890
7. Lee H, Kim S, Chung M et al (2014) A bioengineered array of 3D microvessels for vascular permeability assay. Microvasc Res 91:90–98
8. Kim S, Lee H, Chung M et al (2013) Engineering of functional, perfusable 3D microvascular networks on a chip. Lab Chip 13(8):1489–1500
9. Yeon JH, Ryu HR, Chung M et al (2012) In vitro formation and characterization of a perfusable three-dimensional tubular capillary network in microfluidic devices. Lab Chip 12(16):2815–2822
10. Vickerman V, Blundo J, Chung S et al (2008) Design, fabrication and implementation of a novel multi-parameter control microfluidic platform for three-dimensional cell culture and real-time imaging. Lab Chip 8(9): 1468–1477
11. Young EWK (2013) Advances in microfluidic cell culture systems for studying angiogenesis. J Lab Autom 18(6):427–436
12. Chiu LL, Montgomery M, Liang Y et al (2012) Perfusable branching microvessel bed for vascularization of engineered tissues. Proc Nat Acad Sci U S A 109(50): E3414–E3423
13. Whisler JA, Chen MB, Kamm RD (2014) Control of perfusable microvascular network morphology using a multiculture microfluidic system. Tissue Eng 20(7):543–552

14. Diaz-Santana A, Shan M, Stroock AD (2015) Endothelial cell dynamics during anastomosis in vitro. Integr Biol 7(4):454–466
15. Chan CY, Huang PH, Guo F et al (2013) Accelerating drug discovery via organs-on-chips. Lab Chip 13(24):4697–4710
16. Bhatia SN, Ingber DE (2014) Microfluidic organs-on-chips. Nat Biotechnol 32(8): 760–772
17. Huh D, Torisawa YS, Hamilton GA et al (2012) Microengineered physiological biomimicry: organs-on-chips. Lab Chip 12(12): 2156–2164
18. Hsu YH, Moya ML, Hughes CCW et al (2013) Full range physiological mass transport control in 3D tissue cultures. Lab Chip 13(1): 81–89
19. Hsu YH, Moya ML, Hughes CCW et al (2013) A microfluidic platform for generating large-scale nearly identical human microphysiological system arrays. Lab Chip 13(15): 2990–2998
20. Moya ML, Hsu YH, Lee AP et al (2013) In vitro perfused human capillary networks. Tissue Eng Part C Methods 19(9):730–737
21. Wang X, Phan DTT, Sobrino A et al (2016) Engineering anastomosis between living capillary networks and endothelial cell-lined microfluidic channels. Lab Chip 16(2):282–290
22. Wang X, Phan DTT, Zhao D et al (2016) An on-chip microfluidic pressure regulator that facilitates reproducible loading of cells and hydrogels into microphysiological system platforms. Lab Chip 16(5):868–876

Chapter 25

Human Lung Small Airway-on-a-Chip Protocol

Kambez H. Benam, Marc Mazur, Youngjae Choe, Thomas C. Ferrante, Richard Novak, and Donald E. Ingber

Abstract

Organs-on-chips are microfluidic cell culture devices created using microchip manufacturing techniques that contain hollow microchannels lined by living cells, which recreate specialized tissue–tissue interfaces, physical microenvironments, and vascular perfusion necessary to recapitulate organ-level physiology in vitro. Here we describe a protocol for fabrication, culture, and operation of a human lung "small airway-on-a-chip," which contains a differentiated, mucociliary bronchiolar epithelium exposed to air and an underlying microvascular endothelium that experiences fluid flow. First, microengineering is used to fabricate a multilayered microfluidic device that contains two parallel elastomeric microchannels separated by a thin rigid porous membrane; this requires less than 1 day to complete. Next, primary human airway bronchiolar epithelial cells isolated from healthy normal donors or patients with respiratory disease are cultured on the porous membrane within one microchannel while lung microvascular endothelial cells are cultured on the opposite side of the same membrane in the second channel to create a mucociliated epithelium–endothelium interface; this process take about 4–6 weeks to complete. Finally, culture medium containing neutrophils isolated from fresh whole human blood are flowed through the microvascular channel of the device to enable real-time analysis of capture and recruitment of circulating leukocytes by endothelium under physiological shear; this step requires less than 1 day to complete. The small airway-on-a-chip represents a new microfluidic tool to model complex and dynamic inflammatory responses of healthy and diseased lungs in vitro.

Key words Microfluidics, Organ-on-a-chip, Tissue microengineering, Lung pathophysiology, Small Airway, Microphysiological System

1 Introduction

Chronic respiratory diseases, such as asthma and chronic obstructive pulmonary disease (COPD), are a major cause of morbidity and mortality with a rising prevalence worldwide [1–3]. COPD currently ranks as the third leading cause of death in humans globally with only few pharmacotherapeutics available [1, 4]. Therefore, better understanding of disease biogenesis is crucial to advance drug development strategies, to enable biomarker discovery for early disease detection and to improve clinical prognosis.

Zuzana Koledova (ed.), *3D Cell Culture: Methods and Protocols*, Methods in Molecular Biology, vol. 1612,
DOI 10.1007/978-1-4939-7021-6_25, © Springer Science+Business Media LLC 2017

A number of in vitro and in vivo models have been developed to study human pulmonary diseases. But their applicability for clinical translation is a matter of debate [3–7]. For example, the lung anatomy, immune system, and inflammatory responses of rodents, which are widely used in many laboratories, differ greatly from those in humans [8–10]. Thus, many of these animal models fail to accurately predict drug activities in humans. On the other hand, in vitro models often utilize two-dimensional monolayer cultures of cell lines or undifferentiated primary cells, which lack pathophysiological relevance. Moreover, the greatest limitation of current cell culture systems that mimic airway mucosa, such as ones developed using permeable cell culture inserts, is the inability to recapitulate the dynamic interactions between circulating blood immune cells and endothelium under physiologically relevant vascular shear. More recently, lung organoid cultures containing lung stem cells have been described that mimic lung morphogenesis in three-dimensional (3D) extracellular matrix gels; however, the relevance of these systems for drug development and toxicity analysis is limited because there is no access to the air space and again, microvascular endothelium, flow and cyclic breathing motions that are important for lung physiology are absent.

Advances in microsystems engineering have recently made it possible to create biomimetic microfluidic cell-culture devices, known as organs-on-chips, that contain continuously perfused microchannels lined by living human cells, which recapitulate the multicellular architectures, tissue–tissue interfaces, physicochemical microenvironments, and vascular perfusion of the body [5], potentially offering new opportunities for disease modeling and drug-efficacy assessment [11]. Here, we describe a step-by-step protocol for the fabrication, culture, and operation of a human lung "small airway-on-a-chip" that we recently developed using this microengineering approach [3]. This microfluidic culture device supports differentiation of a columnar, pseudostratified, mucociliary bronchiolar epithelium composed of cells isolated from healthy individuals or people with COPD, which are underlined by a functional microvascular endothelium. Importantly, this platform enables real-time imaging and analysis of capture of circulating leukocytes from flow by endothelium under physiological shear.

2 Materials

2.1 Microdevice Components

1. Poly(dimethylsiloxane) (PDMS).
2. Printed 3D microfluidic molds.
3. Polyester (PET) membranes with 0.4 μm pores.

2.2 Chip Assembly and Fluidics Operational Parts

1. Tygon silicone tubing, formulation 3350 with inner diameter (ID): 0.8 mm, outer diameter (OD): 2.4 mm.
2. 19G 12.7 mm-long blunt straight stainless steel tubing.
3. 19G 9.5 mm-long blunt-tip Luer needle.
4. Nalgene™ wide-mouth LDPE bottle as medium reservoir.
5. MLL Kynar® cap luer connector.
6. MLL-to-1.6 mm Kynar® barbed luer connector.
7. 28 mm-diameter sterile syringe filters, 0.2 μm-pore SFCA-PF membrane.
8. 2-Stop PharMed BPT tubing with ID: 0.25 mm.
9. Devcon 5-min epoxy.
10. 24 mm × 46 mm mounting glass slide.
11. Acetone.
12. 19 mm binder clips.
13. Scalpels.
14. Self-healing cutting pads with grid.
15. Optically clear or colored 3 mm- and 6 mm-thick cast acrylic sheets.
16. 1-mL Luer-Lok disposable syringe.
17. Sterile disposable forceps with blunt tips.
18. Low-temperature self-seal sterilization pouch.
19. Micro binder clips.
20. Premoistened cleanroom 70% ethanol wipes.
21. Acrylic sheets.
22. Heavy duty packing tape.

2.3 Primary Human Lung Cells and Culture Reagents

1. Human small airway epithelial cells from normal donors or patients with COPD.
2. Human lung blood microvascular endothelial cells.
3. Dulbecco's phosphate-buffered saline (PBS), pH 7.4, Ca^{2+}- and Mg^{2+}-free.
4. Trypsin–EDTA solution: 0.05% (wt/vol) trypsin, 0.02% (wt/vol) EDTA.
5. 70% ethanol.
6. Calcein AM dye.
7. Anti-human CD31 (PECAM-1) antibody FITC conjugated.
8. Anti-human LYVE-1 antibody Alexa Fluor 488 conjugated.
9. Neubauer disposable hemocytometer.
10. Tissue culture-treated T-75 flask.

11. 0.4% sterile-filtered Trypan Blue solution.
12. Expansion medium: Small airway epithelial cell growth medium (SAECGM) SupplementPack kit [bovine pituitary extract (BPE), recombinant human epidermal growth factor (rhEGF), recombinant human insulin, hydrocortisone, epinephrine, triiodo-L-thyronine, transferrin, retinoic acid, bovine serum albumin-fatty acid free (BSA-FAF)], 100 U/mL penicillin, and 100 μg/mL streptomycin in small airway epithelial cell basal medium (SAECBM) (*see* **Note 1**).
13. Bovine serum albumin (BSA) stock solution*:* 1.5 mg/mL BSA in Expansion medium (*see* **Note 2**).
14. Retinoic acid solutions: 3 mg/mL and 15 μg/mL all-*trans*-retinoic acid (ATRA) in DMSO (*see* **Note 3**).
15. Submerged medium: 1.5 μg/mL BSA, 100 U/mL penicillin, and 100 μg/mL streptomycin in the 1:1 mix of SAECBM with Dulbecco's Modified Eagle Medium [DMEM; with pyruvate and low (1 g/L) glucose], supplemented with BPE, rhEGF, insulin, hydrocortisone, epinephrine, triiodo-L-thyronine, and transferrin of the SAECGM SupplementPack kit. Fresh before use add ATRA to final concentration 15 ng/mL (*see* **Note 4**).
16. ALI medium: similar to Submerged medium, except the retinoic acid is supplemented to final concentration 3 μg/mL.
17. Endothelial medium: Components of one EGM-2 MV SingleQuot Kit Supplements & Growth Factors kit [rhEGF, vascular endothelial growth factor (VEGF), analog of insulin-like growth factor 1 (R3-IGF-1), ascorbic acid, hydrocortisone, recombinant human fibroblast growth factor beta (rhFGF-β), FBS], 100 U/mL penicillin and 100 μg/mL streptomycin in endothelial basal medium 2 (EBM-2).
18. Extracellular matrix (ECM) coating solution: 300 μg/mL rat tail type I collagen in prechilled tissue culture-grade water.
19. Calcein AM/Wheat Germ Agglutinin (WGA)-Alexa Fluor® 488 Conjugate-containing medium: 1 μg/mL Calcein AM and 5 μg/mL WGA in Endothelium medium (*see* **Note 5**).

2.4 Blood Leukocyte Isolation and Reagents

1. Fresh whole human blood, collected directly into EDTA-containing tubes (*see* **Note 6**).
2. Lymphoprep™ medium.
3. Anti-human CD15 antibody eFluor® 450 conjugated.
4. Anti-human CD16 antibody APC conjugated.
5. 15-mL polypropylene conical tube.
6. 50-mL polypropylene conical tube.
7. 1× Red Blood Cell (RBC) Lysis Buffer (e.g., Affymetrix, 00-4333).

8. Leukocyte medium: 10% (vol/vol) heat-inactivated FBS, 100 U/mL penicillin, and 100 μg/mL streptomycin in RPMI-1640 medium (*see* **Note 7**).
9. Dextran solution: 6% (wt/vol) dextran (molecular weight 450,000–650,000) in 0.9% (wt/vol) NaCl (*see* **Note 8**). Store at 4 °C.
10. Percoll solutions: 80%, 68%, and 55% (vol/vol) Percoll® in Ca^{2+}- and Mg^{2+}-free PBS (*see* **Note 9**).
11. Cell Tracker-containing medium: 1 μg/mL CellTracker™ Red CMPTX Dye in Leukocyte medium (*see* **Note 10**).
12. Hoechst-containing medium: 5 μg/mL Hoechst 33342 in Leukocyte medium (*see* **Note 11**).

2.5 Equipment and Analytical Software

1. Ismatec™ low-speed planetary gear-driven peristaltic pump with 8 or 16 channels.
2. Inverted fluorescent microscope.
3. Fusion 200 touch syringe pump.
4. Plasma machine with oxygen plasma capability.
5. UVO-Ozone cleaner.
6. 60 °C baking oven.
7. Laser cutter.
8. 3D printer.
9. Centrifuge.
10. Flow cytometer.
11. Inverted microscope.
12. Stereo microscope.
13. Zen software.
14. ImageJ software (http://imagej.nih.gov/ij/).
15. Cell Profiler software (http://www.cellprofiler.org).

3 Methods

3.1 Fabrication of the Small Airway Microchip

3.1.1 Fabrication of the Upper and Lower Microchannels (Fig. 1a)

1. Prepare a 10:1 (wt:wt) mixture of PDMS silicone elastomer base and curing agent for molding.
2. Pour the PDMS mixture onto 3D-printed microfluidic molds to create 1000 μm wide by 1000 μm high (upper microchannel), 1000 μm wide by 200 μm high (lower microchannel) channel parts, with a 16.7 mm^2 overlapping channel area. Bottom component thickness is 0.5–1.0 mm; top component thickness is 5.0 mm (*see* Benam et al. [3] for more design details).
3. Degas the PDMS-loaded molds in a vacuum desiccator for 30 min to remove bubbles.

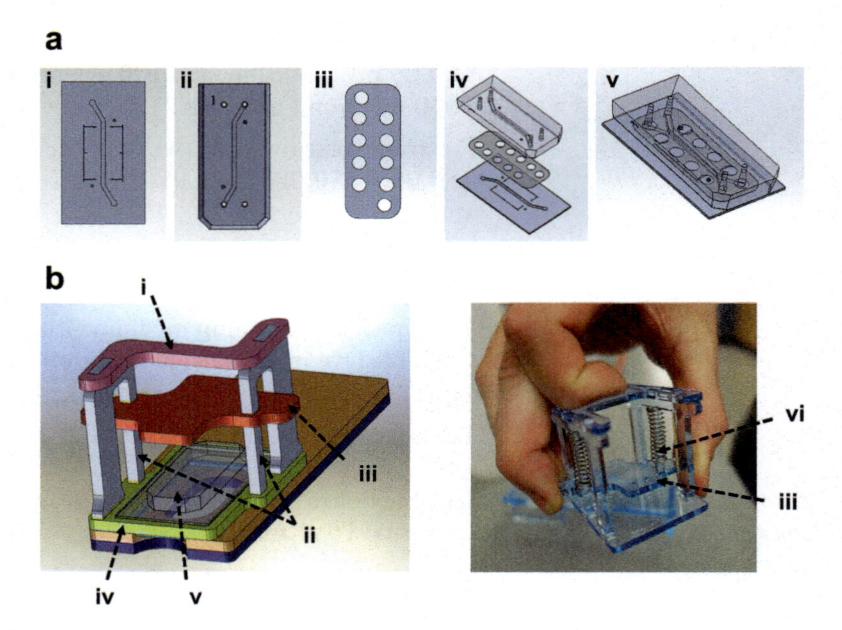

Fig. 1 Fabrication of human lung small airway chip and chip-clamp. (**a**) The microfabricated small airway microdevice consists of top (*i*) and bottom (*ii*) PDMS slabs sandwiching a thin, rigid, porous PET membrane (*iii*). The device recreates airway lumen and subepithelial microvasculature of human lung bronchioles. Lateral circles in the membrane (*iii*) increase surface to surface contact between the PDMS layers. All the three parts are aligned (*iv*) and reversibly bonded (*v*) to form the airway lumen (*upper*) and vascular (*lower*) microchannels. (**b**) Schematic cartoon (*left*) and photograph (*right*) of the chip-clamp. The aim of this system is to press the biochip PDMS layers in a controllable fashion (through the strength of springs) to prevent membrane delamination and enable microchannel-restricted cell culture. The components of the clamp include a top plane (*i*), two vertical supports with spring posts (*ii*), a translational plane (*iii*), a base plane (*iv*), and a cartridge-fitting interface (not visible in this view). A small airway chip mounted on a glass slide (*v*) and springs (*vi*) are also shown at *right*

4. Fully cure the PDMS in a level oven, maintained at 60 °C overnight (*see* **Note 12**).
5. Remove the molds and allow them to cool to room temperature (RT) for 5–10 min.
6. Carefully peel off the fully cured PDMS parts (top and bottom) and if needed, temporarily store in a particle-free environment. For more detailed instructions, *see* Huh et al. [12].

3.1.2 Patterning the PET Membrane and Microdevice Assembly

1. Pattern the PET membrane by laser cutting the 0.4 μm-pore membrane with 2 mm diameter circles at port locations that will allow access to lower microchannel ports and additional lateral 2–3 mm diameter circles to increase surface contact between the top and bottom PDMS layers to provide device

bonding since PDMS and PET do not bond using oxygen plasma. Use a matte steel or aluminum backing to support the membrane during cutting.

2. To align and assemble the microdevice, first clean the PDMS top and bottom layers using packing tape, making sure features are free of debris.
3. Clean the PET membrane by blowing lightly with an air-gun, making sure membrane is free from debris and not wrinkled.
4. Plasma activate the PDMS top layer and the PET membrane, making sure to treat the surfaces that will contact each other.
5. Overlay the PDMS onto the PET membrane and align under a stereo microscope to make a membrane-top component assembly.
6. Plasma activate the membrane-top component assembly and PDMS bottom layer, again making sure to treat the surfaces that will contact each other.
7. Under a stereo microscope, carefully and quickly align the upper and lower microchannels and press gently to seal surfaces. Take care to avoid wrinkling the membrane by applying pressure uniformly over the chip surface.
8. Place the assembled chip in 60 °C oven for at least 2 h.
9. Remove the microdevice and allow to cool down to RT for 5–10 min.

3.2 Fabrication of the Microchip Clamp (Fig. 1b)

The clamp is used to apply approximately 1 kg of force uniformly over the chip surface to seal the chip while enabling imaging and handling. A spring loaded compression plate allows for self-alignment to the chip surface and accommodates variation in chip thicknesses.

1. Cut a 3 mm-thick acrylic sheet into the following parts using a laser cutter:
   (a) A top plane that resembles truncated H-shape (bounding box dimensions: width: 48 mm; length: 38 mm).
   (b) Two vertical supports, with spring posts (bounding box dimensions: height: 48 mm; width: 32 mm).
   (c) A translational plane to compress fitted biochip (bounding box dimensions: width: 52 mm; length: 33 mm).
   (d) A base plane (bounding box dimensions: width: 48 mm; length: 38 mm).
   (e) Cartridge-fitting interface (outer length: 44 mm; inner length: 42 mm; outer width: 25 mm; inner width: 21 mm).
2. Solvent weld the two vertical supports to the top plane and reinforce with 5 min epoxy.

3. After fully bonded, place springs on vertical supports and clamp the base plane to this assembly.
4. Solvent weld the base plane to the vertical supports.
5. Solvent weld the cartridge-fitting interface to the bottom of the base.
6. Reinforce the entire assembly by additional solvent gluing or with epoxy if needed. Allow to dry completely per acrylic glue instructions (typically 24 h).
7. Clean by spraying with 70% ethanol prior to use. The clamps can also be plasma sterilized if desired.

3.3 Fabrication of Chip-Carrying Cartridge (Fig. 2)

1. Cut a 6 mm acrylic sheet into two vertical support pieces (height: 25 mm; width: 38 mm).
2. Cut a 3 mm acrylic sheet into two pieces (length: 108 mm; width: 40 mm):
   (a) One with mating holes for the vertical support pieces.
   (b) One with mating holes for the 3D printed reservoir cap described in the next section.
3. 3D print reservoir cap to enable screwing of Nalgene™ wide-mouth LDPE bottle and which also contained both a tubing and a pressure relief port. Wash to remove any debris.
4. Place the part with mating holes for 3D vertical support (produced in **step 2a**) on top of the vertical support pieces (produced in **step 1**) and overlay with the part with mating holes for the 3D printed reservoir cap (produced in **step 2b**).
5. Solvent-weld together using acetone and reinforce with 5-min epoxy.
6. Place 3D printed reservoir cap into mating holes on the top of this assembly and bond using 5-min epoxy.
7. Sterilize the Chip-carrying Cartridge by oxygen ion plasma treatment using Plasma Etch (run protocol: Plasma Time = 30 s; RF Power = 100 W; O_2 Pressure = 5 sccm (approximately 180 mTorr)) inside a sterilization pouch.

3.4 Fabrication of Cartridge Dock (Fig. 2)

1. Cut a 6 mm-thick acrylic sheet into the following four types of pieces using a laser cutter:
   (a) Horizontal rectangular supports (2 pieces; height: 19 mm; length: 392 mm).
   (b) Vertical supports with handles (3 pieces; height: 70 mm; width: 110 mm).
   (c) Cartridge layer—bottom half with support protrusions (1 piece; width: 120 mm; length: 412 mm).
   (d) Cartridge layer—top half without support protrusions (1 piece; width: 120 mm; length: 412 mm).

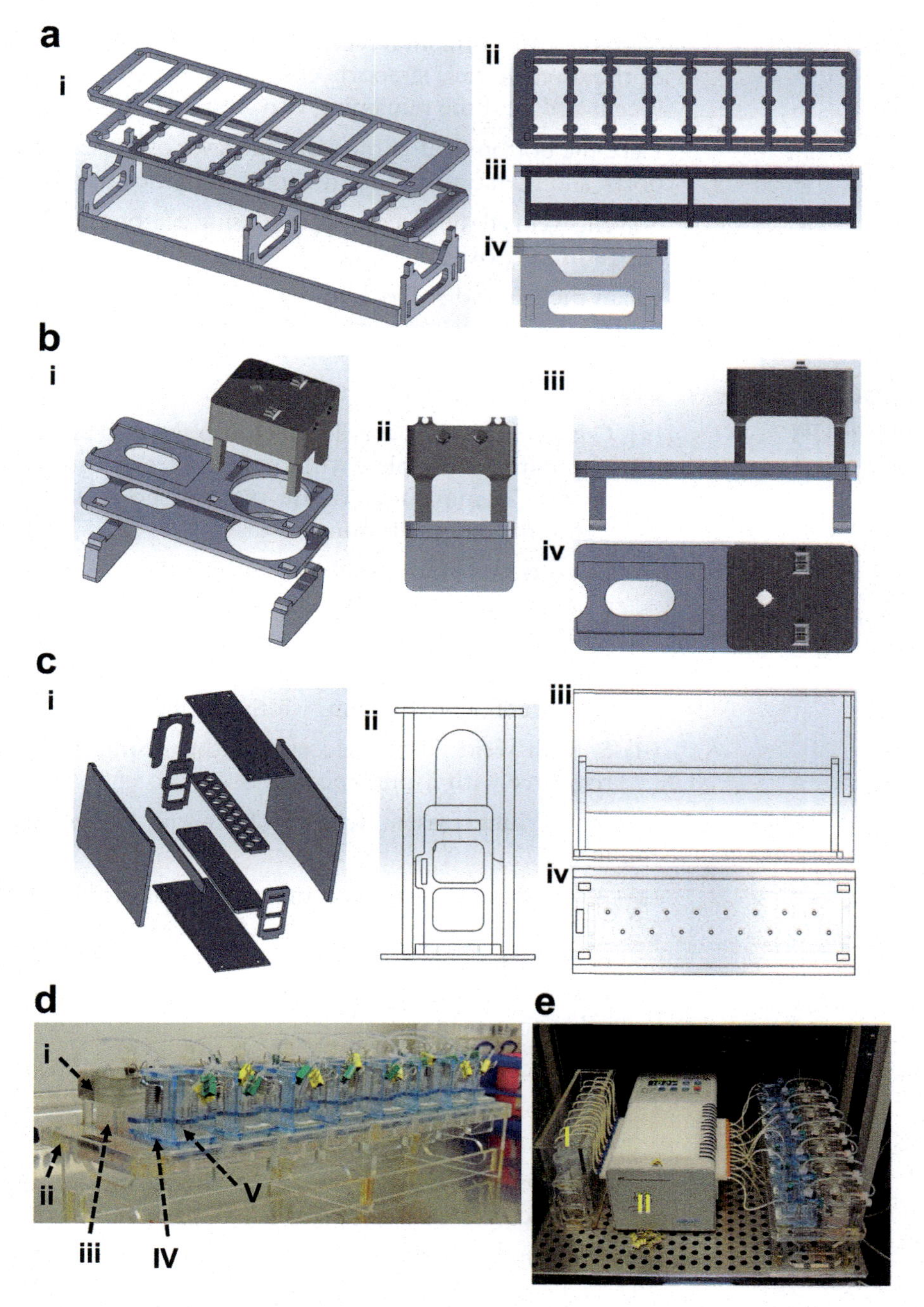

Fig. 2 Design and assembly of cartridge dock, biochip-compatible cartridge and effluent collector for cell culture operation. (**a**) Cartridge dock; aligned parts ready for assembly (*i*), top view (*ii*), side view (*iii*), and lateral view (*iv*) of the dock. (**b**) Biochip-compatible cartridge; aligned parts ready for assembly (*i*), rear view (*ii*), side view (*iii*), and top view (*iv*) of the cartridge. (**c**) Microfluidic chip effluent collector; aligned parts ready for assembly (*i*), lateral view (*ii*), front view (*iii*), and top view (*iv*) of the microfluidic collection system. (**d**) Eight cartridges (*i*) fit side-by-side in a dock (*ii*). A culture medium-containing reservoir (*iii*) is connected to each cartridge, and small airway chips (*iv*) fitted in clamps (*v*) sit on the upper front part of the cartridges. (**e**) The setup in-operation in a humidified 37 °C incubator. Note the tubing (*i*) transferring culture medium out of chip microvascular channel to effluent collector using a peristaltic pump (*ii*) by pull

2. Slide the two horizontal rectangular supports into mating slots of the three vertical supports ensuring that one of the verticals is centered, and the remaining two are flush with the ends.
3. Place the bottom half of the cartridge layer over the vertical supports, and then overlay with the top half of the cartridge layer.
4. Solvent-weld all the pieces together using acetone, and reinforce with 5-min epoxy.
5. Clean the Cartridge Dock by spraying with 70% ethanol prior to use.

3.5 Fabrication of Effluent Collector (Fig. 2)

1. Construct outer assembly as following:
   (a) Cut a 3 mm-thick acrylic sheet into the two pieces using a laser cutter: a top plate with tubing holes (length: 240 mm; width: 74 mm) and a base plate without hole (length: 240 mm; width: 90 mm).
   (b) Cut a 6 mm-thick acrylic sheet into the two pieces using a laser cutter: two side plates (height: 152 mm; width: 240 mm).
   (c) Cut a 6 mm-thick acrylic sheet into a U-shaped endstop support (height: 92 mm; width: 62 mm).
   (d) Solvent-weld all the pieces together using acetone, and reinforce with 5-min epoxy.
2. Construct inner assembly from laser cut 6 mm acrylic as following:
   (a) A top plate (length: 222 mm; width: 38 mm) with sixteen 17.5 mm diameter holes for 15 mL conical tubes, and a base plate (length: 230 mm; width: 50 mm) with sixteen 6 mm-diameter holes for the base of the 15 mL conical tubes to rest in.
   (b) Two side plates (height: 90 mm; width: 44 mm).
   (c) Horizontal rectangular support (height: 19 mm; length: 222 mm).
   (d) Solvent-weld all the pieces together using acetone, and reinforce with 5-min epoxy.
3. Inner assembly can be slid in and out of outer assembly.
4. Clean the Effluent Collector by spraying with 70% ethanol prior to use.

3.6 Microfluidic Culture of Small Airway Epithelial Cells

3.6.1 Preparation of Parts and Cells Prior to Microchip Seeding

1. Preheat a T-75 flask with 20 mL of epithelial Expansion medium in a humidified 37 °C incubator for 30 min.
2. Thaw a frozen vial of epithelial cells and add the cells directly into the T-75 flask and incubate overnight in incubator.
3. Next day, replace the whole medium with 20 mL of fresh prewarmed Expansion medium.

4. Three days after thawing, start replacing 10 mL of the culture medium every 2 days.
5. A day prior to cells reaching 70–80% confluency, place the small airway chip(s) into the UV-ozone cleaner, sterilize the device for 30–40 min, and immediately transfer the sterile devices from the cleaner into a sterile Petri dish to transfer to a biosafety cabinet.
6. Alternatively, the chips can be sterilized by oxygen ion plasma treatment using a plasma machine [run protocol: Plasma Time = 30 s; RF Power = 100 W; O_2 flow rate = 5 sccm (corresponds to approximately 180 mTorr chamber pressure)] inside sterilization pouches.

3.6.2 Clamping Microdevice and Coating with Extracellular Matrix Solution

1. In the biosafety cabinet, place the chip within the pre-sterilized device clamp (Fig. 3) as follows:
   (a) Insert two pre-sterilized 19G blunt straight needles in the inlet and outlet of bottom microfluidic channel of the device.
   (b) Place the clamp upside down and lower the translational plane against the springs.
   (c) Gently transfer the chip with the needles facing down into the clamp and carefully align it by passing the needles through the holes in the translational plane.
   (d) Next, slowly release the pressure on the springs by bringing the translational plane up.
   (e) Place the chip-fitted clamp upright and replace the needles with pre-sterilized 38 mm-long tubings connected to 19G blunt straight needles for all the four inlets and outlets.
2. Slowly introduce freshly prepared 100 μL of the ECM coating solution into each of the upper and lower microfluidic channels, discard the excess liquid, and pinch off the tubings with micro binder clips.
3. Incubate the device overnight at 37 °C.

3.6.3 Seeding, Maintenance and Air-Lifting of Small Airway Epithelial Cells On-Chip

1. Next day, prepare the chips for seeding and culture the epithelial cells as following:
   (a) Inside a biosafety cabinet, aseptically aspirate the ECM coating solution and fill in the microchannels with epithelial Submerged medium.
   (b) Incubate at 37 °C for 15–30 min.
   (c) Remove the culture medium from the T-75 flask containing 70–80% confluent human primary small airway epithelial cells.
   (d) Wash the cells with Ca^{2+}- and Mg^{2+}-free PBS once.

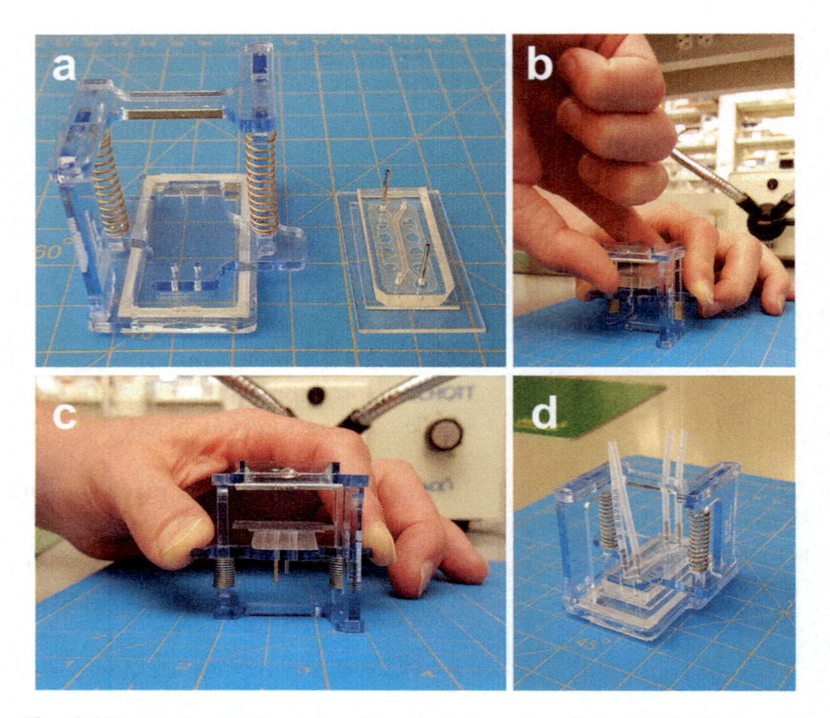

Fig. 3 Microdevice loading into chip-clamp for cell culture. (**a**) A side-by-side photo of the small airway microdevice and clamp. Note two 19G blunt straight needles are placed diagonally in inlet and outlet of the lower vascular channel prior to loading chip into the clamp. (**b**) To initiate loading, place the clamp upside down, lower the translational plane against the springs, and carefully and gently align the chip facing down against the hole of the translational plane. (**c**) Pass the needles through the holes and the chip should now be perfectly aligned across all the inlets and outlets. (**d**) Slowly release the pressure on the springs by bringing the translational plane up, place the chip-fitted clamp upright and replace the needles with 38 mm-long tubings connected to 19G blunt straight needles for all the four inlets and outlets. Note the whole process should be performed in biosafety cabinet using pre-sterilized parts

(e) Add 5 mL of pre-warmed trypsin–EDTA solution to the cell culture flask, incubate at 37 °C for 6–8 min or until the cells are completely detached from the growth surface.

(f) Add 10 mL of Expansion medium to the cell suspension and centrifuge at 200 × *g* for 5 min.

(g) Carefully remove the supernatant and thoroughly resuspend the epithelial cell pellet in 1 mL of Submerged medium culture medium.

(h) Count the cells and bring the cell density to 2–5 × 10^6 cells/mL (centrifuge and resuspend in calculated volume if necessary).

(i) Transfer 40 μL of the cell suspension into upper microchannel of each chip using 100-μL pipette tip (*see* **Note 13**).

(j) Incubate at 37 °C for 4–6 h.

(k) Inspect the device under microscope to ensure cell adhesion throughout to the porous membrane in the top microchannel, and absence of bubble in both top and bottom microchannels.

2. Assemble the cartridge and reservoir for connection to cell cultured chip as following:

(a) Fit in two luer hub 19G needles in the back of a cartridge.

(b) Close one with a 0.2 μm-pore sterile syringe filter and the other with a MLL Kynar® cap luer connector.

(c) Place in a pre-sterilized 50 mm-long tubing connected to a luer hub 19G needle through the top hole on the cartridge.

(d) Transfer the cartridge over a pre-sterilized wide-mouth LDPE bottle to which 10 mL of Submerged medium has been added, and screw tight.

(e) Connect the reservoir medium through the cartridge (top) open luer hub to a pre-sterilized 115 mm-long tubing that has been fitted on both ends with MLL-to-1.6 mm Kynar® barbed luer connectors.

(f) Prime the tubing by aspirating using 200-μL pipette tip against gravity and pinch off the tubing with a micro binder clip.

(g) Fit in a luer hub 19G needle into the barbed luer connector.

(h) Replace the inlet tubing of the chip's bottom microchannel with the cartridge tube.

(i) Replace the top microchannel medium with fresh prewarmed medium and pinch off the inlet and outlet tubings with micro binder clips.

3. Connect the device to a peristaltic pump and perfuse (by pull) the bottom microchannel at a volumetric flow rate of 60 μL/h in a humidified 37 °C incubator with 5% CO_2 (*see* **Note 14**).

4. Replace the upper microchannel medium with fresh prewarmed Submerged medium daily until the cells reach full confluency and demonstrate a cobblestone morphology when observed under a phase-contrast inverted microscope (usually 4–5 days after seeding).

5. In the meantime, add 3 mL of Submerged medium to the chip reservoir every 2 days.

6. Then an air–liquid interface (ALI) can be created by removing the apical medium (*see* **Note 15**).

7. After ALI induction, add 3 mL of ALI medium to the chip reservoir every 2 days throughout the culture, and once a

week replace the whole medium with 10 mL of fresh ALI medium.

8. Monitor the cells every 2 days using phase-contrast microscope (*see* **Note 16**).
9. Wash the epithelial cells once a week by incubating with pre-warmed ALI medium for 5–10 min.
10. Full epithelial differentiation into pseudostratified mucociliated epithelium should occur within 3–5 weeks post-ALI (Fig. 4), with earliest signs of apical secretion and ciliated cell appearance evident within the 7–10 days following ALI.

3.7 Epithelial Coculture with Lung Microvascular Endothelium

1. Preheat a T-75 flask with 20 mL of Endothelial medium in a humidified 37 °C incubator for 30 min (*see* **Note 17**).
2. Thaw a frozen vial of endothelial cells and add the cells directly into the T-75 flask and incubate overnight in incubator.
3. Next day, replace the whole medium with 20 mL of fresh pre-warmed Endothelial medium.
4. Three days after thawing, start replacing 10 mL of the culture medium every 2 days, until cells reach ~80% confluency.
5. Remove the culture medium from the T-75 flask, wash the cells with Ca^{2+}- and Mg^{2+}-free PBS once, and dissociate them by incubating with 5 mL of pre-warmed trypsin–EDTA solution at 37 °C for 4–8 min.
6. Add 10 mL of culture medium to the cell suspension and centrifuge at 200 × *g* for 5 min.
7. Remove the supernatant and thoroughly resuspend the endothelial cell pellet in 1 mL of medium.
8. Count the cells and bring the cell density to 2×10^7 cells/mL (centrifuge and resuspend in calculated volume if necessary).
9. Transfer 20 μL of the cell suspension into lower microchannel of each chip using 100-μL pipette tip (*see* **Note 18**).
10. Incubate at 37 °C for 1.5 h.
11. Replace the lower microchannel medium with fresh pre-warmed medium and connect back to the peristaltic pump and perfuse at volumetric flow rate of 60 μL/h in humidified 37 °C incubator. From now on the Endothelium Medium is supplied to the chip.
12. To ensure full integrity of the endothelial monolayer underneath porous PET membrane, disconnect the chip from flow 2–7 days following cell seeding, live-stain with 20 μL of Calcein AM/Wheat Germ Agglutinin, Alexa Fluor® 488 Conjugate-containing medium by incubating at 37 °C for 30 min, and visualize under inverted fluorescent microscope.

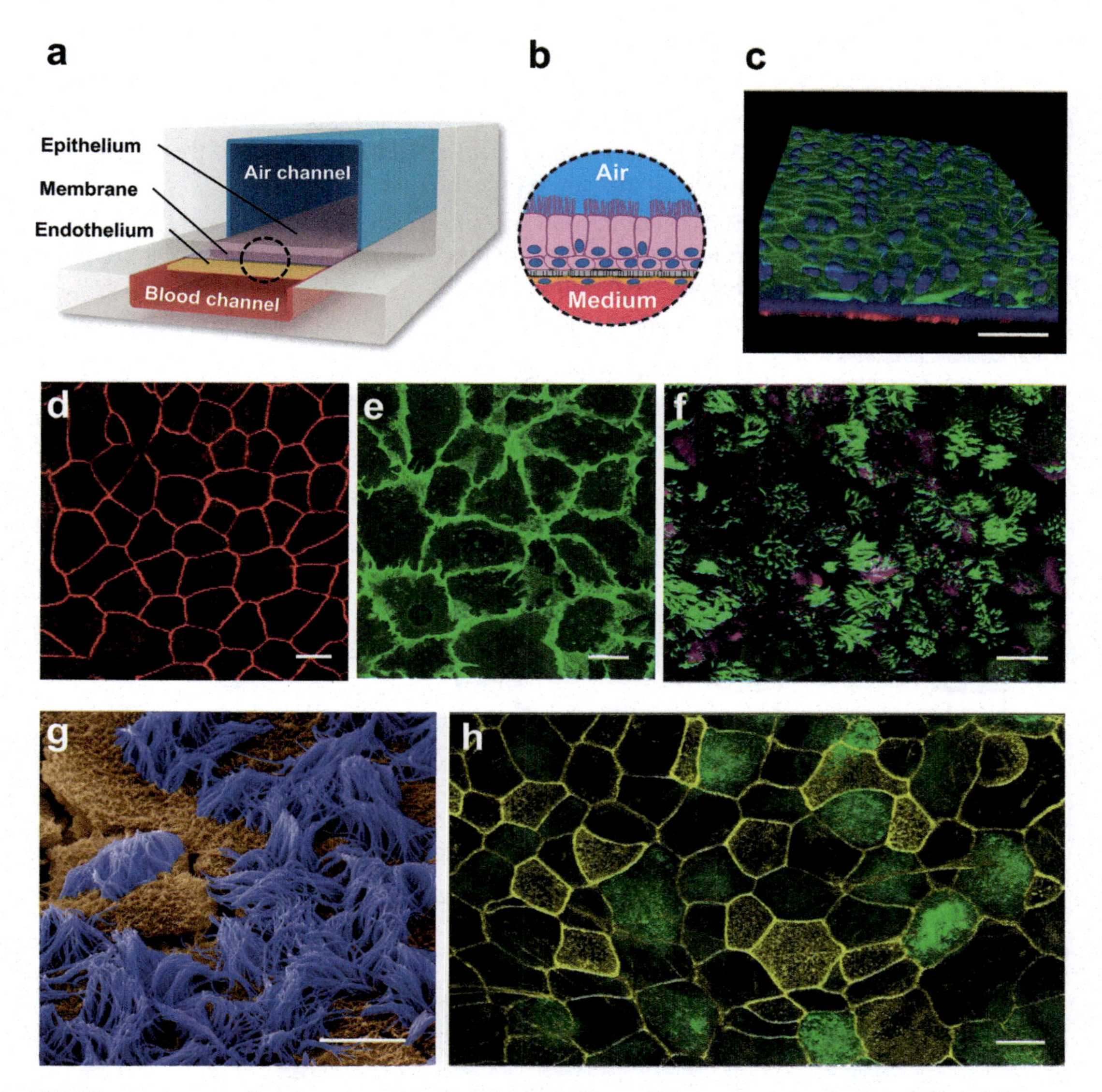

Fig. 4 Human lung small airway-on-a-chip. (**a**) Schematic diagram of a cross section through the small airway-on-a-chip. The *dashed circle* highlights the area depicted in **b**. (**b**) Diagram of the tissue–tissue interface that forms on-chip, showing differentiated airway epithelium (*pink cells*) cultured on top of a porous collagen-coated membrane at an air–liquid interface in the upper channel and the endothelium below (*orange cells*) with flowing medium that feeds both tissue layers. (**c**) A 3D reconstruction showing fully differentiated, pseudostratified airway epithelium formed on-chip by cultured hAECs (*green*, F-actin) with human pulmonary microvascular endothelial cells (*red*, F-actin) on the opposite side of the membrane. *Blue* denotes DAPI-stained nuclei; scale bar, 30 μm. (**d**) The differentiated human small airway epithelium exhibited continuous tight junctional connections on-chip, as demonstrated by ZO1 staining (*red*). Scale bar, 20 μm. (**e**) The human (lung blood microvascular) endothelial monolayer formed on-chip also contained continuous adherens junctions between adjacent cells, as indicated by PECAM-1 staining (*green*). Scale bar, 20 μm. (**f**) Well-differentiated human airway epithelium formed on-chip using hAECs derived from healthy donors demonstrating the presence of high numbers of ciliated cells labeled for β-tubulin IV (*green*) and goblet cells stained with anti-MUC5AC (*magenta*). Scale bar, 20 μm. (**g**) Scanning electron micrograph of cilia (*blue*) on the apical surface of the differentiated airway epithelium formed on-chip (nonciliated cells are in *brown*). Scale bar, 10 μm. (**h**) A confocal immunofluorescence image of club cells in bronchiolar epithelial cells differentiated on-chip (*green*, club cell secretory protein 10; *yellow*, F-actin; scale bar, 20 μm). Reproduced from Benam et al. [3] with permission from Macmillan Publishers Limited (Springer Nature)

3.8 Flowing Immune Cells Through the Small Airway Chip and Analysis of Endothelium–Leukocyte Interaction

3.8.1 Isolation of Mononuclear Leukocytes and Neutrophils from Blood

1. Transfer 15 mL of the blood into a sterile 50-mL conical tube.
2. Add an equal volume of Leukocyte medium to the blood, and mix by gently inverting the tube 4–5 times.
3. Separately, add 15 mL of Ficoll™ solution directly at the bottom of a 50-mL centrifuge tube, and very carefully overlay the Ficoll™ with diluted blood.
4. Centrifuge at RT at 850 × *g* for 20 min with the brakes off.
5. Following density gradient centrifugation, three distinct layers appear: top plasma layer, middle Ficoll™ layer and bottom red blood cell (RBC) layer (Fig. 5a).
6. Peripheral blood mononuclear cells (PBMCs) and granulocytes sit at interfaces of plasma-Ficoll™ layers, and Ficoll™-RBC layers, respectively.

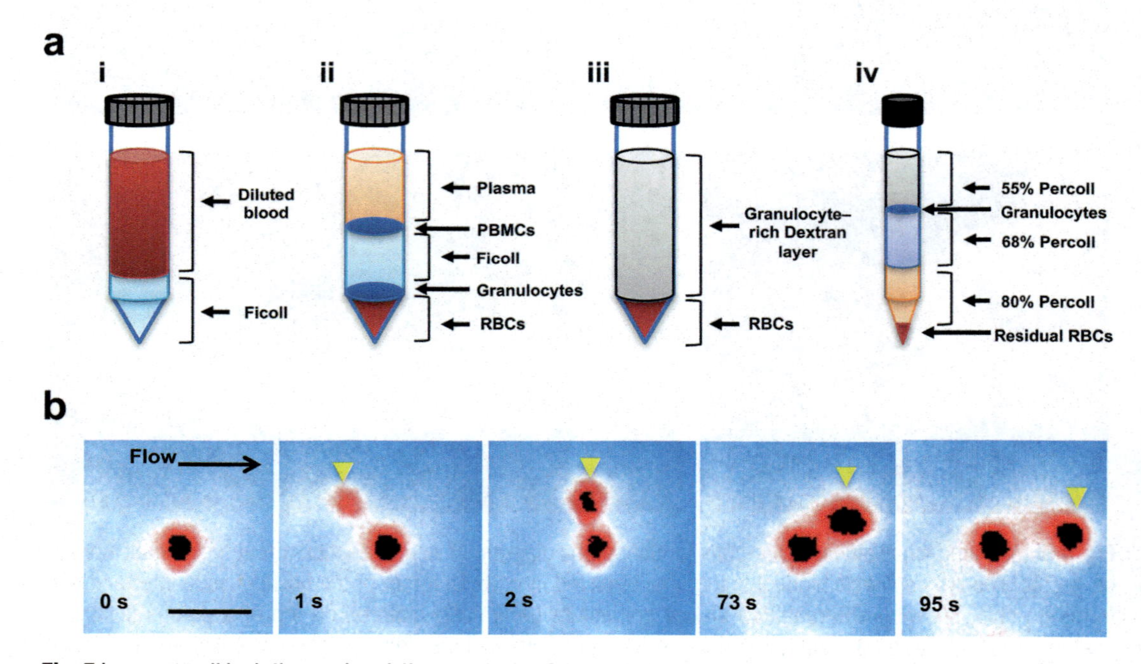

Fig. 5 Immune cell isolation and real-time analysis of dynamic leukocyte–endothelium interactions in the small airway chip. (**a**) Neutrophils are isolated fresh from whole human blood in a sequential manner: first blood is diluted 1:1 with Leukocyte medium and overlaid on (1× volume of blood) Ficoll solution (*i*); the mixture is then centrifuged to create three separate layers [plasma, Ficoll, and red blood cells (RBCs)] and two distinct interfaces [peripheral blood mononuclear cells (PBMCs) on *top* and granulocyte on *bottom*] (*ii*); then 6% dextran solution is added and mixed well with the granulocytes interface and RBCs layer only to sediment the red cells (*iii*); a second density centrifugation of the unsedimented granulocytes using Percoll solutions would yield highly pure neutrophils (*iv*). (**b**) Sequential time-lapse microscopic views showing freshly isolated neutrophils being recruited to the surface of living endothelium under physiological flow and shear stress (1 dyne/cm^2) in the airway chip when stimulated with poly I:C. *Yellow arrowhead* indicates a neutrophil that is captured from the flowing medium in the second image that then adheres and spreads on the endothelial surface in subsequent images, directly adjacent to an already bound neutrophil (neutrophils were live stained with CellTracker Red; *arrow* indicates direction of flow; times are indicated in seconds; scale bar, 20 μm)

7. Collect the PBMCs using a disposable transfer pipette.
8. When needed, use PBMC-derived immune cell populations (e.g., magnetically isolated CD3+ T lymphocytes or CD14+ monocytes) or the whole PBMC for subsequent flow through microfluidic small airway-on-a-chip.
9. Alternatively, when the focus of the studies is neutrophils, PBMCs can be discarded.
10. Gently aspirate and discard the plasma and Ficoll™ layers without disturbing the granulocyte interface.
11. Add 6 mL of Dextran solution and 24 mL of Ca^{2+}- and Mg^{2+}-free PBS to the remaining RBC and granulocyte fractions.
12. Mix gently by inverting the tube 4–5 times.
13. Leave still at RT for 25–30 min.
14. In the meantime, prepare fresh (6 mL) solutions of 80%, 68%, and 55% Percoll.
15. In a 15-mL centrifuge tube, layer 5 mL of the 68% Percoll solution over 5 mL of the 80% Percoll solution, and place the tube still on the side of biosafety cabinet.
16. Following the 30-min incubation, aspirate top leukocyte-rich layer (~25 mL) over precipitated red blood cells and transfer to a new 50-mL conical tube.
17. Add 25 mL of Ca^{2+}- and Mg^{2+}-free PBS and centrifuge at 350 × *g* for 6 min with brakes on at RT.
18. Discard the supernatant and resuspend the cell pellet in 5 mL of the 55% Percoll solution.
19. Gently overlay the cell suspension over the 80% to 68% Percoll mix.
20. Centrifuge at RT at 700 × *g* for 20 min with the brakes off.
21. Collect the neutrophils (located at the interface of top two layers) using a disposable transfer pipette, and transfer to a new 50-mL conical tube.
22. Wash by resuspending the cells in 50 mL of Leukocyte medium followed by centrifugation at 350 × *g* for 6 min (with brakes on) at RT.
23. Optional: lyse any remaining red cells by incubating with 15 mL of 1× RBC lysis buffer at RT for 10 min, stopping the lysis by adding 35 mL of PBS, and spinning the cells at 350 × *g* for 6 min.
24. Wash once more as in **step 22** above.
25. Count the cells and resuspend up to 10^7 viable cells in 5 mL of Cell Tracker- or Hoechst-containing medium.
26. Incubate at 37 °C for 15–30 min.

27. Centrifuge at 350 × *g* for 6 min (with brakes on) at RT.
28. Resuspend the cells in Leukocyte medium to a final density of 10^7 cells/mL and confirm granulocyte purity by flow cytometry analysis using anti-CD15 and anti-CD16 antibodies.
29. Ensure at least 90–95% of the cells are neutrophils.

3.8.2 Live-Imaging and Analysis of Neutrophil-Endothelium Interactions Under Microfluidic Flow

1. Fill a 1-mL syringe with the leukocyte suspension solution, attach to an 18G 15.9 mm-long blunt needle, and load on a syringe pump.
2. Place the unclamped small airway chip upside down on the stage of an inverted fluorescent microscope and secure tight.
3. Connect the inlet tubing of (lower) microvascular channel to the leukocyte-loaded syringe and perfuse (by push) the neutrophils at rates that generate physiological shear stress (*see* **Note 19**).
4. Focus the microscope on the endothelial layer and start imaging every 2 s for 10 min using appropriate filters (DAPI filter for Hoechst or DsRed for CellTracker Red).
5. Stop the neutrophil flow and perfuse the chip with cell-free Leukocyte medium at the same rate to remove unbound immune cells.
6. Capture four to five random micrographs throughout the length of the lengths of the chip.
7. Import the images from **step 4** into ImageJ to generate a video of leukocyte recruitment from flow and its rolling and adhesion over the underlying endothelium (Fig. 5b).
8. Import the images from **step 6** into ImageJ or Cell Profiler to quantify firmly adhered neutrophils to endothelium.

4 Notes

1. Expansion medium is prepared by adding all the components of one SAECGM SupplementPack kit [bovine pituitary extract (BPE), recombinant human epidermal growth factor (rhEGF), recombinant human insulin, hydrocortisone, epinephrine, triiodo-L-thyronine, transferrin, retinoic acid, bovine serum albumin—fatty acid free (BSA-FAF)] along with 5 mL of 100× penicillin–streptomycin solution (10,000 U/mL penicillin and 10 mg/mL streptomycin) to 500 mL of SAECBM.
2. To prepare BSA stock solution, first prepare 30% (wt/vol) BSA in PBS and filter sterilize it. Then mix 30% BSA with Expansion medium at the ratio 50:50, further dilute the resulting (150 mg/mL) solution in Expansion medium at the ratio 1:100 to the final (1.5 mg/mL) solution. Make 500 μL aliquots and freeze at −20 °C.

3. First prepare the 3 mg/mL retinoic acid solution by dissolving ATRA in dimethyl sulfoxide (DMSO), then dilute half of the solution in DMSO at the ratio 1:200 to generate 15 μg/mL ATRA in DMSO. Freeze the 3 mg/mL and 15 μg/mL solutions in 100 μL aliquots at −20 °C.
4. To prepare Submerged medium, mix 250 mL SAECBM with 250 mL of DMEM with pyruvate and low glucose. Add BPE, rhEGF, insulin, hydrocortisone, epinephrine, triiodo-L-thyronine and transferrin of the SAECGM SupplementPack kit, 500 μL of 1.5 mg/mL BSA and 5 mL of 100× penicillin–streptomycin (10,000 U/mL penicillin and 10 mg/mL streptomycin) solution. Before use, add fresh 15 μg/mL ATRA to desired volume at 1:1000 dilution.
5. Calcein AM/Wheat Germ Agglutinin (WGA)-Alexa Fluor® 488 Conjugate-containing medium: Dilute 1 mg/mL Calcein AM 1:1000 and 1 mg/mL WGA solutions 1:200 in Endothelium medium.
6. Ethical approval and consent from patients must be obtained.
7. To prepare Leukocyte medium, add 55 mL of heat-inactivated FBS and 5 mL of 100× penicillin–streptomycin solution to 500 mL of RPMI-1640.
8. For dextran solution, first prepare 0.9% NaCl by dissolving 9 g of NaCl in 1 L of tissue culture-grade water, then filter-sterilize (0.22 μm). To prepare 6% dextran solution, dissolve 6 g of dextran powder in 100 mL of 0.9% NaCl at 37 °C for 1 h. Store at 4 °C until use.
9. First mix 13.5 mL of Percoll® with 1.5 mL of isotonic Ca^{2+}- and Mg^{2+}-free 10× PBS to generate 90% Percoll. Next, prepare 10 mL of each of 80%, 68% and 55% Percoll solutions by diluting the 90% Percoll in Ca^{2+}- and Mg^{2+}-free 1× PBS.
10. Prepare CellTracker™ Red CMPTX Dye stock solution by resuspending 50 μg of the dye in 50 μL of DMSO (store at −20 °C, protected from light). To prepare CellTracker-containing medium dilute 1 μL of the stock CellTracker solution in 1 mL of Leukocyte medium.
11. Hoechst-containing medium is prepared by diluting 1 μL of 10 mg/mL Hoechst 33342 solution in 2 mL of Leukocyte medium.
12. A minimum 4 h is required for PDMS to fully cure at 60 °C; overnight incubation is optional.
13. It is important to visually follow the fluid movement to ensure the whole space in the channel is filled with the cell suspension; collect the displaced medium using premoistened cleanroom 70% ethanol wipes from the outlet tubing.
14. Note that the pump should be purged upon initial connection for 5–10 s to ensure accidental bubbles are removed.

15. It is crucial to ensure the epithelial monolayer is integral and no hole in the culture is present prior to ALI induction.
16. It is easier to visualize the cells under microscope when the upper microchannel is temporarily filled with cell culture medium.
17. We recommend using the primary lung microvascular endothelial cells at passage 2–5, assuming commercial vendor provides the cells at passage 1.
18. It is important to visually follow the fluid movement to ensure the whole space in the channel is filled with the cell suspension; collect the displaced medium using premoistened 70% ethanol wipes from the outlet tubing.
19. Wall shear stress is calculated using the following formula, as previously reported [13]:

$$\tau = \frac{6 \times Q\mu}{h^2 \times w}$$

where, τ is shear stress (dyne/cm^2), Q is volumetric flow rate (mL/s), μ is dynamic fluid viscosity (g/cm.s), h is microchannel height (cm), and w is microchannel width (cm).

Acknowledgments

We thank C. Lucchesi, J. Szabo, R. Villenave, C.D. Hinojosa, and G. Thompson for technical assistance, learning to use instruments and helpful discussions. We also thank the Wyss Microfabrication Team for their help with chip fabrication.

Potential competing interests
D.E. Ingber holds equity in Emulate, Inc. and Opsonix, Inc. and chairs their scientific advisory boards.

References

1. Rennard SI, Drummond MB (2015) Early chronic obstructive pulmonary disease: definition, assessment, and prevention. Lancet 385(9979):1778–1788. doi:10.1016/S0140-6736(15)60647-X
2. Pawankar R (2014) Allergic diseases and asthma: a global public health concern and a call to action. World Allergy Organ J 7(1):12. doi:10.1186/1939-4551-7-12
3. Benam KH, Villenave R, Lucchesi C et al (2016) Small airway-on-a-chip enables analysis of human lung inflammation and drug responses in vitro. Nat Methods 13(2):151–157. doi:10.1038/nmeth.3697
4. Martinez FJ, Donohue JF, Rennard SI (2011) The future of chronic obstructive pulmonary disease treatment—difficulties of and barriers to drug development. Lancet 378(9795):1027–1037. doi:10.1016/S0140-6736(11)61047-7
5. Bhatia SN, Ingber DE (2014) Microfluidic organs-on-chips. Nat Biotechnol 32(8):760–772. doi:10.1038/nbt.2989
6. Benam KH, Dauth S, Hassell B et al (2015) Engineered in vitro disease models. Annu Rev Pathol 10:195–262. doi:10.1146/annurev-pathol-012414-040418
7. Proudfoot AG, McAuley DF, Griffiths MJ et al (2011) Human models of acute lung injury.

Dis Model Mech 4(2):145–153. doi:10.1242/dmm.006213

8. Hyde DM, Hamid Q, Irvin CG (2009) Anatomy, pathology, and physiology of the tracheobronchial tree: emphasis on the distal airways. J Allergy Clin Immunol 124(6 Suppl):S72–S77. doi:10.1016/j.jaci.2009.08.048
9. Kolaczkowska E, Kubes P (2013) Neutrophil recruitment and function in health and inflammation. Nat Rev Immunol 13(3):159–175. doi:10.1038/nri3399
10. Wright JL, Cosio M, Churg A (2008) Animal models of chronic obstructive pulmonary disease. Am J Physiol Lung Cell Mol Physiol 295(1):L1–15. doi:10.1152/ajplung.90200.2008
11. Esch EW, Bahinski A, Huh D (2015) Organs-on-chips at the frontiers of drug discovery. Nat Rev Drug Discov 14(4):248–260. doi:10.1038/nrd4539
12. Huh D, Kim HJ, Fraser JP et al (2013) Microfabrication of human organs-on-chips. Nat Protoc 8(11):2135–2157. doi:10.1038/nprot.2013.137
13. Shao J, Wu L, Wu J et al (2009) Integrated microfluidic chip for endothelial cells culture and analysis exposed to a pulsatile and oscillatory shear stress. Lab Chip 9(21):3118–3125. doi:10.1039/b909312e

Part V

Bioprinting

Chapter 26

Microfluidic Bioprinting of Heterogeneous 3D Tissue Constructs

Cristina Colosi, Marco Costantini, Andrea Barbetta, and Mariella Dentini

Abstract

3D bioprinting is an emerging field that can be described as a robotic additive biofabrication technology that has the potential to build tissues or organs. In general, bioprinting uses a computer-controlled printing device to accurately deposit cells and biomaterials into precise architectures with the goal of creating on demand organized multicellular tissue structures and eventually intra-organ vascular networks. The latter, in turn, will promote the host integration of the engineered tissue/organ in situ once implanted. Existing biofabrication techniques still lay behind this goal. Here, we describe a novel microfluidic printing head—integrated within a custom 3D bioprinter—that allows for the deposition of multimaterial and/or multicellular within a single scaffold by extruding simultaneously different bioinks or by rapidly switching between one bioink and another. The designed bioprinting method effectively moves toward the direction of creating viable tissues and organs for implantation in clinic and research in lab environments.

Key words Bioprinting, Cell-laden scaffolds, Microfluidic dispensing head, Heterogeneous cellular 3D structures, Vascular network, Bioink, Alginate

1 Introduction

The in vitro recapitulation of tissue and organ function represents one of the main objectives of tissue engineering. Developments in this field hold great promises for numerous clinical applications, from alleviating the shortage of organ donation to providing more representative platforms for drug testing. As such, there is an increasingly significant body of literature devoted to the development of better strategies to reconstruct living matter outside of the body [1].

Biological structures are highly organized and heterogeneous, as they are composed of multiple cell types and extracellular matrix (ECM) components that are positioned in precise locations, respectively at the microscale and nanoscale. Complexity of tissue organization has proven to be one roadblock in the way of building functional tissues and organs. Simply injecting or implanting

Zuzana Koledova (ed.), *3D Cell Culture: Methods and Protocols*, Methods in Molecular Biology, vol. 1612,
DOI 10.1007/978-1-4939-7021-6_26, © Springer Science+Business Media LLC 2017

masses of cells in vivo or growing masses of cells in vitro on two dimensional (2D) or three dimensional (3D) surfaces generally fails in recapitulating the specific functions of the engineered tissue or organ. Therefore, alternative methods of organizing cells into proper 3D assembly are needed to facilitate formation of usable tissues. Among them, bioprinting is considered as one of the most promising techniques to recapitulate the complexity of in vivo histo-architectures [2, 3].

Bioprinting consists of the simultaneous deposition of cells and biomaterials, layer by layer, in order to form 3D well-organized structures that can mirror physiologically relevant architectures. The bioprinting process involves the use of a 3D printer (generally renamed *bioprinter*), which can deposit the so-called *bioink*, composed of living cells and ECM-like components. Bioinks can be composed of (1) cell aggregates (i.e., spheroids), (2) cells resuspended in hydrogels or their precursor solutions, or (3) cell-seeded microcarriers. A computer-assisted design is used to guide the placement of the formulated bioink into predetermined geometries that can mimic actual histo-architecture.

Different deposition techniques have been developed to dispose cells and supporting ECM components in 3D, each one with specific advantages and technical limitations. For example, bioprinting techniques based on the extrusion of the bioink through dispensing nozzles can rapidly deposit fibrous 3D structures using very simple setups, but their precision falls in the range of few hundreds micrometers (the typical fiber dimension), which is often inadequate to pattern different types of cells with the necessary accuracy. In contrast, laser-based techniques can deposit droplets of bioink with dimensions as small as 10 μm [4] but their setups are hardly scalable and high-priced.

To bridge the gap between accessibility and performances, we developed a new concept of extrusion-based bioprinting technique, which implements a microfluidic control in the dispensation of the bioink [5]. With the use of microfluidic platforms, it is possible to achieve a precision in the deposition of cells that falls beneath the dimension of the single laid fiber, reaching a level of accuracy that was previously unachievable in extrusion-based systems.

The coupling of microfluidic platforms with the dispensing system is made possible by the use of a coaxial extrusion head that induces the solidification of the bioink in the form of a hydrogel simultaneously to its deposition. In particular, among other components, the bioink contains alginate, whose gelation is induced by exposing it to a cross-linking solution containing calcium ions. The bioink and the cross-linking solution are delivered through the internal and external needles of a coaxial-needles system, respectively. At the ending tip of the dispensing head the two solutions meet causing the immediate solidification of the bioink due to the ionic cross-linking of alginate. In this way, it is possible to deposit hydrogel fibers with dimensions ranging between 150 and 300 μm [6].

Starting from the described coaxial extrusion system, we designed low-viscous, liquid bioinks loadable with high cell densities that, before reaching the dispensing head, can flow in microfluidic chips. Thanks to the laminar flow inside microchannels, it is possible to design specific disposition of different cell types in the stream of liquid bioink exiting from the extruder. The designed cell patterns will be crystallized by the gelation of alginate during the 3D deposition. The printing conditions described above are mild since bioink viscosity is low and cross-linking conditions can be tuned to be harmless toward encapsulated cells. Beside alginate, other macromolecules of biological interest can be dissolved in the bioink and used as ECM components for encapsulated cells. In particular, we presented a bioink containing photo-curable gelatin metacrylate, whose mechanical properties can be freely tuned by dosing the exposition of the printed constructs to UV light. By varying the mechanical properties of the encapsulating matrix, cells can be induced to spread and migrate within the gel.

The described bioprinting technique can thus produce heterogeneous cellular 3D structures, fast and with a simple set up; it has virtually no limits in the chemical characteristics of the encapsulating ECM and can deposit high cell density maintaining high viability for prolonged periods of culture time.

2 Materials

2.1 Microfluidic Chip

1. Computer-Aided Design (CAD) and Manufacturing (CAM) software (e.g., SolidWorks—SolidCAM).
2. Poly(methyl methacrylate) (PMMA) sheets, 3–5 mm thick.
3. Computer Numerical Control (CNC) milling machine (*see* **Note 1**).
4. Bench-drill.
5. High-pressure hot press.

2.2 Microfluidic Coaxial Extruder

1. Two needles that can fit coaxially. We recommend to use a 27 gauge (G) as the internal needle and an 18 G as the external needle. The two needles should have blunt ends. Standard needle with bevel tip can be used after accurately flattening the tip (*see* **Note 2**).
2. Microfluidic chip.
3. Adhesive filler, e.g., two-component epoxy glue.
4. Tools for modeling the needles: metal file, small scissors, or clipper.

2.3 Preparation of the Bioink

1. Heated magnetic stirrer.
2. Brown glass vial (*see* **Note 3**).
3. Low molecular weight Alginate (FMC Biopolymers).
4. Gelatin Methacrylate (gelMA) (*see* **Note 4**).
5. 2-Hydroxy-4′-(2-hydroxyethoxy)-2-methylpropiophenone photoinitiator (Irgacure 2959, Sigma).
6. Bioink buffer: 10% (v/v) fetal bovine serum (FBS) in 25 mM Hepes, pH 7.2–7.5, sterile.
7. Sterile eppendorf tube, 1.5 or 2 mL.
8. Syringe filter, 0.2–0.1 μm.
9. 1 mL syringe with a 19 G needle.
10. Cells of interest, e.g., HUVEC, in dish or flask.
11. Centrifuge.
12. Cell counting chamber.
13. PBS.
14. Trypsin-EDTA solution (1×).
15. Cell Culture Medium suitable for selected cell type, e.g., HUVEC medium: 2% FBS and Vascular Endothelial Growth Factor in Endothelial Basal Medium (EGM-2 BulletKit, Lonza).

2.4 Bioprinting Experiment

1. 3-axis CNC machine or any motorized system that can follow a predetermined printing code in 3D (*see* **Note 5**).
2. Two or more precision syringe pumps (*see* **Note 6**).
3. UV light source (*see* **Note 7**).
4. Cross-linking solution: 300 mM $CaCl_2$ in the bioink buffer, sterile.
5. 1 mL syringes, sterile (*see* **Note 8**).
6. Silicone tubing with 0.5–0.8 mm diameter, sterile.
7. Microscope glass slides.
8. Tweezers, sterile.
9. Spatula, sterile.
10. 70% ethanol.
11. Multiwell plates (*see* **Note 9**).
12. Phosphate buffered saline (PBS).
13. Bioink buffer.
14. Cell culture medium.

2.5 Evaluation of Cell Viability, Fluorescence Microscopy Analysis

1. PBS.
2. LIVE/DEAD® Viability/Cytotoxicity Kit, for mammalian cells (Thermo Fisher Scientific).

3. Confocal fluorescence microscope.
4. Image analysis software (e.g., ImageJ).
5. 4% paraformaldehyde (PFA) in PBS.
6. 0.05% Triton X-100 in PBS.
7. Primary and secondary antibodies and nuclei staining solution (e.g., DAPI) according to the experiment.

3 Methods

3.1 Production of the Microfluidic Chip

1. Generate a 3D design of the desired microfluidic geometry using a CAD software (*see* **Note 10**).
2. Generate the manufacturing instructions for the CNC-milling machine starting from your CAD data, using CAM software (*see* **Note 11**).
3. Secure the PMMA sheet on the working area of the CNC-milling machine.
4. Execute the instructions for the CNC-milling machine generated with the CAM software in order to mill the fluidic channels on top of the PMMA sheet.
5. Drill 0.5–0.8 mm holes in correspondence of the fluidic inlets of microchannels (*see* **Note 12**).
6. Place a blank PMMA sheet on top of the PMMA sheet containing the fluidic channels.
7. Place the two PMMA sheets in the hot-press and let bond for 1 h at 3 bar and 130 °C.
8. Cut the bottom part of the bonded chip crossing the outlet channel to create a fluidic access parallel to the direction of flow, as opposed to inlet ports that are perpendicular to the direction of flow (Fig. 1, step 1).

3.2 Assembling of the Microfluidic Coaxial Extruder

1. Remove the plastic hub from the smaller needle using the scissor or clipper. Check accurately that the needle has blunt ends.
2. Insert the smaller needle inside the outlet channel of the microfluidic chip. Glue the needle watertightly to the microfluidic device with epoxy glue (Fig. 1, step 2).
3. Drill a hole sideways in the plastic hub of the external needle. The size of the hole should fit a small tube (Fig. 1, step 3).
4. Place the external needle coaxially to the inner needle. Pay particular attention to place the two needles as much coaxially as possible. Then, glue the external needle watertightly to the microfluidic device (Fig. 1, step 4) (*see* **Note 13**).

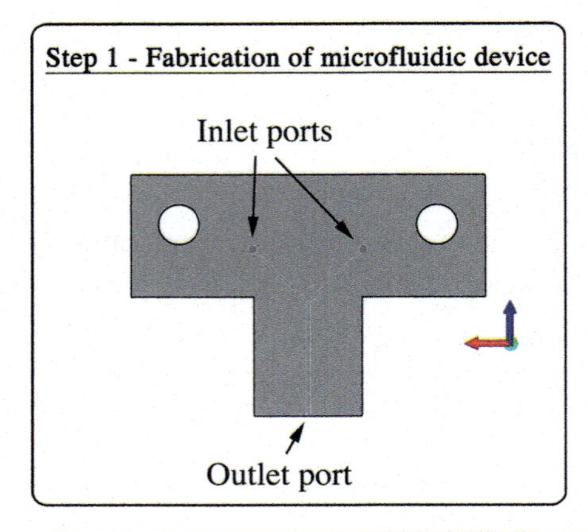

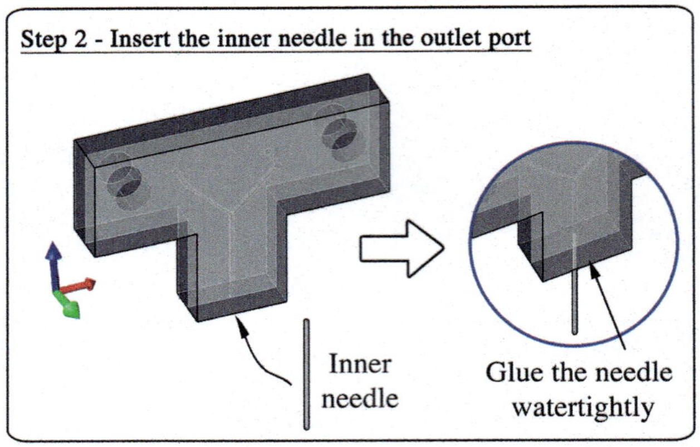

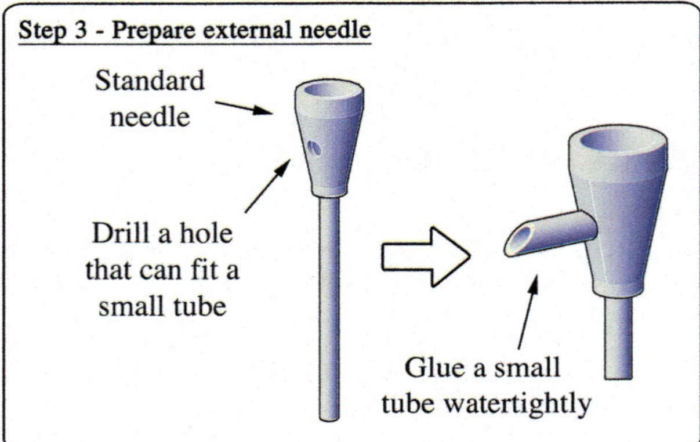

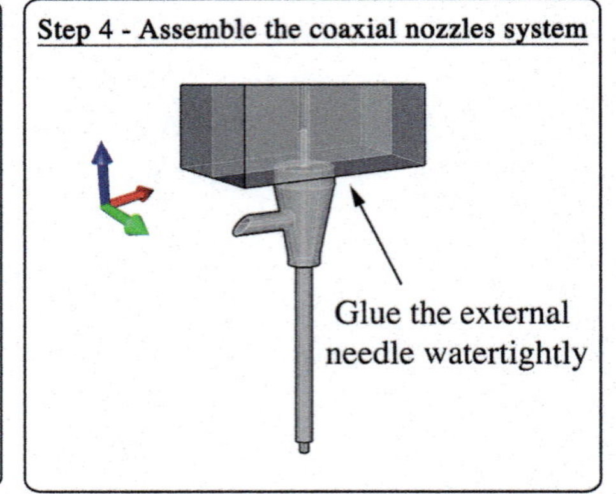

Fig. 1 Schematics of the fabrication process of the microfluidic printing head

3.3 Preparation of the Bioink

3.3.1 Preparation of the Hydrogel Precursor Solution

For each bioprinting experiment, prepare 1 mL of polymer solution. After filtration, it will yield about 600–700 μL of sterile polymer solution.

1. Dissolve the required amount of alginate (for 4% (w/v) solution) and gelMA (for 5% (w/v) solution) in bioink buffer. To accelerate the dissolution of biopolymers, heat the solution to 37 °C.
2. Add photoinitiator to final concentration 0.5% (w/w). Stir on a magnetic stirrer for at least 2 h at 37 °C.
3. Filter-sterilize the polymer solution using a 0.22 μm syringe filter in a laminar flow hood. Collect the filtered solution in a sterile eppendorf tube.
4. Store the filtered solution protected from light at 37 °C and use it within the next few hours (*see* **Note 14**).

3.3.2 Suspension of Cells in the Bioink

Right before the bioprinting procedure, cells are trypsinized and resuspended in the prepared hydrogel solution. All the following steps must be executed inside a sterile laminar flow hood.

1. Place the dish or flask containing confluent cells inside the hood.
2. Wash cells with warm PBS twice (*see* **Note 15**).
3. Add 2–5 mL of Trypsin-EDTA (*see* **Note 15**) and incubate for 5 min inside the incubator.
4. Gently shake the dish-flask to guarantee that most cells detach from it.
5. Add 4–10 mL of cell culture medium in the dish or flask (*see* **Note 16**) and gently suspend cells in the solution.
6. Aspirate the cell suspension and place it in a sterile 50 mL tube.
7. Centrifuge the tube for 5 min at 180 × *g*.
8. Resuspend the cell pellet in 10 mL of cell culture medium.
9. Use the cell counting chamber to determine the concentration of cells in the suspension.
10. Place the volume of cell suspension containing the desired amount of cells in a sterile eppendorf tube, and centrifuge again for 5 min at 180 × *g*.
11. The optimal cell density in the bioink will depend on cell type. For HUVECs, we suggest to prepare pellets containing 5×10^6 cells.
12. Aspirate the supernatant, add 500 μL of the polymer solution and resuspend the cells (this is the bioink).

3.4 Bioprinting Experiment

1. Flow 70% ethanol inside the microfluidic coaxial extruder, both in the bioink and cross-linking solution channels and needles using sterile syringes connected to the extruder with plastic tubing.
2. Flow sterile water inside the microfluidic coaxial extruder as in the previous step.
3. Empty the microfluidic coaxial extruder using a flow of sterile air.
4. Fill a 1 mL syringe with the bioink while avoiding the aspiration of air and formation of bubbles. Fill another 1 mL syringe with the cross-linking solution (Fig. 2, step 1).
5. Connect the syringes to the coaxial extrusion system using tubing.
6. Mount the syringes in the syringe pumps and the coaxial extrusion system in the XYZ motorized system. This step depends on the configuration of the 3D printer setup (Fig. 2, step 2).
7. Set the zero level for Z-axis using a glass slide as deposition plane. According to the 3D printer used, this crucial step might be executed automatically or manually. Pay the highest attention to make a correct zero level as it strongly influences the quality of 3D printed hydrogels fibers (Fig. 2, step 3).

8. Start the flow in all syringe pumps. Typical flow rates are around 5 μL/min (*see* **Note 17**). Wait until gel starts to form at the ending tip of the coaxial extruder.
9. Start the printing code in the bioprinter (*see* **Note 18**) (Fig. 2, step 4).

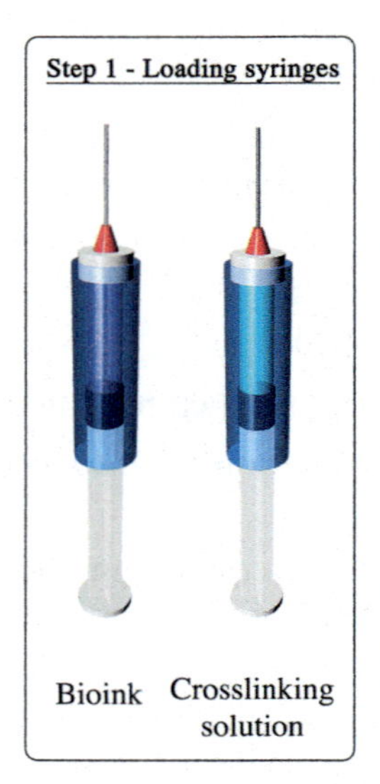

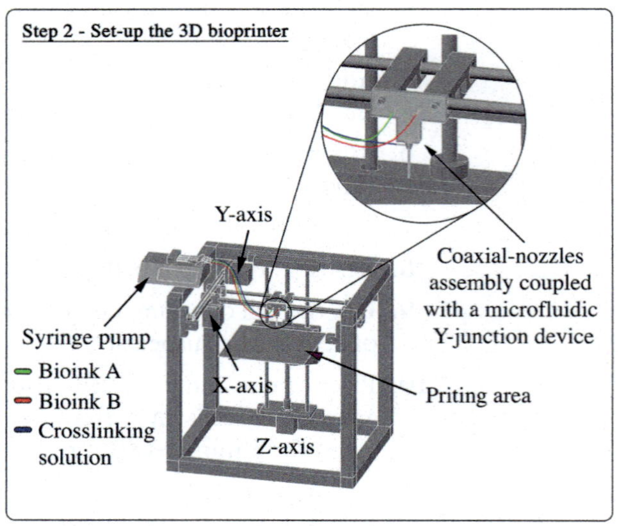

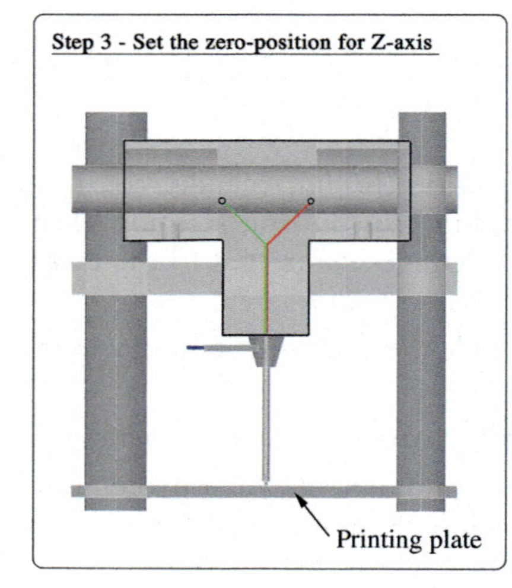

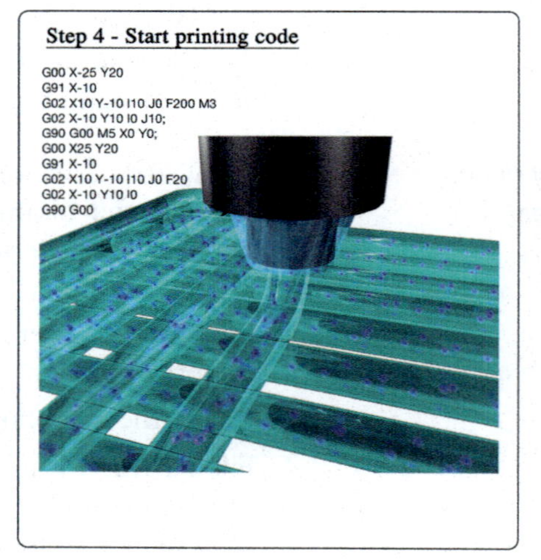

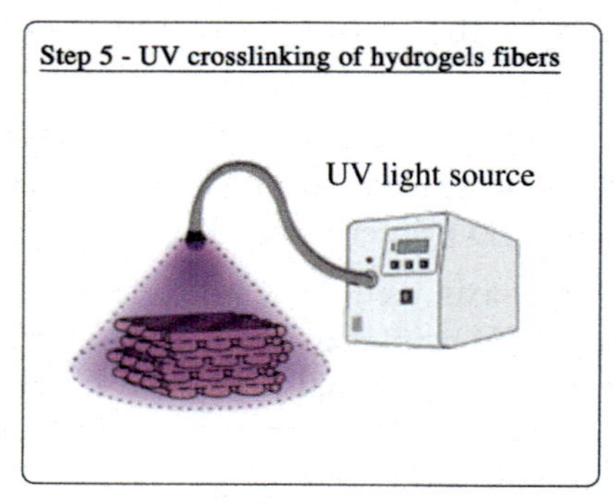

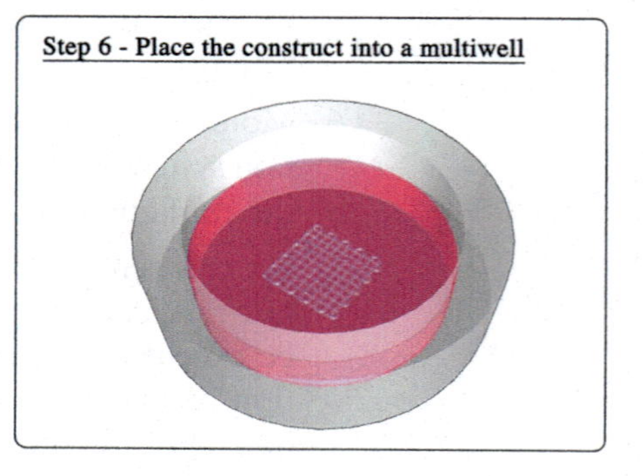

Fig. 2 Schematics of the bioprinting process and of the UV light curing of the cell-laden scaffold

10. Collect the printed construct using a spatula and expose it to UV light. The optimal dose of UV exposure will depend on the type of encapsulated cells, and should be evaluated case-by-case (*see* **Note 19**) (Fig. 2, step 5).
11. Place the cross-linked construct in a multiwell dish. Wash it with bioink buffer and then with PBS, then add 1–2 mL of cell culture medium (Fig. 2, step 6).
12. When all the programmed samples have been printed, place the multiwell plate inside an incubator. Change the medium every other day.

3.5 Evaluation of Cell Viability

For the evaluation of cell viability, we suggest the use of two-color fluorescent probes based on calcein-AM (green, for live cells) and ethidium homodimer-1 (red, for dead cells) in the LIVE/DEAD kit.

1. Calculate the total volume of dyeing solution needed. For each sample in a 12-well plate, approximately 500 μL of the solution is needed.
2. Prepare the dyeing solution by adding 0.2 μL/mL of calcein-AM and 1 μL/mL of ethidium homodimer-1 in PBS. Protect it from light using aluminum foil (*see* **Note 20**).
3. Wash the samples with PBS twice, leaving the well dry.
4. Add 500 μL of dying solution to each sample.
5. Cover the 12-well plate with aluminum foil and incubate it for 30 min inside the incubator.
6. After the incubation time, aspirate the staining solution and wash the samples with PBS twice.
7. Acquire fluorescence pictures of the samples within 1 h after the end of incubation. Ensure to acquire the red and green channels separately.
8. Using image analysis software analyze the green and red channels of the fluorescence pictures individually, counting the number of live (green) and dead (red) cells.
9. Calculate cell survival rate as the proportion of live cells (green) out of the total cell number (green and red).

3.6 Fluorescence Microscopy Analysis of the Bioprinted 3D Cell Construct

Thanks to the optical transparency of the printed hydrogel fibers, the organization of encapsulated cells inside the 3D bioprinted structure can be monitored using bright field, phase contrast or fluorescence microscopy. Given the 3D nature of the construct, the best results will be obtained using confocal microscopy.

F-actin staining (e.g., using fluorescently labelled phalloidin) can be used to evaluate cell disposition and spreading inside the 3D bioprinted construct. Immunohistochemical assays can be performed to analyze the spatial organization of relevant biomolecules in encapsulated cells. As these experiments are strongly dependent on the specific kit elected, we provide general hints to be followed according to specific manufacturer's instructions.

1. Fix the samples using 4% PFA, soaking the sample for approximately 10 min.
2. Permeabilize the cells using 0.05% Triton X-100 in PBS, soaking the sample for approximately 20–30 min.
3. When using fluorescent probes, reevaluate the incubation times with respect to specific kit instructions, as all chemicals need to diffuse through the hydrogel network and reach encapsulated cells in order to react.
4. Calculate the needed volume of dyeing solutions so as to completely soak the 3D construct during incubation.

4 Notes

1. Microfluidic chips can be fabricated using different techniques. We are describing the CNC micromilling process. If you prefer, you may build your chip using soft-lithography. If you do not have access to a clean-room or CNC-milling facility, you can order customized microfluidic chips from specialized companies.
2. Standard needle with bevel tip can be flattened with sandpaper. After that, we recommend to ultrasonicate the needle to remove all burrs and dust.
3. You can use another kind of container to prepare the solution as far as you keep it protected from light after the addition of the photoinitiator.
4. For the preparation of the methacrylated derivative of gelatin, *see* Van Den Bulcke et al. [7]. The reagents needed can all be purchased by Sigma. Instead of GelMA, you can use another macromolecule that can undergo a UV triggered chemical cross-linking.
5. You can use a commercial 3-axis CNC system, like extrusion-based 3D printers, as well as assemble three motorized linear stages in an X–Y–Z configuration and connect it to the computer with a proper electronic board.
6. You will need one syringe pump to control the flow of the cross-linking solution that will be connected to the external needle of the coaxial extruder, and one or more syringe pumps to control the flow of the bioink in the internal needle (one syringe pump) or in your microfluidic chip (using two or more syringe pumps depending on the number of inlets in your chip).
7. We suggest to use a mercury lamp with high-pass filters to limit the emission of the lamp to UV-A portion of the spectrum.

8. You will need one syringe for each bioink solution and another one for the cross-linking solution.
9. Depending on the dimension of your constructs, choose a multiwell plate in which samples can be easily transferred and cultured.
10. The outlet channel of the microfluidic chip must be dimensioned to contain the internal needle of the coaxial extruder, as described in the next paragraph.
11. You need to use an end mill cutter as small as the smallest feature of your microfluidic geometry. A typical feed-rate for the milling of PMMA is 2 mm/min.
12. The dimension of the holes must be selected in accordance with the dimension of the plastic tubing that will be used to connect the microfluidic chip with syringes, as described in the next paragraphs.
13. Make sure that the internal needle protrudes from the external one for approximately 0.5 mm at its inferior. To arrange the two needles at the right relative heights you can shape their length using a metal file, or put spacers between them before fixing with glue.
14. If the solution presents a large amount of bubbles, it can be centrifuged to remove them. This will help in resuspending the cells more homogeneously later.
15. The volumes for washing depend on the cell culture platform selected. For 60 mm dishes, use 3 mL of PBS; for T-75 flasks, use 15 mL of PBS.
16. Cell culture medium neutralizes the action of trypsin. As a general rule, you should use a volume of medium that is twice the volume of trypsin.
17. When using a microfluidic chip, set the flows of different bioinks so as their sum is around 5 μL/min. Flow rates are generally tuned according to the printing speed and fibers diameter desired.
18. You need to finely balance the flow rates, the deposition speed and the printing code to succeed in the 3D deposition experiment. For more information on this topic, *see* Colosi et al. [6].
19. The UV dose will depend on the light intensity and exposure time. For the bioprinting of HUVEC cells, we used 800 mW/cm^2 light intensity, exposure times that ranged between 15 and 60 s, placing the sample at a distance of 7 cm from the light source.
20. You may need to reevaluate the concentration of the two dyes depending on the performances of you fluorescence microscope.

Acknowledgments

The authors thank University of Rome "La Sapienza" for funding this research and Prof. Ali Khademhosseini of the Biomaterials Innovation Research Center Harvard Medical School, Brigham and Women's Hospital, Cambridge, MA 02139, USA for his support in defining the bioprinting protocol.

References

1. Schiele NR, Corr DT, Huang Y et al (2010) Laser-based direct-write techniques for cell printing. Biofabrication 2:032001. doi:10.1088/1758-5082/2/3/032001
2. Malda J, Visser J, Melchels FP et al (2013) 25th anniversary article: engineering hydrogels for biofabrication. Adv Mater 25:5011–5028. doi:10.1002/adma.201302042
3. Melchels FPW, Domingos MAN, Klein TJ et al (2012) Additive manufacturing of tissues and organs. Prog Polym Sci 37:1079–1104. doi:10.1016/j.progpolymsci.2011.11.007
4. Mandrycky C, Wang Z, Kim K et al (2015) 3D bioprinting for engineering complex tissues. Biotechnol Adv 34(4):422–434. doi:10.1016/j.biotechadv.2015.12.011
5. Colosi C, Shin SR, Manoharan V et al (2015) Microfluidic bioprinting of heterogeneous 3D tissue constructs using low-viscosity bioink. Adv Mater:1–8. doi:10.1002/adma.201503310
6. Colosi C, Costantini M, Latini R et al (2014) Rapid prototyping of chitosan-coated alginate scaffolds through the use of a 3D fiber deposition technique. J Mater Chem B 2:6779–6791. doi:10.1039/C4TB00732H
7. Van Den Bulcke AI, Bogdanov B, De Rooze N et al (2000) Structural and rheological properties of methacrylamide modified gelatin hydrogels. Biomacromolecules 1:31–38. doi:10.1021/bm990017d

Chapter 27

Bioprinting of 3D Tissue Models Using Decellularized Extracellular Matrix Bioink

Falguni Pati and Dong-Woo Cho

Abstract

Bioprinting provides an exciting opportunity to print and pattern all the components that make up a tissue—cells and extracellular matrix (ECM) material—in three dimensions (3D) to generate tissue analogues. A large number of materials have been used for making bioinks; however, majority of them cannot represent the complexity of natural ECM and thus are unable to reconstitute the intrinsic cellular morphologies and functions. We present here a method for making of bioink from decellularized extracellular matrices (dECMs) and a protocol for bioprinting of cell-laden constructs with this novel bioink. The dECM bioink is capable of providing an optimized microenvironment that is conducive to the growth of 3D structured tissue. We have prepared bioinks from different tissues, including adipose, cartilage and heart tissues and achieved high cell viability and functionality of the bioprinted tissue structures using our novel bioink.

Key words Bioprinting, Extracellular matrix, Decellularization, Bioink, Cell-laden structure, 3D structured tissue, Thermally induced gelation

1 Introduction

Bioprinting offers wonderful prospects of printing tissue analogues through delivering living cells with appropriate material in a defined and organized manner [1], which is vital for several emerging technologies, including tissue-engineered scaffolds [2, 3], cell-based sensors [4], drug/toxicity screening [5], and tissue or tumor models [6]. The concept of bioprinting is essentially an extension of the idea that uses additive manufacturing (AM) methods for building three-dimensional (3D) tissue structures via a layer-by-layer process [7–10]. A critical requirement for bioprinting is a cytocompatible bioink, which is used for making structures containing live cells. This restriction reduces the choices of materials because of the requirement to work in an aqueous or aqueous-gel environment [1, 11].

Zuzana Koledova (ed.), *3D Cell Culture: Methods and Protocols*, Methods in Molecular Biology, vol. 1612,
DOI 10.1007/978-1-4939-7021-6_27,

In extrusion based printing, bioinks that are gelled either by thermal processes or UV curing or post-print crosslinking are being used for bioprinting [5, 12–15]. Though a number of materials have been used for making bioinks, none of them actually could represent the complexity of natural extracellular matrices (ECMs) and thus are inadequate to recreate a microenvironment that is typical of living tissues. Thus it would be ideal if cells were provided the natural microenvironment similar to their abode. Decellularized extracellular matrix (dECM) could be the best choice for doing so, as it can recapitulate all the features of natural ECM [16]. Moreover, ECM of each tissue is unique in terms of composition and topology, which is generated through dynamic and reciprocal interactions between the resident cells and microenvironment [17].

In this chapter, we describe the procedure for decellularization of various tissues, including heart, cartilage, and adipose tissue [18, 19], by a combination of physical, chemical, and enzymatic methods. We also describe the protocol for making bioinks using those decellularized tissues and their subsequent use for bioprinting of cell-laden construct by extrusion-based 3D printing technology.

2 Materials

Prepare all solutions using ultrapure water (purified deionized water with the sensitivity of 18 MΩ cm at 25 °C) and analytical grade reagents. Prepare and store all reagents at room temperature (unless indicated otherwise).

1. Phosphate buffered saline (PBS), 1×, sterile.
2. DNase and RNase A.
3. Pepsin (P7125, Sigma-Aldrich).
4. Porcine cartilage (*see* **Note 1**).
5. Porcine heart (*see* **Note 1**).
6. Human adipose tissue, obtained by liposuction (*see* **Note 2**).
7. 1% sodium dodecyl sulfate (SDS) solution: 1% SDS in PBS.
8. 0.5% SDS solution: 0.5% SDS in water.
9. Hypotonic Tris–HCl buffer: 10 mM Tris–HCl, pH 8.0 in water.
10. Trypsin solution: 0.25% trypsin in PBS.
11. Hypertonic buffer: 1.5 M NaCl, 50 mM Tris–HCl, pH 7.6.
12. Nuclease solution: 50 U/mL deoxyribonuclease and 1 U/mL ribonuclease A in 10 mM Tris–HCl, pH 7.5.
13. Peracetic acid (PAA) solution: 0.1% PAA in 4% ethanol.
14. Distilled water, sterile.
15. 0.5 M acetic acid.

16. 10 M NaOH.
17. Stainless steel sieve, pore size ~250 μm.
18. 250 mm Nitex filters (Sefar America Inc.).
19. Digestion solution: 0.3% collagenase type I, 10 mg/mL bovine serum albumin and 100 U/mL penicillin and 100 μg/mL streptomycin in DMEM/F12.
20. Maintenance medium: 10% FBS, 100 U/mL penicillin, and 100 μg/mL streptomycin in DMEM/F12.
21. 0.05% trypsin solution: 0.05% trypsin in 0.53 mM EDTA.
22. 10× DMEM.
23. CAD software, e.g., Solid Edge or Solidworks.
24. Bioprinter, e.g., extrusion-based bioprinter.
25. Polycaprolactone (PCL) granules.

3 Methods

Carry out all procedures at room temperature unless otherwise specified.

3.1 Decellularization of Tissues

3.1.1 Decellularization of Heart

1. Isolate the left ventricle from the whole porcine heart (*see* **Note 1**).
2. Using a scalpel, cut the tissue into pieces of about 1–2 mm^3.
3. Stir the chopped heart tissue in 1% SDS solution for 48 h.
4. Next, treat the tissue with 1% Triton X-100 for 30 min.
5. Wash the decellularized heart tissue using sterile PBS for at least 3 days, while changing the PBS every day.
6. To sterilize the decellularized tissue, treat it with PAA for 4 h (*see* **Note 3**).
7. Wash the tissue several times with sterile PBS and sterile distilled water.
8. Strain the dECMs using a stainless steel sieve, lyophilize, and store at −20 °C.

3.1.2 Decellularization of Cartilage

1. Separate the hyaline cartilage from articular cartilage and mince the hyaline cartilage into small pieces of about 1–2 mm^3 using a sharp scalpel.
2. Place the minced cartilage into hypotonic Tris–HCl buffer in a 50 mL centrifuge tube and conduct 6 cycles of freeze (at −80 °C) and thaw (at 37 °C) (*see* **Note 4**).
3. Homogenize the cartilage slurry by vortexing and treat it with trypsin solution for 24 h at 37 °C with vigorous agitation, while replacing the trypsin solution with a fresh one every 4 h

by straining the cartilage tissue with a stainless steel sieve and resuspending in fresh trypsin solution.

4. Wash the trypsinized cartilage slurry with hypertonic buffer.
5. Treat the cartilage slurry with nuclease solution at 37 °C for 4 h with gentle agitation.
6. To remove all the enzymes, wash the enzyme-treated cartilage slurry with the hypotonic Tris–HCl solution for 20 h.
7. Treat the cartilage slurry with 1% Triton X-100 for 24 h.
8. Finally, to remove all detergent, wash the decellularized cartilage tissue thoroughly for at least 3 days with distilled water, while changing the water every 12 h.
9. To sterilize the decellularized tissue, treat it with PAA for 4 h (*see* **Note 3**).
10. Wash the tissue several times with sterile PBS and sterile distilled water.
11. Strain the dECMs using a stainless steel sieve, lyophilize, and store at −20 °C.

3.1.3 Decellularization of Adipose Tissue

1. Put the liposuction material in a 50 mL centrifuge tube.
2. Centrifuge the adipose tissue at 500 × *g* for 10 min to separate the oil and blood from the tissue.
3. Discard the top layer containing oil and interphase between top and middle layer. Collect the middle layer containing adipose tissue. Also, discard the bottom layer containing blood and serum (Fig. 1). Collect the middle layer containing adipose tissue and wash with PBS for at least three times.

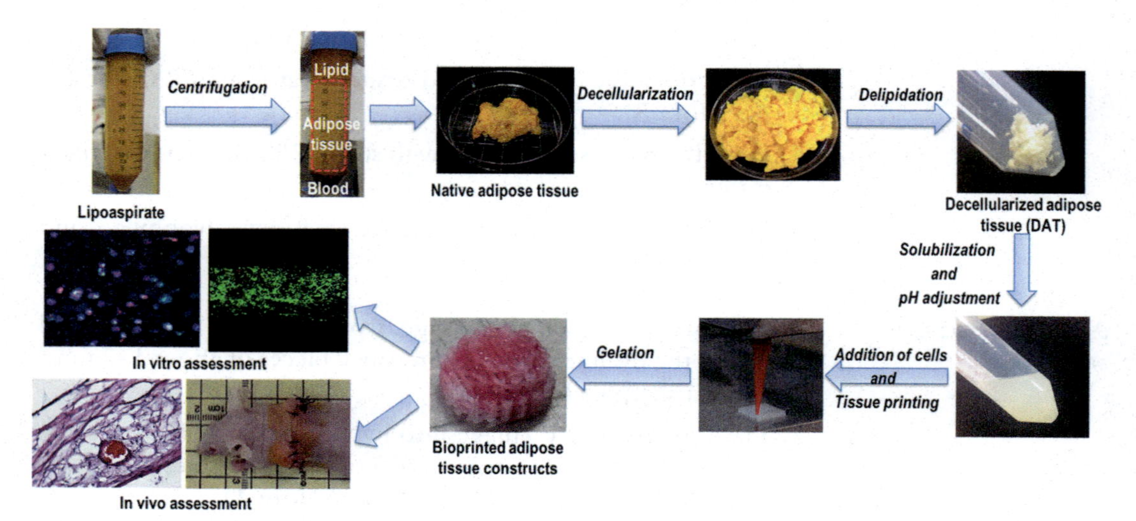

Fig. 1 Schematic elucidating preparation of decellularized adipose tissue (DAT) bioink and printing of dome-shaped adipose tissue constructs and their in vitro and in vivo evaluation. Reproduced from Pati et al., 2015 [19] with permission from Elsevier B.V

4. Decellularize the tissue with 0.5% SDS solution for 48 h, while changing the solution every 12 h.
5. Treat the decellularized adipose tissue with isopropanol to remove lipids for 48 h, change the isopropanol every 12 h (Fig. 1).
6. Wash the tissue several times with PBS.
7. Treat the decellularized and delipidated tissue with PAA (*see* **Note 3**).
8. Wash the tissue several times with sterile PBS and sterile distilled water.
9. Strain the dECM, lyophilize, and store at −20 °C.

3.2 Preparation of dECM Solution and Pre-Gel

1. Pulverize the lyophilized dECM into fine powder using a mortar and pestle with the help of liquid N_2. This process can also be carried out using some mini mills.
2. Prepare 3% dECM solution by digesting about 3 g of pulverized dECM in 100 mL of 0.5 M acetic acid with 30 mg of pepsin for 48 h.
3. After complete solubilization of dECM, centrifuge the solution at 500 × *g* for 10 min to remove the particulate residues if any (*see* **Note 5**).
4. Adjust the pH to ~7 (*see* **Note 6**) by dropwise addition of cold 10 M NaOH solution while maintaining the temperature of the dECM below 10 °C to avoid premature gelation of the dECM (*see* **Note 7**).
5. The pH-adjusted dECM pre-gel can be stored in a refrigerator at 4 °C for about a week.

3.3 Isolation of Human Adipose-Derived Mesenchymal Stem Cells (hASCs)

1. Digest the adipose tissue in digestion solution at 37 °C for 6 h with intermittent shaking (*see* **Note 8**).
2. Filter the resulting suspensions (fat) using Nitex filters to remove debris.
3. Centrifuge at 400 × *g* for 8 min.
4. Resuspend the cell pellet in maintenance medium.
5. Plate the primary cells in tissue culture flasks and incubate in a humidified atmosphere at 37 °C with 5% CO_2 for 48 h.
6. Discard the medium and wash the culture dishes with PBS to remove non-adherent cells.
7. Add fresh maintenance medium and change the medium every other day, maintaining cells at subconfluent levels (passage 0).
8. For subculturing, detach the cultured cells from the tissue culture plate by treatment with 0.05% trypsin solution for 2 min.
9. Check under microscope if about 90–95% of cells are detached. Then stop trypsinization (neutralize trypsin) by adding maintenance medium.

10. Transfer the dislodged cells along with the medium into a new centrifuge tube.
11. Centrifuge the mixture at 500 × *g* for 3 min.
12. Discard the supernatant and resuspend the cell pellet in fresh maintenance medium.
13. Count the cells using a counting chamber or a cell counter.
14. Seed at a density of 5×10^3 cells/cm^2, and culture in maintenance medium in a humidified atmosphere at 37 °C with 5% CO_2.
15. Use cells between passages 4 and 5 for making bioink and printing.

3.4 Cell Encapsulation in the dECM Pre-Gel

1. Conduct all the processes on ice to prevent gelation of the bioink.
2. Prepare about 10 mL of pH-adjusted dECM pre-gel for bioprinting, which would be sufficient for printing ~5 3D constructs of $10 \times 10 \times 5$ mm^3 size. Add 1 mL of 10× DMEM to 9 mL of pH-adjusted dECM pre-gel (i.e., 1/10 of final volume) (*see* **Note 9**).
3. Prepare cell suspension at the density 10^6–10^7 cells/mL (*see* **Note 10**) and mix it thoroughly with the pH-adjusted ECM pre-gel. Mix about 1/20th volume of cell suspension to that of pH-adjusted ECM pre-gel (*see* **Note 11**).
4. The resulting mixture of cells and dECM is called bioink. Use the bioink immediately for bioprinting.

3.5 Printing of Cell-Laden Construct

1. Draw 3D structure with lattice geometry using a CAD software.
2. Save the CAD file in STL. format and generate g-code using any open source software.
3. Load the prepared bioink into a sterilized syringe and fasten it to the bioprinter. The temperature of the printing head has to be maintained below 15 °C to avoid gelation of the bioink within the syringe (*see* **Note 12**).
4. Load PCL granules to the other syringe and heat it to 80 °C to melt the PCL. Wait until the PCL is molten completely (~ 1–2 h) (*see* **Note 13**).
5. Set a culture dish into which the structure will be printed.
6. Open the software interface of the 3D bioprinter and load the g-code of the 3D structure. Use the command print to direct the 3D printer to print the specified structure.
7. The printer will deposit the molten PCL in the first layer according to the CAD drawing. For deposition of molten PCL, a pneumatic pressure of 650 kPa is required [20].

8. The bioink should be then infused only once by pneumatic pressure into the PCL canal. The cell-laden bioink could also be dispensed using a plunger-based low-dosage dispensing system. Print the structure directly in a six-well plate or petri dish (*see* **Notes 14** and **15**).
9. Repeat the same process to produce a 3D hybrid scaffold filled with the bioink solution (Fig. 2).

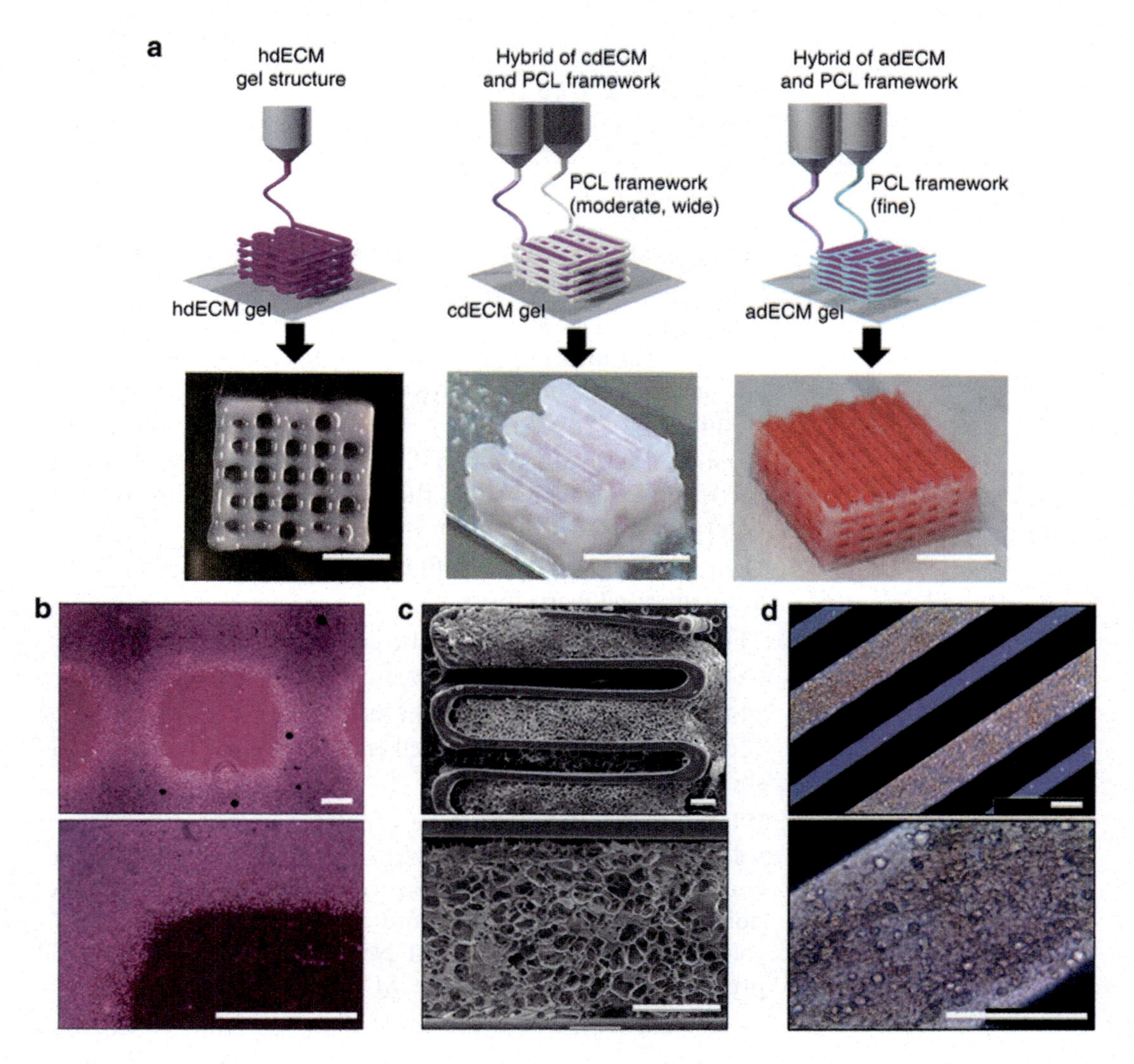

Fig. 2 Printing process of particular tissue constructs with dECM bioink. (**a**) Heart tissue construct was printed with only heart dECM (hdECM). Cartilage and adipose tissues were printed with cartilage dECM (cdECM) and adipose dECM (adECM), respectively, and in combination with PCL framework (*scale bar*, 5 mm). (**b**) Representative microscopic images of hdECM construct (*scale bar*, 400 μm). (**c**) Scanning electron microscopy images of hybrid structure of cdECM with PCL framework (*scale bar*, 400 μm). (**d**) Images of cell-printed structure of adECM with PCL framework (*scale bar*, 400 μm). Reproduced from Pati et al., 2014 [18] with permission from Macmillan Publishers Limited (Springer Nature)

10. Maintain the temperature of the printing stage at 37 °C while printing (*see* **Note 15**).
11. Soon after printing, add warm cell culture medium (maintenance cell culture medium or other, dependent on the experiment) to the printed structure from the side without disturbing the structure and place in an incubator (37 °C, 5% CO_2).
12. Replace the medium after 2–4 h with fresh warm medium and culture the printed construct in the cell culture incubator (*see* **Note 16**) with changing the medium every alternative day.
13. Check the viability of the cells within the bioprinted cell-laden construct (*see* **Note 17**).

4 Notes

1. Porcine cartilage and heart should be collected from the nearby slaughter house.
2. The adipose tissue should be collected from hospital after liposuction with informed consent from the persons undergoing liposuction and under approval from the institutional ethical committee.
3. Treatment with PAA is very important to sterilize the tissue. After this treatment handle the tissue carefully with sterilized utensils and tools.
4. Gently massage the tissue after every cycle with your finger to facilitate decellularization.
5. This step is optional and can be omitted if the solution seems to be homogeneous and no particulate material is seen.
6. The pH of dECM solution is in the range of 2.8–3 at this stage and thus cannot be used for cell encapsulation. Hence, the pH should be adjusted to ~7.
7. While adjusting the pH, take care to avoid warming up the solution above ~10 °C. Use cold 10 M NaOH solution (refrigerated overnight or cooled on ice for 1 h before the procedure). Perform the procedure while keeping the dECM solution (as well as the 10 M NaOH stock) on ice to avoid premature gelation of the dECM.
8. Please follow the literature for detail about the cell isolation procedure [21].
9. Addition of 10× DMEM is necessary to prepare the final bioink with 1× DMEM.
10. The concentration of cells used can be chosen from 10^6–10^7 cells/mL, depending upon the application.
11. In our study, we used two syringes for constructing each structure, one containing PCL and maintained at 80 °C, the other

one loaded with cell-laden dECM bioink and maintained at temperatures below 15 °C. The applied pneumatic pressure was in the range of 400–650 kPa and 10–20 kPa for dispensing PCL and bioink, respectively. In the process of printing synthetic polymer, temperature of the molten PCL comes down very fast in the air gap before reaching the stage and hence the process is not deleterious to the cells.

12. Addition of more than 1/20th volume of cell suspension to the pH adjusted dECM pre-gel can cause reduction in viscosity that is not suitable for printing.
13. Make provision in the CAD modeling for the printer to build three layers of PCL to reach the required height and also to make canals for subsequent bioink infusion [20]. Hence, each layer of PCL actually consists of three PCL fine layers. If the CAD model is perfectly prepared, the 3D printer will follow the instructions to print the required structure.
14. Make provision in the CAD modeling for positioning of the bioink syringe to the middle of the pore. The nozzle end of the bioink syringe should locate at the middle of a pore placed between PCL lines to infuse the bioink solution into the scaffold [20].
15. Thermally induced gelation mechanism is used for solidification of the dECM structure. Hence, the temperature of the stage should be kept at 37 °C.
16. Culture time of the cells can be varied depending upon application. Change the medium every 2 days with fresh warm medium.
17. Live/Dead assay should be conducted after printing to check the cell viability. Use Live/Dead assay kit for this assay and follow the manufacturer's protocol.

Acknowledgments

This work was partially supported by the Early Career Research (ECR) grant awarded by Science and Engineering Research Board, Department of Science and Technology, Government of India (ECR/2015/000458) and National Research Foundation (NRF) of Korea grant funded by the Korean government (MSIP) (No. 2010-0018294).

References

1. Derby B (2012) Printing and prototyping of tissues and scaffolds. Science 338:921–926
2. Griffith LG, Naughton G (2002) Tissue engineering—current challenges and expanding opportunities. Science 295: 1009–1014
3. Gaetani R, Doevendans PA, Metz CHG et al (2012) Cardiac tissue engineering using tissue printing technology and human cardiac progenitor cells. Biomaterials 33:1782–1790
4. Falconnet D, Csucs G, Michelle Grandin H et al (2006) Surface engineering approaches to

micropattern surfaces for cell-based assays. Biomaterials 27:3044–3063
5. Chang R, Nam J, Sun W (2008) Direct cell writing of 3D microorgan for in vitro pharmacokinetic model. Tissue Eng Part C Methods 14:157–166
6. Fischbach C, Chen R, Matsumoto T et al (2007) Engineering tumors with 3D scaffolds. Nat Methods 4:855–860
7. Derby B (2008) Bioprinting: inkjet printing proteins and hybrid cell-containing materials and structures. J Mater Chem 18:5717–5721
8. Kundu J, Shim J-H, Jang J et al (2013) An additive manufacturing-based PCL–alginate–chondrocyte bioprinted scaffold for cartilage tissue engineering. J Tissue Eng Regen Med 9:1286–1297
9. Jung JW, Kang H-W, Kang T-Y et al (2012) Projection image-generation algorithm for fabrication of a complex structure using projection-based Microstereolithography. Int J Precis Eng Manuf 13:445–449
10. Seol Y-J, Kang T-Y, Cho D-W (2012) Solid freeform fabrication technology applied to tissue engineering with various biomaterials. Soft Matter 8:1730–1735
11. Ferris C, Gilmore K, Wallace G et al (2013) Biofabrication: an overview of the approaches used for printing of living cells. Appl Microbiol Biotechnol 97:4243–4258
12. Billiet T, Gevaert E, De Schryver T et al (2014) The 3D printing of gelatin methacrylamide cell-laden tissue-engineered constructs with high cell viability. Biomaterials 35:49–62
13. Wang X, Yan Y, Pan Y et al (2006) Generation of three-dimensional hepatocyte/gelatin structures with rapid prototyping system. Tissue Eng 12:83–90
14. Yan Y, Wang X, Xiong Z et al (2005) Direct construction of a three-dimensional structure with cells and hydrogel. J Bioact Compat Polym 20:259–269
15. Pati F, Gantelius J, Svahn HA (2016) 3D bioprinting of tissue/organ models. Angew Chem Int Ed Engl 55:4650–4665
16. Sellaro TL, Ranade A, Faulk DM et al (2010) Maintenance of human hepatocyte function in vitro by liver-derived extracellular matrix gels. Tissue Eng Part A 16:1075–1082
17. Frantz C, Stewart KM, Weaver VM (2010) The extracellular matrix at a glance. J Cell Sci 15:4195–4200
18. Pati F, Jang J, Ha D-H et al (2014) Printing three dimensional tissue analogues with decellularized extracellular matrix bioink. Nat Commun 5:3935
19. Pati F, Ha D-H, Jang J et al (2015) Biomimetic 3D tissue printing for soft tissue regeneration. Biomaterials 62:164–175
20. Jin-Hyung S, Jong Young K, Min P et al (2011) Development of a hybrid scaffold with synthetic biomaterials and hydrogel using solid freeform fabrication technology. Biofabrication 3:034102
21. Park I-S, Han M, Rhie J-W et al (2009) The correlation between human adipose-derived stem cells differentiation and cell adhesion mechanism. Biomaterials 30:6835–6843

Chapter 28

Bioprinting Cartilage Tissue from Mesenchymal Stem Cells and PEG Hydrogel

Guifang Gao, Karen Hubbell, Arndt F. Schilling, Guohao Dai, and Xiaofeng Cui

Abstract

Bioprinting based on thermal inkjet printing is one of the most attractive enabling technologies for tissue engineering and regeneration. During the printing process, cells, scaffolds, and growth factors are rapidly deposited to the desired two-dimensional (2D) and three-dimensional (3D) locations. Ideally, the bio-printed tissues are able to mimic the native anatomic structures in order to restore the biological functions. In this study, a bioprinting platform for 3D cartilage tissue engineering was developed using a commercially available thermal inkjet printer with simultaneous photopolymerization. The engineered cartilage demonstrated native zonal organization, ideal extracellular matrix (ECM) composition, and proper mechanical properties. Compared to the conventional tissue fabrication approach, which requires extended UV exposure, the viability of the printed cells with simultaneous photopolymerization was significantly higher. Printed neocartilage demonstrated excellent glycosaminoglycan (GAG) and collagen type II production, which was consistent with gene expression profile. Therefore, this platform is ideal for anatomic tissue engineering with accurate cell distribution and arrangement.

Key words Cartilage, Inkjet printing, Human mesenchymal stem cells, Hydrogel, Photopolymerization, Tissue engineering

1 Introduction

Bioprinting based on thermal inkjet printing is one of the most promising enabling technologies in the field of tissue engineering and regenerative medicine. With digital control and high throughput printing capacity, cells, scaffolds, and growth factors can be precisely deposited to the desired two-dimensional (2D) and three-dimensional (3D) positions rapidly. Combining with simultaneous photopolymerization and photosensitive biomaterials, this technology is able to fix the cells and other printed substances to the initially deposited locations. Numerous successful applications have been achieved using this technology in tissue engineering and regenerative medicine [1–13]. In this study, a bioprinting platform

Zuzana Koledova (ed.), *3D Cell Culture: Methods and Protocols*, Methods in Molecular Biology, vol. 1612,
DOI 10.1007/978-1-4939-7021-6_28,

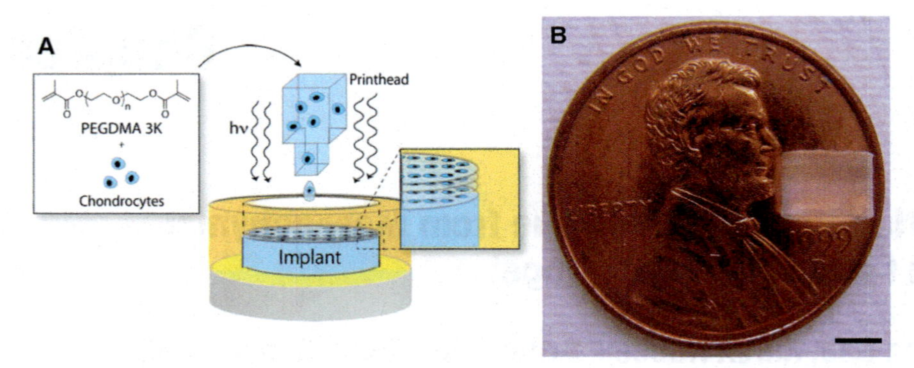

Fig. 1 Printed neocartilage tissue. (**a**) Schematic of cartilage bioprinting with simultaneous photopolymerization and layer-by-layer assembly. (**b**) A printed neocartilage tissue with 4 mm in diameter and 4 mm in height. *Scale bar*, 2 mm. Reproduced from Cui et al., 2015 [5] with permission from John Wiley & Sons, Inc

was established with a modified Hewlett-Packard (HP) Deskjet 500 thermal inkjet printer and a simultaneous photopolymerization system. Synthetic hydrogels formulated from poly(ethylene glycol) (PEG) have shown the capacity of maintaining cell viability and promoting chondrogenic ECM production [14, 15]. In addition, photo-cross-linkable PEG is highly soluble in water with low viscosity, which is ideal for simultaneous polymerization during 3D bioprinting.

In this chapter, we describe the procedure for precise printing of human mesenchymal stem cells (hMSCs) suspended in PEG diacrylate (PEGDA; MW 3400) to construct neocartilage, layer-by-layer with 1400 dpi in 3D resolution (Fig. 1). Using this protocol, a construct with homogeneous distribution of deposited cells in a 3D scaffold is achieved (Fig. 2a), that represents cartilage tissue with excellent mechanical properties and enhanced ECM production. By contrast, in manual fabrication the cells would accumulate at the bottom of the gel instead of their initially deposited positions due to gravity and slower polymerization process, and this would lead to inhomogeneous cartilage formation after culture (Fig. 2b) [4, 5]. This cell accumulation was also observed in previous reports of manual fabrication of cartilage tissue [16, 17]. The printed cells in 3D PEG hydrogel demonstrated chondrogenic phenotype with gradually increased proteoglycan production during culture (Fig. 3) [5].

2 Materials

Prepare all solutions using deionized (DI) water and analytical grade reagents. Prepare and store all reagents at room temperature (unless otherwise indicated). Diligently follow all waste disposal regulations when disposing waste materials.

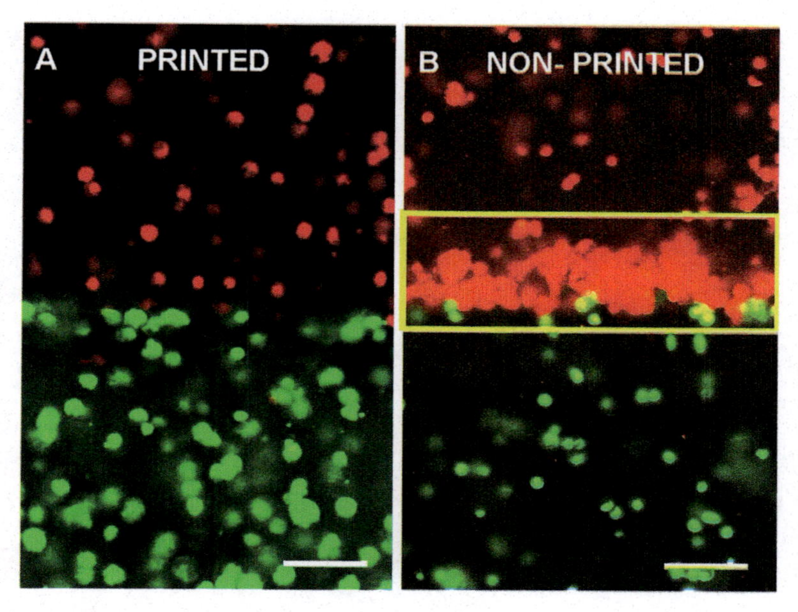

Fig. 2 Cells labeled with *green* and *orange* fluorescent dyes demonstrate the zonal cartilage bioprinting feasibility. (**a**) Printed cells maintained their initially deposited positions in the 3D hydrogel. The printing and photopolymerization process was completed in 4 min with cell viability of 90% ($n = 3$). (**b**) Cells accumulated to the bottom or interface due to gravity without simultaneous photopolymerization. It took 10 min of UV exposure to gel the construct with the same size as in A, resulting cell viability was 63% ($n = 3$). *Scale bars*, 100 μm. Reproduced from Cui et al., 2015 [4] with permission from Mary Ann Liebert, Inc. publishers

1. HP Deskjet 500 thermal inkjet printer.
2. HP 51626A black ink cartridge.
3. UVP B-100AP ultraviolet (UV) lamp.
4. UV light meter.
5. Sterile DI water.
6. Adobe Photoshop software.
7. Bone-marrow derived hMSCs.
8. hMSC culture medium: 10% fetal bovine serum (FBS), 100 U/mL penicillin, 100 μg/mL streptomycin, 4 mM L-glutamine in Dulbecco's Modified Eagle's Medium (DMEM). Store at 4 °C up to 1 month.
9. Sterile phosphate buffer saline (PBS).
10. 0.05% trypsin–EDTA.
11. 10% FBS solution: 10% FBS in PBS.
12. PEGDA solution: 10% (w/v) PEGDA and 0.05% I-2959 in PBS, filter sterilized. Prepared freshly every time before use.
13. hMSC chondrogenic medium: 1× insulin–transferrin–selenium with sodium pyruvate (ITS-A), 0.1 mM ascorbic acid,

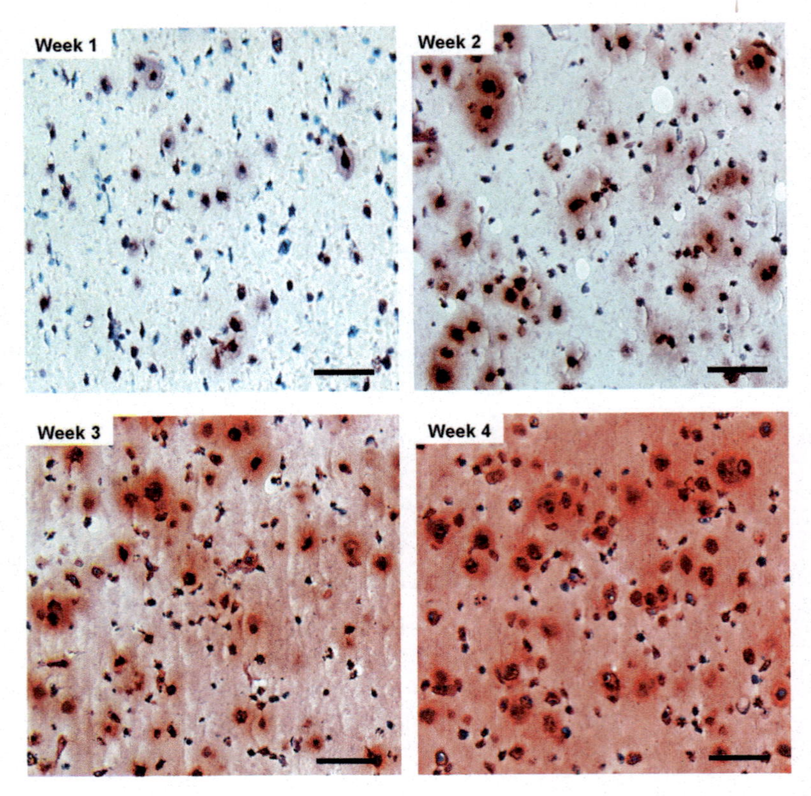

Fig. 3 Safranin-O staining of printed cells in PEG hydrogel shows increased proteoglycan production during the tissue construct culture. *Scale bars*, 100 μm

1.25 mg/mL human serum albumin, 0.1 μM dexamethasone, 10 ng/mL TGF-β3 in L-glutamine-supplemented DMEM. TGF-β3 needs to be added freshly every time before feeding the cells and the complete chondrogenic medium should be used within 12 h.

14. LIVE/DEAD Viability/Cytotoxicity kit.
15. Confocal microscope, e.g., Zeiss LSM 510 laser scanning confocal microscope.

3 Methods

3.1 Bioprinting Platform Establishment

The printer modification was based on a HP Deskjet 500 thermal inkjet printer and HP 51626A black ink cartridge.

1. Remove the top plastic cover of the printer and carefully detach the control panel from the cover.
2. Detach the three cable connections between the printer top portion and base. Remove the printer top portion from the base.

3. On the printer top portion, remove the small plastic and rubber accessories (printhead cleaning system) at the right hand side under the ink cartridge.
4. Remove the base of the paper tray with springs.
5. Remove the metallic plate coving the plastic paper feeding bar.
6. Cut off the plastic paper feeding bar at the middle feeding wheel position using a hand saw or other cutting tool.
7. Remove the two paper feeding wheels exposed after the previous step (*see* **Note 1**).
8. Clean the dust and debris using canned air and ethanol wipes.
9. Attach the printer top portion to the base.
10. UV-sterilize the modified printer for at least 2 h in a laminar flow hood before use (*see* **Note 2**).
11. Cut off the cap of a HP 51626a ink cartridge using a hand saw or other cutting tool (*see* **Note 3**).
12. Discard the ink and remove the steel filter that covers the bottom well reservoir of the cartridge (*see* **Note 4**).
13. Rinse the cartridge thoroughly using running tap water.
14. Clean the cartridge in ultrasonic bath in DI water for 10 min to remove the residual ink.
15. Examine the cartridge to make sure that all the ink has been removed. To disinfect the cartridge, rinse or spray it thoroughly with 70% ethanol, followed by sterilized DI water.
16. Set up a long-wavelength UV lamp over the printing platform to provide simultaneous photopolymerization capacity.
17. Measure UV intensity at the printing platform using a UV light meter. Adjust the distance between the UV lamp and the printer platform so that the intensity at the printing subject is between 4 and 8 mW/cm^2 (*see* **Note 5**).

3.2 hMSC Culture, Expansion and Bioink Preparation

1. hMSCs are maintained in hMSC culture medium in cell culture incubator at 37 °C with humidified air containing 5% CO_2, with cell culture medium change every 3 days and subculture as needed.
2. To expand hMSCs, wash the cell culture dish with monolayer hMSCs twice with PBS and treat with 0.05% Trypsin/EDTA for 5 min.
3. After the cells have detached, neutralize the trypsin using 10% FBS solution.
4. Pipette the cell suspension repeatedly to create a single-celled suspension, then transfer it to a centrifuge tube and centrifuge at 300 × *g* for 5 min.
5. Count the cells using a hemocytometer or similar device.

6. Plate 5 × 10^6 hMSCs per a T175 tissue culture flask in hMSC culture medium.
7. Incubate the cells at 37 °C, 5% CO_2 until they reach 85% confluence. Change the culture medium every 3 days as necessary.
8. For bioink preparation, use the cells from the same passage at 85% confluence.
9. Trypsinize and count the cells as described above (**steps 2–5**).
10. Prepare the bioink by resuspending hMSC in PEGDA solution at 5 × 10^6 cells/mL (*see* **Note 6**).

3.3 Cartilage Tissue Printing

1. Turn on the printer and the printer-controlling laptop/computer (*see* **Note 7**).
2. Create a printing pattern of a solid circle with 4 mm in diameter using Microsoft Word or Adobe Photoshop.
3. Adjust the position of the pattern and make sure it will print exactly to the desired position (*see* **Note 8**).
4. Calculate the number of prints needed to reach the desired thickness of scaffold. For 4 mm in height, 220 prints are required to create the desired scaffold.
5. Load the bioink into the ink cartridge (*see* **Note 9**). Cover the cartridge with the sterilized aluminum foil to protect it from the direct UV exposure during printing.
6. Send printing command to the printer. Pull the paper sensor when the printer starts to print (*see* **Notes 10** and **11**). The whole printing process should take less than 4 min for a scaffold with 4 mm in diameter and 4 mm in height.
7. Transfer the printed neocartilage to a 24-well plate and add 1.5 mL hMSC chondrogenic differentiation medium to each well.

3.4 Cell Viability Evaluation in 3D Scaffold

1. Incubate the neocartilage in LIVE/DEAD Viability/Cytotoxicity working solution at room temperature for 15 min in dark.
2. Cut the cell-laden hydrogel in half and take fluorescent images of the cutting area.
3. Count live (green) and dead (red) cells by a blinded observer in five randomly taken images.
4. Calculate cell viability by dividing the number of live cells by the total number of cells counted (live plus dead cells).

4 Notes

1. When modifying the printer, electronic saw will be handy to remove the hard paper-feeding wheels.
2. In addition to UV exposure, gamma irradiation or ethylene oxide can also be used to sterilize the printer platform and the printing cartridges.

3. When cutting off the plastic cap of the ink cartridge, pay careful attention to not destroy the electric connectors at the back of the cartridge.
4. The steel cover inside, at the bottom of the ink cartridge must be removed for further cleaning and cell seeding.
5. When the distance between the printing platform and the UV light is set at approximately 25 cm, the ideal UV intensity can be reached.
6. Before printing, calculate the volume of bioink to prepare according to the number and volume of the constructs to print. Use prepared PEGDA solution to resuspend the hMSCs to avoid the extra steps or bringing the inaccuracy of the final cell concentration. Before adding the PEGDA solution, remove as much medium as possible from the centrifuged tube.
7. All the printing process should be performed in a laminar flow hood to avoid contamination.
8. During the printing process, it will be helpful to print the initial pattern using an unmodified ink cartridge to locate the position for bioink deposition. When switching back to the bioink, be careful to keep the printer platform steady.
9. Add 100–200 μL bioink into the cartridge each time. This amount should be sufficient to print multiple constructs. Do not add too much bioink into the cartridge each time, otherwise the cells will accumulate to the bottom and block the printhead.
10. A flexible string can be attached to the paper sensor so it is easier to operate during the printing process.
11. During the bioprinting process, the printer will perform the printhead cleaning process routinely. During this process, paper trigger need to be pulled until the cleaning process is completed.

Acknowledgments

This work was supported by the Fundamental Research Funds for the Central Universities (WUT: 2015IB004, 2017IB004).

References

1. Gao G, Cui X (2016) Three-dimensional bioprinting in tissue engineering and regenerative medicine. Biotechnol Lett 38:203–211
2. Gao G, Schilling AF, Hubbell K et al (2015) Improved properties of bone and cartilage tissue from 3D inkjet-bioprinted human mesenchymal stem cells by simultaneous deposition and photocrosslinking in PEG-GelMA. Biotechnol Lett 37:2349–2355
3. Cui X, Boland T (2009) Human microvasculature fabrication using thermal inkjet printing technology. Biomaterials 30:6221–6227

4. Cui X, Breitenkamp K, Finn MG et al (2012) Direct human cartilage repair using three-dimensional bioprinting technology. Tissue Eng Part A 18:1304–1312
5. Cui X, Breitenkamp K, Lotz M et al (2012) Synergistic action of fibroblast growth factor-2 and transforming growth factor-beta1 enhances bioprinted human neocartilage formation. Biotechnol Bioeng 109:2357–2368
6. Cui X, Breitenkamp K, Finn MG et al (2011) Direct human cartilage repair using thermal inkjet printing technology. Osteoarthr Cartil 19:S47–S48
7. Cui X, Dean D, Ruggeri ZM et al (2010) Cell damage evaluation of thermal inkjet printed Chinese hamster ovary cells. Biotechnol Bioeng 106:963–969
8. Cui X, Hasegawa A, Lotz M et al (2012) Structured three-dimensional co-culture of mesenchymal stem cells with meniscus cells promotes meniscal phenotype without hypertrophy. Biotechnol Bioeng 109:2369–2380
9. Cui X, Gao G, Qiu Y (2013) Accelerated myotube formation using bioprinting technology for biosensor applications. Biotechnol Lett 35:315–321
10. Cui X, Boland T, D'Lima D et al (2012) Thermal inkjet printing in tissue engineering and regenerative medicine. Recent Pat Drug Deliv Formul 6:149–155
11. Gao G, Yonezawa T, Hubbell K et al (2015) Inkjet-bioprinted acrylated peptides and PEG hydrogel with human mesenchymal stem cells promote robust bone and cartilage formation with minimal printhead clogging. Biotechnol J 10:1568–1577
12. Gao G, Schilling AF, Yonezawa T et al (2014) Bioactive nanoparticles stimulate bone tissue formation in bioprinted three-dimensional scaffold and human mesenchymal stem cells. Biotechnol J 9:1304–1311
13. Cui X, Gao G, Yonezawa T et al (2014) Human cartilage tissue fabrication using three-dimensional inkjet printing technology. J Vis Exp 88:51294
14. Bryant SJ, Anseth KS (2002) Hydrogel properties influence ECM production by chondrocytes photoencapsulated in poly(ethylene glycol) hydrogels. J Biomed Mater Res 59:63–72
15. Elisseeff J, McIntosh W, Anseth K et al (2000) Photoencapsulation of chondrocytes in poly (ethylene oxide)-based semi-interpenetrating networks. J Biomed Mater Res 51:164–171
16. Kim TK, Sharma B, Williams CG et al (2003) Experimental model for cartilage tissue engineering to regenerate the zonal organization of articular cartilage. Osteoarthr Cartil 11:653–664
17. Sharma B, Williams CG, Kim TK et al (2007) Designing zonal organization into tissue-engineered cartilage. Tissue Eng 13: 405–414

Part VI

Imaging and Image Analysis of 3D Cell Cultures

Chapter 29

Real-Time Cell Cycle Imaging in a 3D Cell Culture Model of Melanoma

Loredana Spoerri*, Kimberley A. Beaumont*, Andrea Anfosso, and Nikolas K. Haass

Abstract

Aberrant cell cycle progression is a hallmark of solid tumors; therefore, cell cycle analysis is an invaluable technique to study cancer cell biology. However, cell cycle progression has been most commonly assessed by methods that are limited to temporal snapshots or that lack spatial information. Here, we describe a technique that allows spatiotemporal real-time tracking of cell cycle progression of individual cells in a multicellular context. The power of this system lies in the use of 3D melanoma spheroids generated from melanoma cells engineered with the fluorescent ubiquitination-based cell cycle indicator (FUCCI). This technique allows us to gain further and more detailed insight into several relevant aspects of solid cancer cell biology, such as tumor growth, proliferation, invasion, and drug sensitivity.

Key words Fluorescent ubiquitination-based cell cycle indicator (FUCCI), Real-time imaging, 3D spheroid, Tumor heterogeneity, Tumor microenvironment, Cancer drug resistance, Migration, Invasion

1 Introduction

To understand cancer cell biology, especially the process of metastasis, it is necessary to assess the proliferative and invasive behavior in the context of the location within the tumor and its complex microenvironment [1, 2]. In comparison to 2D cell culture models, in vitro 3D tumor models mirror in vivo tumor biology and drug response more faithfully [3–5]. They are therefore an important tool in the field of cancer research and a good compromise between the lack of a microenvironment encountered under 2D culture conditions and the great complexity of in vivo animal models.

Within the range of available in vitro 3D tumor models, spheroids present the advantage of being the relatively quickest and technically simplest method to recapitulate multiple characteristics

*These authors contributed equally to this work.

Zuzana Koledova (ed.), *3D Cell Culture: Methods and Protocols*, Methods in Molecular Biology, vol. 1612,
DOI 10.1007/978-1-4939-7021-6_29, © Springer Science+Business Media LLC 2017

of solid tumors at once: Spheroids mimic physiological tumor behavior in terms of growth, proliferation, invasion, cell-cell and cell-matrix interactions, molecule diffusion, oxygen/nutrient gradients—with a hypoxic zone and a central necrosis—as well as drug sensitivity [3, 5]. Spheroids can be generated from cell lines but also from primary patient-derived cells [6].

Since uncontrolled proliferation is a hallmark of malignancies [7, 8], understanding cell cycle behavior in detail is crucial in cancer research. Quantitation of DNA content by flow cytometry of cells stained with fluorescent markers and DNA pulse incorporation and detection of nucleotide analogues are probably the most commonly used techniques for cell cycle analysis. Even though the latter approach provides some chronological insight into the studied process, both methods provide information limited to snapshots taken at specific time points. However, cell cycle analysis has been recently revolutionized by a new method: fluorescent ubiquitination-based cell-cycle indicator (FUCCI) [9]. This genetically encoded system allows spatial and temporal real-time tracking of cell cycle progression of individual cells in a multicellular context. The FUCCI technology is based on the overexpression of two modified cell cycle-dependent proteins, Geminin and Cdt1, each respectively fused to the green and red fluorescence emitting proteins Azami Green [mAG-hGem(1–110)] and monomeric Kusabira Orange 2 [mKO2-hCdt1(30–120)]. Synthesis and degradation of Cdt1 and Geminin during cell cycle progression results in the nucleus of FUCCI-expressing cells to appear red in G_1 phase, yellow in early S phase, and green in late S, G_2, and M phases. Immediately following cytokinesis and for a brief period of time at the very beginning of G_1 phase, the cell nucleus does not display any fluorescence.

We have incorporated this real-time cell cycle tracking system into a melanoma 3D in vitro model by generating spheroids initiated from FUCCI melanoma cell lines [10]. We have used this system to study dynamic heterogeneity in cell cycle behavior and invasion [10], drug sensitivity [11] and acquired multidrug tolerance in melanoma [12]. We have demonstrated the presence and specific distribution of sub-compartments of cells with different cell cycle behavior. More specifically, cycling cells were found in the spheroid periphery, while G_1-arrested cells located in more internal zones. This specific distribution correlated with oxygen and nutrient accessibility. These characteristics reflected those observed in vivo in mouse xenograft tumors generated from corresponding FUCCI melanoma cell lines, confirming the physiological relevance of this in vitro cell-based model. Furthermore, we showed that invading cells are actively cycling and that cells arrested in G_1, due to their specific location within the spheroid or to drug treatment, were able to resume proliferation and invasion when re-exposed to the same favorable conditions experienced by the cycling cells [10].

The method described in this chapter allows us to study the spatiotemporal cell cycle dynamics of individual cells within the 3D structure of spheroids in real time, and therefore to gain insight into cancer relevant processes such as proliferation and invasion [13]. Briefly, cultured melanoma cells are transduced with the FUCCI cell cycle indicator system. The red and green double-positive cells are isolated by fluorescence-activated cell sorting to obtain optimal and comparable fluorescence intensity of both the green and red fluorescence. Spheroids are then generated using these cells based on a non-adherent surface method [5, 14], embedded in a collagen matrix and their individual cell behavior monitored by confocal fluorescent time-lapse microscopy. 3D reconstruction of FUCCI cell distribution in spheroids can be performed using multiphoton microscopy and 3D stitching. The physical separation of the tumor sub-compartments by Hoechst dye diffusion and subsequent fluorescence-activated cell sorting has been described recently [10, 14] and is beyond the scope of this methods chapter.

2 Materials

2.1 Generation of FUCCI Expressing Melanoma Cell Lines

1. CO_2 incubator.
2. 6-well and 96-well plates.
3. HEK293T cells and the melanoma cell line(s) of choice to be transduced.
4. Distilled H_2O (dH_2O) and Milli-Q H_2O (MQ H_2O).
5. Trypsinization solution: 0.05% Trypsin, 0.5 mM EDTA in PBS without Ca^{2+} and Mg^{2+}.
6. HEK293T medium: 10% fetal bovine serum (FBS) in Dulbecco's Modified Eagle Medium.
7. Melanoma cell medium ("Tu4% medium"): 80% MCDB-153 medium, 20% L-15 medium (Leibovitz), 4% FBS, 5 μg/mL insulin, 1.68 mM $CaCl_2$. Dissolve the whole content of an MCDB-153 vial (17.6 g of powder) in 800 mL deionized H_2O, add 15.7 mL of 7.5% sodium bicarbonate, adjust pH to 7.2–7.4 using NaOH pellets or concentrated NaOH (*see* **Note 1**), and top up with deionized H_2O to 1 L. Filter sterilize using a 0.2 μm filter, remove 200 mL, add 200 mL of L-15 medium (Leibovitz), 40 mL of FBS, 500 μL of 10 mg/mL insulin, and 1120 μL of 1.5 M $CaCl_2$.
8. FUCCI constructs: mKO2-hCdt1 (30–120) and mAG-hGem (1–110) [9], sub-cloned into a replication-defective, self-inactivating lentiviral expression vector system [15].
9. Lentiviral vectors: pMDLg/pRRE, pRSV-Rev and pCMV-VSV-G [15].
10. 2× HeBS (HEPES buffered saline): 50 mM HEPES, 10 mM KCl, 12 mM Dextrose, 280 mM NaCl, 1.5 mM Na_2HPO_4. Adjust to pH 7.1. Filter with a 0.22 μm filter, aliquot and store at −20 °C.

11. 2.5 M $CaCl_2$: To prepare 40 mL, dissolve 11.025 g of $CaCl_2$ in 30 mL of dH_2O. Filter with a 0.22 μm filter, aliquot and store at −20 °C.
12. Sterile phosphate buffered solution (PBS): 137 mM NaCl, 2.7 mM KCl, 10 mM Na_2HPO_4, 1.8 mM KH_2PO_4 in H_2O.
13. Centrifuge.
14. 0.22 μm and 0.45 μm filters.
15. −80 °C and −20 °C freezer.
16. Polybrene, 100 mg/mL in sterile H_2O. Filter and store at −20 °C.
17. Inverted fluorescence microscope.
18. Hemocytometer or automated cell counter.
19. Sorting medium 10% FBS in PBS.
20. FACSAria cell sorter.
21. Inverted phase-contrast microscope.

2.2 Spheroid Formation and Embedding

1. Well-coating solution: 1.5% agarose in sterile PBS. Mix 0.45 g of tissue culture agarose in 30 mL of sterile PBS (*see* **Note 2**) and microwave until the agarose has completely dissolved (*see* **Note 3**). Prepare freshly before plate coating.
2. Trypsin neutralizing medium: 10% FBS in L-15 medium (Leibovitz).
3. Melanoma cell medium.
4. 200 mM L-Glutamine.
5. 10× Eagle's Minimum Essential Medium (EMEM).
6. Bovine type I collagen (R&D Systems), 5 mg/mL.
7. FBS.
8. 7.5% (w/v) $NaHCO_3$.
9. Sterile dH_2O.
10. 96-well plate.
11. Hemocytometer or automated cell counter.
12. CO_2 incubator.
13. Inverted phase-contrast microscope.

2.3 FUCCI-Spheroid Imaging

1. Time-lapse confocal microscope, e.g., Leica SPF confocal inverted microscope equipped with eight laser lines and five filter-free PMT detectors (488 and 561 nm lasers to excite mAG and mKO2), low magnification and long working distance objectives (10× PL FLUOTAR objective (NA 0.3) or a 20× PL FLUOTAR objective (NA 0.5)) as well as an incubator chamber to maintain the standard cell culture conditions (i.e., humidity, 37 °C and 5% CO_2).

2. Multiphoton microscope, e.g., a custom-built upright TriM Scope two-photon microscope (LaVision BioTec) equipped with a diode-pumped, wideband mode-locked Ti:Sapphire femtosecond laser (MaiTai, SpectraPhysics) and an APE Optical Parametric Oscillator (OPO), Photomultiplier tubes (Hamamatsu Photonics), 520 nm dichroic and bandpass filter 560/40 (Chroma Technology) and a water-dipping 20× objective (Olympus XLUMPlanFL IR coated, NA 0.95).
3. Software: Imaris (Bitplane), Image J/Fiji (open source), Volocity (Perkin Elmer).

3 Methods

3.1 Transfection of HEK293T Cells

1. Harvest exponentially growing HEK293T cells (*see* **Note 4**) the day before the transfection, centrifuge at 300 × *g* for 5 min, count and plate them in a six-well plate, so the cells will be 70% confluent the next day.
2. On the day of transfection, replace old medium with 2 mL of fresh medium 4 h before transfection.
3. Prepare 500 μL of calcium phosphate-DNA mix per each transfection in MQ H_2O. Co-transfect HEK293T cells with 0.5 μg each of the packaging defective helper construct (pMDLg/pRRE), the nuclear localization signal Rev plasmid (pRSV-Rev), a plasmid coding for a heterologous (pCMV-VSV-G) envelope protein, 8.5 μg of the vector construct harboring either mKO2-hCdt1 (30–120) or mAG-hGem (1–110) and 50 μL of 2.5 M $CaCl_2$.
4. Add 500 μL of the transfection solution drop by drop to 500 μL of the 2× HeBS solution while vortexing (*see* **Note 5**).
5. Let the solution stand at room temperature for 15 min, vortex every 5 min.
6. Slowly add the 1 mL of solution to each well while gently swirling the plate.
7. After 4 h remove the medium, wash with PBS, add fresh medium, and incubate at 37 °C, 5% CO_2.
8. After 48 h harvest the supernatant containing the lentiviral particles and centrifuge at 300 × *g* for 5 min (*see* **Note 6**).
9. Filter the supernatant with a 0.45 μm filter.
10. Aliquot in cryovials and store the supernatant at −80 °C or proceed with the transduction of the melanoma cells (*see* **Note 7**).

3.2 Transduction of Melanoma Cells

1. Plate melanoma cell line in a six-well plate so that they will be in the exponential grow phase during the transduction the next day (*see* **Notes 8–10**).

2. Replace old medium with 2 mL of fresh medium 4 h before transfection.
3. Thaw the lentiviral stock on ice.
4. Prepare the polybrene working solution at 8 mg/mL in culture medium.
5. Prepare 400 μL of transduction solution by mixing 200 μL of each FUCCI lentivirus (mKO2-hCdt1 (30–120) and mAG-hGem (1–110)) and add 0.4 μL of 8 mg/mL polybrene (final concentration of 8 μg/mL) (*see* **Notes 11** and **12**).
6. Add the transduction solution drop by drop while gently swirling the plate and return plate to the incubator.
7. After 72 h check the transduction efficiency under the fluorescence microscope.

3.3 Sorting of FUCCI-Expressing Single-Cell Clones

1. Trypsinize melanoma cells in the exponential growth phase (*see* **Notes 8** and **13**).
2. Centrifuge at 300 × *g* for 5 min.
3. Remove the supernatant and count the cells.
4. Resuspend the cells in sorting medium at 1 × 10^6 cells/mL.
5. Using a FACSAria cell sorter, acquire using the 488 nm laser and the B530 detector for mAG, and B575 for mKO2.
6. Sort double-positive cells, highly expressing mKO2-hCdt1 (30–120) and mAG-hGem (1–100) into a 96-well plate prefilled with medium.
7. Check at the phase-contrast microscope that no more than one cell is present in each well (*see* **Note 14**).
8. Return to the incubator.
9. Check every day to spot the first dividing clones (*see* **Note 15**).
10. Trypsinize cells when confluent and split to a 24-well plate. Continue to expand cells until there are enough to freeze down in liquid nitrogen for further use.

3.4 Spheroid Formation

1. To prepare one 96-well plate for spheroid formation, freshly prepare 30 mL of well-coating solution and immediately dispense 100 μL of the solution in each well (*see* **Note 16**). Let the plate rest at room temperature for 15 min to allow the agarose to cool down and harden.
2. In the meantime, trypsinize the FUCCI-expressing cells (*see* **Note 17**): Wash the cells once with PBS and then treat with 0.05% Trypsin/EDTA for 5 min at 37 °C.
3. Neutralize the trypsin by adding trypsin neutralizing medium and pipet repeatedly to create a single-cell suspension.

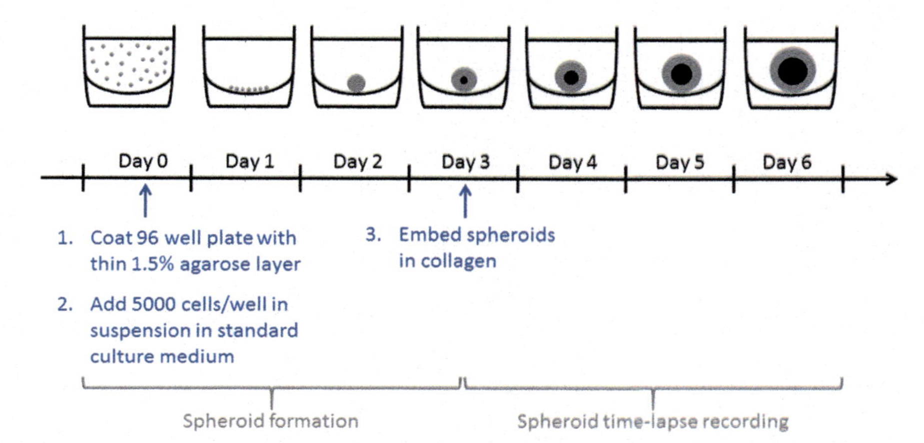

Fig. 1 Spheroid formation: a day-by day schematic illustration of the different stages of spheroid formation and the main steps involved in this procedure

4. Centrifuge at 300 × *g* for 5 min.
5. Remove supernatant and resuspend the cells in cell culture medium.
6. Count the cells using a hemocytometer or an automatic cell counter and resuspend 600,000 cells in 24 mL of cell culture medium to obtain a solution at a concentration of 2.5×10^4 cells/mL.
7. Using a multichannel pipette add 200 μL of the cell suspension to each well so that 5000 cells are seeded per well (*see* **Note 18**).
8. Place the plate in the incubator for the following 72 h, allowing the spheroids to form (Fig. 1) (*see* **Note 19**).
9. Use a phase-contrast microscope to inspect the spheroids and mark the wells where the spheroids have formed properly (*see* **Note 20**).

3.5 Spheroid Embedding in Collagen Matrix

1. To prepare 1 mL of the collagen embedding solution, mix 100 μL of 10× EMEM, 10 μL of 200 mM L-Glutamine, and 400 μL of bovine type I collagen (*see* **Notes 21** and **22**). Work on ice (*see* **Note 23**).
2. Titrate the solution with 7.5% $NaHCO_3$ until the solution turns from a yellow-orange to a peach-pink color (the volume required for this change in pH should be around 15–30 μL).
3. Add 100 μL of FBS and top up to 1 mL with H_2O.
4. Mix well while taking care to avoid generating bubbles and keep on ice.
5. Add 40 μL in each well of a new flat-bottom 96-well plate (*see* **Note 24**) and incubate plate for 5 min at 37 °C to allow the collagen to polymerize (*see* **Note 25**).

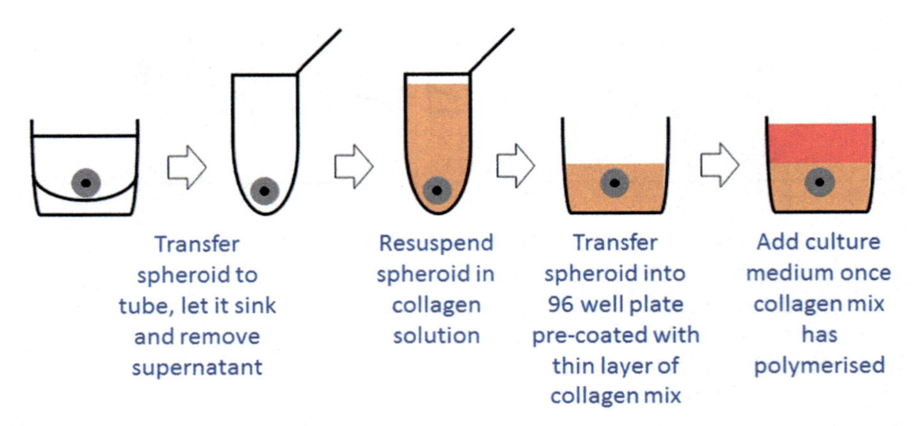

Fig. 2 Spheroid embedding: schematic illustrating the main steps involved in the procedure for embedding spheroids in collagen matrix

6. Transfer spheroids from the 96-well plate where they have been cultured to sterile microcentrifuge tubes using a pipette and pipette tips with enlarged aperture (*see* **Note 26**). Transfer one spheroid per tube (Fig. 2) (*see* **Note 27**).
7. Let spheroids settle by gravity to the tube's bottom and remove the supernatant taking care to not disturb the spheroid (*see* **Note 28**).
8. Gently resuspend each spheroid in 60 μL of collagen solution (use pipette and pipette tips with enlarged aperture) and transfer into the 96-well plate that was previously coated with 40 μL per well of collagen solution (**step 5**).
9. Incubate plate for 15 min at 37 °C to allow the collagen to polymerize.
10. With the use of a phase-contrast microscope inspect spheroid integrity and distribution in the well (if multiple spheroids were seeded per well).
11. Add 200 μL per well of appropriate medium and incubate at 37 °C until the setup of the spheroid time-lapse confocal microscopy analysis.

3.6 FUCCI-Spheroid Time-Lapse Imaging Using Confocal Microscopy

1. Set the microscope to a low magnification and turn on the laser to excite the Azami Green or Kusabira Orange 2 proteins (*see* **Note 29**).
2. Move to the well containing the spheroid to be imaged, locate the spheroid, and center it in the field of view.
3. Set the microscope to a magnification that allows visualization of the whole spheroid (*see* **Note 30**).
4. Position the plane of view in the center of the spheroid (*see* **Note 31**). To do so, adjust the *z* variable starting with the view

plane at the bottom of the spheroid then moving it deeper into the spheroid. During this process the diameter of the spheroid should increase until the center of the spheroid has been reached (*see* **Note 32**).

5. Adjust the necessary settings (exposure time, gain, laser power, etc.) to achieve optimal signal without overexposed zones.
6. Set up four optical slices on each side of the central plane, each separated by 5 μm. This results in a total of nine optical slices (including the central slice) encompassing a thickness of 40 μm, 20 μm above the central plane and 20 μm below the central plane of the spheroid (*see* **Note 33**).
7. Set up the total recording duration to at least 24–48 h and the intervals' time to 10–20 min (*see* **Note 34**).
8. Start the time-lapse recording (Fig. 3).
9. The obtained images are processed using Imaris, Volocity, or Image J. Obtain cell parameters such as speed by automated or manual cell tracking in the aforementioned software programs. Convert time-lapse image sequences into AVI or MOV format for display in common media players.

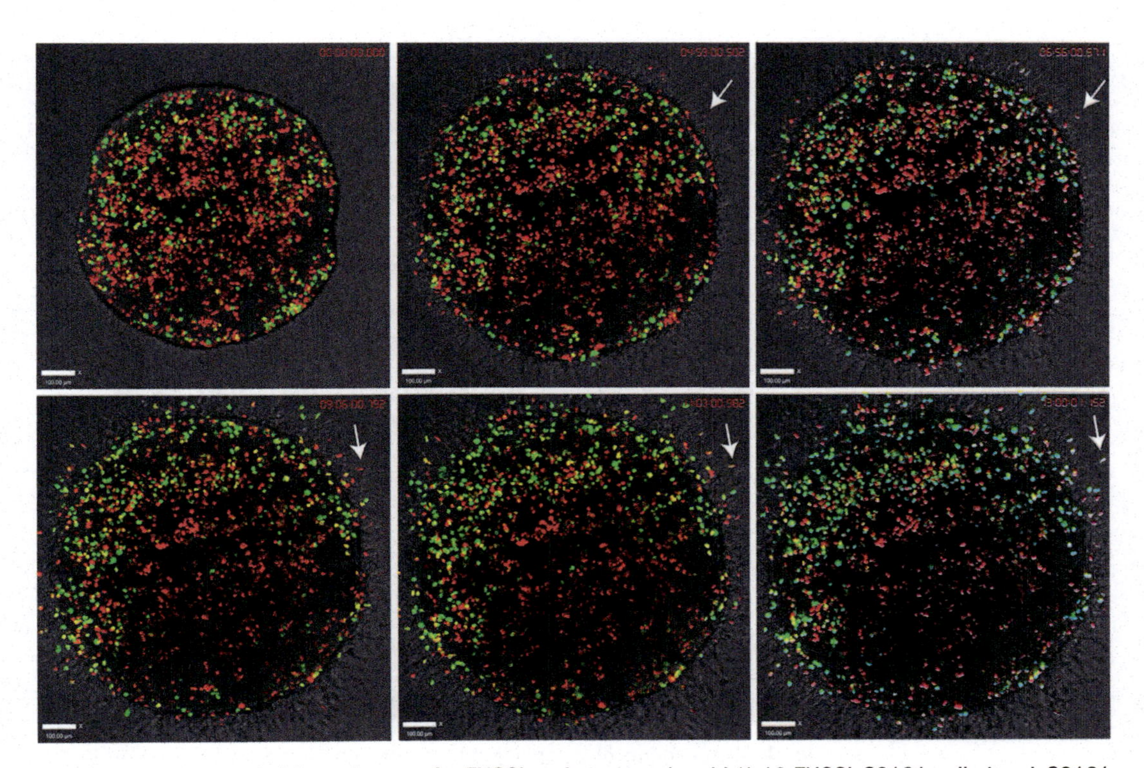

Fig. 3 Confocal extended focus image of a FUCCI-melanoma spheroid (1:10 FUCCI-C8161 cells to wt-C8161 cells, from 0 to 13 h after collagen embedding. The top few *z*-slices were removed to reveal the *red* G1-arrested cells inside the spheroid. *White arrows* indicate a single invading cell tracked over time. Note the cell cycle change of the invading cell from G1 (*red*) to S/G2/M (*green*). Scale bar, 100 μm

3.7 Spheroid Imaging Using Multiphoton Microscopy and 3D Stitching

1. Visualize the spheroid using a 20× water-dipping objective, lowered into the medium (*see* **Note 35**).
2. Excite the FUCCI-spheroid using a wavelength of 920 nm (Ti:Sapphire laser) and 1060 nm (OPO).
3. Use a 465 dichroic to remove second-harmonic generation (SHG) signals originating from the Ti:Sapphire laser (*see* **Note 36**).
4. Use a 520 nm dichroic to split the mKO from the mAG emission, and a bandpass filter 560/40 to further refine the mKO signal and exclude SHG light originating from the OPO.
5. Collect mAG and mKO fluorescence in photomultiplier tubes.
6. Acquire three-dimensional images (*x*, *y*, *z*) of spheroids using the LaVision acquisition software. Use a *z* spacing of 4 μm, with montaging to cover the entire XY area of the spheroid. Use a *z*-depth of approximately 600 μm or until the signal fades, in order to image as much of the spheroid depth as possible.
7. Stitch *z*-stack images in FIJI or ImageJ using the Stich Grid of Images Plugin [16], and perform 3D volume rendering using Volocity software (Fig. 4) (*see* **Note 37**).

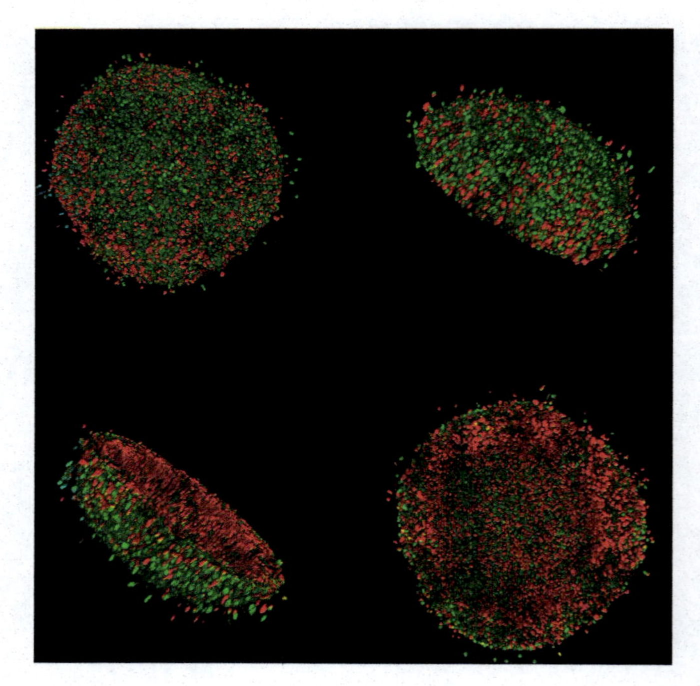

Fig. 4 3D rendering of a multiphoton microscopy image *z*-stack of a FUCCI-melanoma spheroid (C8161) 24 h after collagen implantation

4 Notes

1. To start pH should be around 6.8 and should need 2–3 NaOH pellets to reach the right pH and change color to red. Notice that pH might increase slightly when topping up with H_2O to 1 L.
2. It is advisable to prepare at least 20 mL extra in addition to the required volume. Due to its viscosity this solution tends indeed to stick to the surfaces of the container.
3. Coating plates with 1.5% agarose will prevent adhesion of the cells to the bottom of the well and at the same time creates a meniscus that allows grouping of the cells at the center of the well.
4. HEK293T is the embryonic kidney 293 cell line (HEK293) carrying the SV40 T-antigen. The T-antigen makes these cells highly transfectable with vectors carrying the SV40 origin of replication resulting in high viral titers.
5. Add the 2× HeBS dropwise to the $CaCl_2$/plasmid DNA solution, to achieve optimal calcium-phosphate/DNA precipitate.
6. Gloves and lab coat must be used when working with lentiviral vectors. The transduction must be carried out in a PC2 laboratory equipped with a certified Class II Biosafety cabinet and a tissue culture incubator. The buckets in the centrifuge must be equipped with aerosol-tight covers. Decontaminate the surface of the Biosafety cabinet with 10% bleach at the end of the session.
7. Lentiviral particles can be used fresh, or stored at −80 °C. After thawing and re-freezing, there will be a loss of virus potency.
8. The number of cells to be seeded depends on the cell line used.
9. This step is very important as actively dividing cells give a higher transduction rate than nondividing cells.
10. Low cell confluence is desired to allow growth for 2 days without splitting.
11. Remember to perform also the single color and the mock transduction as later flow cytometry sorting will be performed for which single-color signal compensation is required.
12. Calculation of the virus MOI is not required as expression levels of each construct can be measured by flow cytometry and used for sorting to achieve comparable fluorescence intensity of the FUCCI fluorophores.
13. This is important because confluent cells arrest in G_1 and accumulate Cdt1. This will result in stronger intensity of the red fluorescence compared to actively dividing cells that pass through G_1.

14. It is possible that no cell is visible at this stage.
15. Usually, cells grow close to the edge of the wells. Two cells close together are likely a single-cell clone. Two cells far apart within the well are likely two independent clones and should therefore be disregarded.
16. This procedure needs to be performed quickly to prevent hardening of the agarose before it has been dispensed in the wells. Use a multichannel pipette and change tips between each 96-well plate (if preparing multiple plates).
17. In order to facilitate easier cell tracking, untransduced parental cells (non-fluorescent) may be mixed with FUCCI cells in a ratio of 10:1 or even 20:1 when forming spheroids. This will reduce problems with cell separation in imaging software due to density.
18. It is advisable to use cells that were grown at a density of 80% or lower. We have observed in the past that cells at higher confluency resulted in poor spheroid formation.
19. The cells will initially group in close contact to each other as a non-adherent single layer in the center of the well. Over time they will initiate cell-cell contact and begin to form a 3D structure. Once cell-cell contact is initiated (usually after 24 h), the cells will restart to proliferate allowing the spheroid to expand.
20. A properly formed spheroid should have a reasonably round and regular shape with a necrotic core observable by phase-contrast microscopy as a darker central zone, although the latter might not be visible in all cell lines. The success of spheroid formation depends on the specific cell lines used to form spheroids as well as on the integrity of the agarose coating, i.e., spheroids can form properly in some well but not in others. Some cell lines require seeding of higher number of wells compared to others to obtain the same number of properly formed spheroids.
21. The volumes given here are for 1 mL of collagen solution which is enough to embed 9 spheroids, each in a well of a 96-well plate (100 μL per spheroid plus minimal manipulation-related loss of volume). To embed spheroids in larger wells (48- or 24-well plates), adjust the volumes proportionally to the well's surface area. The number of spheroids embedded per well can also be adjusted accordingly (e.g., three spheroids in a well of a 24-well plate), however take into consideration that multiple spheroids seeded in a single well might embed in close contact to each other, preventing/compromising their experimental use.
22. In addition, other cell types such as fibroblasts and/or endothelial cells can be added to the collagen gel to mimic the stroma better. Examples of this technique are outlined in [17–19].

23. All work must be undertaken on ice to avoid collagen polymerization, unless otherwise desired, i.e., during and following incubation at 37 °C.
24. A normal plastic tissue culture plate may be used—but to improve image quality, and if a smaller working distance objective is needed, an imaging plate (either glass or thin plastic bottom) should be used.
25. This is important to prevent contact between the transferred spheroid and the bottom of the well. Contact will allow cells that are part of the spheroids to adhere to the plastic and proliferate as a monolayer on the well's surface.
26. With a pair of scissors, enlarge the opening of the *pipette tip* by *cutting* approximately 2–3 mm *off* the *tip*. Before use, clean and disinfect the blades of the scissors with 80% ethanol.
27. To successfully aspirate a spheroid, immerse the end of the pipette tip in the supernatant and move as close as possible to the spheroid without touching it, then gently but assertively retract the plunger to generate a fluid flow with which the spheroid will move into the tip. Gently release the spheroid into the tube avoiding to damage it.
28. Remove the majority of the supernatant using a pipette attached to a vacuum pump and then carefully remove as much as possible of the remaining supernatant using a manual pipette with a gel-loading tip.
29. In the case of untreated spheroids the green and red fluorescence signals are expected to have similar intensity and their use to locate the spheroid is interchangeable. If the intensity of one of the two fluorescence signals is decreased (e.g., due to drug treatment), the channel for the fluorophore with the strongest intensity should be used to locate the spheroid.
30. The starting size and growth rate of the spheroid, as well as whether visualization of cell invasion is desired, are factors to consider when choosing the magnification. If visualization of cell invasion of a fast growing spheroid is desired, the visualization field at the start of the experiment should include sufficient empty space surrounding the spheroid accounting for the increase in size of the spheroid as well as cell moving out of the spheroid.
31. The ideal position to monitor cell cycle behavior in FUCCI melanoma spheroids is in the middle of the spheroid, as this allows observing both the inner arrested zone and the outer cycling layer of cells (based on the specific distribution of subcompartments of cells with different cell cycle behavior outlined in the introduction paragraph).
32. Due to the limited sample penetration depth of confocal microscopy, it might not be possible to visualize the central

plane of certain spheroids without experiencing loss of signal. This depends on the spheroid's size, density as well as other factors. The loss of signal might be uneven throughout the plane and give rise to artifacts. The plane of view should be localized as deep as possible without experiencing loss of signal. This compromise might be adopted where setting the plane of view slightly away from the center of the spheroid still provides the desired information. To confirm the lack of artifacts, the same experiment can be repeated for specific time points where sections of the fixed spheroid (instead of the whole spheroid) are imaged using the same microscopy technique. This method prevents potential artifacts generated by limited imaging penetration depth.

33. Collecting *z*-stack images allows the creation of a maximum intensity projection image to enhance signal intensity when signal from a single optical slice is too weak. The thickness of the *z*-stack and the number of optical slices it contains should be adjusted according to spheroid's characteristics and in particular to spheroid size. For instance, in melanoma FUCCI spheroids, an excessive *z*-stack thickness might result in signal contamination from the outer cycling layer into the inner arrested zone of cells. Removal of the bottom z-slices is therefore necessary for optimal outcome.
34. These factors depend on the time necessary for the phenomenon of interest to be captured and the desired level of details. For monitoring changes in cell cycle behavior using the FUCCI system, the total recording duration and intervals' time need to be adjusted according to the cell line's cell cycle time.
35. An inverted multiphoton microscope may be preferable if repeated imaging is to take place, or time lapse imaging. This will avoid potential contamination due to the need to remove the tissue culture plate lid.
36. SHG may also be used to image type I collagen fibers [20, 21].
37. Several other imaging software have 3D stitching plugins, and FIJI also has additional stitching plugins. If this FIJI plugin fails, one of the other methods should be tried. An alternative is to start with two tiles that do stitch together, and then to build up the larger image iteratively by gradually adding more tiles. For this method the maximum intensity blending method must be used.

Acknowledgments

We thank Prof. Meenhard Herlyn, The Wistar Institute, Philadelphia, PA; Prof. Keiran Smalley, Moffitt Cancer Center, Tampa, FL; Prof. Wolfgang Weninger & Dr. Ben Roediger, Centenary Institute,

Sydney, NSW; Dr. Crystal Tonnessen, UQ Diamantina Institute, Brisbane, QLD, and the imaging facilities of the Centenary and UQ Diamantina Institutes for their contribution to optimizing this protocol over the years. We thank Prof. Atsushi Miyawaki, RIKEN, Wako-city, Japan, for providing the FUCCI constructs. N.K.H. is a Cameron fellow of the Melanoma and Skin Cancer Research Institute, Australia. K.A.B. is a fellow of Cancer Institute New South Wales (13/ECF/1-39). The work leading to this protocol was supported by project grants RG 09-08 and RG 13-06 (Cancer Council New South Wales), 570778 and 1051996 (Priority-driven collaborative cancer research scheme/Cancer Australia/Cure Cancer Australia Foundation), 08/RFG/1-27 (Cancer Institute New South Wales), APP1003637 and APP1084893 (National Health and Medical Research Council).

References

1. Brandner JM, Haass NK (2013) Melanoma's connections to the tumour microenvironment. Pathology 45(5):443–452. doi:10.1097/PAT.0b013e328363b3bd
2. Villanueva J, Herlyn M (2008) Melanoma and the tumor microenvironment. Curr Oncol Rep 10(5):439–446
3. Beaumont KA, Mohana-Kumaran N, Haass NK (2014) Modeling melanoma in vitro and in vivo. Healthcare 2(1):27–46. doi:10.3390/healthcare2010027
4. Santiago-Walker A, Li L, Haass NK et al (2009) Melanocytes: from morphology to application. Skin Pharmacol Physiol 22(2):114–121. doi:10.1159/000178870
5. Smalley KS, Lioni M, Noma K et al (2008) In vitro three-dimensional tumor microenvironment models for anticancer drug discovery. Expert Opin Drug Discovery 3(1):1–10. doi:10.1517/17460441.3.1.1
6. Wroblewski D, Mijatov B, Mohana-Kumaran N et al (2013) The BH3-mimetic ABT-737 sensitizes human melanoma cells to apoptosis induced by selective BRAF inhibitors but does not reverse acquired resistance. Carcinogenesis 34(2):237–247. doi:10.1093/carcin/bgs330
7. Hanahan D, Weinberg RA (2000) The hallmarks of cancer. Cell 100(1):57–70. doi:10.1016/S0092-8674(00)81683-9
8. Hanahan D, Weinberg RA (2011) Hallmarks of cancer: the next generation. Cell 144(5):646–674. doi:10.1016/j.cell.2011.02.013
9. Sakaue-Sawano A, Kurokawa H, Morimura T et al (2008) Visualizing spatiotemporal dynamics of multicellular cell-cycle progression. Cell 132(3):487–498. doi:10.1016/j.cell.2007.12.033
10. Haass NK, Beaumont KA, Hill DS et al (2014) Real-time cell cycle imaging during melanoma growth, invasion, and drug response. Pigment Cell Melanoma Res 27(5):764–776. doi:10.1111/pcmr.12274
11. Beaumont KA, Hill DS, Daignault SM et al (2016) Cell cycle phase-specific drug resistance as an escape mechanism of melanoma cells. J Invest Dermatol. doi:10.1016/j.jid.2016.02.805
12. Ravindran Menon D, Das S, Krepler C et al (2015) A stress-induced early innate response causes multidrug tolerance in melanoma. Oncogene 34(34):4448–4459. doi:10.1038/onc.2014.372
13. Haass NK (2015) Dynamic tumor heterogeneity in melanoma therapy: how do we address this in a novel model system? Melanoma Manag 2(2):93–95. doi:10.2217/mmt.15.1
14. Beaumont KA, Anfosso A, Ahmed F et al (2015) Imaging- and flow cytometry-based analysis of cell position and the cell cycle in 3D melanoma spheroids. J Vis Exp 106:e53486. doi:10.3791/53486
15. Smalley KS, Brafford P, Haass NK et al (2005) Up-regulated expression of zonula occludens protein-1 in human melanoma associates with N-cadherin and contributes to invasion and adhesion. Am J Pathol 166(5):1541–1554. doi:10.1016/S0002-9440(10)62370-X
16. Preibisch S, Saalfeld S, Tomancak P (2009) Globally optimal stitching of tiled 3D microscopic image acquisitions. Bioinformatics 25(11):1463–1465. doi:10.1093/bioinformatics/btp184
17. Flach EH, Rebecca VW, Herlyn M et al (2011) Fibroblasts contribute to melanoma tumor

growth and drug resistance. Mol Pharm 8(6):2039–2049. doi:10.1021/mp200421k

18. Haass NK, Sproesser K, Nguyen TK et al (2008) The mitogen-activated protein/extracellular signal-regulated kinase kinase inhibitor AZD6244 (ARRY-142886) induces growth arrest in melanoma cells and tumor regression when combined with docetaxel. Clin Cancer Res 14(1):230–239. doi:10.1158/1078-0432.CCR-07-1440
19. Velazquez OC, Snyder R, Liu ZJ et al (2002) Fibroblast-dependent differentiation of human microvascular endothelial cells into capillary-like 3-dimensional networks. FASEB J 16(10):1316–1318. doi:10.1096/fj.01-1011fje
20. Kirkpatrick ND, Hoying JB, Botting SK et al (2006) In vitro model for endogenous optical signatures of collagen. J Biomed Opt 11(5):054021. doi:10.1117/1.2360516
21. Tong PL, Qin J, Cooper CL et al (2013) A quantitative approach to histopathological dissection of elastin-related disorders using multiphoton microscopy. Br J Dermatol 169(4):869–879. doi:10.1111/bjd.12430

Chapter 30

Revealing 3D Ultrastructure and Morphology of Stem Cell Spheroids by Electron Microscopy

Josef Jaros, Michal Petrov, Marketa Tesarova, and Ales Hampl

Abstract

Cell culture methods have been developed in efforts to produce biologically relevant systems for developmental and disease modeling, and appropriate analytical tools are essential. Knowledge of ultrastructural characteristics represents the basis to reveal in situ the cellular morphology, cell-cell interactions, organelle distribution, niches in which cells reside, and many more. The traditional method for 3D visualization of ultrastructural components, serial sectioning using transmission electron microscopy (TEM), is very labor-intensive due to contentious TEM slice preparation and subsequent image processing of the whole collection. In this chapter, we present serial block-face scanning electron microscopy, together with complex methodology for spheroid formation, contrasting of cellular compartments, image processing, and 3D visualization. The described technique is effective for detailed morphological analysis of stem cell spheroids, organoids, as well as organotypic cell cultures.

Key words 3D visualization, Image reconstruction, Image segmentation, Morphology, Organoid, SBF-SEM, Scanning electron microscopy, Serial block-face, Spheroid, Stem cell, Ultrastructure

1 Introduction

Stem cells are broadly studied and envisioned for use as model systems in regenerative medicine, developmental and disease modeling, as well as for evaluating the effectiveness and safety of therapeutic drugs. However, conventionally used planar cell cultivation (monolayer) lacks spatial organization of cells, which leads to aberrant behaviors: flattened shape, abnormal polarization, altered response to pharmaceutical reagents, and loss of differentiated phenotype [1]. Therefore, three-dimensional (3D) cell culture methods have been developed in efforts to produce the most possible biologically relevant models in vitro. The complex 3D environment provides the cell-cell and cell-extracellular matrix (ECM) signaling, and also

Electronic supplementary material: The online version of this chapter (doi: 10.1007/978-1-4939-7021-6_30) contains supplementary material, which is available to authorized users.

Zuzana Koledova (ed.), *3D Cell Culture: Methods and Protocols*, Methods in Molecular Biology, vol. 1612,
DOI 10.1007/978-1-4939-7021-6_30, © Springer Science+Business Media LLC 2017

establishes structural features, such as morphology and spatial organization of cells and subcellular compartments [2, 3]. Still, 3D cell culture is not yet optimized due to lack of standardized analytical tools for quantifying biological response [4].

Utilization of microscopy and imaging for morphological characterization of aggregates, individual cells, and ultrastructural compartments with subsequent 3D visualization could overcome some of these limits [5–7]. Both light microscopy and transmission electron microscopy have been used to determine architectural parameters, but they have technical limitations and/or they are very labor-intensive and require a high degree of skill and training. Then, high-resolution 3D imaging of cellular ultrastructure remains mostly elusive. Dramatic improvement in electron microscopy was achieved by developing two methods, which utilize automated sectioning of biological specimen. First, serial block-face imaging scanning electron microscopy (SBF-SEM) uses the automated ultramicrotome located inside the SEM chamber and removes sections (≥20 nm thick) from the block face [8], while the second method applies focused ion beam (FIB) to mill away sections (~5 nm) of substrate of investigated sample [9]. The principal of imaging is then the same for both, SEM image of truncated surface is acquired after each section.

Here, we describe utilization of the SBF-SEM method which provides scanning of relatively large overviews (typically 500–5000 μm^2) and volumes in comparison to FIB, which is limited by the size of the field that can be milled and imaged, usually less than 800 μm^2 [10].

We have employed pluripotent stem cells due to their high proliferation potential, differentiation capacity, and colony-forming behavior during cultivation in 3D matrices. Staining procedures were suited to electron dense contrasting of DNA and membranes for 3D modeling and visualization of chromatin, membranes, and organelles (e.g., Golgi apparatus, mitochondria). Further image processing was performed by commercial software in our case (VG studio Max, Avizo), however freeware (e.g., Fiji, The Visualization Toolkit) or other commercial software (Gatan Microscopy Suite, Autodesk, etc.) is available.

In the present protocol, we demonstrate the methodology for revealing the 3D cellular ultrastructure of stem cell spheroids and their inner compartments, which is efficiently applicable for any of 3D cell culture models.

2 Materials

Prepare all solutions using ultrapure water (ddH_2O; prepared by purifying deionized water to attain a sensitivity of 18 MΩcm at 25 °C) and analytical grade reagents. Prepare and store all reagents at room temperature (RT) unless indicated otherwise. Thoroughly follow all waste disposal regulations when disposing waste materials, especially in case of toxic reagents, which are used for electron microscopy contrasting.

2.1 Cell Culture and 3D Spheroid Formation

All materials and solutions must be sterile.

1. Human embryonic stem cells (hESCs) grown on Matrigel, e.g., H9, CCTL-14, or CCTL-12 (*see* **Note 1**).
2. Matrigel hESC-qualified matrix (Corning, 354277).
3. Conditioned medium (CM): Knockout serum replacement medium (20% KSR, 2 mM nonessential amino acids, 2 mM L-glutamine, 50 U/mL penicillin, 50 μg/mL streptomycin, 0.1 mM β-mercaptoethanol, and 4 ng/mL of FGF2 in DMEM/F12) collected from irradiated mouse embryonic fibroblast culture every 24 h for 5 days, freshly supplemented with 4 ng/mL of FGF2 and 2 mM L-glutamine and sterile-filtered before use (*see* **Note 2**).
4. Phosphate buffered saline (PBS).
5. TrypLE Express (1×).
6. Fetal bovine serum (FBS), heat inactivated.
7. Glass Pasteur pipette with flamed opening to create a bevel.
8. 60 mm cell culture-treated tissue culture dishes.
9. 60 mm nontreated tissue culture dish.
10. 2 mL round-bottom centrifuge vial.
11. Tissue culture equipment, such as cell culture incubator with temperature, humidity and CO_2 control, laminar flow cabinet, inverted optical microscope, centrifuge, water bath, etc.

2.2 Fixation and Contrasting of Spheroids for Electron Microscopy

Most of the fixatives and contrasting agents are irritants or toxic on inhalation or skin contact. Work in a fume hood and wear gloves; nitrile, PVA, or Barrier-brand gloves are preferred for work with contrasting agents.

1. CacoB working solution: 0.1 M sodium cacodylate in ddH_2O. Prepare as 2× CacoB stock solution (0.2 M sodium cacodylate buffer) by adding 4.28 g of sodium cacodylate to 50 mL of ddH_2O. Adjust pH by adding 5 mL of 0.2 N HCl and fill the volume to 100 mL with ddH_2O. Aliquot (10 mL) and store at −20 °C. For working solution, dilute 1 part of 2× CacoB stock solution in 1 part of ddH_2O.
2. GA stock solution: 6% glutaraldehyde in ddH_2O (*see* **Note 3**). Store at −20 °C.
3. FA stock solution: 4% formaldehyde in 0.2 M CacoB stock solution (*see* **Note 3**). To prepare 10 mL, add 1 mL of 40%

formaldehyde to 9 mL of 0.2 M CacoB. Store at −20 °C up to 6 months.

4. GAFA fixative: Mix GA stock solution and FA stock solutions in the ratio 1:1 (v/v).
5. Tannin-GAFA fixative: 1% (w/v) tannin in GAFA fixative.
6. OsO_4 fixative: 4% OsO_4 in ddH_2O (*see* **Note 4**). OsO_4 is obtained commercially as 1 g crystals sealed in breakable glass ampoules. Score the glass ampoule with a diamond pencil in upright position (if it is not prescored by the manufacturer) above an opened 50 mL glass stoppered reagent bottle. Drop the ampoule immediately into the reagent bottle, add 25 mL of ddH_2O water, insert the stopper, and then gently jiggle the bottle until the glass ampoule is lying flat at the bottom without any air bubbles. After dissolving crystals at RT, place the reagent bottle to store at 4 °C (*see* **Note 5**).
7. FerroOS postfix: 1.5% (w/v) potassium ferrocyanide in 1% OsO_4. Add 30 mg of potassium ferrocyanide to 1 mL of CacoB stock solution. Vortex the vial to dissolve and add 0.5 mL of OsO_4 post-fixative with 0.5 mL of ddH_2O. Filter with a 0.22 μm filter before use to remove coagulates. Prepare freshly or store in a double container at 4 °C for 1–2 days (*see* **Note 6**).
8. UA stock: 5% (w/v) uranyl acetate solution. To prepare 10 mL, add 0.5 g of uranyl acetate to 10 mL of ddH_2O. Cover with aluminum foil and stir overnight (*see* **Note 7**). Filter with a 0.22 μm filter. The solution is stable for at least 6 months at 4 °C.
9. UA stain. 1% uranyl acetate. Dissolve 1 part of UA stock in 4 parts of ddH_2O. Store at 4 °C. If possible, use freshly prepared staining solution and filter out or centrifuge the undissolved crystals of uranyl acetate before use.
10. Razor blades (0.009 single edge).
11. Paintbrush, thin.
12. Weighing spatula.

2.3 Embedding, Electron Microscopy, and Image Processing

1. London Resin White (LRW). Prior to mounting, equilibrate 1.98 g of catalyst and 100 g of LRW (Medium Grade) to RT for 15 min, then mix together and shake thoroughly for 30 min with a magnetic stirrer. The catalyst will take full 24 h in a refrigerator to dissolve completely. During this time it is most helpful if the mixture can be shaken from time to time. Once mixed and fully dissolved, the resin must be stored at 4 °C to maintain its shelf life. Once catalyzed, LRW's shelf life is 12 months if stored in a refrigerator or it can be aliquoted and stored in −80 °C. Allow the resin to reach RT at least 15 min before use (*see* **Note 8**).

2. 100% ethanol.
3. Flat embedding coffin mold (or gelatin capsules).
4. Paraffin film (or Aclar film, or polyethylene-based film).
5. Oven for 65 °C (or UV lamp, 365 nm, 4 W).
6. Conductive epoxy glue.
7. Tescan MAIA3 XMU SEM with integrated Gatan 3View2XP ultramicrotome.
8. Flat screwdriver and hex key.
9. Image processing and 3D visualization software—VG studio Max 2.2 (Volume Graphics GmbH) and Avizo 7.0 (FEI).

3 Methods

3.1 hESC Culture

All cell handling should be performed under sterile conditions in a Class II Biological Hazard Flow Hood, and all biologically contaminated material should be disposed of properly. All cell cultures should be maintained at 37 °C in a humidified atmosphere of 5% CO_2 in air.

1. Prior to each passaging, remove overtly differentiated hESC colonies to maintain the general undifferentiated and homogeneous state of the culture. Under a dissecting microscope within a laminar flow cabinet, mechanically excise differentiated regions from the dish using a Pasteur pipette with bevel and aspirate them.
2. To passage hESCs, wash the cells with PBS and add TrypLE (1 mL per 60 mm petri dish). Incubate for 4 min at 37 °C, and release the cells by tapping the dish.
3. Add an equal volume of CM medium to dilute the TrypLE and pipette the medium three to four times across the dish to wash off remaining adherent cells.
4. Transfer the cells gently to a 15 mL conical tube and centrifuge at 100 × *g* for 4 min at 4 °C.
5. Aspirate supernatant and resuspend pellet in 1 mL of CM medium.
6. Count the cells using Trypan blue and hemocytometer and dilute to concentration of 1×10^5 cells per mL.
7. Seed hESCs at 20,000 cells/cm^2 on Matrigel-coated tissue culture dishes. Cultivate in a cell culture incubator for 3–5 days with daily change of medium.

3.2 3D Spheroid Formation

1. After cells reach 80% of confluency, prepare the matrix for 3D formation. Place an aliquot (270–350 μL) of Matrigel into ice for at least 15 min to thaw. Also, place the 200 μL pipette tips

and centrifuge tubes for at least 15 min at −20 °C to chill prior to work with Matrigel.

2. In the meantime, harvest the cells as described in Subheading 3.1, **steps 2–6**.
3. Withdraw the volume with 1000 cells, drop it to the bottom of cold centrifuge tube placed in ice, and resuspend the cells in cold CM to the final volume of 50 μL.
4. Withdraw 50 μL of Matrigel using a chilled tip and gently mix it with the cell suspension. Avoid bubble formation.
5. Slowly pipette 5 μL of the suspension into drops or 10 μL of the suspension into lines onto the hydrophobic surface of a nontreated culture dish (*see* **Note 9**).
6. Allow the Matrigel-cell suspension to solidify for 5 min in the cell culture incubator. Then slowly add CM pre-warmed to 37 °C (*see* **Note 9**).
7. Culture for 3–6 days. Change the medium daily.

3.3 Fixation and Contrasting for EM

1. Aspirate culture medium and wash samples with PBS pre-warmed to 37 °C.
2. Pour carefully Tannin-GAFA fixative to the dish and let it sit covered in a fume hood for 1 h at RT.
3. Wash two times for 5 min each in 0.1 M CacoB.
4. Drain off the buffer and circumscribe the Matrigel drops/lines with cotton swabs or hydrophobic pen to draw a hydrophobic barrier (*see* **Note 10**).
5. Drop freshly filtered FerroOS postfix on samples and incubate them covered for 1 h in a fume hood (*see* **Note 11**).
6. Wash three times with 0.1 M CacoB for 5 min each (*see* **Note 12**).
7. With a razor blade, cut the samples into pieces of approximately 1 × 1 × 1 mm (Fig. 1a), or separate the appropriate spheroids on optical microscope. Transfer the samples to 2 mL centrifuge vials.
8. Rinse with ddH_2O two times for 5 min each. In the meantime, prepare fresh UA stain.
9. Continue by contrasting samples with 1% UA stain for 20 min at RT in a dark place to protect from light degradation.
10. Wash three times quickly in ddH_2O.

3.4 Embedding to LRW

1. Dehydrate the samples in an ascending ice-cold alcohol series [30, 50, 70, 80, 90% (v/v)] for 10 min at each concentration at 4 °C.
2. Infiltrate the samples stepwise with LRW resin. Incubate the specimen in a mixture containing 50:50 (v/v) LRW/EtOH

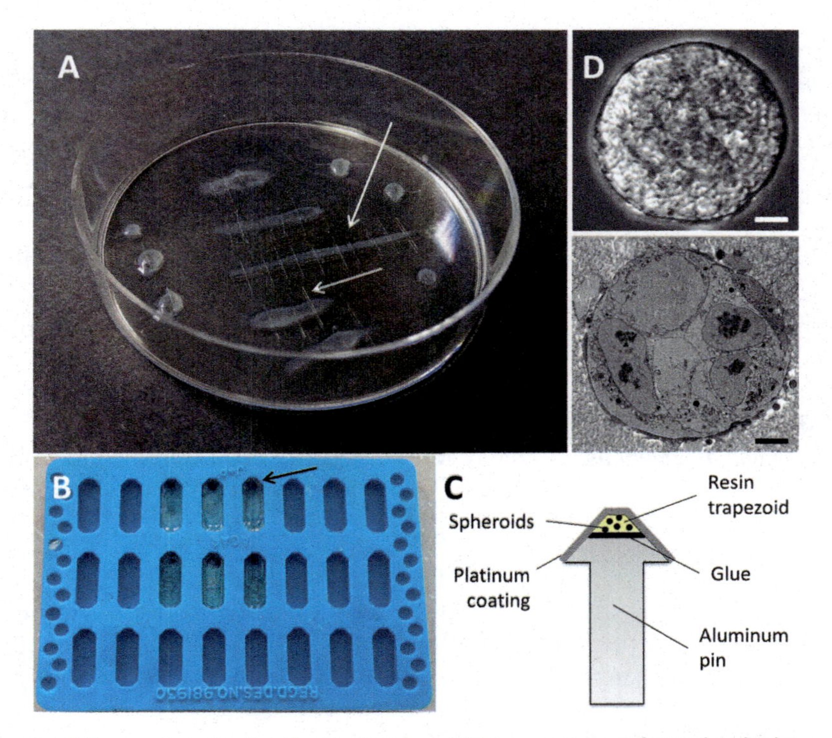

Fig. 1 Processing of cell culture in Matrigel. (**a**) Lines and drops formed on hydrophobic surface of a 60 mm nontreated Petri dish. *White arrows* show cutting of specimen after fixation. (**b**) Embedding of samples with LRW in coffin mold. *Black arrow* indicates orientation of samples in block. (**c**) Schematic illustration of trimmed block glued to pin for serial block face imaging. (**d**) Representative images of hESC spheroids organized within Matrigel after 4 days of cultivation observed in transmitted optical microscope (*upper panel*) and scanning electron microscope (*lower panel*). Scale bar, 10 μm

for 30 min at 4 °C. Then exchange the mixture for 75:25 (v/v) LRW/EtOH for 30 min at 4 °C. Finally, cover samples with 100% LRW resin and leave overnight at 4 °C.

3. Next day, allow vials to reach RT for 10 min. Using a metal-weighing spatula (palette knife) and a paintbrush, transfer the samples to an embedding coffin mold (or a gelatin capsule in a rack) filled with 100% LRW. Ensure that there are no bubbles in the resin.
4. Gently orient the samples to the end of the embedding mold using a paintbrush. Resin must be slightly over-filled in the cavities of mold (Fig. 1b).
5. Cut a piece of paraffin film to the size of the mold, and place it right on the top of mold to exclude oxygen.
6. Polymerize the resin in an oven at 65 °C for 2 days (*see* **Note 13**).
7. Remove hardened blocks from the mold using a razor blade or spatula.

8. Cut excess resin with a razor blade and trim the block carefully until you reach the sample with spheroids, which should be visible as black.
9. Trim the block into a 3 mm high trapezoid (pyramid shape) with the spheroid located on top and cut it off from the rest of the block (*see* **Note 14**).
10. Stick the trapezoid on the top of an aluminum pin with conductive epoxy glue and leave to harden for 5–10 min (Fig. 1c).
11. Coat the sample with 5 nm platinum (or gold) in a sputter coater.

3.5 Serial Block Face Imaging of Spheroids

Once the sample is trimmed, glued to the aluminum pin and sputtered, it has to be properly positioned in a pin holder, the removable part of Gatan 3View2 ultramicrotome assembled in SEM.

1. Put the pin in the pin holder and tighten the side screw (*see* **Note 15**).
2. Release the hex screw from the pin holder bottom. Place the holder under an optical microscope on the table and by rotating concentric ring with fastened pin, position the sample into the center (*see* **Note 16**).
3. Once the positioning is done, fasten the hex screw from bellow of the pin holder.
4. Vent the SEM chamber and open the door of electron microscope.
5. Place the optical microscope onto the door of electron microscope (Fig. 2).

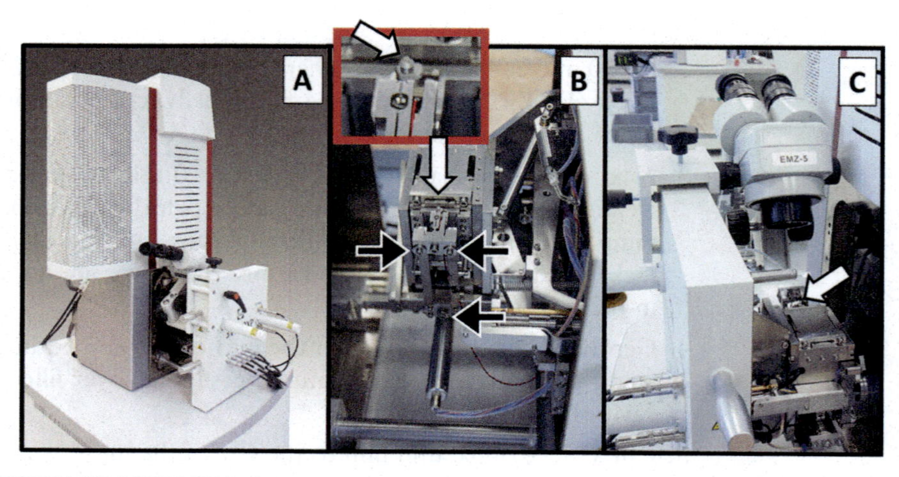

Fig. 2 TESCAN MAIA3 XMU SEM with integrated Gatan 3View2XP ultramicrotome. (**a**) The optical microscope on the SEM door serves for precise sample positioning with respect to the knife. (**b**) A detail of the Gatan ultramicrotome. The *white arrows* indicate where the pin with sample is located and *black arrows* show screws for a knife movement adjustment. (**c**) The ultramicrotome integrated in the door of SEM, rear view

6. Turn on the LED light.
7. In Gatan software set the X, Y position to zero and lower the Z position of ultramicrotome to the minimum.
8. Check if the knife is in the position "clear." If not set it in the knife control panel.
9. Insert the pin holder with sample to the ultramicrotome. Using the screw that controls the coarse sample adjustment, lower the holder so that the top of the sample is under the knife edge (*see* **Note 17**).
10. Once the knife does not collide with the sample, release the knife movement by setting the "near knife" in the Gatan software.
11. Hold the knife above the sample and raise it slowly toward the edge using the screw that controls the coarse sample adjustment. Check the distance by an optical microscope (*see* **Note 18**).
12. After approaching the sample, fasten the two hex screws to fix the pin (*see* **Note 19**).
13. Perform several (e.g., 5–10) cutting cycles and visually check whether the knife goes across the whole sample area.
14. The range of the knife movement can be slightly adjusted by tightening or loosening the appropriate screw. Also the displacement of the knife can be adjusted perpendicularly to the direction of knife movement (*see* **Note 20**).
15. Turn off the LED light, remove the optical microscope from the SEM door, close and pump the chamber of the electron microscope.
16. Run the wide field imaging mode of the electron microscope and focus the sample. Set the cutting speed and thickness of ultramicrotome to the highest value to approach fast to the sample.
17. Once the knife touches the sample, stop the knife approach and adjust parameters of cutting. These are cutting speed, thickness, amount of cutting cycles, and knife oscillation as the minimum. Besides, different imaging modes can be chosen: current (actual field of view), multiple regions of interest, and stitching (*see* **Note 21**).
18. Control the processing of SBF imaging during the first 20 section cuts and at least once after a hundred section cuts (*see* **Note 22**).

3.6 Image Processing and 3D Visualization of Ultrastructures

The organization of cells in the organoid is visualized and organelles are segmented in the selected cells. Image processing is performed using a combination of commercial software VG studio Max 2.2 and Avizo 7.0. For the segmentation of the cells and organelles, image filtration is necessary. However, despite the filtration, in most cases, the segmentation cannot be done fully automatically because of the stained protein matrix surrounding spheroids. Therefore, the 3D modeling is performed by combination of manual segmentation and automated interpolation.

3.6.1 Filtration of the Data and Preparation for Manual Segmentation

1. Import images in tiff format into the VG Studio software. Select pixel size for dimensions *x* and *y* (e.g., 40 nm). The parameter *z* determines the distance between slices (e.g., 70 nm).
2. Apply image filters to get rid of a pattern that was introduced by sample preparation (cutting the samples for the scanning). Use Median filter and Adaptive Gauss: Smoothing 0.8, Edge Threshold 0.1 (*see* **Note 23**).
3. Export DICOM image stack.

3.6.2 Manual Segmentation

1. Import DICOM image stack into Avizo software: *Open data*—mark all DICOMs—option *Read complete volume into memory*.
2. Loaded volume of the data appears in a file format *.vol. Select this file for all processes with raw data.
3. Create three orthogonal views on the data: Right click on the file *.vol—*Orthoslice*—and choose three different orientations *xy*, *xz*, *yz*.
4. Start manual segmentation in special window: Right click on file *.vol—*Image segmentation*—option *Edit new label field*. Each label field represents one object (cell) or a type of objects (nucleus, organelles, etc.) (Fig. 3).

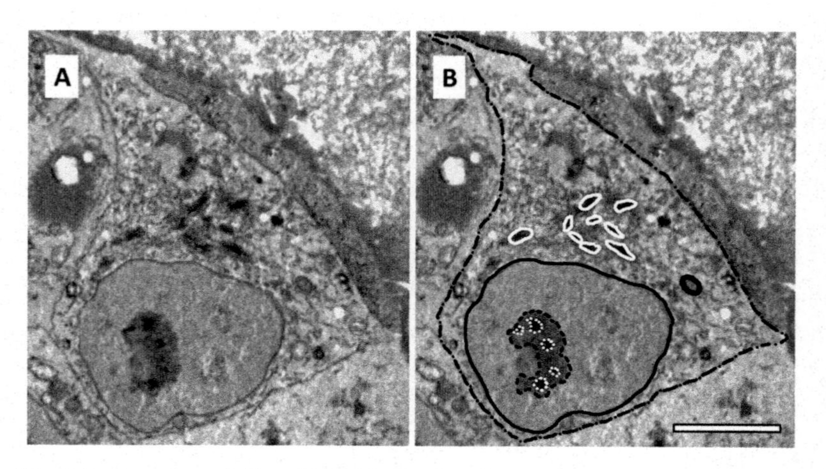

Fig. 3 Image segmentation of individual cell and its organelles within a spheroid. Cytoplasmic membrane, *black dot-and-dash line*; nucleus, *solid line*; nucleolus, *dash line*; pars fibrosa, *white dotted line*; dictyosomes of Golgi complex, *white solid*; secondary lysosomes, *black thick solid line*. Scale bar, 10 μm. (**a**) original image, (**b**) image with marked organelles

5. Set histogram and adjust contrast in the right panel. It is possible to change gray value intervals for each orthogonal view separately.
6. Manual segmentation is performed by drawing around the selected area by using software tools:
   (a) *Brush*—manual drawing; Size of the brush can be changed according to the dimensions of the segmented object.
   (b) *2D lasso*—useful for filling borders of larger, inhomogeneous areas.
   (c) *Blow tool*—uses gradient method for areas with distinct borders and homogenous filling inside. The segmented area spreads until it reaches a threshold represented by a significant gray value.
7. Perform manual segmentation for approximately one in five or ten slices (depending on the diversity of the object). Compute the rest by interpolation (use: ctrl + I). Check the interpolated slices (because of occasional incompleteness).
8. Create 3D view on the segmented area: *Add selected voxels*. Then move back to project scene tree—mark item *labels—Display* > *Isosurface* > *Apply*.
9. Construct STL model after the manual segmentation is done: Mark selected item of *labels—Compute* > *Generate Surface* > *Apply*. First, *Smoothing labels* is computed, and then *Computing triangulation* is performed (*see* **Note 24**). Computed surface appears in the right panel as file *.surf.
10. Export the STL model by marking item file *.surf: *Save Data As* > select suffix STL ascii.

3.6.3 Polygon Rendering and Final Visualization

1. Import STL models to VG studio software into raw data. *File* > *Import* > *Polygon*.
2. Transfer polygon to region of interest in the volumetric data. Software function *ROI from polygon/CAD* in the left panel.
3. Use the function *Smoothing* if distinct inaccuracies occurred during interpolation of manual segmentation (*see* **Note 25**).
4. Create a surface from finished and fixed region of interest by *Surface determination* (top panel). Use the *advanced mode* for determining starting contour—It can be defined in histogram or by finished region of interest. For determining only the border of the region of interest and ignoring gray values, use *expert mode* > select *Multi material* and set *Edge threshold* to the highest possible value (255 for 8-bit image and 65,535 for 16-bit images) > *Finish*.
5. Export new STL from surface determination or just put new polygons into project scene tree by function *Surface*

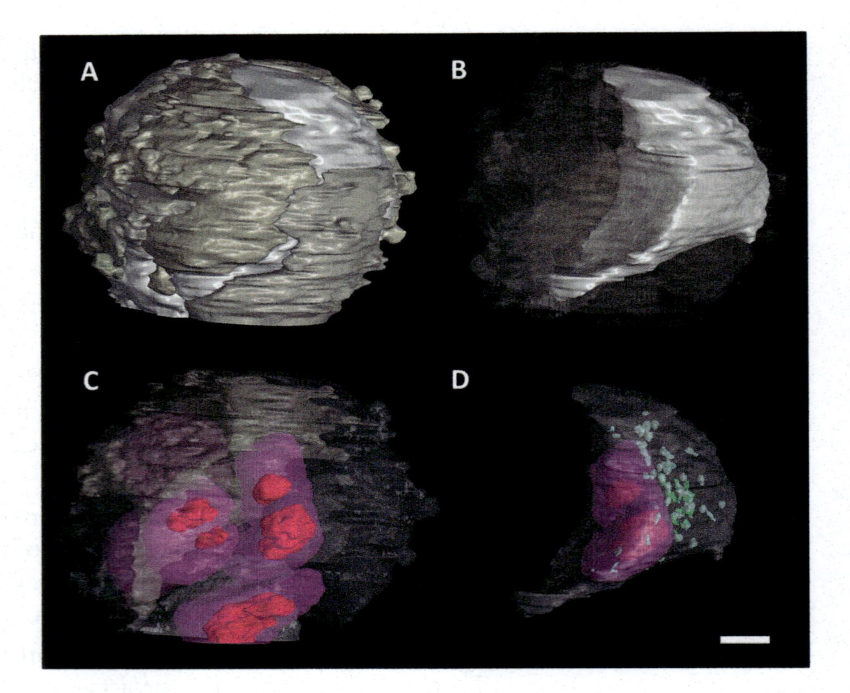

Fig. 4 3D visualization of stem cell spheroid. (**a**) Individual cells distinguished by *brownish-gray colors*. (**b**) Entombment of individual cell and its distribution within spheroid. (**c**) Visualization of subcellular components. Nuclei and nucleoli enhanced. (**d**) 3D visualization of a single cell in a spheroid, its morphology, and subcellular components. Cytoplasmic membrane, *grey*; nucleus, *magenta*; nucleolus, *red*; dictyosomes of Golgi complex, *green*; secondary lysosome, *blue*. Scale bar, 10 μm. Animation of 3D visualization of the spheroid is presented in Supplementary Video 1

extraction. *Preset selection* determines the mesh density. *Normal* is usually a sufficient choice.

6. Adjust the created model by *Polygon rendering* (top panel). To change opacity, switch the rendering settings to *Raytracing renderer mode*. Changing colors of the front face and the back face is possible in the *Material* section.
7. Export and save final images (Fig. 4) or animations (Supplementary Video 1) in the format required.

4 Notes

1. Human pluripotent stem cells are usually grown on feeder support layer. Their adaptation to feeder-free conditions for monotypic cultivation and preparation of pluripotent stem cell culture was described [11]. Coating of tissue culture plastic with Matrigel was performed according to manufacturer's instructions.

2. Alternatively, xeno-free medium can be used instead of conditioned medium. mTeSR-1 is commercially available or can be prepared as described in Ludwig and Thomson, 2007 [12].
3. Caution! The compound is toxic on inhalation and skin contact.
4. Caution! The vapors are extremely toxic. All work must be done in a fume hood. Use nitrile gloves (not latex); double-gloving is recommended when working with pure osmium tetroxide or concentrated solutions. Check gloves for contamination. Inspect all equipment for osmium tetroxide contamination after work. If contamination is found, immediately cover the contaminated area with high polyunsaturated vegetable oil or milk powder to neutralize the osmium. After neutralization, discard into appropriate waste container. Do not leave it in laboratory waste.
5. Because osmium vapors may escape in fridge, it is the best to store the reagent bottle in a second air-tight container in a dedicated refrigerator. Protect it from prolonged light exposure.
6. Same as for OsO_4.
7. Uranyl acetate precipitates when ageing and exposed to light, especially UV. Addition of glacial acetic acid [0.3% (v/v)] prevents photodegradation and increases long-term storage; however, it decreases staining effect. Moreover, UA is both toxic and mildly radioactive, but as long as the material remains external to the body, it is not harmful. Avoid the possibility of inhalation or ingestion of the material.
8. Usually, the resin is supplied as non-catalyzed for long-term storage together with the catalyst (benzyol peroxide). Do not attempt to heat the resin to speed the dissolution of the catalyst. All methacrylates (acrylics) should be considered hazardous. Direct contact and inhalation should be strongly avoided. While moderately toxic and allergenic, high concentrations may be very harmful to tissue. In addition, the methacrylates are combustible and vapors may be explosive. All acrylics should be stored in airtight vessels and used in chemical fume hoods with the appropriate protective clothing (gloves and goggles). Vinyl, latex and nitrile gloves alone are quickly compromised by contact with methacrylates. Recommended gloves are polyvinyl alcohol or Barrier-brand.
9. Use required size of culture dish. For 60 mm Petri dish use 5 mL of medium, for a well of a six-well plate use 3 mL of medium.
10. Hydrophobic barrier allows significant reduction of volume of chemicals. If strips of cell/Matrigel suspension detach from

the culture dish during longer cultivation, draw the barrier for fixation with Tannin-GAFA already.

11. During manipulation with osmium, work in a fume hood. Open vials only for the interval and keep samples covered. Avoid inhalation. Longer incubation is possible, but be aware that delicate details can be darkened. You can also use shorter incubation time if the background increases for your samples.
12. If necessary, it is possible to leave samples in the third bath of CacoB at 4 °C overnight; however, it is recommended to continue in processing.
13. The polymerization of resin can be done also by UV lamp at 4 °C (in a cold room or refrigerator) in the distance of 10–15 cm from the lamp for 2 days. The aerobic conditions lead to pro-longed polymerization and shrinkage of blocks in both conditions—heating, as well UV light.
14. When gelatin capsules are used for embedding, initial trimming of the block with a jeweler's saw is faster and the final trim with the razor blade is then much easier.
15. If the sample has a rectangular shape, position it so that the longer side of the sample is parallel to the side screw.
16. The side screw that holds the pin must stay on the left or right side with respect to the pin holder so that it does not block the ultramicrotome knife later. Precise positioning of the sample in the holder is recommended to perform using an optical stereo microscope or macroscope.
17. The side screw that holds the pin must stay perpendicular to the knife movement direction.
18. Slow knife swinging helps to better estimate the sample-knife distance.
19. After tightening screws once again carefully check and eventually correct the distance between the knife and the surface of the sample.
20. If the knife movement does not cover the whole sample area even after all possible adjustments, the sample has to be removed, its position in the pin holder corrected, and its placement repeated.
21. Adjustment of the electron optics has to be done when the knife is in the clear position.
22. Melting of the sample is usually visible after 20–30 cuts. In this case, it is advised to change the amount of energy deposited to the volume. This can be done multiple ways, e.g., changing the scanning speed, the field of view, the cutting thickness, and cutting speed.

23. Median filtering is a good choice most of the time. Alternatively, 2D filtering can be used if the images are not too noisy.
24. The process *Computing triangulation* lasts longer in comparison to the previous processes (*Add selected voxels*, *Smoothing labels*, etc.).
25. The strength of smoothing is set by parameter from interval 1 to 100. The yellow line shows the future smoothed border of the object in the raw data. Most suitable value for smoothing is usually from interval 5 to 30 (it differs by resolution, size of objects, etc.).

Acknowledgment

This study was supported by: the project FP7 Regpot ICRC-ERA-HumanBridge no. 316345, the project no. LQ1605 from the National Program of Sustainability II (MEYS CR), Czech Science Foundation (GA16-02702S), the project HistoPARK from the Centre for Analysis and Modeling of Tissues and Organs (CZ.1.07/2.3.00/20.0185, European Social Fund in the Czech Republic), funds from the Faculty of Medicine of Masaryk University (MUNI/M/1050/2013), and the project CEITEC 2020 no. LQ1601 by the Ministry of Education, Youth and Sports of the Czech Republic.

References

1. Caliari SR, Burdick JA (2016) A practical guide to hydrogels for cell culture. Nat Methods 13:405–414
2. Edmondson R, Broglie JJ, Adcock AF et al (2014) Three-dimensional cell culture systems and their applications in drug discovery and cell-based biosensors. Assay Drug Dev Technol 12:207–218
3. Pasquinelli G, Tazzari P, Ricci F et al (2007) Ultrastructural characteristics of human mesenchymal stromal (stem) cells derived from bone marrow and term placenta. Ultrastruct Pathol 31:23–31
4. Rimann M, Graf-Hausner U (2012) Synthetic 3D multicellular systems for drug development. Curr Opin Biotechnol 23:803–809
5. Bock DD, Lee W-CA, Kerlin AM et al (2011) Network anatomy and in vivo physiology of visual cortical neurons. Nature 471:177–182
6. Mishchenko Y (2009) Automation of 3D reconstruction of neural tissue from large volume of conventional serial section transmission electron micrographs. J Neurosci Methods 176:276–289
7. Takemura S, Bharioke A, Lu Z et al (2013) A visual motion detection circuit suggested by drosophila connectomics. Nature 500: 175–181
8. Denk W, Horstmann H (2004) Serial block-face scanning electron microscopy to reconstruct three-dimensional tissue nanostructure. PLoS Biol 2:e329
9. Bushby AJ, P'ng KMY, Young RD et al (2011) Imaging three-dimensional tissue architectures by focused ion beam scanning electron microscopy. Nat Protoc 6:845–858
10. Kremer A, Lippens S, Bartunkova S et al (2015) Developing 3D SEM in a broad biological context. J Microsc 259:80–96
11. Kunova M, Matulka K, Eiselleova L et al (2013) Adaptation to robust monolayer expansion produces human pluripotent stem cells with improved viability. Stem Cells Transl Med 2:246–254
12. Ludwig T, Thomson JA (2007) Defined, feeder-independent medium for human embryonic stem cell culture. Curr Protoc Stem Cell Biol Chapter 1:Unit 1C.2

Chapter 31

Quantitative Phenotypic Image Analysis of Three-Dimensional Organotypic Cultures

Malin Åkerfelt, Mervi Toriseva, and Matthias Nees

Abstract

Glandular epithelial cells differentiate into three-dimensional (3D) multicellular or acinar structures, particularly when embedded in laminin-rich extracellular matrix (ECM). The spectrum of different multicellular morphologies formed in 3D is a reliable indicator for the differentiation potential of normal, non-transformed cells compared to different stages of malignant progression. Motile cancer cells may actively invade the matrix, utilizing epithelial, mesenchymal, or mixed modes of motility. Dynamic phenotypic changes involved in 3D tumor cell invasion are also very sensitive to small-molecule inhibitors that, e.g., target the actin cytoskeleton. Our strategy is to recapitulate the formation and the histology of complex solid cancer tissues in vitro, based on cell culture technologies that promote the intrinsic differentiation potential of normal and transformed epithelial cells, and also including stromal fibroblasts and other key components of the tumor microenvironment. We have developed a streamlined stand-alone software solution that supports the detailed quantitative phenotypic analysis of organotypic 3D cultures. This approach utilizes the power of automated image analysis as a phenotypic readout in cell-based assays. AMIDA (Automated Morphometric Image Data Analysis) allows quantitative measurements of a large number of multicellular structures, which can form a multitude of different organoid shapes, sizes, and textures according to their capacity to engage in epithelial differentiation programs or not. At the far end of this spectrum of tumor-relevant differentiation properties, there are highly invasive tumor cells or multicellular structures that may rapidly invade the surrounding ECM, but fail to form higher-order epithelial tissue structures. Furthermore, this system allows us to monitor dynamic changes that can result from the extraordinary plasticity of tumor cells, e.g., epithelial-to-mesenchymal transition in live cell settings. Furthermore, AMIDA supports an automated workflow, and can be combined with quality control and statistical tools for data interpretation and visualization. Our approach supports the growing needs for user-friendly, straightforward solutions that facilitate cell-based organotypic 3D assays in basic research, drug discovery, and target validation.

Key words Morphology, Organoid, Extracellular matrix, Image data analysis, Phenotypic screening, Drug discovery, Target validation

1 Introduction

The culture of glandular epithelial cells in matrix, such as collagen, hydrogels, or Matrigel, was first established more than two decades ago [1]. The miniaturized organotypic 3D cultures described here in

Zuzana Koledova (ed.), *3D Cell Culture: Methods and Protocols*, Methods in Molecular Biology, vol. 1612,
DOI 10.1007/978-1-4939-7021-6_31,

this protocol have been developed in our laboratory and previously described [2], followed by a detailed description of the imaging and image analysis strategies [3]. This cell culture protocol is based on Ibidi Angiogenesis products, namely the Angiogenesis 15-well μ-Slides (Ibidi GmbH). These slides feature a well-in-a-well geometry that consists of two compartments, with the smaller well residing within the center of the larger well. This simple, but effective, design allows cells to be embedded in a defined and narrow focal plane between two layers of matrix. The choice of ECM is completely open, and a broad spectrum of synthetic hydrogels, alginates, agarose or soft agar, methyl-cellulose, or biologically active matrices such as collagen type I and Matrigel can be used. Approximately 30 prostate cancer (PrCa) cell lines, derivatives, and primary prostate cells have been tested with this standardized 3D cell culture platform in our laboratory [2]. Most tumor cell lines form either round, irregular, or invasive multicellular organoids. One of the most interesting and dynamic tumor cell lines is the PrCa cell line PC-3, which is characterized by extreme epithelial plasticity. PC-3 organoids initially form well-differentiated, polarized, and hollow acinar structures that strikingly resemble well-differentiated normal acini, with a complete basement membrane (BM) and devoid of any invasive properties. However, after 7–10 days, these multicellular organoids spontaneously transform into organoids with overtly stellate morphology, indicating local cell invasion with a characteristic collective streaming mode of cell motility [4], and combining epithelial with mesenchymal features (e.g., reorganization of the cortical actin cytoskeleton into a contractile acto-myosin system or stress fibers that promote cell motility) [5]. PC-3 cells represent a particularly suitable model to demonstrate the complex and highly dynamic nature of epithelial-to-mesenchymal transformation (EMT) and spontaneous differentiation versus de-differentiation in 3D organotypic cultures (Fig. 1). This cellular trans-differentiation is concomitant with striking morphometric changes and therefore an outstanding example for the potential of phenotypic screening strategies, particularly in cancer.

Advanced phenotypic image analysis tools are now widely established for single cells and 2D cultures, but are only emerging in the 3D field. ImageJ represents the most widely used open-source image analysis software. Both CellC [7, 8] and CellProfiler [9, 10] are also open-source software programs, specifically tailored for high-content analyses of microscopic images, but mainly intended for single cells and two-dimensional conditions. In addition, several commercial software packages have been recently released, mainly for the pharmaceutical high content screening and imaging market, including VoxelView (SGI Vital Images; www.vitalimages.com), Imaris (www.bitplane.com), Metamorph (Molecular Devices; www.moleculardevices.com), Definiens Developer XD (www.definiens.com), Analysis (Olympus, OSIS; www.soft-imaging.de), and Volocity (PerkinElmer; www.perkinelmer.com) [11, 12]. However, all of these programs are

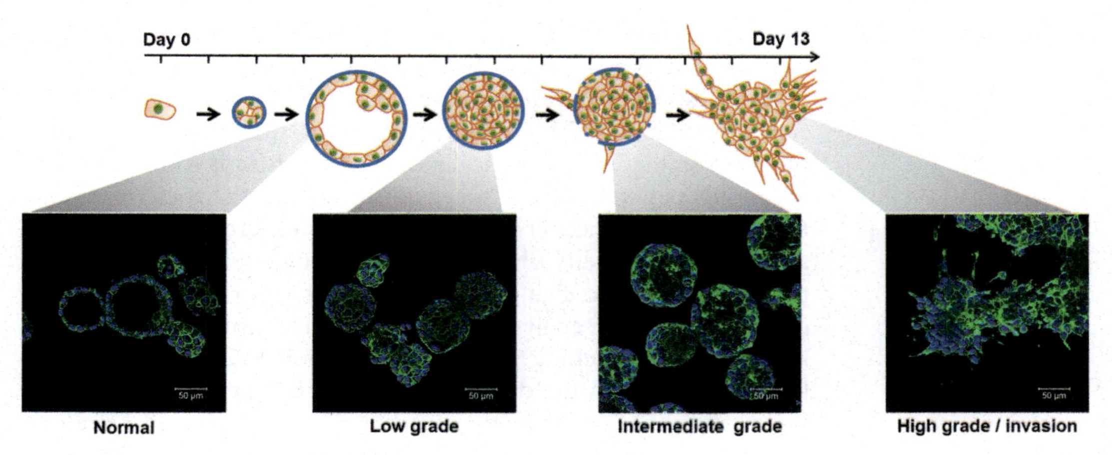

Fig. 1 Illustration of PrCa PC-3 cells grown in 3D culture in Matrigel. This allows organotypic organoid formation, acinar differentiation, and invasion. In the schematic illustration, the basement membrane formed by the organoid is depicted in *blue*. Our models specifically address the complex biology of human tissues, and faithfully recapitulate morphologic features such as invasion. Example confocal microscopy images show PC3 organoids, stained with filamentous actin (F-actin; *green*) and DAPI (4',6-diamidino-2-phenylindole) to label nuclei (*blue*), at different stages of 3D organotypic culture. Reproduced from Björk et al., 2016 [6] with permission from Macmillan Publishers Limited (Springer Nature)

specialized on the detailed analysis of a few selected 3D organoids or tissue-like histology. None of these programs was designed for the systematic analysis of multiple structures formed in a single well, nor across many wells as would be required for the purpose of phenotypic screening. As morphologies formed in 3D also vary massively in shape, size, geometry, density, surface features, and textures, none of the existing programs is specifically optimized for the heterogeneous and dynamic features observed in organotypic 3D cultures.

Many 3D model systems are based on the formation of a single spheroid per well, and basic image analysis platforms, as those mentioned above, may be suitable to extract the limited number of features formed under non-adherent conditions, e.g., in hanging drop plates. In contrast, the model system described here is inspired by the spontaneous formation of complex, tissue-like structures after embedding normal or tumor cells as single cells into supportive ECM, and the spectrum of morphometric features formed is dramatically increased in these settings compared to spheroids. Therefore, it is necessary to quantitate the characteristic heterogeneity and the multiple dynamic changes that can be observed in tissue-like 3D cultures such as invasion and differentiation. These tasks are orders of magnitudes more complex than simple spheroid size and shape estimations, and the needs are more closely related to the analysis of tissue architecture or histology in clinical pathology. The reliable evaluation of tissue-like features in in vitro model systems requires a very specific and accurate measurement of the most biologically relevant morphometric features.

In our laboratory, we have developed AMIDA, a light multi-parametric image analysis program, designed for high-content analysis of complex and heterogeneous 3D organoid cultures. AMIDA enables rapid segmentation and quantitative measurements of single images, typically containing dozens to hundreds of organoids per well, with a spectrum of different shapes, sizes, and textures. AMIDA covers the spectrum of different morphologies and the heterogeneity observed within a single well, but also variations and dynamic changes between wells. Accordingly, AMIDA can be used for the analysis of large numbers of images and thus for high-content screening. However, additional functions such as batch operation required for this task have been excluded from the basic software version openly available. The computational readout represents a logical compromise between exceedingly detailed imaging of a vast spectrum of morphological details, and fast experimental throughput [3], as would be required for high-content screening purposes. In practice, the AMIDA program first identifies individual multicellular structures for image segmentation, and then assigns numerical values for a panel of selected, particularly cancer-relevant parameters to the objects. These numeric values are then exported as an Excel file, with detailed data available for each single structure. Moreover, for speed of analysis and ease of interpretation, the values of the often large number of organoids covered even in a single well are subsequently averaged and can be plotted based on statistical parameters.

AMIDA has primarily been designed to retrieve information from 3D confocal image stacks [3]. However, due to the special 3D cell culture design used in Ibidi angiogenesis slides, there is often very little spatial overlap of multicellular structures in the Z-axis. Therefore, we have restricted the quantitative analysis in practise to 2D maximum intensity projections of the 3D images, a decision that contributes significantly to the reduction of data analyzed per well and thus speeding up the statistical analysis of larger series of experiments. AMIDA has been successfully used in many studies from our laboratory [2, 3, 6, 13–15].

By examining the morphologies of organoids formed by prostate-derived cell lines [2], we devised 19 phenotypic parameters that were considered as most informative and most directly linked to physiologically relevant cancer biology (e.g., growth, invasion, differentiation). The morphometric parameters implemented in AMIDA can be divided into three classes: general features, morphological features, and functional aspects (Table 1). Image data can be analyzed in three channels (red: R, green: G, and blue: B), where green and red are the most frequently used in live cell imaging to detect overall morphology of the multicellular structures and to visualize dead cells, respectively. Values for general and morphological parameters are always derived from the same RGB channel and used for structural segmentation. General parameters include information related to the size (*area*) of any segmented object, its relation to neighbors (number

Table 1
Morphometric parameters produced by the AMIDA image analysis program

Parameter	Explanation	Unit
General		
Area	Area of the segmented structure	Pixels
Neighbors	The number of neighboring structures touching the segmented structure	Numbers
SharedBound	The length of the shared boundary of all neighbors for the structure	Pixels
Closest	The distance of the closest neighbor of the segmented structure from the center point to the center point	Pixels
CellRatio	Portion of segmented area in the image	%
Morphological		
Roundness	Roundness of the segmented structure	%
FiltRound	Roundness of the segmented structure after filtering the object with the opening operator	%
RoundDiff	Difference between the Roundness and Filtered Roundness	%
AppIndex	Index for severity of appendages of the segmented structure	No unit
MaxApp	Estimate for the maximum length of appendages of the segmented structure	Pixels
MedApp	Estimate for the median length of appendages of the s egmented structure	Pixels
Roughness	Roughness of the surface of the segmented structure	%
AppNumber	Estimate for the number of appendages in the segmented structure	Numbers
Functional		
DensityR/G/B	Density of the segmented structure for each channel. Sum of pixel intensities divided by the number of pixels	Gray levels/ pix
DeviationR/G/B	Standard deviation of the pixel intensities in the segmented structure for each channel	No unit
AreaRatioR/G/B	Ratio of substructures of a certain color inside the segmented structure	%
HollownessR/G/B	Estimate of the hollowness of the segmented structure for each channel	%
CellNumberR/G/B	Estimate of the number of cells in the red channel inside the segmented structure	Numbers
AveAreaR/G/B	Average area of the cells inside the segmented structure for each channel	Pixels

of *neighbors*, *shared boundaries* with neighbors, *closest* neighbors), and the amount of cellular matter in relation to the local background (*cell ratio*, *average ratio*). Morphological parameters include measures for features typically associated with the phenotype of multicellular organoids such as symmetry (*roundness*), small surface contour features (*roughness*), and measures that indicate invasive processes (*appendages*, incl. number, maximum and median length of invasive structures). For each of the three channels, functional parameters can assess a structure's signal *density*, the number of cells per structure (*cell number*), polarization of cells within the organoid (*hollowness*), and the cell area relative to the size of the entire organoid (*average area*). Currently, all morphometric calculations are based on key biological events highly relevant for cancer research, such as growth, multicellular differentiation and formation of glandular structures (acini), cell death or apoptosis, and invasion. These simple morphometric measures are intuitive, and directly and functionally relate to the underlying biological processes. For example, "*roundness*" has been shown to relate to the formation of a complete BM or basal lamina that typically defines the round nature of well-differentiated acini or organoids. In cell lines that fail to form a complete BM like LNCaP, organoid structures often become irregular and display reduced roundness. They may also become prone to dynamic changes over time or upon drug treatments, such as emerging invasive structures. For example, when PC-3 cells spontaneously transform themselves from round organoids into invasive structures, the BM is actively degraded by proteolytic processes. Nevertheless, descriptive mathematical metrics can be implemented, thus opening the possibility to quantitate additional structural aspects that are currently beyond human recognition. Our image analysis approach supports the growing needs for user-friendly, straightforward, and fast solutions that facilitate large-scale, cell-based organotypic 3D assays in basic research, drug discovery, and target validation. The complete pipeline for the phenotypic image-based analysis of the multicellular structures in 3D cell culture system is presented in Fig. 2.

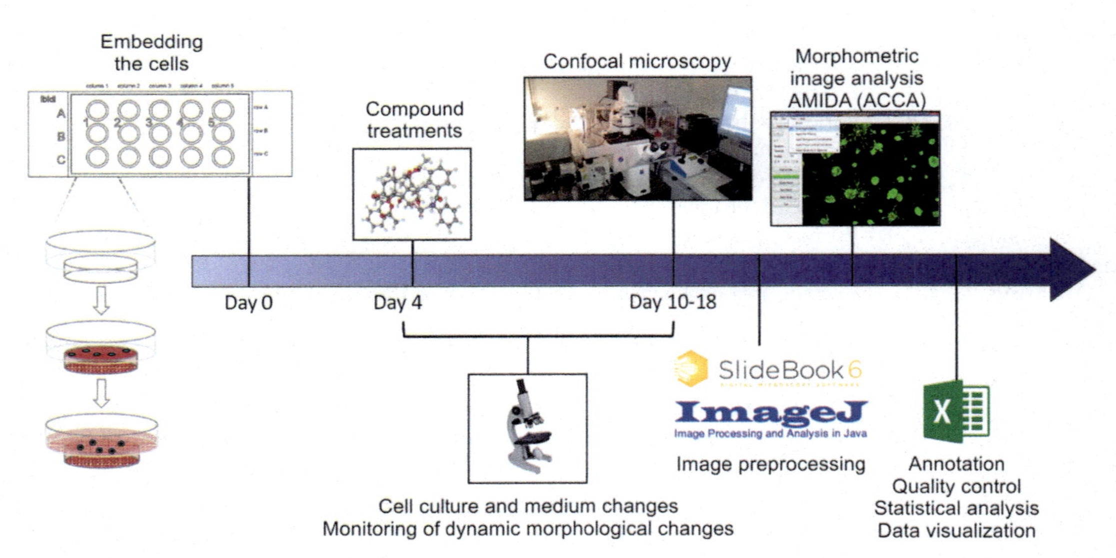

Fig. 2 Working pipeline for morphology and image-based phenotypic analysis of multicellular organoids in 3D cell culture. The concept can be used for compound screens or target validation. It is based on an optimized organotypic cell culture platform complemented with a user-friendly proprietary image analysis program AMIDA. Ibidi Angiogenesis μ-Slides have a unique design that facilitates 3D cell culturing between two layers of extracellular matrix, on a very narrow focal plane. The operation schedule for a typical experiment from cell seeding to image analysis and morphometric data visualization is presented. Reproduced from Härmä et al., 2014 [3] with permission under the terms of the Creative Commons Attribution License

2 Materials

1. μ-Slide Angiogenesis (Ibidi GmbH) (*see* **Note 1**).
2. Matrigel Matrix Growth Factor Reduced (BD Biosciences) (*see* **Note 2**).
3. Phosphate buffered saline (PBS).
4. Trypsin.
5. Normal and cancer epithelial cell lines as desired, e.g., PC3, LNCaP, EP156T, NCI-H660, and other described in Härmä et al. [2].
6. Normal epithelial cell culture (NEC) medium: 12.5 mg/L bovine pituitary extract, 1.25 mg/L EGF, 100 U/mL penicillin, 100 μg/mL streptomycin, and 2 mM L-glutamine in Keratinocyte Serum-Free medium.
7. Epithelial cancer cell line (ECC) medium: 5–10% fetal bovine serum (FBS), 100 U/mL penicillin, 100 μg/mL streptomycin, and 2 mM L-glutamine in RPMI-1640.
8. 1 mM Calcein AM in anhydrous DMSO (*see* **Note 3**).
9. 1 mM Ethidium homodimer-2 (EthD-2) in DMSO (*see* **Note 4**).
10. Confocal microscope Zeiss Axiovert-200 M microscope equipped with Yokogawa CSU22 spinning disc confocal unit and with a Zeiss Plan-Neofluar 5× objective.
11. SlideBook digital microscopy software (3i Intelligent Imaging Innovations Inc.) (*see* **Note 5**).
12. ImageJ (NIH) (*see* **Note 6**).
13. AMIDA software (previously called ACCA) can be downloaded freely from the publication site (AMIDA Program S1) (*see* ref. 3) (*see* **Note 7**) http://journals.plos.org/plosone/article?id=10.1371/journal.pone.0096426#s5
14. Excel (Microsoft).

3 Methods

3.1 Miniaturized 3D Organotypic Cultures

1. Perform the 3D organotypic culture experiments in 15-well μ-Slides Angiogenesis.
2. Use growth factor-reduced Matrigel as ECM to promote epithelial differentiation.
3. Thaw Matrigel on ice at 4 °C. Prepare a standardized Matrigel stock solution of 8 mg/mL diluted with cell culture medium, as the Matrigel concentration varies between batches (*see* **Note 8**).
4. Vortex the Matrigel briefly to ensure that the matrix is evenly dispersed. Always keep Matrigel on ice to avoid solidification.

5. Fill bottom wells of Ibidi Angiogenesis μ-Slides with 10 μL/well of 50% Matrigel-medium mixture (4 mg/mL Matrigel) (*see* **Note 9**).
6. Incubate the μ-Slides filled with Matrigel at 37 °C for 30 min and allow to polymerize.
7. Meanwhile, prepare normal epithelial or cancer cells cultured in 2D conditions to be used in the experiment, such as normal epithelial cells and derivatives cultured in NEC medium or PrCa cell lines cultivated in ECC medium (*see* **Note 10**).
8. Detach the epithelial cells from the plate with trypsin, wash once in medium.
9. Suspend the detached cells in medium (for 3D culture of normal epithelial cells, use NEC medium with 2% FBS; for 3D culture of PrCa cell lines use the same medium as for 2D) at 20,000 cells/mL and place 50 μL of cells mixed with medium on top of the polymerized bottom gel (i.e., at a density of 1000 cells/well).
10. Incubate at 37 °C for 1 h (*see* **Note 11**) to allow cells to attach.
11. Aspirate the medium carefully, and cover the adherent cells with 20 μL of 25% Matrigel in medium mixture (i.e., 2 mg/mL Matrigel) (*see* **Note 12**).
12. Humidify the μ-Slides by adding 25 μL of sterile PBS between the wells and by surrounding them. Add new PBS to the slide periphery every time the medium is changed.
13. Allow the upper gel to polymerize at 37 °C in standard cell culture conditions (5% CO_2, and 95% humidity), for at least 3 h, but preferably overnight.
14. Fill the wells with medium, and change medium every second to third day (*see* **Note 13**).

3.2 Compound Treatments

1. Add experimental compounds upon medium change on day 4 of 3D culture, when the cells have established in their new environment.
2. Dissolve all compounds in appropriate vehicle (e.g., DMSO, ethanol, or PBS).
3. Perform all drug exposures at least in triplicates, including vehicle controls.
4. Continue the 3D cultures and treatments for 6 more days, and stop the experiment on day 10 (*see* **Note 14**).

3.3 Live Cell Staining, Image Acquisition and Preprocessing

1. Stain multicellular organotypic structures with live/dead cell dyes, 2 μM Calcein-AM (for viability) and 1 μM EthD-2 (for cell death) in cell culture medium, and incubate for 30 min in 37 °C (*see* **Note 15**).

2. Acquire 3D confocal image stacks with a confocal spinning disc microscope, equipped with 5× objective (*see* **Note 16**).
3. Scan all 15 wells of the Ibidi μ-Slide, by acquiring four stacks of images per well, a total of 60 images/μ-Slide, with the spinning disc confocal (*see* **Note 17**).
4. The x and y dimensions for a single field are approximately 4.4 mm × 3.3 mm, and z ranging between 300 and 800 μm with 20–40 μm image plane intervals.
5. Create maximum intensity projections with SlideBook.
6. Remove background noise by normalization, using either SlideBook and/or ImageJ programs (*see* **Note 18**). Reduce background noise by, e.g., excluding the 5 and 95 percentiles of the image distribution. If necessary, perform additional background reduction.

3.4 Image Segmentation Using AMIDA

1. Open an image file in AMIDA by clicking "select image data".
2. Browse image layers with the arrow buttons (*see* **Note 19**).
3. Adjust the two basic parameters to assure proper segmentation quality: sensitivity and threshold. The dimension-less sensitivity parameter controls the splitting of segmented multicellular regions (organoids) in the analyzed image, and can be chosen for a range between 2 and 50. Choosing a smaller value generally leads to smaller segmented regions; and optimal settings can be empirically determined. The equally dimension-less threshold parameter controls the cut-off value of the histogram for image segmentation, in the range of 1–5. Choosing higher threshold values leads to a more stringent segmentation.
4. Select red, green, and/or blue channels to be analyzed by checking the boxes "R," "G," and/or "B."
5. Perform the actual image segmentation and quantitative analysis of the structures in the image by clicking on "Analyze Data." The segmented image will open up in a new window.
6. Visualize the results by clicking "Display Results."
7. Save the segmented image by clicking "Save Results" (Fig. 3).
8. By using the "Batch Mode" function, the numerical data from each segmented structure in the image can be saved as a Microsoft Excel file: xxx_anal.xls. Manual visual inspection of segmented images is advisable.

3.5 Additional Tools for Viewing or Filtering of Images in AMIDA

1. The "View" drop down tab opens up ways to view the segmented image, either as a single color channel, or as a binary or core-binary image.

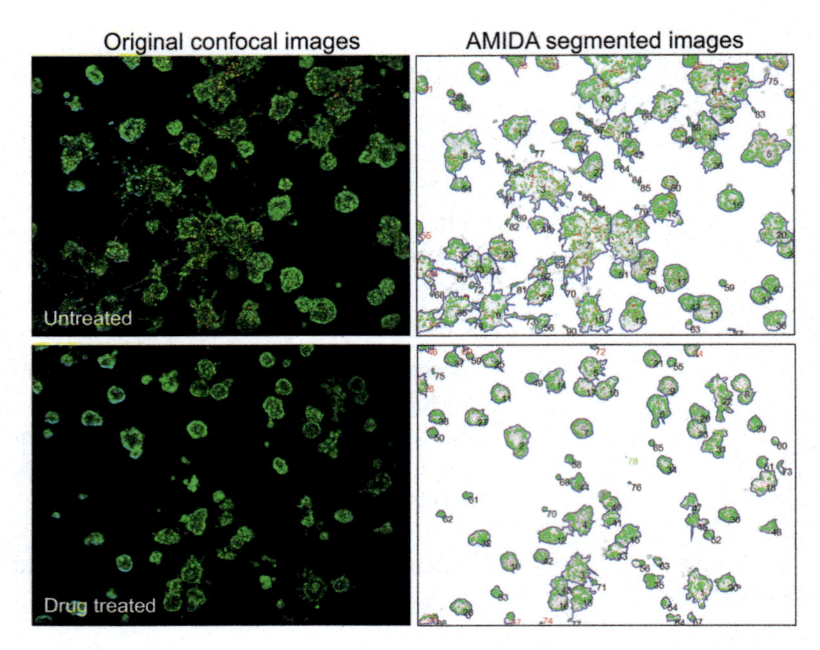

Fig. 3 Quantification of key parameters analyzed by AMIDA. Representative PC-3 organoids, either untreated or treated with an invasion blocking drug, stained for viable cells with Calcein AM (*green*), and dead cells with EthD-2 (*red*). The organoids were imaged with a spinning disk confocal microscope (5× objective), and then segmented, numbered, and analyzed with AMIDA (sensitivity: 20, threshold: 1, smallest: 10, channels: *green* and *red*)

2. The "Tool" drop down tab opens up other additional adjustment options, described below.
3. "Zoom" opens an image zoom tool that allows the user to inspect magnified image areas.
4. "Only Segmentation" turns the morphometric analysis off for faster segmentation, which is helpful when aiming for optimal sensitivity and threshold settings.
5. "Apply Pre-filtering" activates several pre-filtering modes (median, Gaussian, edge enhancement) that can be explored on a "trial and error" basis, aiming for optimized results.
6. "Apply Background Compensation" switches background compensation on.
7. AMIDA can also be used for the segmentation of black and white phase contrast images. "Apply phase-contrast conversion" turns on phase-contrast conversion mode (*see* **Note 20**).
8. "Select Sensitivity of Measures" fine-tunes the sensitivity of morphometric measures (options "low," "medium," and "high" sensitivity).
9. The "Help" drop down tab describes the current version of the software.

3.6 Quantitative Phenotypic Analysis Using AMIDA

1. Open the numerical data file: xxx_anal.xls and it is advisable to perform proper annotation and quality control in Excel (*see* **Note 21**). Data annotation refers to providing necessary experimental information, e.g., name of cell lines, experimental conditions, time-points, compounds, and concentrations for drug treatments. Data quality control refers to removing erroneously segmented structures, cell debris, staining/imaging irregularities, noise, and other artifacts. Quality control assessment can be performed manually.
2. Numerical values are appointed by AMIDA for selected morphological features, *see* Table 1.
3. The data can be readily plotted in Excel (*see* **Note 21**), e.g., we use box-and-whisker plots with p-values indicated for experimental data as compared to control.

4 Notes

1. The 15-well μ-Slides were designed for investigation of angiogenesis in tube formation assays, but are perfectly suited for 3D cell culture of other cell types including cancer cells and cell-based imaging of cancer organoids.
2. Matrigel is especially suited for the culture of epithelial cells in 3D as it effectively promotes differentiation. Nevertheless, there are no technical restrictions with this protocol to the use of other biological or synthetic matrix preparations.
3. Calcein AM is a non-fluorescent, hydrophobic compound that actively permeates cell membranes. In living cells, the hydrolysis of Calcein AM by intracellular esterases generates free Calcein, a hydrophilic, strongly fluorescent and non-toxic compound that is retained in the cytoplasm and is not able to leave the cell by diffusion. Calcein AM thus provides a very simple, but effective method to demonstrate cell viability.
4. EthD-2 is a high-affinity nucleic acid stain that is only weakly fluorescent until bound to double-stranded DNA, upon which it emits a strong red fluorescence. The dye is not able to permeate intact membranes of living cells, and thus provides a simple method to identify dead cells in real time, live cell cultures.
5. SlideBook is a digital microscopy software that comes as a standard and basic image processing program that controls the confocal microscope.
6. ImageJ is an open platform for scientific image analysis.
7. A collection of exemplary images is available for testing of the AMIDA software (AMIDA Data S1), *see* Härmä et al., 2014 [3], http://journals.plos.org/plosone/article?id=10.1371/journal.pone.0096426#s5.

8. Smaller Matrigel aliquots can be made directly from the stock and should be stored at −20 °C.
9. Always prepare some extra Matrigel-medium mixture (for a few extra wells), since the gel tends to stick to pipette tips.
10. For MDA-PCa-2b cells use Ham's F12 supplemented with 20% FBS. For NCI-H660 cells use RPMI supplemented with 5% FBS, 5 μg/mL insulin, 10 μg/mL transferrin, 30 nM sodium selenite, 10 nM hydrocortisone, 10 nM beta-estradiol, 4 mM L-glutamine, and 10 ng/mL EGF.
11. Depending on the growth and invasiveness of the cell line, 700–1500 cells/well should be used. This warrants that all organoids are generated from a single cell (clonal approach), and avoids merging of organoids already during early time points in organotypic cell culture. Some cell lines attach more slowly to the matrix and may require >1.5 h to adhere properly.
12. Never put the aspiration tip into the bottom well to avoid accidental removal of the cells.
13. The wells have to be completely filled with medium, which has to be changed regularly in order for the cells to grow optimally.
14. For cell lines that show a strong invasive phenotype (e.g., PC-3, PC-3 M, ALVA31), drug treatments are typically initiated already on day 2, and emerging structures including invasive "appendages" are imaged on days 8–9. More slowly growing cell lines devoid of any invasive or dynamic properties (e.g., LAPC-4, EP156T, primary prostate epithelial cells) are incubated for 8 days to form organoids of significant size (e.g., >50 μm diameter) prior to treatments.
15. Many other live cell dyes can be used for staining of the organotypic 3D cultures, but potential cytotoxicity and optimal concentrations have to be empirically evaluated.
16. Low magnification (e.g., 4–5×) should be used to support reasonable throughput (n).
17. The image resolution is on purpose kept at a relatively low range (672 × 512 pixels) to promote fast image acquisition. The typical scan time of two channels, all wells of an Ibidi 15-well μ-Slide, is 10 min. After the live cell staining, the same wells can be fixed and used for immunofluorescent stainings.
18. It is important that the image contrast is sufficient and that all background fluorescence is removed, in order to assure reliable and robust image segmentation.
19. In practice, AMIDA automatically applies an intensity projection algorithm to generate simple 2D raster graphics. As this may increase the overall time required for image analysis, the user can also convert 3D images into intensity projections

using any other image processing program of choice (e.g., ImageJ, CellProfiler), before uploading of images to AMIDA.

20. Since phase-contrast pictures differ significantly from confocal images, they must be preprocessed differently prior to segmentation, and converted to fit the customized thresholding technique described. *See* Härmä et al. [3] for detailed information.
21. Microsoft Excel or other corresponding programs for handling numerical data can be used for annotation, plotting, and graphical visualization of the results.

Acknowledgments

This work was supported by Academy of Finland (grant no. 267326), and K. Albin Johannson's Foundation.

References

1. Streuli CH, Bailey N, Bissell MJ (1991) Control of mammary epithelial differentiation: basement membrane induces tissue-specific gene expression in the absence of cell-cell interaction and morphological polarity. J Cell Biol 115:1383–1395
2. Härmä V, Virtanen J, Mäkelä R et al (2010) A comprehensive panel of three-dimensional models for studies of prostate cancer growth, invasion and drug responses. PLoS One 5:e10431
3. Härmä V, Schukov HP, Happonen A et al (2014) Quantification of dynamic morphological drug responses in 3D organotypic cell cultures by automated image analysis. PLoS One 9:e96426
4. Friedl P, Locker J, Sahai E et al (2012) Classifying collective cancer cell invasion. Nat Cell Biol 14:777–783
5. Yamazaki D, Kurisu S, Takenawa T (2005) Regulation of cancer cell motility through actin reorganization. Cancer Sci 96:379–386
6. Björk JK, Åkerfelt M, Joutsen J et al (2016) Heat-shock factor 2 is a suppressor of prostate cancer invasion. Oncogene 35:1770–1784
7. Selinummi J, Sarkanen JR, Niemisto A et al (2006) Quantification of vesicles in differentiating human SH-SY5Y neuroblastoma cells by automated image analysis. Neurosci Lett 396:102–107
8. Selinummi J, Seppälä J, Yli-Harja O et al (2005) Software for quantification of labeled bacteria from digital microscope images by automated image analysis. Biotechniques 39:859–863
9. Carpenter AE, Jones TR, Lamprecht MR et al (2006) CellProfiler: Image analysis software for identifying and quantifying cell phenotypes. Genome Biol 7:R100
10. Kamentsky L, Jones TR, Fraser A et al (2011) Improved structure, function and compatibility for CellProfiler: modular high-throughput image analysis software. Bioinformatics 27:1179–1180
11. Megason SG, Fraser SE (2007) Imaging in systems biology. Cell 130:784–795
12. Megason SG, Fraser SE (2003) Digitizing life at the level of the cell: high-performance laser-scanning microscopy and image analysis for in toto imaging of development. Mech Dev 120:1407–1420
13. Härmä V, Knuuttila M, Virtanen J et al (2012) Lysophosphatidic acid and sphingosine-1-phosphate promote morphogenesis and block invasion of prostate cancer cells in three-dimensional organotypic models. Oncogene 31:2075–2089
14. Härmä V, Haavikko R, Virtanen J et al (2015) Optimization of invasion-specific effects of betulin derivatives on prostate cancer cells through lead development. PLoS One 10:e0126111
15. Åkerfelt M, Bayramoglu N, Robinson S et al (2015) Automated tracking of tumor-stroma morphology in microtissues identifies functional targets within the tumor microenvironment for therapeutic intervention. Oncotarget 6:30035–30056

INDEX

Zuzana Koledova (ed.), *3D Cell Culture: Methods and Protocols*, Methods in Molecular Biology, vol. 1612,
DOI 10.1007/978-1-4939-7021-6, © Springer Science+Business Media LLC 2017

D

E

F

N

O

P

Q

R

S

T

U

V

W

Y

Z

Printed by Printforce, the Netherlands